WHAT WORKS IN CONSERVATION

What Works in Conservation

2021

Edited byWilliam J. Sutherland, Lynn V. Dicks, Silviu O. Petrovan and Rebecca K. Smith

OpenBook Publishers

http://www.openbookpublishers.com

Digital material and resources associated with this volume are available at https://www.openbookpublishers.com/product/1490#resources and http://www.conservationevidence.com

What Works in Conservation Series | ISSN: 2059-4232 (Print); 2059-4240 (Online)

ISBN Paperback: 978-1-80064-272-0
ISBN Hardback: 978-1-80064-273-7
ISBN Digital (PDF): 978-1-80064-274-4
ISBN Digital ebook (epub): 978-1-80064-275-1
ISBN Digital ebook (mobi): 978-1-80064-276-8
DOI: 10.11647/OBP.0267

Funded by Arcadia, DEFRA, ESRC, MAVA Foundation, NERC, Natural England, Robert Bosch Stiftung, Synchronicity Earth, South West Water and Waitrose Ltd.

Cover image: A close up shot of the underside of a Dwarf Cavendish (*Musa acuminata*) by Ben Clough, CC BY-SA 3.0. Wikimedia http://commons.wikimedia.org/wiki/File:Dwarf_cavendish_leaf_2.jpg. Cover design: Heidi Coburn

Contents

Introduction

This book has been created to help you make decisions about practical conservation management by providing an assessment, from the available scientific evidence, of what works and what does not work in conservation. It also tells you if no evidence has been found about whether or not a conservation intervention is effective. This is the 2021 edition of *What Works in Conservation*, which was first published in 2015 and is updated annually.

Who is *What Works in Conservation* for?

This book is for people who have to make decisions about how best to support or conserve biodiversity. These include land managers, conservationists in the public or private sector, farmers, campaigners, advisors or consultants, policymakers, researchers or people taking action to protect local wildlife. *What Works in Conservation* and the associated synopses summarize scientific evidence relevant to conservation objectives and the actions that could be taken to achieve them. *What Works in Conservation* also provides an assessment of the effectiveness of interventions based on available evidence.

We do not aim to make decisions for people, but to support decision-making by providing what evidence there is (or is not) about the effects that your planned actions could have. It is important that you read the full details of the evidence, freely available online at www.conservationevidence.com, before making any decisions about implementing an intervention.

The Conservation Evidence project

The Conservation Evidence project has four parts, all of which are available from our website conservationevidence.com:

1. An ever-expanding searchable **database of over 7,800 summaries** of previously published scientific papers, reports, reviews or systematic reviews that document the effects of interventions.

 https://doi.org/10.11647/OBP.0267.16

2. **Synopses** of the evidence captured in part 1) relating to particular species groups, habitats or conservation issues. Synopses bring together the evidence and background for all possible interventions. Most synopses are also available to purchase in printed book form, and all of them can be downloaded for free as electronic material.
3. ***What Works in Conservation*** provides an assessment of the effectiveness of interventions based on available evidence. It contains both the key messages from the evidence for each conservation intervention from the relevant synopses, and an assessment of the effectiveness of each intervention by expert panels.
4. An online, **open access journal**, the *Conservation Evidence Journal* that publishes new pieces of research on the effects of conservation management interventions. All our papers are written by, or in conjunction with, those who carried out the conservation work and include some monitoring of its effects.

Alongside this project, the Centre for Evidence-Based Conservation (http://www.cebc.bangor.ac.uk) and the Collaboration for Environmental Evidence (http://www.environmentalevidence.org) carry out and compile systematic reviews of evidence on the effectiveness of particular conservation interventions. We recommend carrying out a systematic review, which is more comprehensive than our summaries of evidence, when decisions have to be made with particularly important consequences. Systematic reviews are included in the Conservation Evidence database.

Which conservation interventions are included?

Lists of interventions for each synopsis are developed and agreed in partnership with an advisory board made up of international conservationists and academics with expertise in the subject. We aim to include all actions that have been carried out or advised for the conservation of the specific group of species or habitat or for the specific conservation issue.

The lists of interventions are organized into categories based on the International Union for the Conservation of Nature (IUCN) classifications of direct threats and conservation actions (https://www.iucnredlist.org/resources/classification-schemes). Interventions are primarily grouped according to the relevant direct threats. However, some interventions can be used in response to many different threats and so these have been grouped according to conservation action.

How we review the literature

We gather evidence by searching relevant scientific journals from volume one through to the most recent volume. Thirty general conservation journals are regularly searched by Conservation Evidence. Specialist journals are also searched for each synopsis (330 have been searched so far) as well as over 300 non-English journals. We also search reports, unpublished literature and evidence provided by our advisory boards. Two of the synopses used systematic mapping exercises undertaken by, or in partnership with, other institutions. Systematic mapping uses a rigorous search protocol (involving an array of specified search terms) to retrieve studies from several scientific databases. Evidence published in languages other than English is included when it is identified. Evidence from all around the world is included in synopses. One exception is farmland conservation, which only covers northern Europe (all European countries west of Russia, but not those south of France, Switzerland, Austria, Hungary and Romania). Any apparent bias towards evidence from some regions in a particular synopsis reflects the current biases in published research papers available to Conservation Evidence.

The criteria for inclusion of studies in the Conservation Evidence database are as follows:

- A conservation intervention must have been carried out.
- The effects of the intervention must have been monitored quantitatively.

These criteria exclude studies examining the effects of specific interventions without actually doing them. For example, predictive modelling studies and studies looking at species distributions in areas with long-standing management histories (correlative studies) are excluded. Such studies can suggest that an intervention could be effective, but do not provide direct evidence of a causal relationship between the intervention and the observed biodiversity pattern.

For each study we summarise the results that are relevant to each intervention. Unless specifically stated, results reflect statistical tests performed on the data within the papers.

What does *What Works in Conservation* include?

What Works in Conservation includes **only the key messages from each synopsis**, which provide a rapid overview of the evidence. These messages are condensed from the summary text for each intervention within each synopsis. **For the full text and references see www.conservationevidence.com**

Panels of experts have assessed the collated evidence for each intervention to determine effectiveness, certainty of the evidence and, in most cases, whether there are negative side-effects (harms). Using these assessments, interventions are categorized based on a combination of effectiveness (the size of benefit or harm) and certainty (the strength of the evidence). The following categories are used: Beneficial, Likely to be beneficial, Trade-off between benefit and harms, Unknown effectiveness, Unlikely to be beneficial, Likely to be ineffective or harmful (for more details see below).

Expert assessment of the evidence

The average of several experts' opinions has been shown to be a more reliable and accurate assessment than the opinion of a single expert. We therefore ask a panel of experts to use their judgement to assess whether evidence within the synopsis indicates that an intervention is effective or not. They are also asked to assess how certain they are of the effectiveness given the quality of evidence available for that intervention (certainty of the evidence). Negative side-effects described in the collated evidence are also assessed (harms). They base their assessment solely on the evidence in the synopsis. We use a modified Delphi method to quantify the effectiveness and certainty of evidence of each intervention, based on the summarized evidence. The Delphi method is a structured process that involves asking a panel of experts to state their individual opinion on a subject by scoring anonymously. They can then revise their own scores after seeing a summary of scores and comments from the rest of the panel. Final scores are then collated. Scores and comments are kept anonymous throughout the process so that participants are not overly influenced by any single member of the panel.

For each intervention, experts are asked to read the summarized evidence in the synopsis and then score to indicate their assessment of the following:

Effectiveness: 0 = no effect, 100% = always effective.

The score uses an assessment by independent experts of the effectiveness of this action based on the summarized evidence (0% = not effective, 100% = highly effective). This score is based on the direction and size of the effects reported in each study. Actions with high scores typically have large, desirable effects on the target species/habitat in each study. There is some variation between actions, e.g. 100% effectiveness in adding underpasses under roads for bat conservation will likely have different impacts to 100% effectiveness in restoring marsh habitat. The effectiveness score does not consider the quantity or quality of studies; a single, poorly designed study could generate a high effectiveness score. The effectiveness score is combined with the certainty and harms scores to determine the overall effectiveness category (for more details see https://www.conservationevidence.com/content/page/79)

Certainty of the evidence: 0 = no evidence, 100% = high quality evidence; complete certainty. This is certainty of effectiveness of intervention, not of harms.

An assessment by independent experts of the certainty of the evidence for this action based on the summarized evidence (0% = no evidence, 100% = high quality evidence). How certain can we be that the effectiveness score applies to all targets of the intervention (e.g. all birds for an action in the bird synopsis)? This score is based on the number, quality and coverage (species, habitats, geographical locations) of studies. Actions with high scores are supported by lots of well-designed studies with a broad coverage relative to the scope of the intervention. However, the definition of "lots" and "well-designed" will vary between interventions and synopses depending on the breadth of the subject. The certainty score is combined with the effectiveness and harms scores to determine the overall effectiveness category.

Harms: 0 = none, 100% = major negative side-effects to the group of species/habitat of concern.

An assessment by independent experts of the harms of this action to the target group of species/habitat, based on the summarized evidence (0% = none, 100% = major undesirable effects). Undesirable effects on other groups of species/habitats are not considered in this score. The harms score is combined with the effectiveness and certainty scores to determine the overall effectiveness category.

Categorization of interventions

After one or two rounds of initial scoring, interventions are categorized by their effectiveness, as assessed by the expert panel. The median score from all the experts' assessments is calculated for the effectiveness, certainty and harms for each intervention. Categorization is based on these median values i.e. on a combination of the size of the benefit and harm and the strength of the evidence. The table and figure overleaf show how interventions are categorized using the median scores. There is an important distinction between lack of benefit and lack of evidence of benefit.

Once interventions are categorized, experts are given the chance to object if they believe an intervention has been categorized incorrectly. Interventions that receive a specified number (depending on the size of the panel) of strong objections from experts are re-scored by the expert panel and re-categorized accordingly. Experts did not see the categories for the farmland synopsis or for the 'Reduce predation by other species' section of the bird synopsis and so those categories are based on the second round of scoring.

How to use *What Works in Conservation*

Please remember that the categories provided in this book are meant as a guide and a starting point in assessing the effectiveness of conservation interventions. The assessments are based on the available evidence for the target group of species for each intervention and may therefore refer to different species or habitat to the one(s) you are considering. Before making any decisions about implementing interventions it is vital that you read the more detailed accounts of the evidence, in order to assess their relevance to your species or system. Full details of the evidence are available at www.conservationevidence.com.

There may also be significant negative side-effects on the target groups or other species or communities that have not been identified in our assessment. A lack of evidence means that we have been unable to assess whether or not an intervention is effective or has any harmful impacts.

Table of categories of effectiveness

Category	Description	General criteria	Thresholds
Beneficial	Effectiveness has been demonstrated by clear evidence. Expectation of harms is small compared with the benefits	High median benefit score High median certainty score Low median harm score	Effectiveness: >60% Certainty: >60% Harm: <20%
Likely to be beneficial	Effectiveness is less well established than for those listed under 'beneficial' **OR** There is clear evidence of medium effectiveness	High benefit score Lower certainty score Low harm score **OR** Medium benefit score High certainty score Low harm score	Effectiveness: >60% Certainty: 40–60% Harm: <20% **OR** Effectiveness: 40–60% Certainty: ≥40% Harm: <20%
Trade-off between benefit and harms	Interventions for which practitioners must weigh up the beneficial and harmful effects according to individual circumstances and priorities	Medium benefit and medium harm scores **OR** High benefit and high harm scores High certainty score	Effectiveness: ≥40% Certainty: ≥40% Harm: ≥20%
Unknown effectiveness (limited evidence)	Currently insufficient data, or data of inadequate quality	Low certainty score	Effectiveness: Any Certainty: <40% Harm: Any
Unlikely to be beneficial	Lack of effectiveness is less well established than for those listed under 'likely to be ineffective or harmful'	Low benefit score Medium certainty score and/or some variation between experts	Effectiveness: <40% Certainty: 40–60% Harm: <20%
Likely to be ineffective or harmful	Ineffectiveness or harmfulness has been demonstrated by clear evidence	Low benefit score High certainty score (regardless of harms) **OR** Low benefit score High harm score (regardless of certainty of effectiveness)	Effectiveness: <40% Certainty: >60% Harm: Any **OR** Effectiveness: <40% Certainty: ≥ 40% Harm: ≥20%

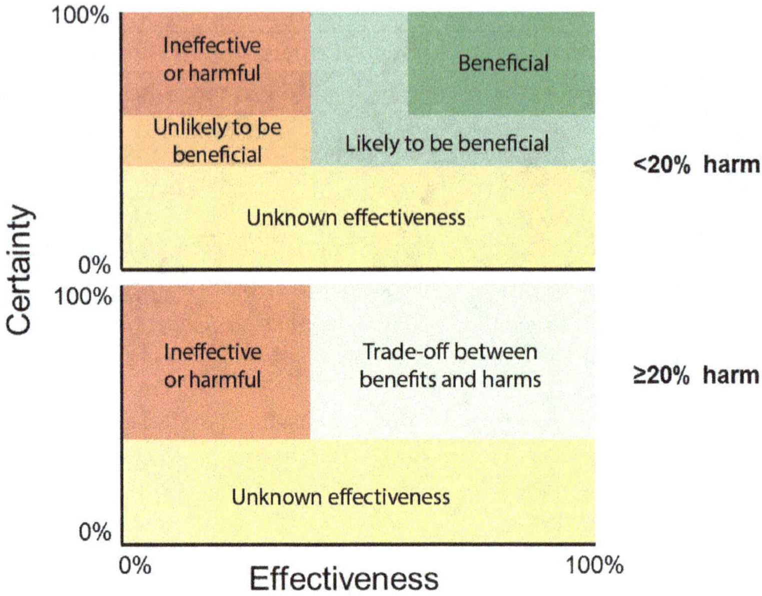

Categories of effectiveness based on a combination of effectiveness (the size of the benefit and harm) and certainty (the strength of the evidence). The top graph refers to interventions with harms <20% and the bottom graph to interventions with harms ≥20%.

1. AMPHIBIAN CONSERVATION

Rebecca K. Smith, Helen Meredith & William J. Sutherland

Expert assessors

Ariadne Angulo, Co-Chair of the Amphibian Specialist Group, Peru
Robert Brodman, Saint Joseph's College, Indiana, USA
Andrew Cunningham, Institute of Zoology, Zoological Society of London, UK
Jeff Dawson, Durrell Wildlife Conservation Trust, UK
Rob Gandola, University of Southampton, UK
Jaime García Moreno, International Union for Conservation of Nature, The Netherlands
Trent Garner, Institute of Zoology, Zoological Society of London, UK
Richard Griffiths, Durrell Institute of Conservation and Ecology, UK
Sergei Kuzmin, Russian Academy of Sciences
Michael Lanoo, Indiana University, USA
Michael Lau, WWF-Hong Kong
James Lewis, Amphibian Survival Alliance/Global Wildlife Conservation, USA
An Martel, Ghent University, Belgium
LeGrand Nono Gonwouo, Cameroon Herpetology-Conservation Biology Foundation
Deanna Olson, US Forest Service
Timo Paasikunnas, Curator of Conservation at Helsinki Zoo, Finland
Frank Pasmans, Ghent University, Belgium
Silviu Petrovan, Froglife, UK
Carlos Martínez Rivera, Philadelphia Zoo, USA
Gonçalo Rosa, Institute of Zoology, Zoological Society of London, UK
David Sewell, Durrell Institute of Conservation and Ecology, UK
Rebecca K. Smith, University of Cambridge, UK
Ben Tapley, Herpetology Department, Zoological Society of London, UK
Jeanne Tarrant, Endangered Wildlife Trust, South Africa
Karthikeyan Vasudevan, Wildlife Institute of India
Victor Wasonga, National Museums of Kenya
Ché Weldon, North-West University, South Africa
Sally Wren, Amphibian Specialist Group Programme Officer, New Zealand

Scope of assessment: for native wild amphibian species across the world.
Assessed: 2014.
Effectiveness measure is the median % score for effectiveness.
Certainty measure is the median % certainty of evidence for effectiveness, determined by the quantity and quality of the evidence in the synopsis.
Harm measure is the median % score for negative side-effects to the group of species of concern.

 https://doi.org/10.11647/OBP.0267.01

This book is meant as a guide to the evidence available for different conservation interventions and as a starting point in assessing their effectiveness. The assessments are based on the available evidence for the target group of species for each intervention. The assessment may therefore refer to different species or habitat to the one(s) you are considering. Before making any decisions about implementing interventions it is vital that you read the more detailed accounts of the evidence in order to assess their relevance for your study species or system.

Full details of the evidence are available at**www. conservationevidence.com**

There may also be significant negative side-effects on the target groups or other species or communities that have not been identified in this assessment.

A lack of evidence means that we have been unable to assess whether or not an intervention is effective or has any harmful impacts.

1.1 Threat: Residential and commercial development

Based on the collated evidence, what is the current assessment of the effectiveness of interventions for residential and commercial development?	
Unknown effectiveness (limited evidence)	• Legal protection of species
No evidence found (no assessment)	• Protect brownfield or ex-industrial sites • Restrict herbicide, fungicide and pesticide use on and around ponds on golf courses

Unknown effectiveness (limited evidence)

Legal protection of species

Three reviews, including one systematic review, in the Netherlands and UK found that legal protection of amphibians was not effective at protecting populations during development. Two reviews found that the number of great crested newt mitigation licences issued in England and Wales increased over 10 years. *Assessment: unknown effectiveness — limited evidence (effectiveness 10%; certainty 35%; harms 7%).*

http://www.conservationevidence.com/actions/779

No evidence found (no assessment)

We have captured no evidence for the following interventions:

- Protect brownfield or ex-industrial sites
- Restrict herbicide, fungicide and pesticide use on and around ponds on golf courses

1.2 Threat: Agriculture

1.2.1 Engage farmers and other volunteers

Based on the collated evidence, what is the current assessment of the effectiveness of interventions for engaging farmers and other volunteers?	
Likely to be beneficial	• Engage landowners and other volunteers to manage land for amphibians • Pay farmers to cover the costs of conservation measures

Likely to be beneficial

Engage landowners and other volunteers to manage land for amphibians

Three studies, including one replicated and one controlled study, in Estonia, Mexico and Taiwan found that engaging landowners and other volunteers in habitat management increased amphibian populations and axolotl weight. Six studies in Estonia, the USA and UK found that up to 41,000 volunteers were engaged in habitat restoration programmes for amphibians and restored up to 1,023 ponds or 11,500 km^2 of habitat. *Assessment: likely to be beneficial (effectiveness 70%; certainty 55%; harms 5%).*

http://www.conservationevidence.com/actions/777

Pay farmers to cover the costs of conservation measures

Four of five studies, including two replicated studies, in Denmark, Sweden and Taiwan found that payments to farmers increased amphibian populations, numbers of species or breeding habitat. One found that amphibian habitat was not maintained. *Assessment: likely to be beneficial (effectiveness 70%; certainty 53%; harms 10%).*

http://www.conservationevidence.com/actions/818

1.2.2 Terrestrial habitat management

Based on the collated evidence, what is the current assessment of the effectiveness of interventions for terrestrial habitat management in agricultural systems?	
Unknown effectiveness (limited evidence)	• Manage cutting regime • Manage grazing regime
No evidence found (no assessment)	• Maintain or restore hedges • Plant new hedges • Reduced tillage

Manage silviculture practices in plantations

Studies investigating the effects of silviculture practices are discussed in 'Threat: Biological resource use — Logging and wood harvesting'.

Unknown effectiveness (limited evidence)

Manage cutting regime

One before-and-after study in Australia found that restoration that included reduced mowing increased numbers of frog species. *Assessment for 'Change mowing regime' from 'Habitat restoration and creation' section: unknown effectiveness — limited evidence (effectiveness 50%; certainty 30%; harms 0%).*

http://www.conservationevidence.com/actions/788

Manage grazing regime

Two studies, including one replicated, controlled study, in the UK and USA found that grazed plots had lower numbers of toads than ungrazed plots and that grazing, along with burning, decreased numbers of amphibian species. Five studies, including four replicated studies, in Denmark, Estonia and the UK found that habitat management that included reintroduction of grazing maintained or increased toad populations. *Assessment: unknown effectiveness — limited evidence (effectiveness 45%; certainty 39%; harms 10%).*

http://www.conservationevidence.com/actions/780

No evidence found (no assessment)

We have captured no evidence for the following interventions:

- Maintain or restore hedges
- Plant new hedges
- Reduced tillage

1.2.3 Aquatic habitat management

Based on the collated evidence, what is the current assessment of the effectiveness of interventions for aquatic habitat management in agricultural systems?	
Likely to be beneficial	• Manage ditches
Likely to be ineffective or harmful	• Exclude domestic animals or wild hogs from ponds by fencing

Likely to be beneficial

Manage ditches

One controlled, before-and-after study in the UK found that managing ditches increased toad numbers. One replicated, site comparison study in the Netherlands found that numbers of amphibians and species were higher in ditches managed under agri-environment schemes compared to those managed conventionally. *Assessment: likely to be beneficial (effectiveness 71%; certainty 60%; harms 0%).*

http://www.conservationevidence.com/actions/749

Likely to be ineffective or harmful

Exclude domestic animals or wild hogs from ponds by fencing

Four replicated studies, including one randomized, controlled, before-and-after study, in the USA found that excluding livestock from streams or ponds did not increase overall numbers of amphibians, species, eggs or larval survival, but did increase larval and metamorph abundance. One before-and-after study in the UK found that pond restoration that included livestock exclusion increased pond use by breeding toads. *Assessment: likely to be ineffective or harmful (effectiveness 31%; certainty 50%; harms 25%).*

http://www.conservationevidence.com/actions/746

1.3 Threat: Energy production and mining

Based on the collated evidence, what is the current assessment of the effectiveness of interventions for energy production and mining?	
Unknown effectiveness (limited evidence)	• Artificially mist habitat to keep it damp

Unknown effectiveness (limited evidence)

Artificially mist habitat to keep it damp

One before-and-after study in Tanzania found that installing a sprinkler system to mitigate against a reduction of river flow did not maintain a population of Kihansi spray toads. *Assessment: unknown effectiveness — limited evidence (effectiveness 24%; certainty 20%; harms 0%).*

http://www.conservationevidence.com/actions/755

1.4 Threat: Transportation and service corridors

Based on the collated evidence, what is the current assessment of the effectiveness of interventions for transportation and service corridors?	
Likely to be beneficial	• Close roads during seasonal amphibian migration • Modify gully pots and kerbs
Trade-off between benefit and harms	• Install barrier fencing along roads • Install culverts or tunnels as road crossings
Unknown effectiveness (limited evidence)	• Use signage to warn motorists
Unlikely to be beneficial	• Use humans to assist migrating amphibians across roads

Likely to be beneficial

Close roads during seasonal amphibian migration

Two studies, including one replicated study, in Germany found that road closure sites protected large numbers of amphibians from mortality during breeding migrations. *Assessment: likely to be beneficial (effectiveness 85%; certainty 50%; harms 0%).*

http://www.conservationevidence.com/actions/842

Modify gully pots and kerbs

One before-and-after study in the UK found that moving gully pots 10 cm away from the kerb decreased the number of great crested newts that fell in by 80%. *Assessment: likely to be beneficial (effectiveness 80%; certainty 40%; harms 0%).*

http://www.conservationevidence.com/actions/782

Trade-off between benefit and harms

Install barrier fencing along roads

Seven of eight studies, including one replicated and two controlled studies, in Germany, Canada and the USA found that barrier fencing with culverts decreased amphibian road deaths, in three cases depending on fence design. One study found that few amphibians were diverted by barriers. *Assessment: trade-offs between benefits and harms (effectiveness 65%; certainty 68%; harms 23%).*

http://www.conservationevidence.com/actions/756

Install culverts or tunnels as road crossings

Thirty-two studies investigated the effectiveness of installing culverts or tunnels as road crossings for amphibians. Six of seven studies, including three replicated studies, in Canada, Europe and the USA found that installing culverts or tunnels decreased amphibian road deaths. One found no effect on road deaths. Fifteen of 24 studies, including one review, in Australia, Canada, Europe and the USA found that tunnels were used by amphibians. Four found mixed effects depending on species, site or culvert type. Five found that culverts were not used or were used by less than 10% of amphibians. Six studies, including one replicated, controlled study, in Canada, Europe and the USA investigated the use of culverts with flowing water. Two found that they were used by amphibians. Three found that they were rarely or not used. Certain culvert designs were found not to be suitable for amphibians. *Assessment: trade-offs between benefits and harms (effectiveness 60%; certainty 75%; harms 25%).*

http://www.conservationevidence.com/actions/884

Unknown effectiveness (limited evidence)

Use signage to warn motorists

One study in the UK found that despite warning signs and human assistance across roads, some toads were still killed on roads. *Assessment: unknown effectiveness — limited evidence (effectiveness 10%; certainty 10%; harms 0%).*

http://www.conservationevidence.com/actions/841

Unlikely to be beneficial

Use humans to assist migrating amphibians across roads

Three studies, including one replicated study, in Italy and the UK found that despite assisting toads across roads during breeding migrations, toads were still killed on roads and 64–70% of populations declined. Five studies in Germany, Italy and the UK found that large numbers of amphibians were moved across roads by up to 400 patrols. *Assessment: unlikely to be beneficial (effectiveness 35%; certainty 40%; harms 3%).*

http://www.conservationevidence.com/actions/784

1.5 Threat: Biological resource use

1.5.1 Hunting and collecting terrestrial animals

Based on the collated evidence, what is the current assessment of the effectiveness of interventions for hunting and collecting terrestrial animals?	
Likely to be beneficial	• Reduce impact of amphibian trade
Unknown effectiveness (limited evidence)	• Use legislative regulation to protect wild populations
No evidence found (no assessment)	• Commercially breed amphibians for the pet trade • Use amphibians sustainably

Likely to be beneficial

Reduce impact of amphibian trade

One review found that reducing trade through legislation allowed frog populations to recover from over-exploitation. *Assessment: likely to be beneficial (effectiveness 76%; certainty 40%; harms 0%).*

http://www.conservationevidence.com/actions/824

Unknown effectiveness (limited evidence)

Use legislative regulation to protect wild populations

One review found that legislation to reduce trade resulted in the recovery of frog populations. One study in South Africa found that the number of permits issued for scientific and educational use of amphibians increased from 1987 to 1990. *Assessment: unknown effectiveness — limited evidence (effectiveness 60%; certainty 30%; harms 5%).*

http://www.conservationevidence.com/actions/785

No evidence found (no assessment)

We have captured no evidence for the following interventions:

- Commercially breed amphibians for the pet trade
- Use amphibians sustainably

1.5.2 Logging and wood harvesting

Based on the collated evidence, what is the current assessment of the effectiveness of interventions for logging and wood harvest?	
Likely to be beneficial	• Retain riparian buffer strips during timber harvest • Use shelterwood harvesting instead of clearcutting
Trade-off between benefit and harms	• Leave coarse woody debris in forests
Unknown effectiveness (limited evidence)	• Use patch retention harvesting instead of clearcutting
Unlikely to be beneficial	• Leave standing deadwood/snags in forests • Use leave-tree harvesting instead of clearcutting
Likely to be ineffective or harmful	• Harvest groups of trees instead of clearcutting • Thin trees within forests

Likely to be beneficial

Retain riparian buffer strips during timber harvest

Six replicated and/or controlled studies in Canada and the USA compared amphibian numbers following clearcutting with or without riparian buffer strips. Five found mixed effects and one found that abundance was higher with riparian buffers. Two of four replicated studies, including one randomized, controlled, before-and-after study, in Canada and the USA found that numbers of species and abundance were greater in wider buffer strips. Two found no effect of buffer width. *Assessment: likely to be beneficial (effectiveness 50%; certainty 61%; harms 10%).*

http://www.conservationevidence.com/actions/747

Use shelterwood harvesting instead of clearcutting

Three studies, including two randomized, replicated, controlled, before-and-after studies, in the USA found that compared to clearcutting, shelterwood harvesting resulted in higher or similar salamander abundance. One meta-analysis of studies in North America found that partial harvest, which included shelterwood harvesting, resulted in smaller reductions in salamander populations than clearcutting. *Assessment: likely to be beneficial (effectiveness 40%; certainty 57%; harms 10%).*

http://www.conservationevidence.com/actions/851

Trade-off between benefit and harms

Leave coarse woody debris in forests

Two replicated, controlled studies in the USA found that abundance was similar in clearcuts with woody debris retained or removed for eight of nine amphibian species, but that the overall response of amphibians was more negative where woody debris was retained. Two replicated, controlled studies in the USA and Indonesia found that the removal of coarse woody debris from standing forest did not affect amphibian diversity or overall amphibian abundance, but did reduce species richness. One replicated, controlled study in the USA found that migrating amphibians used clearcuts where woody debris was retained more than where it was removed. One replicated, site comparison study in the USA found that within clearcut forest, survival of juvenile amphibians was significantly higher within piles of woody debris than in open areas. *Assessment: trade-offs between benefits and harms (effectiveness 40%; certainty 60%; harms 26%).*

http://www.conservationevidence.com/actions/843

Unknown effectiveness (limited evidence)

Use patch retention harvesting instead of clearcutting

We found no evidence for the effect of retaining patches of trees rather than clearcutting on amphibian populations. One replicated study in Canada found that although released red-legged frogs did not move towards retained tree patches, large patches were selected more and moved out of less than small patches. *Assessment: unknown effectiveness — limited evidence (effectiveness 20%; certainty 25%; harms 0%).*

http://www.conservationevidence.com/actions/847

Unlikely to be beneficial

Leave standing deadwood/snags in forests

One randomized, replicated, controlled, before-and-after study in the USA found that compared to total clearcutting, leaving dead and wildlife trees did not result in higher abundances of salamanders. One randomized, replicated, controlled study in the USA found that numbers of amphibians and species were similar with removal or creation of dead trees within forest. *Assessment: unlikely to be beneficial (effectiveness 5%; certainty 58%; harms 2%).*

http://www.conservationevidence.com/actions/845

Use leave-tree harvesting instead of clearcutting

Two studies, including one randomized, replicated, controlled, before-and-after study, in the USA found that compared to clearcutting, leaving a low density of trees during harvest did not result in higher salamander abundance. *Assessment: unlikely to be beneficial (effectiveness 10%; certainty 48%; harms 11%).*

http://www.conservationevidence.com/actions/846

Likely to be ineffective or harmful

Harvest groups of trees instead of clearcutting

Three studies, including two randomized, replicated, controlled, before-and-after studies, in the USA found that harvesting trees in small groups resulted in similar amphibian abundance to clearcutting. One meta-analysis and one randomized, replicated, controlled, before-and-after study in North America and the USA found that harvesting, which included harvesting groups of trees, resulted in smaller reductions in salamander populations than clearcutting. *Assessment: likely to be ineffective or harmful (effectiveness 33%; certainty 60%; harms 23%).*

http://www.conservationevidence.com/actions/844

Thin trees within forests

Six studies, including five replicated and/or controlled studies, in the USA compared amphibians in thinned to unharvested forest. Three found that thinning had mixed effects and one found no effect on abundance. One found that amphibian abundance increased following thinning but the body condition of ensatina salamanders decreased. One found a negative overall response of amphibians. Four studies, including two replicated, controlled studies, in the USA compared amphibians in thinned to clearcut forest. Two found that thinning had mixed effects on abundance and two found higher amphibian abundance or a less negative overall response of amphibians following thinning. One meta-analysis of studies in North America found that partial harvest, which included thinning, decreased salamander populations, but resulted in smaller reductions than clearcutting. *Assessment: likely to be ineffective or harmful (effectiveness 35%; certainty 60%; harms 40%).*

http://www.conservationevidence.com/actions/852

1.6 Threat: Human intrusions and disturbance

Based on the collated evidence, what is the current assessment of the effectiveness of interventions for human intrusions and disturbance?	
No evidence found (no assessment)	• Use signs and access restrictions to reduce disturbance

No evidence found (no assessment)

We have captured no evidence for the following intervention:

- Use signs and access restrictions to reduce disturbance

1.7 Threat: Natural system modifications

Based on the collated evidence, what is the current assessment of the effectiveness of interventions for natural system modifications?	
Beneficial	• Regulate water levels
Unknown effectiveness (limited evidence)	• Mechanically remove mid-storey or ground vegetation
Likely to be ineffective or harmful	• Use herbicides to control mid-storey or ground vegetation • Use prescribed fire or modifications to burning regime: forests • Use prescribed fire or modifications to burning regime: grassland

Beneficial

Regulate water levels

Three studies, including one replicated, site comparison study, in the UK and USA found that maintaining pond water levels, in two cases with other habitat management, increased or maintained amphibian populations or increased breeding success. One replicated, controlled study in Brazil found that keeping rice fields flooded after harvest did not change amphibian abundance or numbers of species, but changed species composition. One replicated, controlled study in the USA found that draining ponds increased abundance and numbers of amphibian species. *Assessment: beneficial (effectiveness 70%; certainty 65%; harms 10%).*

http://www.conservationevidence.com/actions/833

Unknown effectiveness (limited evidence)

Mechanically remove mid-storey or ground vegetation

One randomized, replicated, controlled study in the USA found that mechanical understory reduction increased numbers of amphibian species, but not amphibian abundance. *Assessment: unknown effectiveness — limited evidence (effectiveness 40%; certainty 30%; harms 0%).*

http://www.conservationevidence.com/actions/781

Likely to be ineffective or harmful

Use herbicides to control mid-storey or ground vegetation

Three studies, including two randomized, replicated, controlled studies, in the USA found that understory removal using herbicide had no effect or negative effects on amphibian abundance. One replicated, site comparison study in Canada found that following logging, abundance was similar or lower in stands with herbicide treatment and planting compared to those left to regenerate naturally. *Assessment: likely to be ineffective or harmful (effectiveness 10%; certainty 50%; harms 50%).*

http://www.conservationevidence.com/actions/778

Use prescribed fire or modifications to burning regime (forests)

Eight of 15 studies, including three randomized, replicated, controlled studies, in Australia, North America and the USA found no effect of prescribed forest fires on amphibian abundance or numbers of species. Four found that fires had mixed effects on abundance. Four found that abundance, numbers of species or hatching success increased and one that abundance decreased. *Assessment: likely to be ineffective or harmful (effectiveness 30%; certainty 58%; harms 40%).*

http://www.conservationevidence.com/actions/877

Use prescribed fire or modifications to burning regime (grassland)

Two of three studies, including one replicated, before-and-after study, in the USA and Argentina found that prescribed fires in grassland decreased amphibian abundance or numbers of species. One found that spring, but not autumn or winter burns in grassland, decreased abundance. *Assessment: likely to be ineffective or harmful (effectiveness 10%; certainty 40%; harms 70%).*

http://www.conservationevidence.com/actions/862

1.8 Threat: Invasive and other problematic species

1.8.1 Reduce predation by other species

Based on the collated evidence, what is the current assessment of the effectiveness of interventions for reducing predation by other species?	
Beneficial	• Remove or control fish by drying out ponds
Likely to be beneficial	• Remove or control fish population by catching • Remove or control invasive bullfrogs • Remove or control invasive viperine snake • Remove or control mammals
Trade-off between benefit and harms	• Remove or control fish using Rotenone
Unknown effectiveness (limited evidence)	• Exclude fish with barriers
No evidence found (no assessment)	• Encourage aquatic plant growth as refuge against fish predation • Remove or control non-native crayfish

Beneficial

Remove or control fish by drying out ponds

One before-and-after study in the USA found that draining ponds to eliminate fish increased numbers of amphibian species. Four studies, including one review, in Estonia, the UK and USA found that pond drying to eliminate fish, along with other management activities, increased amphibian abundance, numbers of species and breeding success. *Assessment: beneficial (effectiveness 80%; certainty 66%; harms 3%).*

http://www.conservationevidence.com/actions/826

Likely to be beneficial

Remove or control fish population by catching

Four of six studies, including two replicated, controlled studies, in Sweden, the USA and UK found that removing fish by catching them increased amphibian abundance, survival and recruitment. Two found no significant effect on newt populations or toad breeding success. *Assessment: likely to be beneficial (effectiveness 50%; certainty 52%; harms 0%).*

http://www.conservationevidence.com/actions/827

Remove or control invasive bullfrogs

Two studies, including one replicated, before-and-after study, in the USA and Mexico found that removing American bullfrogs increased the size and range of frog populations. One replicated, before-and-after study in the USA found that following bullfrog removal, frogs were found out in the open more. *Assessment: likely to be beneficial (effectiveness 79%; certainty 60%; harms 0%).*

http://www.conservationevidence.com/actions/825

Remove or control invasive viperine snake

One before-and-after study in Mallorca found that numbers of Mallorcan midwife toad larvae increased after intensive, but not less intensive, removal of viperine snakes. *Assessment: likely to be beneficial (effectiveness 50%; certainty 40%; harms 0%).*

http://www.conservationevidence.com/actions/830

Remove or control mammals

One controlled study in New Zealand found that controlling rats had no significant effect on numbers of Hochstetter's frog. Two studies, one of which was controlled, in New Zealand found that predator-proof enclosures enabled or increased survival of frog species. *Assessment: likely to be beneficial (effectiveness 50%; certainty 40%; harms 0%).*

http://www.conservationevidence.com/actions/839

Trade-off between benefit and harms

Remove or control fish using Rotenone

Three studies, including one replicated study, in Sweden, the UK and USA found that eliminating fish using rotenone increased numbers of amphibians, amphibian species and recruitment. One review in Australia, the UK and USA found that fish control that included using rotenone increased breeding success. Two replicated studies in Pakistan and the UK found that rotenone use resulted in frog deaths and negative effects on newts. *Assessment: trade-offs between benefits and harms (effectiveness 65%; certainty 60%; harms 52%).*

http://www.conservationevidence.com/actions/828

Unknown effectiveness (limited evidence)

Exclude fish with barriers

One controlled study in Mexico found that excluding fish using a barrier increased weight gain of axolotls. *Assessment: unknown effectiveness — limited evidence (effectiveness 30%; certainty 20%; harms 0%).*

http://www.conservationevidence.com/actions/829

No evidence found (no assessment)

We have captured no evidence for the following interventions:

- Encourage aquatic plant growth as refuge against fish predation
- Remove or control non-native crayfish.

1.8.2 Reduce competition with other species

Based on the collated evidence, what is the current assessment of the effectiveness of interventions for reducing competition with other species?	
Unknown effectiveness (limited evidence)	• Reduce competition from native amphibians • Remove or control invasive Cuban tree frogs
No evidence found (no assessment)	• Remove or control invasive cane toads

Unknown effectiveness (limited evidence)

Reduce competition from native amphibians

One replicated, site comparison study in the UK found that common toad control did not increase natterjack toad populations. *Assessment: unknown effectiveness — limited evidence (effectiveness 10%; certainty 23%; harms 0%).*

http://www.conservationevidence.com/actions/821

Remove or control invasive Cuban tree frogs

One before-and-after study in the USA found that removal of invasive Cuban tree frogs increased numbers of native frogs. *Assessment: unknown effectiveness — limited evidence (effectiveness 65%; certainty 30%; harms 0%).*

http://www.conservationevidence.com/actions/822

No evidence found (no assessment)

We have captured no evidence for the following interventions:

- Remove or control invasive cane toads.

1.8.3 Reduce adverse habitat alteration by other species

Based on the collated evidence, what is the current assessment of the effectiveness of interventions for reducing adverse habitat alteration by other species?	
Likely to be beneficial	• Control invasive plants
No evidence found (no assessment)	• Prevent heavy usage/exclude wildfowl from aquatic habitat

Likely to be beneficial

Control invasive plants

One before-and-after study in the UK found that habitat and species management that included controlling swamp stonecrop, increased a population of natterjack toads. One replicated, controlled study in the USA found that more Oregon spotted frogs laid eggs in areas where invasive reed canarygrass was mown. *Assessment: likely to be beneficial (effectiveness 60%; certainty 47%; harms 0%).*

http://www.conservationevidence.com/actions/823

No evidence found (no assessment)

We have captured no evidence for the following intervention:

- Prevent heavy usage/exclude wildfowl from aquatic habitat.

1.8.4 Reduce parasitism and disease – chytridiomycosis

Based on the collated evidence, what is the current assessment of the effectiveness of interventions for reducing chytridiomycosis?	
Likely to be beneficial	• Use temperature treatment to reduce infection
Trade-off between benefit and harms	• Use antifungal treatment to reduce infection
Unknown effectiveness (limited evidence)	• Add salt to ponds • Immunize amphibians against infection • Remove the chytrid fungus from ponds • Sterilize equipment when moving between amphibian sites • Treating amphibians in the wild or pre-release • Use gloves to handle amphibians
Unlikely to be beneficial	• Use antibacterial treatment to reduce infection • Use antifungal skin bacteria or peptides to reduce infection
No evidence found (no assessment)	• Use zooplankton to remove zoospores

Likely to be beneficial

Use temperature treatment to reduce infection

Four of five studies, including four replicated, controlled studies, in Australia, Switzerland and the USA found that increasing enclosure or water temperature to 30–37°C for over 16 hours cured amphibians of chytridiomycosis. One found that treatment did not cure frogs. *Assessment: likely to be beneficial (effectiveness 60%; certainty 70%; harms 10%).*

http://www.conservationevidence.com/actions/770

Trade-off between benefit and harms

Use antifungal treatment to reduce infection

Twelve of 16 studies, including four randomized, replicated, controlled studies, in Europe, Australia, Tasmania, Japan and the USA found that antifungal treatment cured or increased survival of amphibians with chytridiomycosis. Four studies found that treatments did not cure chytridiomycosis, but did reduce infection levels or had mixed results. Six of the eight studies testing treatment with itraconazole found that it was effective at curing chytridiomycosis. One found that it reduced infection levels and one found mixed effects. Six studies found that specific fungicides caused death or other negative side effects in amphibians. *Assessment: trade-offs between benefits and harms (effectiveness 71%; certainty 70%; harms 50%).*

http://www.conservationevidence.com/actions/882

Unknown effectiveness (limited evidence)

Add salt to ponds

One study in Australia found that following addition of salt to a pond containing the chytrid fungus, a population of green and golden bell frogs remained free of chytridiomycosis for over six months. *Assessment: unknown effectiveness — limited evidence (effectiveness 41%; certainty 25%; harms 50%).*

http://www.conservationevidence.com/actions/762

Immunize amphibians against infection

One randomized, replicated, controlled study in the USA found that vaccinating mountain yellow-legged frogs with formalin-killed chytrid fungus did not significantly reduce chytridiomycosis infection rate or mortality. *Assessment: unknown effectiveness — limited evidence (effectiveness 0%; certainty 25%; harms 0%).*

http://www.conservationevidence.com/actions/765

Remove the chytrid fungus from ponds

One before-and-after study in Mallorca found that drying out a pond and treating resident midwife toads with fungicide reduced levels of infection but did not eradicate chytridiomycosis. *Assessment: unknown effectiveness — limited evidence (effectiveness 25%; certainty 25%; harms 0%).*

http://www.conservationevidence.com/actions/766

Sterilize equipment when moving between amphibian sites

We found no evidence for the effects of sterilizing equipment when moving between amphibian sites on the spread of disease between amphibian populations or individuals. Two randomized, replicated, controlled study in Switzerland and Sweden found that Virkon S disinfectant did not affect survival, mass or behaviour of eggs, tadpoles or hatchlings. However, one of the studies found that bleach significantly reduced tadpole survival. *Assessment: unknown effectiveness — limited evidence (effectiveness 10%; certainty 30%; harms 40%).*

http://www.conservationevidence.com/actions/768

Treating amphibians in the wild or pre-release

One before-and-after study in Mallorca found that treating wild toads with fungicide and drying out the pond reduced infection levels but did not eradicate chytridiomycosis. *Assessment: unknown effectiveness — limited evidence (effectiveness 27%; certainty 30%; harms 0%).*

http://www.conservationevidence.com/actions/767

Use gloves to handle amphibians

We found no evidence for the effects of using gloves on the spread of disease between amphibian populations or individuals. A review for Canada and the USA found that there were no adverse effects of handling 22 amphibian species using disposable gloves. However, three replicated studies in Australia and Austria found that deaths of tadpoles were caused by latex, vinyl and nitrile gloves for 60–100% of species tested. *Assessment: unknown effectiveness — limited evidence (effectiveness 9%; certainty 35%; harms 65%).*

http://www.conservationevidence.com/actions/769

Unlikely to be beneficial

Use antibacterial treatment to reduce infection

Two studies, including one randomized, replicated, controlled study, in New Zealand and Australia found that treatment with chloramphenicol antibiotic, with other interventions in some cases, cured frogs of chytridiomycosis. One replicated, controlled study found that treatment with trimethoprim-sulfadiazine increased survival time but did not cure infected frogs. *Assessment: unlikely to be beneficial (effectiveness 38%; certainty 45%; harms 10%).*

http://www.conservationevidence.com/actions/763

Use antifungal skin bacteria or peptides to reduce infection

Three of four randomized, replicated, controlled studies in the USA found that introducing antifungal bacteria to the skin of chytrid infected amphibians did not reduce infection rate or deaths. One found that it prevented infection and death. One randomized, replicated, controlled study in the USA found that adding antifungal skin bacteria to soil significantly reduced chytridiomycosis infection rate in salamanders. One randomized, replicated, controlled study in Switzerland found that treatment with antimicrobial skin peptides before or after infection with chytridiomycosis did not increase toad survival. *Assessment: unlikely to be beneficial (effectiveness 29%; certainty 50%; harms 10%).*

http://www.conservationevidence.com/actions/764

No evidence found (no assessment)

We have captured no evidence for the following intervention:

- Use zooplankton to remove zoospores

1.8.5 Reduce parasitism and disease – ranaviruses

Based on the collated evidence, what is the current assessment of the effectiveness of interventions for reducing ranaviruses?	
No evidence found (no assessment)	• Sterilize equipment to prevent ranaviruses

No evidence found (no assessment)

We have captured no evidence for the following intervention:

- Sterilize equipment to prevent ranaviruses.

1.9 Threat: Pollution

1.9.1 Agricultural pollution

Based on the collated evidence, what is the current assessment of the effectiveness of interventions for agricultural pollution?	
Unknown effectiveness (limited evidence)	• Create walls or barriers to exclude pollutants • Plant riparian buffer strips • Reduce pesticide, herbicide or fertilizer use
No evidence found (no assessment)	• Prevent pollution from agricultural lands or sewage treatment facilities entering watercourses

Unknown effectiveness (limited evidence)

Create walls or barriers to exclude pollutants

One controlled study in Mexico found that installing filters across canals to improve water quality and exclude fish increased weight gain in axolotls. *Assessment: unknown effectiveness — limited evidence (effectiveness 35%; certainty 29%; harms 0%).*

http://www.conservationevidence.com/actions/771

Plant riparian buffer strips

One replicated, controlled study in the USA found that planting buffer strips along streams did not increase amphibian abundance or numbers of species. *Assessment: unknown effectiveness — limited evidence (effectiveness 0%; certainty 30%; harms 0%).*

http://www.conservationevidence.com/actions/819

Reduce pesticide, herbicide or fertilizer use

One study in Taiwan found that halting pesticide use, along with habitat management, increased a population of frogs. *Assessment: unknown effectiveness — limited evidence (effectiveness 71%; certainty 26%; harms 0%).*

http://www.conservationevidence.com/actions/832

No evidence found (no assessment)

We have captured no evidence for the following intervention:

- Prevent pollution from agricultural lands or sewage treatment facilities entering watercourses

1.9.2 Industrial pollution

Based on the collated evidence, what is the current assessment of the effectiveness of interventions for industrial pollution?	
Trade-off between benefit and harms	• Add limestone to water bodies to reduce acidification
No evidence found (no assessment)	• Augment ponds with ground water to reduce acidification

Trade-off between benefit and harms

Add limestone to water bodies to reduce acidification

Five before-and-after studies, including one controlled, replicated study, in the Netherlands and UK found that adding limestone to ponds resulted in establishment of one of three translocated amphibian populations, a temporary increase in breeding and metamorphosis by natterjack toads and increased egg and larval survival of frogs. One replicated, site comparison study in the UK found that habitat management that included adding limestone to ponds increased natterjack toad populations. However, two before-and-after studies, including one controlled study, in the UK found that adding limestone to ponds resulted in increased numbers of abnormal eggs, high tadpole mortality and pond abandonment. *Assessment: trade-offs between benefits and harms (effectiveness 47%; certainty 50%; harms 50%).*

http://www.conservationevidence.com/actions/748

No evidence found (no assessment)

We have captured no evidence for the following intervention:

- Augment ponds with ground water to reduce acidification.

1.10 Threat: Climate change and severe weather

Based on the collated evidence, what is the current assessment of the effectiveness of interventions for climate change and severe weather?	
Beneficial	• Deepen ponds to prevent desiccation (deepen, de-silt or re-profile)
Unknown effectiveness (limited evidence)	• Use irrigation systems for amphibian sites (artificially mist habitat)
No evidence found (no assessment)	• Artificially shade ponds to prevent desiccation • Protect habitat along elevational gradients • Provide shelter habitat

Create microclimate and microhabitat refuges

Studies investigating the effects of creating refuges are discussed in 'Habitat restoration and creation' and 'Threat: Biological resource use — Leave coarse woody debris in forests'.

Maintain ephemeral ponds

Studies investigating the effects of regulating water levels and deepening ponds are discussed in 'Threat: Natural system modifications — Regulate water levels' and 'Habitat restoration and creation — Deepen, de-silt or re-profile ponds'.

Beneficial

Deepen ponds to prevent desiccation

Four studies, including one replicated, controlled study, in France, Denmark and the UK found that pond deepening and enlarging or re-profiling resulted in establishment or increased populations of amphibians. Four before-and-after studies in Denmark and the UK found that pond deepening, along with other interventions, maintained newt or increased toad populations. *Assessment for 'Deepen, de-silt or re-profie ponds' from* 'Habitat restoration and creation' section: *beneficial (effectiveness 71%; certainty 65%; harms 0%).*

http://www.conservationevidence.com/actions/806

Unknown effectiveness (limited evidence)

Use irrigation systems for amphibian sites

One before-and-after study in Tanzania found that installing a sprinkler system to mitigate against a reduction of river flow did not maintain a population of Kihansi spray toads. *Assessment for 'Artificially mist habitat to keep it damp' from 'Threat: Energy production and mining' section: unknown effectiveness — limited evidence (effectiveness 24%; certainty 20%; harms 0%).*

http://www.conservationevidence.com/actions/804

No evidence found (no assessment)

We have captured no evidence for the following interventions:

- Artificially shade ponds to prevent desiccation
- Protect habitat along elevational gradients
- Provide shelter habitat.

1.11 Habitat protection

Based on the collated evidence, what is the current assessment of the effectiveness of interventions for habitat protection?	
Trade-off between benefit and harms	• Retain buffer zones around core habitat
Unknown effectiveness (limited evidence)	• Protect habitats for amphibians • Retain connectivity between habitat patches

Trade-off between benefit and harms

Retain buffer zones around core habitat

Two studies, including one replicated, controlled study, in Australia and the USA found that retaining unmown buffers around ponds increased numbers of frog species, but had mixed effects on tadpole mass and survival. One replicated, site comparison study in the USA found that retaining buffers along ridge tops within harvested forest increased salamander abundance, body condition and genetic diversity. However, one replicated study in the USA found that 30 m buffer zones around wetlands were not sufficient to protect marbled salamanders. *Assessment: trade-offs between benefits and harms (effectiveness 50%; certainty 50%; harms 25%).*

http://www.conservationevidence.com/actions/850

Unknown effectiveness (limited evidence)

Protect habitats for amphibians

One replicated, site comparison study in the UK found that statutory level habitat protection helped protect natterjack toad populations. One before-and-after study in the UK found that protecting a pond during development had mixed effects on populations of amphibians. *Assessment: unknown effectiveness — limited evidence (effectiveness 51%; certainty 31%; harms 9%).*

http://www.conservationevidence.com/actions/820

Retain connectivity between habitat patches

One before-and-after study in Australia found that retaining native vegetation corridors maintained populations of frogs over 20 years. *Assessment: unknown effectiveness — limited evidence (effectiveness 60%; certainty 31%; harms 0%).*

http://www.conservationevidence.com/actions/853

1.12 Habitat restoration and creation

1.12.1 Terrestrial habitat

Based on the collated evidence, what is the current assessment of the effectiveness of interventions for terrestrial habitat restoration and creation?	
Beneficial	• Replant vegetation
Likely to be beneficial	• Clear vegetation • Create artificial hibernacula or aestivation sites • Create refuges • Restore habitat connectivity
Unknown effectiveness (limited evidence)	• Change mowing regime
No evidence found (no assessment)	• Create habitat connectivity

Beneficial

Replant vegetation

Four studies, including one replicated study, in Australia, Spain and the USA found that amphibians colonized replanted forest, reseeded grassland and seeded and transplanted upland habitat. Three of four studies, including two replicated studies, in Australia, Canada, Spain and the USA found that areas planted with trees or grass had similar amphibian abundance or community composition to natural sites and one found similar or lower abundance compared to naturally regenerated forest. One found that wetlands within reseeded grasslands were used less than those in natural grasslands. One before-and-after study in Australia found that numbers of frog species increased following restoration that included planting shrubs and trees. *Assessment: beneficial (effectiveness 70%; certainty 63%; harms 3%).*

http://www.conservationevidence.com/actions/849

Likely to be beneficial

Clear vegetation

Seven studies, including four replicated studies, in Australia, Estonia and the UK found that vegetation clearance, along with other habitat management and in some cases release of amphibians, increased or maintained amphibian populations or increased numbers of frog species. However, great crested newt populations were only maintained for six years, but not in the longer term. *Assessment: likely to be beneficial (effectiveness 60%; certainty 54%; harms 10%).*

http://www.conservationevidence.com/actions/761

Create artificial hibernacula or aestivation sites

Two replicated studies in the UK found that artificial hibernacula were used by two of three amphibian species and along with other terrestrial habitat management maintained populations of great crested newts. *Assessment: likely to be beneficial (effectiveness 50%; certainty 44%; harms 0%).*

http://www.conservationevidence.com/actions/759

Create refuges

Two replicated, controlled studies, one of which was randomized, in the USA and Indonesia found that adding coarse woody debris to forest floors had no effect on the number of amphibian species or overall abundance, but had mixed effects on abundance of individual species. One before-and-after study in Australia found that restoration that included reintroducing coarse woody debris to the forest floor increased frog species. Three studies, including two replicated studies, in New Zealand, the UK and USA found that artificial refugia were used by amphibians and, along with other interventions, maintained newt populations. *Assessment: likely to be beneficial (effectiveness 45%; certainty 55%; harms 0%).*

http://www.conservationevidence.com/actions/772

Restore habitat connectivity

One before-and-after study in Italy found that restoring habitat connectivity by raising a road on a viaduct significantly decreased amphibian deaths. *Assessment: likely to be beneficial (effectiveness 75%; certainty 40%; harms 0%).*

http://www.conservationevidence.com/actions/840

Unknown effectiveness (limited evidence)

Change mowing regime

One before-and-after study in Australia found that restoration that included reduced mowing increased numbers of frog species. *Assessment: unknown effectiveness — limited evidence (effectiveness 50%; certainty 30%; harms 0%).*

http://www.conservationevidence.com/actions/783

No evidence found (no assessment)

We have captured no evidence for the following intervention:

- Create habitat connectivity.

1.12.2 Aquatic habitat

Based on the collated evidence, what is the current assessment of the effectiveness of interventions for aquatic habitat restoration and creation?	
Beneficial	• Create ponds (amphibians in general) • Create ponds: frogs • Create ponds: natterjack toads • Create ponds: salamanders (including newts) • Create wetlands • Deepen, de-silt or re-profile ponds • Restore wetlands

Likely to be beneficial	• Create ponds: great crested newts • Create ponds: green toads • Create ponds: toads • Remove specific aquatic plants (invasive species) • Restore ponds
Unknown effectiveness (limited evidence)	• Remove tree canopy to reduce pond shading
No evidence found (no assessment)	• Add nutrients to new ponds as larvae food source • Add specific plants to aquatic habitats • Add woody debris to ponds • Create refuge areas in aquatic habitats

Beneficial

Create ponds (amphibians in general)

Twenty-eight studies investigated the colonization of created ponds by amphibians in general, all of which found that amphibians used all or some of the created ponds. Five of nine studies in Australia, Canada, Spain, the UK and USA found that numbers of species were similar or higher in created compared to natural ponds. Nine studies in Europe and the USA found that amphibians established stable populations, used or reproduced in created ponds. Four found that species composition differed, and abundance, juvenile productivity or size in created ponds depended on species. One study found that numbers of species were similar or lower in created ponds. Sixteen studies in Europe and the USA found that created ponds were used or colonized by up to 15 naturally colonizing species, up to 10 species that reproduced or by captive-bred amphibians. Five studies in Europe and the USA found that pond creation, with restoration in three cases, maintained and increased populations or increased species. *Assessment: beneficial (effectiveness 80%; certainty 80%; harms 0%).*

http://www.conservationevidence.com/actions/869

Create ponds (frogs)

Six of nine studies in Australia, Italy, Spain, the UK and USA found that frogs established breeding populations or reproduced in created ponds. One study in Denmark found that frogs colonized created ponds. One study in the Netherlands found that pond creation, along with vegetation clearance, increased frog populations. One study in the USA found that survival increased with age of created ponds. *Assessment: beneficial (effectiveness 75%; certainty 70%; harms 0%).*

http://www.conservationevidence.com/actions/865

Create ponds (natterjack toads)

Five studies in the UK and Denmark found that pond creation, along with other interventions, maintained or increased populations at 75–100% of sites. One study in the UK found that compared to natural ponds, created ponds had lower tadpole mortality from desiccation, but higher mortality from predation by invertebrates. *Assessment: beneficial (effectiveness 75%; certainty 70%; harms 10%).*

http://www.conservationevidence.com/actions/866

Create ponds (salamanders including newts)

Three studies in France, Germany and the USA found that alpine newts, captive-bred smooth newts and translocated spotted salamanders established stable breeding populations in 20–100% of created ponds. Three studies in France, China and the USA found that alpine newts, Chinhai salamanders and translocated spotted salamanders, but not tiger salamanders, reproduced in created ponds. *Assessment: beneficial (effectiveness 70%; certainty 65%; harms 0%).*

http://www.conservationevidence.com/actions/867

Create wetlands

Fifteen studies, including one review and seven replicated studies, in Australia, Kenya and the USA, investigated the effectiveness of creating wetlands for amphibians. Six studies found that created wetlands had similar amphibian abundance, numbers of species or communities as natural wetlands or in one case adjacent forest. Two of those studies found that created wetlands had fewer amphibians, amphibian species and different communities compared to natural wetlands. One global review and two other studies combined created and restored wetlands and found that amphibian abundance and numbers of species were similar or higher compared to natural wetlands. Five of the studies found that up to 15 amphibian species used created wetlands. One study found that captive-bred frogs did not establish in a created wetland. *Assessment: beneficial (effectiveness 75%; certainty 70%; harms 0%).*

http://www.conservationevidence.com/actions/880

Deepen, de-silt or re-profile ponds

Four studies, including one replicated, controlled study, in France, Denmark and the UK found that pond deepening and enlarging or re-profiling resulted in establishment or increased populations of amphibians. Four before-and-after studies in Denmark and the UK found that pond deepening, along with other interventions, maintained newt or increased toad populations. *Assessment: beneficial (effectiveness 71%; certainty 65%; harms 0%).*

http://www.conservationevidence.com/actions/817

Restore wetlands

Seventeen studies, including one review and 11 replicated studies, in Canada, Taiwan and the USA, investigated the effectiveness of wetland restoration for amphibians. Seven of ten studies found that amphibian abundance, numbers of species and species composition were similar in restored and natural wetlands. Two found that abundance or numbers of species were lower and species composition different to natural wetlands. One found mixed results. One global review found that in 89% of cases, restored and created wetlands had similar or higher amphibian abundance or numbers of species to natural wetlands. Seven of nine studies found that wetland restoration increased numbers of amphibian species, with breeding populations establishing in some cases, and maintained or increased abundance of individual species. Three found that amphibian abundance or numbers of species did not increase with restoration. Three of the studies found that restored wetlands were colonized by up to eight amphibian species. *Assessment: beneficial (effectiveness 80%; certainty 73%; harms 0%).*

http://www.conservationevidence.com/actions/879

Likely to be beneficial

Create ponds (great crested newts)

Three studies in Germany and the UK found that great crested newts established breeding populations in created ponds. One systematic review in the UK found that there was no conclusive evidence that mitigation, which often included pond creation, resulted in self-sustaining populations. Four studies in the UK found that great crested newts colonized up to 88% of, or reproduced in 38% of created ponds. *Assessment: likely to be beneficial (effectiveness 60%; certainty 61%; harms 0%).*

http://www.conservationevidence.com/actions/863

Create ponds (green toads)

Two studies in Denmark found that pond creation, along with other interventions, significantly increased green toad populations. One study in Sweden found that green toads used or reproduced in 41–59% of created ponds. *Assessment: likely to be beneficial (effectiveness 73%; certainty 59%; harms 0%).*

http://www.conservationevidence.com/actions/864

Create ponds (toads)

Five studies in Germany, Switzerland, the UK and USA found that toads established breeding populations or reproduced in 16–100% of created ponds. Two studies in Denmark and Switzerland found that wild but not captive-bred toads colonized 29–100% of created ponds. One study in Denmark found that creating ponds, along with other interventions, increased toad populations. *Assessments: likely to be beneficial (effectiveness 70%; certainty 60%; harms 0%).*

http://www.conservationevidence.com/actions/868

Remove specific aquatic plants

One before-and-after study in the UK found that habitat and species management that included controlling swamp stonecrop, increased a population of natterjack toads. One replicated, controlled study in the USA found that more Oregon spotted frogs laid eggs in areas where invasive reed canarygrass was mown. *Assessment for 'Control invasive plants' from* 'Threat: Invasive alien and other problematic species': *likely to be beneficial (effectiveness 60%; certainty 47%; harms 0%).*

http://www.conservationevidence.com/actions/815

Restore ponds

Fifteen studies investigated the effectiveness of pond restoration for amphibians. Three studies, including one replicated, controlled, before-and-after study in Denmark, the UK and USA found that pond restoration did not increase or had mixed effects on population numbers and hatching success. One replicated, before-and-after study in the UK found that restoration increased pond use. One replicated study in Sweden found that only 10% of restored ponds were used for breeding. Three before-and-after studies, including one replicated, controlled study, in Denmark and Italy found that restored and created ponds were colonized by up to seven species. Eight of nine studies, including one systematic review, in Denmark, Estonia, Italy and the UK found that pond restoration, along with other habitat management, maintained or increased populations, or increased pond occupancy, ponds with breeding success or numbers of amphibian species. One found that numbers of species did not increase and one found that great crested newt populations did not establish. *Assessment: likely to be beneficial (effectiveness 60%; certainty 63%; harms 0%).*

http://www.conservationevidence.com/actions/878

Unknown effectiveness (limited evidence)

Remove tree canopy to reduce pond shading

One before-and-after study in the USA found that canopy removal did not increase hatching success of spotted salamanders. One before-and-after study in Denmark found that following pond restoration that included canopy removal, translocated toads established breeding populations. *Assessment: unknown effectiveness — limited evidence (effectiveness 30%; certainty 25%; harms 0%).*

http://www.conservationevidence.com/actions/758

No evidence found (no assessment)

We have captured no evidence for the following interventions:

- Add nutrients to new ponds as larvae food source
- Add specific plants to aquatic habitats
- Add woody debris to ponds
- Create refuge areas in aquatic habitats.

1.13 Species management

Strict protocols should be followed when carrying out these interventions to minimise potential spread of disease-causing agents such as chytrid fungi and Ranavirus.

1.13.1 Translocate amphibians

Based on the collated evidence, what is the current assessment of the effectiveness of translocations?	
Likely to be beneficial	• Translocate amphibians (amphibians in general) • Translocate amphibians (great crested newts) • Translocate amphibians (natterjack toads) • Translocate amphibians (salamanders including newts) • Translocate amphibians (toads) • Translocate amphibians (wood frogs)
Trade-off between benefit and harms	• Translocate amphibians (frogs)

Likely to be beneficial

Translocate amphibians (amphibians in general)

Overall, three global reviews and one study in the USA found that 65% of amphibian translocations that could be assessed resulted in established breeding populations or substantial recruitment to the adult population. A further two translocations resulted in breeding and one in survival following release. One review found that translocations of over 1,000 animals were more successful, but that success was not related to the source of animals (wild or captive), life-stage, continent or reason for translocation. *Assessment: likely to be beneficial (effectiveness 60%; certainty 60%; harms 19%).*

http://www.conservationevidence.com/actions/854

Translocate amphibians (great crested newts)

Four of six studies in the UK found that translocated great crested newts maintained or established breeding populations. One found that populations survived at least one year in 37% of cases, but one found that within three years breeding failed in 48% of ponds. A systematic review of 31 studies found no conclusive evidence that mitigation that included translocations resulted in self-sustaining populations. One review found that newts reproduced following 56% of translocations, in some cases along with other interventions. *Assessment: likely to be beneficial (effectiveness 50%; certainty 50%; harms 10%).*

http://www.conservationevidence.com/actions/858

Translocate amphibians (natterjack toads)

Three studies in France and the UK found that translocated natterjack toad eggs, tadpoles, juveniles or adults established breeding populations at some sites, although head-started or captive-bred animals were also released at some sites. Re-establishing toads on dune or saltmarsh habitat was more successful than on heathland. One study in the UK found that repeated translocations of wild rather than captive-bred toads were more successful. *Assessment: likely to be beneficial (effectiveness 60%; certainty 56%; harms 10%).*

http://www.conservationevidence.com/actions/859

Translocate amphibians (salamanders including newts)

Four studies in the UK and USA found that translocated eggs or adults established breeding populations of salamanders or smooth newts. One study in the USA found that one of two salamander species reproduced following translocation of eggs, tadpoles and metamorphs. One study in the USA found that translocated salamander eggs hatched and tadpoles had similar survival rates as in donor ponds. *Assessment: likely to be beneficial (effectiveness 70%; certainty 55%; harms 0%).*

http://www.conservationevidence.com/actions/860

Translocate amphibians (toads)

Two of four studies in Denmark, Germany, the UK and USA found that translocating eggs and/or adults established common toad breeding populations. One found populations of garlic toads established at two of four sites and one that breeding populations of boreal toads were not established. One study in Denmark found that translocating green toad eggs to existing populations, along with habitat management, increased population numbers. Four studies in Germany, Italy, South Africa and the USA found that translocated adult toads reproduced, survived up to six or 23 years, or some metamorphs survived over winter. *Assessment: likely to be beneficial (effectiveness 60%; certainty 56%; harms 10%).*

http://www.conservationevidence.com/actions/855

Translocate amphibians (wood frogs)

Two studies in the USA found that following translocation of wood frog eggs, breeding populations were established in 25–50% of created ponds. One study in the USA found that translocated eggs hatched and up to 57% survived as tadpoles in pond enclosures. *Assessment: likely to be beneficial (effectiveness 40%; certainty 50%; harms 0%).*

http://www.conservationevidence.com/actions/856

Trade-off between benefit and harms

Translocate amphibians (frogs)

Eight of ten studies in New Zealand, Spain, Sweden, the UK and USA found that translocating frog eggs, juveniles or adults established breeding populations. Two found that breeding populations went extinct within five years or did not establish. Five studies in Canada, New Zealand and the USA found that translocations of eggs, juveniles or adults resulted in little or no breeding at some sites. Five studies in Italy, New Zealand and the USA found that translocated juveniles or adults survived the winter or up to eight years. One study in the USA found that survival was lower for Oregon spotted frogs translocated as adults compared to eggs. Two studies in the USA found that 60–100% of translocated frogs left the release site and 35–73% returned to their original pond within 32 days. Two studies in found that frogs either lost or gained weight after translocation. *Assessment: trade-offs between benefits and harms (effectiveness 58%; certainty 65%; harms 20%).*

http://www.conservationevidence.com/actions/861

1.13.2 Captive breeding, rearing and releases

Based on the collated evidence, what is the current assessment of the effectiveness of captive breeding, rearing and releases?	
Likely to be beneficial	• Release captive-bred individuals (amphibians in general) • Release captive-bred individuals: frogs
Trade-off between benefit and harms	• Breed amphibians in captivity: frogs • Breed amphibians in captivity: harlequin toads • Breed amphibians in captivity: Mallorcan midwife toad • Breed amphibians in captivity: salamanders (including newts) • Breed amphibians in captivity: toads • Head-start amphibians for release • Release captive-bred individuals: Mallorcan midwife toads • Release captive-bred individuals: toads • Use artificial fertilization in captive breeding • Use hormone treatment to induce sperm and egg release
Unknown effectiveness (limited evidence)	• Release captive-bred individuals: salamanders (including newts)
Unlikely to be beneficial	• Freeze sperm or eggs for future use
Likely to be ineffective or harmful	• Release captive-bred individuals: green and golden bell frogs

Likely to be beneficial

Release captive-bred individuals (amphibians in general)

One review found that 41% of release programmes of captive-bred or head-started amphibians showed evidence of breeding in the wild for multiple generations, 29% showed some evidence of breeding and 12% evidence of survival following release. *Assessment: likely to be beneficial (effectiveness 55%; certainty 50%; harms 10%).*

http://www.conservationevidence.com/actions/871

Release captive-bred individuals (frogs)

Five of six studies in Europe, Hong Kong and the USA found that captive-bred frogs released as tadpoles, juveniles or adults established breeding populations and in some cases colonized new sites. Three studies in Australia and the USA found that a high proportion of frogs released as eggs survived to metamorphosis, some released tadpoles survived the first few months, but few released froglets survived. Four studies in Australia, Italy, the UK and USA found that captive-bred frogs reproduced at 31–100% of release sites, or that breeding was limited. *Assessment: likely to be beneficial (effectiveness 60%; certainty 60%; harms 15%).*

http://www.conservationevidence.com/actions/870

Trade-off between benefit and harms

Breed amphibians in captivity (frogs)

Twenty-three of 33 studies across the world found that amphibians produced eggs in captivity. Seven found mixed results, with some species or populations reproducing successfully, but with other species difficult to maintain or raise to adults. Two found that frogs did not breed successfully or died in captivity. Seventeen of the studies found that captive-bred frogs were raised successfully to hatching, tadpoles, froglets or adults in captivity. Four studies in Canada, Fiji, Hong Kong and Italy found that 30–88% of eggs hatched, or survival to metamorphosis was 75%, as froglets was 17–51% or to adults was 50–90%. *Assessment: trade-offs between benefits and harms (effectiveness 60%; certainty 68%; harms 30%).*

http://www.conservationevidence.com/actions/835

Breed amphibians in captivity (harlequin toads)

Four of five studies in Colombia, Ecuador, Germany and the USA found that harlequin toads reproduced in captivity. One found that eggs were only produced by simulating a dry and wet season and one found that breeding was difficult. One found that captive-bred harlequin toads were raised successfully to metamorphosis in captivity and two found that most toads died before or after hatching. *Assessment: trade-offs between benefits and harms (effectiveness 44%; certainty 50%; harms 28%).*

http://www.conservationevidence.com/actions/836

Breed amphibians in captivity (Mallorcan midwife toad)

Two studies in the UK found that Mallorcan midwife toads produced eggs that were raised to metamorphs or toadlets in captivity. However, clutches dropped by males were not successfully maintained artificially. One study in the UK found that toads bred in captivity for nine or more generations had slower development, reduced genetic diversity and predator defence traits. *Assessment: trade-offs between benefits and harms (effectiveness 69%; certainty 55%; harms 40%).*

http://www.conservationevidence.com/actions/837

Breed amphibians in captivity (salamanders including newts)

Four of six studies in Japan, Germany, the UK and USA found that eggs were produced successfully in captivity. Captive-bred salamanders were raised to yearlings, larvae or adults. One review found that four of five salamander species bred successfully in captivity. Four studies in Germany, Mexico and the USA found that egg production, larval development, body condition and survival were affected by water temperature, density or enclosure type. *Assessment: trade-offs between benefits and harms (effectiveness 60%; certainty 50%; harms 25%).*

http://www.conservationevidence.com/actions/838

Breed amphibians in captivity (toads)

Ten studies in Germany, Italy, Spain, the UK and USA found that toads produced eggs in captivity. Eight found that toads were raised successfully to tadpoles, toadlets or adults in captivity. Two found that most died after hatching or metamorphosis. Two reviews found mixed results with four species of toad or 21% of captive populations of Puerto Rican crested toads breeding successfully. Four studies in Germany, Spain and the USA found that reproductive success was affected by tank location and humidity. *Assessment: trade-offs between benefits and harms (effectiveness 65%; certainty 60%; harms 25%).*

http://www.conservationevidence.com/actions/848

Head-start amphibians for release

Twenty-two studies head-started amphibians from eggs and monitored them after release. A global review and six of 10 studies in Europe and the USA found that released head-started tadpoles, metamorphs or juveniles established breeding populations or increased existing populations. Two found mixed results with breeding populations established in 71% of studies reviewed or at 50% of sites. Two found that head-started metamorphs or adults did not establish a breeding population or prevent a population decline. An additional 10 studies in Australia, Canada, Europe and the USA measured aspects of survival or breeding success of released head-started amphibians and found mixed results. Three studies in the USA only provided results for head-starting in captivity. Two of those found that eggs could be reared to tadpoles, but only one successfully reared adults. *Assessment: trade-offs between benefits and harms (effectiveness 60%; certainty 60%; harms 25%).*

http://www.conservationevidence.com/actions/881

Release captive-bred individuals (Mallorcan midwife toad)

Three studies in Mallorca found that captive-bred midwife toads released as tadpoles, toadlets or adults established breeding populations at 38–100% of sites. One study in the UK found that predator defences were maintained, but genetic diversity was reduced in a captive-bred population. *Assessment: trade-offs between benefits and harms (effectiveness 68%; certainty 58%; harms 20%).*

http://www.conservationevidence.com/actions/873

Release captive-bred individuals (toads)

Two of three studies in Denmark, Sweden and the USA found that captive-bred toads released as tadpoles, juveniles or metamorphs established populations. The other found that populations were not established. Two studies in Puerto Rico found that survival of released captive-bred Puerto Rican crested toads was low. *Assessment: trade-offs between benefits and harms (effectiveness 40%; certainty 50%; harms 20%).*

http://www.conservationevidence.com/actions/875

Use artificial fertilization in captive breeding

Three replicated studies, including two randomized studies, in Australia and the USA found that the success of artificial fertilization depended on the type and number of doses of hormones used to stimulate egg production. One replicated study in Australia found that 55% of eggs were fertilized artificially, but soon died. *Assessment: trade-offs between benefits and harms (effectiveness 40%; certainty 40%; harms 20%).*

http://www.conservationevidence.com/actions/834

Use hormone treatment to induce sperm and egg release

One review and nine of 10 replicated studies, including two randomized, controlled studies, in Austria, Australia, China, Latvia, Russia and the USA found that hormone treatment of male amphibians stimulated or increased sperm production, or resulted in successful breeding. One found that hormone treatment of males and females did not result in breeding. One review and nine of 14 replicated studies, including six randomized and/ or controlled studies, in Australia, Canada, China, Ecuador, Latvia and the USA found that hormone treatment of female amphibians had mixed results, with 30–71% of females producing viable eggs following treatment, or with egg production depending on the combination, amount or number of doses of hormones. Three found that hormone treatment stimulated egg production or successful breeding. Two found that treatment did not stimulate or increase egg production. *Assessment: trade-offs between benefits and harms (effectiveness 50%; certainty 65%; harms 30%).*

http://www.conservationevidence.com/actions/883

Unknown effectiveness (limited evidence)

Release captive-bred individuals (salamanders including newts)

One study in Germany found that captive-bred great crested newts and smooth newts released as larvae, juveniles and adults established stable breeding populations. *Assessment: unknown effectiveness — limited evidence (effectiveness 70%; certainty 30%; harms 0%).*

http://www.conservationevidence.com/actions/874

Unlikely to be beneficial

Freeze sperm or eggs for future use

Ten replicated studies, including three controlled studies, in Austria, Australia, Russia, the UK and USA found that following freezing, viability of amphibian sperm, and in one case eggs, depended on species, cryoprotectant used, storage temperature or method and freezing or thawing rate. One found that sperm could be frozen for up to 58 weeks. *Assessment: unlikely to be beneficial (effectiveness 35%; certainty 50%; harms 10%).*

http://www.conservationevidence.com/actions/876

Likely to be ineffective or harmful

Release captive-bred individuals (green and golden bell frogs)

Three studies in Australia found that captive-bred green and golden bell frogs released mainly as tadpoles did not established breeding populations, or only established breeding populations in 25% of release programmes. One study in Australia found that some frogs released as tadpoles survived at least 13 months. *Assessment: likely to be ineffective or harmful (effectiveness 20%; certainty 50%; harms 20%).*

http://www.conservationevidence.com/actions/872

1.14 Education and awareness raising

Based on the collated evidence, what is the current assessment of the effectiveness of interventions for education and awareness raising?	
Likely to be beneficial	• Engage volunteers to collect amphibian data (citizen science) • Provide education programmes about amphibians • Raise awareness amongst the general public through campaigns and public information

Likely to be beneficial

Engage volunteers to collect amphibian data (citizen science)

Five studies in Canada, the UK and USA found that amphibian data collection projects engaged up to 10,506 volunteers and were active in 16–17 states in the USA. Five studies in the UK and USA found that volunteers surveyed up to 7,872 sites, swabbed almost 6,000 amphibians and submitted thousands of amphibian records. *Assessment: likely to be beneficial (effectiveness 66%; certainty 60%; harms 0%).*

http://www.conservationevidence.com/actions/760

Provide education programmes about amphibians

One study in Taiwan found that education programmes about wetlands and amphibians, along with other interventions, doubled a population of Taipei frogs. Four studies, including one replicated study, in Germany,

Mexico, Slovenia, Zimbabwe and the USA found that education programmes increased the amphibian knowledge of students. *Assessment: likely to be beneficial (effectiveness 58%; certainty 55%; harms 0%).*

http://www.conservationevidence.com/actions/776

Raise awareness amongst the general public through campaigns and public information

Two studies, including one replicated, before-and-after study, in Estonia and the UK found that raising public awareness, along with other interventions, increased amphibian breeding habitat and numbers of toads. One before-and-after study in Mexico found that raising awareness in tourists increased their knowledge of axolotls. However, one study in Taiwan found that holding press conferences had no effect on a frog conservation project. *Assessment: likely to be beneficial (effectiveness 60%; certainty 51%; harms 0%).*

http://www.conservationevidence.com/actions/831

2. BAT CONSERVATION

Anna Berthinussen, Olivia C. Richardson & John D. Altringham

Expert assessors

Levente Barti, Myotis Bat Conservation Group, Romania
Manisha Bhardwaj, Swedish University of Agricultural Sciences, Sweden
Johnny De Jong, Swedish University of Agricultural Sciences, Sweden
Jasja Dekker, Jasja Dekker Dierecologie, Netherlands
Brock Fenton, University of Western Ontario, Canada
Winifred Frick, Bat Conservation International, United States
Anita Glover, Vincent Wildlife Trust, United Kingdom
Alice Hughes, Xishuangbanna Tropical Botanical Garden, China, P.R.
Csaba Jere, Romanian Bat Protection Association, Romania
Julia Lins Luz, Universidade Federal do Rio de Janeiro, Brazil
Christoph Meyer, University of Salford, United Kingdom
Silviu Petrovan, University of Cambridge, United Kingdom
Guido Reiter, Austrian Coordination Centre for Bat Conservation & Research, Austria
Luisa Rodrigues, Institute for Nature Conservation and Forests, Portugal
Danilo Russo, Università degli Studi di Napoli Federico II, Italy
Uttam Saikia, Zoological Survey of India, India
Mbeng Donatus Waghiiwimbom, University of Maroua, Cameroon
Carol Williams, Bat Conservation Trust, United Kingdom

Scope of assessment: for native wild bat species across the world.

Assessed: 2021. For previous assessments and expert panels please check *What Works in Conservation 2020*.

Effectiveness measure is the median % score for effectiveness.

Certainty measure is the median % certainty of evidence for effectiveness, determined by the quantity and quality of the evidence in the synopsis.

Harm measure is the median % score for negative side-effects to the group of species of concern.

 https://doi.org/10.11647/OBP.0267.02

This book is meant as a guide to the evidence available for different conservation interventions and as a starting point in assessing their effectiveness. The assessments are based on the available evidence for the target group of species for each intervention. The assessment may therefore refer to different species or habitat to the one(s) you are considering. Before making any decisions about implementing interventions it is vital that you read the more detailed accounts of the evidence in order to assess their relevance for your study species or system.

Full details of the evidence are available at
www.conservationevidence.com

There may also be significant negative side-effects on the target groups or other species or communities that have not been identified in this assessment.

A lack of evidence means that we have been unable to assess whether or not an intervention is effective or has any harmful impacts.

2.1 Threat: Residential and commercial development

Based on the collated evidence, what is the current assessment of the effectiveness of interventions for residential and commercial development?	
Likely to be beneficial	• Retain existing bat roosts and access points within developments
Unknown effectiveness	• Change timing of building work • Create alternative bat roosts within developments • Create or restore bat foraging habitat in urban areas • Exclude bats from roosts during building work • Legally protect bats during development • Protect brownfield or ex-industrial sites • Relocate access points to bat roosts within developments
No evidence found (no assessment)	• Educate homeowners about building and planning laws relating to bats to reduce disturbance to bat roosts • Increase semi-natural habitat within gardens • Install sound-proofing insulation between bat roosts and areas occupied by humans within developments • Plant gardens with night-scented flowers • Protect greenfield sites or undeveloped land in urban areas

Likely to be beneficial

Retain existing bat roosts and access points within developments

Three studies evaluated the effects of retaining existing bat roosts and access points within developments on bat populations. Two studies were in the UK and one was in Ireland.

COMMUNITY RESPONSE (0 STUDIES)

POPULATION RESPONSE (0 STUDIES)

BEHAVIOUR (3 STUDIES)

Use (3 studies): One before-and-after study in Ireland found similar numbers of brown long-eared bats roosting within an attic after existing access points were retained during renovations. One replicated, before-and-after study in the UK found that four of nine bat roosts retained within developments were used as maternity colonies, in two cases by similar or greater numbers of bats after development had taken place. One review in the UK found that bats used two-thirds of retained and modified bat roosts after development, and retained roosts were more likely to be used than newly created roosts.

Assessment: likely to be beneficial (effectiveness 77%; certainty 60%; harms 0%).

https://www.conservationevidence.com/actions/947

Unknown effectiveness

Change timing of building work

One study evaluated the effects of changing the timing of building work on bat populations. The study was in Ireland.

COMMUNITY RESPONSE (0 STUDIES)

POPULATION RESPONSE (0 STUDIES)

BEHAVIOUR (1 STUDY)

Use (1 study): One before-and-after study in Ireland found that carrying out roofing work outside of the bat maternity season, along with retaining bat access points, resulted in a similar number of brown long-eared bats continuing to use a roost within an attic.

Assessment: unknown effectiveness (effectiveness 50%; certainty 12%; harms 0%).

https://www.conservationevidence.com/actions/950

Create alternative bat roosts within developments

Eleven studies evaluated the effects of creating alternative bat roosts within developments on bat populations. Nine studies were in Europe and two were in the USA.

COMMUNITY RESPONSE (0 STUDIES)

POPULATION RESPONSE (0 STUDIES)

BEHAVIOUR (11 STUDIES)

Use: (11 studies): Two replicated studies in the USA and UK found that bats did not use any of the alternative roosts provided in bat houses or a purpose-built bat wall after exclusion from buildings. Three studies (two replicated) in the USA and UK and one review in the UK found that bat boxes or bat lofts/barns were used by bats at 13–74% of development sites, and bat lofts/barns were used by maternity colonies at one of 19 development sites. Three of five before-and-after studies in Portugal, Ireland, Spain and the UK found that bat colonies used purpose-built roosts in higher or similar numbers after the original roosts were destroyed. The other two studies found that bats used purpose-built roosts in lower numbers than the original roost. One review in the UK found that new bat boxes/lofts built to replace destroyed roosts were four times less likely to be used by returning bats than roosts retained during development.

Assessment: unknown effectiveness (effectiveness 45%; certainty 35%; harms 0%).

https://www.conservationevidence.com/actions/949

Create or restore bat foraging habitat in urban areas

Three studies evaluated the effects of creating or restoring bat foraging habitat in urban areas on bat populations. One study in each of the UK and USA evaluated green roofs and one study in the USA evaluated restored forest fragments.

COMMUNITY RESPONSE (1 STUDY)

Richness/diversity (1 study): One replicated, controlled, site comparison study in the USA found no difference in species richness over green roofs and conventional unvegetated roofs.

POPULATION RESPONSE (3 STUDIES)

Abundance (3 studies): One site comparison study in the USA found higher bat activity (relative abundance) in two of seven restored forest fragments in urban areas than in two unrestored forest fragments. One replicated, controlled, site comparison study in the UK found greater bat activity over 'biodiverse' green roofs than conventional unvegetated roofs, but not over 'sedum' green roofs. One replicated, controlled, site comparison study in the USA found greater bat activity for three of five bat species over green roofs than over conventional unvegetated roofs.

BEHAVIOUR (0 STUDIES)

Assessment: unknown effectiveness (effectiveness 60%; certainty 36%; harms 0%).

https://www.conservationevidence.com/actions/954

Exclude bats from roosts during building work

One study evaluated the effects of excluding bats from roosts during building work on bat populations. The study was in the UK.

COMMUNITY RESPONSE (0 STUDIES)

POPULATION RESPONSE (0 STUDIES)

BEHAVIOUR (1 STUDY)

Behaviour change (1 study): One replicated, before-and-after study in the UK found that excluding bats from roosts within buildings did not change roost switching frequency, core foraging areas or foraging preferences of soprano pipistrelle colonies.

Assessment: unknown effectiveness (effectiveness 45%; certainty 23%; harms 17%).

https://www.conservationevidence.com/actions/1930

Legally protect bats during development

Four studies evaluated the effects of legally protecting bats by issuing licences during development on bat populations. The four studies were in the UK.

COMMUNITY RESPONSE (0 STUDIES)

POPULATION RESPONSE (0 STUDIES)

BEHAVIOUR (2 STUDIES)

Change in human behaviour (2 studies): One review in the UK found that the number of development licences for bats more than doubled over three years in Scotland. One review in the UK found that 81% of licensees did not carry out post-development monitoring to assess whether bats used the roost structures installed.

OTHER (3 STUDIES)

Impact on bat roost sites (3 studies): One review in the UK found that licenced activities during building developments had a negative impact on bat roosts, with 68% of roosts being destroyed. One replicated, before-and-after study in the UK found that five of 28 compensation roosts provided under licence were used, and two by similar or greater numbers of bats after development. One review in the UK found that 31–67% of compensation roosts provided under licence were used by bats.

Assessment: unknown effectiveness (effectiveness 45%; certainty 35%; harms 10%).

https://www.conservationevidence.com/actions/1935

Protect brownfield or ex-industrial sites

One study evaluated the effects of protecting brownfield or ex-industrial sites on bat populations. The study as in the USA.

COMMUNITY RESPONSE (1 STUDY)

Richness/diversity (1 study): One study in the USA found that five bat species were recorded within a protected urban wildlife refuge on an abandoned manufacturing site.

POPULATION RESPONSE (0 STUDIES)

BEHAVIOUR (0 STUDIES)

Assessment: unknown effectiveness (effectiveness 40%; certainty 20%; harms 0%).

https://www.conservationevidence.com/actions/953

Relocate access points to bat roosts within developments

Two studies evaluated the effects of relocating access points to bat roosts within building developments on bat populations. One study was in Ireland and one in the UK.

COMMUNITY RESPONSE (0 STUDIES)

POPULATION RESPONSE (0 STUDIES)

BEHAVIOUR (2 STUDIES)

Use (2 studies): One before-and-after study in Ireland found that fewer brown long-eared bats used a roost after the access points were relocated, and no bats were observed flying through them. One before-and-after study in the UK found that few lesser horseshoe bats used an alternative access point with a 'bend' design to re-enter a roost in a building development, but the number of bats using the roost increased after an access point with a 'straight' design was installed.

Assessment: unknown effectiveness (effectiveness 45%; certainty 32%; harms 10%).

https://www.conservationevidence.com/actions/946

No evidence found (no assessment)

We have captured no evidence for the following interventions:

- Educate homeowners about building and planning laws relating to bats to reduce disturbance to bat roosts
- Increase semi-natural habitat within gardens
- Install sound-proofing insulation between bat roosts and areas occupied by humans within developments
- Plant gardens with night-scented flowers
- Protect greenfield sites or undeveloped land in urban areas.

2.2 Threat: Agriculture

2.2.1 All farming systems

Based on the collated evidence, what is the current assessment of the effectiveness of interventions for all farming systems?	
Likely to be beneficial	• Retain or plant native trees and shrubs amongst crops (agroforestry) • Use organic farming instead of conventional farming
Unknown effectiveness	• Create tree plantations on agricultural land • Engage farmers and landowners to manage land for bats • Manage hedges to benefit bats • Plant field margins with a diverse mix of plant species • Reduce field size (or maintain small fields) • Retain riparian buffers on agricultural land • Retain unmown field margins
Unlikely to be beneficial	• Pay farmers to cover the costs of conservation measures (e.g. agri-environment schemes)
No evidence found (no assessment)	• Increase the proportion of semi-natural habitat in the farmed landscape • Manage ditches to benefit bats • Plant in-field trees • Plant new hedges • Provide or retain set-aside areas in farmland • Retain existing in-field trees • Retain remnant forest or woodland on agricultural land

Likely to be beneficial

Retain or plant native trees and shrubs amongst crops (agroforestry)

Eight studies evaluated the effects of retaining or planting native trees and shrubs amongst crops on bat populations. Four studies were in Mexico, three were in South America and one was in Tanzania.

COMMUNITY RESPONSE (7 STUDIES)

Community composition (1 study): One replicated, site comparison study in Tanzania found different compositions of bat species in coffee plantations with different amounts and types of shade cover.

Richness/diversity (7 studies): Four of six replicated, site comparison studies in Columbia, Mexico and Costa Rica found a similar number of bat species in shaded and unshaded coffee plantations, and in coffee plantations with different amounts and types of shade cover. The two other studies found more bat species and higher bat diversity in coffee, cacao and banana plantations with varied shade cover, than in plantations with a single shade species or no shade. One replicated, site comparison study in Tanzania found more bat species in shaded coffee plantations than in traditional mixed agroforestry systems with natural forest vegetation.

POPULATION RESPONSE (6 STUDIES)

Abundance (5 studies): Two replicated, site comparison studies in Mexico captured more bats in coffee plantations with varied shade cover than in plantations with a single shade species. One replicated, site comparison study in Mexico found higher activity (relative abundance) of forest bat species in plantations with a varied shade cover than in plantations with a single shade species, but the opposite was true for open habitat bat species. One replicated, site comparison study in Costa Rica found no difference in the number of bats captured between cacao and banana shade plantations and unshaded monocultures. One replicated, site comparison study in Tanzania found greater bat occurrence in shaded coffee plantations than in traditional mixed agroforestry systems with natural forest vegetation.

Condition (1 study): One replicated, site comparison study in Columbia found that great fruit-eating bats captured in 'silvopastoral' areas that used agroforestry, along with no chemicals, had higher body weights and body condition scores than those in conventional farming areas.

BEHAVIOUR (0 STUDIES)

Assessment: likely to be beneficial (effectiveness 60%; certainty 53%; harms 0%).

https://www.conservationevidence.com/actions/963

Use organic farming instead of conventional farming

Twelve studies evaluated the effects of using organic farming instead of conventional farming on bat populations. Eight studies were in Europe, two in the USA, one in Canada and one in Chile.

COMMUNITY RESPONSE (7 STUDIES)

Community composition (1 study): One replicated, paired sites study in the USA found that the composition of bat species did not differ between organic and non-organic farms.

Richness/diversity (7 studies): Five of seven replicated, paired sites or site comparison studies in Europe, the USA, Canada and Chile found that the number of bat species did not differ between organic and non-organic farms. The other two studies found more bat species on organic farms than non-organic farms.

POPULATION RESPONSE (12 STUDIES)

Abundance (12 studies): Five of nine replicated, paired sites or site comparison studies in Europe, the USA, Canada and Chile found that overall bat activity (relative abundance) and common pipistrelle activity did not differ between organic and non-organic farms. The other four studies found higher overall bat activity, bat feeding activity, Brazilian free-tailed bat activity, and activity of four of seven bat species on organic farms than non-organic farms. Two replicated, paired sites and site comparison studies in the UK found higher activity of Myotis species over water and rivers on organic farms than non-organic farms, but no differences were found for other species or habitats. One replicated, site comparison study in France found higher activity for two of three bat species over organic fields than two of three types of conventionally managed fields.

BEHAVIOUR (0 STUDIES)

Assessment: likely to be beneficial (effectiveness 60%; certainty 60%; harms 0%).

https://www.conservationevidence.com/actions/961

Unknown effectiveness

Create tree plantations on agricultural land

Three studies evaluated the effects of creating tree plantations on agricultural land on bat populations. The three studies were in Australia.

COMMUNITY RESPONSE (3 STUDIES)

Richness/diversity (3 studies): Three replicated, site comparison studies in Australia found no difference in the number of bat species in agricultural areas with and without plantations of native trees.

POPULATION RESPONSE (3 STUDIES)

Abundance (3 studies): Two of three replicated, site comparison studies in Australia found no difference in bat activity (relative abundance) in agricultural areas with and without plantations of native trees. The other study found higher bat activity in plantations next to remnant native vegetation than in isolated plantations or over grazing land. In all three studies, bat activity was lower in plantations compared to original forest and woodland remnants.

BEHAVIOUR (0 STUDIES)

Assessment: unknown effectiveness (effectiveness 30%; certainty 30%; harms 0%).

https://www.conservationevidence.com/actions/958

Engage farmers and landowners to manage land for bats

One study evaluated the effects of engaging farmers and landowners to manage land for bats on bat populations. The study was in the UK.

COMMUNITY RESPONSE (0 STUDIES)

POPULATION RESPONSE (1 STUDY)

Abundance (1 study): One study in the UK found that during a five-year project to engage farmers and landowners to manage land for bats, the overall population of greater horseshoe bats at four maternity roosts in the area increased (but see summary below).

BEHAVIOUR (1 STUDY)

Change in human behaviour (1 study): One study in the UK found that a landowner engagement project resulted in 77 bat-related management agreements covering approximately 6,536 ha of land.

Assessment: unknown effectiveness (effectiveness 55%; certainty 20%; harms 0%).

https://www.conservationevidence.com/actions/1936

Manage hedges to benefit bats

Two studies evaluated the effects of managing hedges to benefit bat populations. Both studies were in the UK.

COMMUNITY RESPONSE (1 STUDY)

Richness/diversity (1 study): One replicated, site comparison study in the UK found that hedges trimmed ≥3 years prior had more bat species recorded along them than hedges trimmed during the previous winter.

POPULATION RESPONSE (2 STUDIES)

Abundance (2 studies): One replicated, paired sites study in the UK found that pipistrelle activity (relative abundance) did not differ between hedges managed for wildlife on agri-environment scheme farms and hedges on

conventional farms. One replicated, site comparison study in the UK found that hedges trimmed ≥3 years prior had higher activity of two of eight bat species/species groups than hedges trimmed during the previous winter.
BEHAVIOUR (0 STUDIES)
Assessment: unknown effectiveness (effectiveness 50%; certainty 36%; harms 0%).

https://www.conservationevidence.com/actions/1943

Plant field margins with a diverse mix of plant species

One study evaluated the effects of planting field margins with a diverse mix of plant species on bats populations. The study was in the UK.
COMMUNITY RESPONSE (0 STUDIES)
POPULATION RESPONSE (1 STUDY)
Abundance (1 study): One replicated, site comparison study in the UK found that the activity (relative abundance) of soprano pipistrelles and barbastelle bats increased with a greater diversity of plant species within field margins, but there was no effect on common pipistrelle activity nor on the occurrence of any of the six bat species studied.
BEHAVIOUR (0 STUDIES)
Assessment: unknown effectiveness (effectiveness 50%; certainty 30%; harms 0%).

https://www.conservationevidence.com/actions/1941

Reduce field size (or maintain small fields)

One study evaluated the effects of maintaining small fields on bat populations. The study was in Canada.
COMMUNITY RESPONSE (0 STUDIES)
POPULATION RESPONSE (1 STUDY)
Abundance (1 study): One replicated, site comparison study in Canada found that agricultural landscapes with smaller fields had higher activity (relative abundance) of six of seven bat species than landscapes with larger fields.
BEHAVIOUR (0 STUDIES)
Assessment: unknown effectiveness (effectiveness 60%; certainty 30%; harms 0%).

https://www.conservationevidence.com/actions/1939

Retain riparian buffers on agricultural land

One study evaluated the effects of retaining riparian buffers on agricultural land on bat populations. The study was in the UK.
COMMUNITY RESPONSE (0 STUDIES)
POPULATION RESPONSE (1 STUDY)

Abundance (1 study): One replicated, paired sites study in the UK found that pipistrelle activity (relative abundance) did not differ along waterways with buffers of vegetation on agri-environment scheme farms and waterways on conventional farms.

BEHAVIOUR (0 STUDIES)

Assessment: unknown effectiveness (effectiveness 40%; certainty 20%; harms 0%).

https://www.conservationevidence.com/actions/2284

Retain unmown field margins

One study evaluated the effects of retaining unmown field margins on bats populations. The study was in the UK.

COMMUNITY RESPONSE (0 STUDIES)

POPULATION RESPONSE (1 STUDY)

Abundance (1 study): One replicated, paired sites study in the UK found that pipistrelle activity (relative abundance) did not differ between unmown field margins managed for wildlife on agri-environment scheme farms and field margins on conventional farms.

BEHAVIOUR (0 STUDIES)

Assessment: unknown effectiveness (effectiveness 0%; certainty 10%; harms 0%).

https://www.conservationevidence.com/actions/1940

Unlikely to be beneficial

Pay farmers to cover the costs of conservation measures (e.g. agri-environment schemes)

Three studies evaluated the effects of agri-environment schemes on bat populations. The three studies were in the UK.

COMMUNITY RESPONSE (0 STUDIES)

POPULATION RESPONSE (3 STUDIES)

Abundance (3 studies): Two of three replicated, paired sites studies in the UK found that overall bat activity (relative abundance) or the occurrence of six bat species did not differ significantly between farms managed under agri-environment schemes and those managed conventionally. One of the studies found that agri-environment scheme farms had similar activity of five bat species, and lower activity of one bat species, compared to conventional farms. The other study found lower overall bat activity and activity of pipistrelle species on agri-environment scheme farms than conventional farms.

BEHAVIOUR (0 STUDIES)

Assessment: unlikely to be beneficial (effectiveness 27%; certainty 42%; harms 0%).

https://www.conservationevidence.com/actions/962

No evidence found (no assessment)

We have captured no evidence for the following interventions:

- Increase the proportion of semi-natural habitat in the farmed landscape
- Manage ditches to benefit bats
- Plant in-field trees
- Plant new hedges
- Provide or retain set-aside areas in farmland
- Retain existing in-field trees
- Retain remnant forest or woodland on agricultural land.

2.2.2 Livestock farming

Based on the collated evidence, what is the current assessment of the effectiveness of interventions for livestock farming?	
Likely to be beneficial	• Manage livestock water troughs as a drinking resource for bats
No evidence found (no assessment)	• Avoid the use of antiparasitic drugs for livestock • Manage grazing regimes to increase invertebrate prey • Replace culling of bats with non-lethal methods of preventing vampire bats from spreading rabies to livestock

Likely to be beneficial

Manage livestock water troughs as a drinking resource for bats

Two studies evaluated the effects of managing livestock water troughs as a drinking resource for bats. Both studies were in the USA.

COMMUNITY RESPONSE (0 STUDIES)
POPULATION RESPONSE (0 STUDIES)
BEHAVIOUR (2 STUDIES)
Use (2 studies): One replicated, paired sites study in the USA found that removing livestock modifications from water troughs resulted in bats drinking from them more frequently. One paired sites study in the USA found that livestock water tanks that were larger, full of water or surrounded by sparse vegetation had more bats drinking from them than smaller, half full tanks surrounded by no or dense vegetation.
Behaviour change (1 study): One replicated, paired sites study in the USA found that when livestock modifications were removed from water troughs, bats approached troughs fewer times before successfully drinking from them.
Assessment: likely to be beneficial (effectiveness 67%; certainty 45%; harms 0%).

https://www.conservationevidence.com/actions/1951

No evidence found (no assessment)

We have captured no evidence for the following interventions:

- Avoid the use of antiparasitic drugs for livestock
- Manage grazing regimes to increase invertebrate prey
- Replace culling of bats with non-lethal methods of preventing vampire bats from spreading rabies to livestock.

2.2.3 Perennial, non-timber crops

Based on the collated evidence, what is the current assessment of the effectiveness of interventions for perennial, non-timber crops?	
Likely to be beneficial	• Use non-lethal measures to prevent bats from accessing fruit in orchards to reduce human-wildlife conflict
No evidence found (no assessment)	• Introduce certification for bat-friendly crop harvesting regimes • Prevent culling of bats around fruit orchards • Restore and manage abandoned orchards for bats

Likely to be beneficial

Use non-lethal measures to prevent bats from accessing fruit in orchards to reduce human-wildlife conflict

Two studies evaluated the effects of using non-lethal measures to prevent bats from accessing fruit in orchards to reduce human-wildlife conflict. The studies were in Madagascar and Mauritius.

COMMUNITY RESPONSE (0 STUDIES)

POPULATION RESPONSE (0 STUDIES)

BEHAVIOUR (0 STUDIES)

OTHER (2 STUDIES)

Human-wildlife conflict (2 studies): Two replicated, controlled studies (including one randomized study) in Madagascar and Mauritius found that using an organic deterrent spray, hanging plastic flags in trees, or covering individual tree branches with nylon net bags reduced damage to lychees caused by Madagascan flying foxes or Mauritius fruit bats. One of the studies found that ringing bells in lychee trees deterred most Madagascan flying foxes.

Assessment: likely to be beneficial (effectiveness 70%; certainty 50%; harms 0%).

https://www.conservationevidence.com/actions/1953

No evidence found (no assessment)

We have captured no evidence for the following interventions:

- Introduce certification for bat-friendly crop harvesting regimes
- Prevent culling of bats around fruit orchards
- Restore and manage abandoned orchards for bats.

2.3 Threat: Energy production

2.3.1 Wind turbines

Based on the collated evidence, what is the current assessment of the effectiveness of interventions for wind turbines?	
Beneficial	• Increase the wind speed at which turbines become operational ('cut-in speed') • Prevent turbine blades from turning at low wind speeds ('feathering')
Likely to be beneficial	• Automatically reduce turbine blade rotation when bat activity is high • Slow rotation of turbine blades at low wind speeds
Unknown effectiveness	• Apply textured coating to turbines • Deter bats from turbines using ultrasound
No evidence found (no assessment)	• Close off potential access points on turbines to prevent roosting bats • Deter bats from turbines using low-level ultraviolet light • Deter bats from turbines using radar • Modify turbine placement to reduce bat fatalities • Paint turbines to reduce insect attraction • Reduce rotor diameter • Reduce turbine height • Remove turbine lighting to reduce bat and insect attraction • Retain a buffer between turbines and habitat features used by bats

Beneficial

Increase the wind speed at which turbines become operational ('cut-in speed')

Twelve studies evaluated the effects of increasing the wind speed at which turbines become operational ('cut-in speed') on bat populations. Ten studies were in the USA and two were in Canada.

COMMUNITY RESPONSE (0 STUDIES)

POPULATION RESPONSE (12 STUDIES)

Survival (12 studies): Ten of 12 studies (including 10 replicated, randomized, controlled studies and one before-and-after study) in the USA and Canada found that increasing the wind speed at which turbines become operational ('cut-in speed'), or increasing the cut-in speed along with preventing turbine blades from turning at low wind speeds ('feathering') resulted in fewer bat fatalities than at conventionally operated turbines. The other two studies found that increasing cut-in speeds did not reduce bat fatalities, but sample sizes were small or treatments were applied for short periods only.

BEHAVIOUR (0 STUDIES)

Assessment: beneficial (effectiveness 80%; certainty 72%; harms 0%).

https://www.conservationevidence.com/actions/1960

Prevent turbine blades from turning at low wind speeds ('feathering')

Six studies evaluated the effects of preventing turbine blades from turning at low wind speeds on bat populations. Five studies were in the USA and one was in Canada.

COMMUNITY RESPONSE (0 STUDIES)

POPULATION RESPONSE (6 STUDIES)

Survival (6 studies): Five of six studies (including five replicated, controlled studies and one before-and-after study) in the USA and Canada found that preventing turbine blades from turning at low wind speeds ('feathering'), or feathering along with increasing the wind speed at which turbines become operational ('cut-in speed') resulted in fewer bat fatalities than at conventionally operated turbines. The other study found that automatically feathering turbine blades at low wind speeds did not reduce bat fatalities.

BEHAVIOUR (0 STUDIES)

Assessment: beneficial (effectiveness 82%; certainty 70%; harms 0%).

https://www.conservationevidence.com/actions/970

Likely to be beneficial

Automatically reduce turbine blade rotation when bat activity is high

Two studies evaluated the effects of automatically reducing turbine blade rotation when bat activity is high on bat populations. One study was in Germany, and one in the USA.

COMMUNITY RESPONSE (0 STUDIES)

POPULATION RESPONSE (2 STUDIES)

Survival (2 studies): Two replicated studies (one randomized, controlled and one paired sites study) in Germany and the USA found that automatically reducing the rotation speed of wind turbine blades when bat activity is predicted to be high resulted in fewer bat fatalities for all bat species combined and for five bat species.

BEHAVIOUR (0 STUDIES)

Assessment: likely to be beneficial (effectiveness 80%; certainty 60%; harms 0%).

https://www.conservationevidence.com/actions/971

Slow rotation of turbine blades at low wind speeds

One study evaluated the effects of slowing the rotation of turbine blades at low wind speeds on bat populations. The study was in Canada.

COMMUNITY RESPONSE (0 STUDIES)

POPULATION RESPONSE (1 STUDY)

Survival (1 study): One replicated, randomized, controlled study in Canada found that bat fatalities were reduced when turbine blades were slowed at low wind speeds.

BEHAVIOUR (0 STUDIES)

Assessment: likely to be beneficial (effectiveness 60%; certainty 40%; harms 0%).

https://www.conservationevidence.com/actions/2939

Unknown effectiveness

Apply textured coating to turbines

One study evaluated the effects of applying a textured coating to turbines on bat populations. The study was in the USA.

COMMUNITY RESPONSE (0 STUDIES)

POPULATION RESPONSE (1 STUDY)

Abundance (1 study): One paired sites study in the USA found that applying a textured coating to a turbine did not reduce the activity of four bat species or the number of bats observed.

BEHAVIOUR (0 STUDIES)

Assessment: unknown effectiveness (effectiveness 0%; certainty 22%; harms 0%).

https://www.conservationevidence.com/actions/1957

Deter bats from turbines using ultrasound

Four studies evaluated the effects of deterring bats from wind turbines using ultrasound on bat populations. The four studies were in the USA.

COMMUNITY RESPONSE (0 STUDIES)

POPULATION RESPONSE (3 STUDIES)

Survival (3 studies): Three replicated, randomized, controlled studies (one with a before-and-after trial in the second year) in the USA found mixed results. In the first year of one study, 21-51% fewer bats were killed at turbines with an ultrasonic deterrent fitted than at control turbines, but in the second year, from 2% more to 64% fewer bats were killed at turbines with ultrasonic deterrents fitted. One study found that using an ultrasonic deterrent emitting a constant or pulsed signal had mixed effects on the fatality rates of three bat species. One study found that using ultrasonic deterrents resulted in fewer fatalities for two of three bat species.

BEHAVIOUR (1 STUDY)

Behaviour change (1 study): One paired sites study in the USA found fewer bats flying near one of two wind turbines with an ultrasonic deterrent compared to turbines without, but the effect of the deterrent overall was not significant.

Assessment: unknown effectiveness (effectiveness 40%; certainty 35%; harms 10%).

https://www.conservationevidence.com/actions/968

No evidence found (no assessment)

We have captured no evidence for the following interventions:

- Close off potential access points on turbines to prevent roosting bats
- Deter bats from turbines using low-level ultraviolet light
- Deter bats from turbines using radar
- Modify turbine placement to reduce bat fatalities
- Paint turbines to reduce insect attraction
- Remove turbine lighting to reduce bat and insect attraction
- Retain a buffer between turbines and habitat features used by bats.

2.3.2 Mining

Based on the collated evidence, what is the current assessment of the effectiveness of interventions for mining?	
Trade-off between benefit and harms	• Install and maintain gates at mine entrances to restrict public access
Unknown effectiveness	• Maintain microclimate in closed/abandoned mines • Restore bat foraging habitat at ex-quarry sites
No evidence found (no assessment)	• Exclude bats from roosts prior to mine reclamation • Provide artificial subterranean bat roosts to replace roosts in reclaimed mines • Relocate bats from reclaimed mines to alternative subterranean roost sites • Reopen entrances to closed mines and make suitable for roosting bats • Retain access points for bats following mine closures

Trade-off between benefit and harms

Install and maintain gates at mine entrances to restrict public access

Nine studies evaluated the effects of installing gates at mine entrances on bat populations. Eight studies were in the USA and one in Australia.

COMMUNITY RESPONSE (1 STUDY)

Richness/diversity (1 study): One replicated, before-and-after study in the USA found that fewer bat species entered mines after gates were installed.

POPULATION RESPONSE (3 STUDIES)

Abundance (3 studies): Two replicated, site comparison or before-and-after studies in the USA and Australia found fewer bats in mines or at mine entrances after gates were installed. One replicated, controlled, before-and-after study in the USA found that bat activity (relative abundance) remained stable or increased at five of seven gated mines, and decreased at two gated mines.

BEHAVIOUR (6 STUDIES)

Use (2 studies): One before-and-after study in the USA found that 43 of 47 mines continued to be used 12 years after gates were installed, however bats abandoned four mines with 'ladder' design gates. One replicated study in the USA found that gate design and time since gate installation had varied effects on the presence of four bat species.

Behaviour change (4 studies): Four replicated, before-and-after or site comparison studies in the USA and Australia found that bats at mine entrances circled more and entered mines less after gates were installed.

OTHER (2 STUDIES)

Collisions with gates (1 study): One replicated, controlled, before-and-after study in the USA found that up to 7% of bats at mine entrances collided with mine gates.

Assessment: trade-off between benefit and harms (effectiveness 50%; certainty 50%; harms 46%).

https://www.conservationevidence.com/actions/1963

Unknown effectiveness

Maintain microclimate in closed/abandoned mines

One study evaluated the effects of maintaining the microclimate in an abandoned mine on bat populations. The study was in the USA.

COMMUNITY RESPONSE (0 STUDIES)

POPULATION RESPONSE (1 STUDY)

Abundance (1 study): One before-and-after study in the USA found that modifying the microclimate of an abandoned mine by closing a man-made entrance resulted in a greater number of bats hibernating within the mine.

BEHAVIOUR (0 STUDIES)

Assessment: unknown effectiveness (effectiveness 45%; certainty 20%; harms 0%).

https://www.conservationevidence.com/actions/1964

Restore bat foraging habitat at ex-quarry sites

One study evaluated the effects of restoring bat foraging habitat at ex-quarry sites on bat populations. The study was in France.

COMMUNITY RESPONSE (0 STUDIES)

POPULATION RESPONSE (1 STUDY)

Abundance (1 study): One replicated, site comparison study in France found that gravel-sand pits had higher overall bat activity (relative abundance) 10 years after restoration than gravel-sand pit sites before or during quarrying.

BEHAVIOUR (0 STUDIES)

Assessment: unknown effectiveness (effectiveness 50%; certainty 35%; harms 0%).

https://www.conservationevidence.com/actions/2286

No evidence found (no assessment)

We have captured no evidence for the following interventions:

- Exclude bats from roosts prior to mine reclamation
- Provide artificial subterranean bat roosts to replace roosts in reclaimed mines
- Relocate bats from reclaimed mines to alternative subterranean roost sites
- Reopen entrances to closed mines and make suitable for roosting bats
- Retain access points for bats following mine closures.

2.4 Threat: Transportation and service corridors

2.4.1 Roads

Based on the collated evidence, what is the current assessment of the effectiveness of interventions for roads?	
Likely to be beneficial	• Create spaces for roosting bats in road/railway bridges and culverts • Install overpasses as road/railway crossing structures for bats • Install underpasses or culverts as road/railway crossing structures for bats • Maintain bat roosts in road/railway bridges and culverts
Unknown effectiveness	• Divert bats to safe crossing points over or under roads/railways with plantings or fencing • Install green bridges as road/railway crossing structures for bats • Provide alternative bat roosts during maintenance work at road/railway bridges and culverts
Unlikely to be beneficial	• Install bat gantries or bat bridges as road/railway crossing structures for bats

Based on the collated evidence, what is the current assessment of the effectiveness of interventions for roads?	
No evidence found (no assessment)	• Avoid planting fruit trees alongside roads/railways in areas with fruit bats • Change timing of maintenance work at road/railway bridges and culverts • Deter bats from roads/railways using lighting • Deter bats from roads/railways using ultrasound • Exclude bats from roosts during maintenance work at road/railway bridges and culverts • Install hop-overs as road/railway crossing structures for bats • Manage vegetation along utility and service line corridors to increase foraging habitat for bats • Minimize road lighting to reduce insect attraction • Replace or improve habitat for bats around roads/railways • Replace or improve roosting habitat for bats along utility and service line corridors

Likely to be beneficial

Create spaces for roosting bats in road/railway bridges and culverts

One study evaluated the effects of creating spaces for roosting bats in road bridges. The study was in the USA.

COMMUNITY RESPONSE (0 STUDIES)

POPULATION RESPONSE (0 STUDIES)

BEHAVIOUR (1 STUDY)

Use (1 study): One review in the USA found that spaces created in road bridges for roosting bats to replace those lost during bridge works were recolonized by bats in similar or greater numbers to the original roosts at four of eight sites.

Assessment: likely to be beneficial (effectiveness 60%; certainty 40%; harms 0%).

https://www.conservationevidence.com/actions/1967

Install overpasses as road/railway crossing structures for bats

Four studies evaluated the effects of installing overpasses as road crossing structures for bats. Three studies were in Europe and one in Australia.

COMMUNITY RESPONSE (1 STUDY)

Richness/diversity (1 study): One site comparison study in Australia found that the same number of bat species were recorded at an overpass and in nearby forest and bushland.

POPULATION RESPONSE (0 STUDIES)

BEHAVIOUR (3 STUDIES)

Use (3 studies): Two replicated studies (including one site comparison) in Ireland and France found that two or three bat species/species groups used overpasses but up to three-quarters of bats crossed the road below at traffic height or crossed at other nearby locations. One study in the UK found that an overpass with planters was used by two-thirds of crossing bats, and an unvegetated overpass with a paved road over it was not used by crossing bats.

Assessment: likely to be beneficial (effectiveness 50%; certainty 46%; harms 0%).

https://www.conservationevidence.com/actions/977

Install underpasses or culverts as road/railway crossing structures for bats

Eight studies evaluated the effects of installing underpasses or culverts as road crossing structures for bats. Seven studies were in Europe and one in Australia.

COMMUNITY RESPONSE (0 STUDIES)

POPULATION RESPONSE (0 STUDIES)

BEHAVIOUR (8 STUDIES)

Use (8 studies): Eight studies (including six replicated studies) in Germany, Ireland, the UK, Australia and France found that bats used underpasses and culverts below roads, and crossed over the roads above them, in varying proportions. One replicated, site comparison study in Australia found that bat species adapted to cluttered habitats used small culverts and underpasses more than bat species adapted to open or edge habitats. One replicated, site comparison study in France found that the use of underpasses by five bat species was influenced by underpass type and height, road width, and the amount of forest and hedgerows in the surrounding landscape.

Assessment: likely to be beneficial (effectiveness 60%; certainty 60%; harms 0%).

https://www.conservationevidence.com/actions/976

Maintain bat roosts in road/railway bridges and culverts

Two studies evaluated the effects of maintaining bat roosts within road bridges on bat populations. One study was in Ireland and one in the USA.

COMMUNITY RESPONSE (0 STUDIES)

POPULATION RESPONSE (0 STUDIES)

BEHAVIOUR (2 STUDIES)

Use (2 studies): One before-and-after study in Ireland found that a maternity colony of Daubenton's bats continued to roost in a road bridge over a river in similar numbers after crevices were retained during repair work. One review in the USA found that when bat roosts were maintained during bridge replacement works, Yuma myotis and Mexican free-tailed bats recolonised most roosts in similar numbers to before the works, but pallid bats did not return.

Assessment: likely to be beneficial (effectiveness 70%; certainty 45%; harms 0%).

https://www.conservationevidence.com/actions/1966

Unknown effectiveness

Divert bats to safe crossing points over or under roads/railways with plantings or fencing

One study evaluated the effects of diverting bats using an artificial hedgerow on bat populations. The study was in Switzerland.

COMMUNITY RESPONSE (0 STUDIES)

POPULATION RESPONSE (0 STUDIES)

BEHAVIOUR (1 STUDY)

Use (1 study): One controlled, before-and-after study in Switzerland found that up to one fifth of lesser horseshoe bats within a colony flew along an artificial hedgerow to commute.

Assessment: unknown effectiveness (effectiveness 10%; certainty 10%; harms 5%).

https://www.conservationevidence.com/actions/981

Install green bridges as road/railway crossing structures for bats

One study evaluated the effects of installing green bridges as road crossing structures for bats. The study was in the UK.

COMMUNITY RESPONSE (0 STUDIES)

POPULATION RESPONSE (0 STUDIES)

BEHAVIOUR (1 STUDY)
Use (1 study): One study in the UK found that a green bridge was used by 97% of bats crossing a road.
Assessment: unknown effectiveness (effectiveness 70%; certainty 27%; harms 0%).

https://www.conservationevidence.com/actions/979

Provide alternative bat roosts during maintenance work at road/railway bridges and culverts

One study evaluated the effects of providing alternative bat roosts during maintenance work at road bridges. The study was in the USA.
COMMUNITY RESPONSE (0 STUDIES)
POPULATION RESPONSE (0 STUDIES)
BEHAVIOUR (1 STUDY)
Use (1 study): One review in the USA found that bat houses provided as alternative roosts during bridge replacement works were used by fewer Mexican free-tailed bats than the original roost at one site and were not used by bats at all at three sites.
Assessment: unknown effectiveness (effectiveness 20%; certainty 22%; harms 0%).

https://www.conservationevidence.com/actions/2942

Unlikely to be beneficial

Install bat gantries or bat bridges as road/railway crossing structures for bats

Three studies evaluated the effects of installing bat gantries as road crossing structures for bats. Two studies were in the UK and one in France.
COMMUNITY RESPONSE (0 STUDIES)
POPULATION RESPONSE (0 STUDIES)
BEHAVIOUR (3 STUDIES)
Use (3 studies): Two replicated studies (including one site comparison) in the UK found that fewer bats used bat gantries than crossed the road below at traffic height, and one bat gantry was not used at all. One replicated study in France found that a temporary bat gantry was used by three bat species/species groups, but almost half of crossing bats flew over the road at other locations.
Assessment: unlikely to be beneficial (effectiveness 20%; certainty 42%; harms 0%).

https://www.conservationevidence.com/actions/978

No evidence found (no assessment)

We have captured no evidence for the following interventions:

- Avoid planting fruit trees alongside roads/railways in areas with fruit bats
- Change timing of maintenance work at road/railway bridges and culverts
- Deter bats from roads/railways using lighting
- Deter bats from roads/railways using ultrasound
- Exclude bats from roosts during maintenance work at road/railway bridges and culverts
- Install hop-overs as road/railway crossing structures for bats
- Manage vegetation along utility and service line corridors to increase foraging habitat for bats
- Minimize road lighting to reduce insect attraction
- Replace or improve habitat for bats around roads/railways
- Replace or improve roosting habitat for bats along utility and service line corridors.

2.5 Threat: Biological resource use

2.5.1 Hunting

Based on the collated evidence, what is the current assessment of the effectiveness of interventions for hunting?	
Unknown effectiveness	• Inform local communities about disease risks from hunting and eating bat meat to reduce killing of bats • Inform local communities about the negative impacts of bat hunting to reduce killing of bats
No evidence found (no assessment)	• Encourage online vendors to remove bat specimens for sale • Enforce regulations to prevent trafficking and trade of bats • Introduce alternative treatments to reduce the use of bats in traditional medicine • Introduce and enforce legislation to control hunting of bats • Introduce other food sources to replace bat meat • Introduce other income sources to replace bat trade • Replace culling of bats with non-lethal methods of preventing vampire bats from spreading rabies to humans • Restrict the collection of bat specimens for research • Strengthen cultural traditions that discourage bat harvesting

Unknown effectiveness

Inform local communities about disease risks from hunting and eating bat meat to reduce killing of bats

One study evaluated the effects of informing local communities about disease risks from hunting and eating bat meat to reduce killing of bats on bat populations. The study was in Ghana.

COMMUNITY RESPONSE (0 STUDIES)

POPULATION RESPONSE (0 STUDIES)

BEHAVIOUR (1 STUDY)

Behaviour change (1 study): One before-and-after study in Ghana found that fewer hunters intended to hunt bats in future after they were provided with education about the risks of diseases carried by bats.

Assessment: unknown effectiveness (effectiveness 50%; certainty 25%; harms 0%).

https://www.conservationevidence.com/actions/1974

Inform local communities about the negative impacts of bat hunting to reduce killing of bats

One study evaluated the effects of informing local communities about the negative impacts of bat hunting to reduce killing of bats on bat populations. The study was in Ghana.

COMMUNITY RESPONSE (0 STUDIES)

POPULATION RESPONSE (0 STUDIES)

BEHAVIOUR (1 STUDY)

Behaviour change (1 study): One before-and-after study in Ghana found that after providing education about the ecological roles of bats fewer hunters intended to hunt bats in the future.

Assessment: unknown effectiveness (effectiveness 50%; certainty 20%; harms 0%).

https://www.conservationevidence.com/actions/1973

No evidence found (no assessment)

We have captured no evidence for the following interventions:

- Encourage online vendors to remove bat specimens for sale
- Enforce regulations to prevent trafficking and trade of bats
- Introduce alternative treatments to reduce the use of bats in traditional medicine

- Introduce and enforce legislation to control hunting of bats
- Introduce other food sources to replace bat meat
- Introduce other income sources to replace bat trade
- Replace culling of bats with non-lethal methods of preventing vampire bats from spreading rabies to humans
- Restrict the collection of bat specimens for research
- Strengthen cultural traditions that discourage bat harvesting.

2.5.2 Guano harvesting

Based on the collated evidence, what is the current assessment of the effectiveness of interventions for guano harvesting?	
No evidence found (no assessment)	• Introduce and enforce legislation to regulate harvesting of bat guano

No evidence found (no assessment)

We have captured no evidence for the following interventions:

- Introduce and enforce legislation to regulate harvesting of bat guano.

2.5.3 Logging and wood harvesting

Based on the collated evidence, what is the current assessment of the effectiveness of interventions for logging and wood harvesting?	
Likely to be beneficial	• Retain forested corridors in logged areas • Thin trees within forest and woodland • Use selective or reduced impact logging instead of conventional logging
Unknown effectiveness	• Manage forest and woodland to encourage understorey growth • Retain residual tree patches in logged areas • Use shelterwood cutting instead of clearcutting

Based on the collated evidence, what is the current assessment of the effectiveness of interventions for logging and wood harvesting?	
No evidence found (no assessment)	• Change timing of forestry operations • Coppice woodland • Encourage natural regeneration in former plantations • Maintain forest and woodland edges for foraging bats • Protect roost trees during forest operations • Replant native trees in logged areas • Retain buffers around roost trees in logged areas • Retain riparian buffers in logged areas • Strengthen cultural traditions such as sacred groves that prevent timber harvesting • Train arborists and forestry operatives to identify potential bat roosts

Likely to be beneficial

Retain forested corridors in logged areas

Three studies evaluated the effects of retaining forested corridors in logged areas on bat populations. The three studies were in the USA.

COMMUNITY RESPONSE (0 STUDIES)

POPULATION RESPONSE (1 STUDY)

Abundance (1 studies): One replicated, site comparison study in the USA found that bat activity (relative abundance) was significantly higher along the edges of forested corridors than in corridor interiors or in adjacent logged stands, which had similar activity levels.

BEHAVIOUR (2 STUDIES)

Use (2 studies): One replicated, site comparison study in the USA found more Seminole bats roosting in forested corridors than logged stands or mature forest. One replicated, site comparison study in the USA found more male but fewer female evening bats roosting in forested corridors than logged stands.

Assessment: likely to be beneficial (effectiveness 55%; certainty 40%; harms 0%).

https://www.conservationevidence.com/actions/996

Thin trees within forest and woodland

Eleven studies evaluated the effects of thinning trees within forest and woodland on bat populations. Six studies were in the USA, four were in Australia and one was in Canada.

COMMUNITY RESPONSE (2 STUDIES)

Richness/diversity (2 studies): One replicated, site comparison study in Australia recorded the same bat species in thinned and unthinned forest, except for the chocolate wattled bat, which was not recorded in forests with unthinned regrowth. One replicated, site comparison study in Australia found that forest thinned up to 20 years previously had higher bat diversity than unthinned forest, but sites thinned more than 20 years previously did not differ.

POPULATION RESPONSE (11 STUDIES)

Abundance (11 studies): Five of six replicated, site comparison studies (including two paired sites studies and one controlled study) in the USA and Australia found higher overall bat activity (relative abundance) in thinned or thinned and burned forest than unthinned forest. The other study found similar overall bat activity in thinned and unthinned stands. One replicated, randomized, site comparison study in the USA found higher overall bat activity for three of four types of thinning and burning treatments. One replicated, site comparison study in Australia found that forest thinned up to eight years previously or more than 20 years previously had higher bat activity than unthinned forest, but sites thinned 8–20 years previously did not differ. Three replicated, controlled studies (including one site comparison and one before-and-after study) in Canada and Australia found that thinning increased the activity of some bat species but not others.

BEHAVIOUR (0 STUDIES)

Assessment: likely to be beneficial (effectiveness 60%; certainty 60%; harms 0%).

https://www.conservationevidence.com/actions/991

Use selective or reduced impact logging instead of conventional logging

Four studies evaluated the effects of using selective or reduced impact logging instead of conventional logging on bat populations. Two studies were in the Neotropics, one study was in Italy and one in Germany.

COMMUNITY RESPONSE (1 STUDY)

Community composition (1 study): One replicated, controlled, site comparison study in Trinidad found that the composition of bat species differed between selectively logged and conventionally logged forest.
Richness/diversity (1 study): One replicated, site comparison study in Germany found similar bat diversity in selectively logged and conventionally logged forest.
POPULATION RESPONSE (3 STUDIES)
Abundance (3 studies): One replicated, site comparison study in Germany found similar overall bat activity (relative abundance) in selectively logged and conventionally logged forest. One review of 41 studies in the Neotropics found that reduced impact logging had a smaller effect on bat abundance than conventional logging. One replicated, site comparison study in Italy found greater bat activity at two of three sites that used selective logging techniques to open up the forest canopy rather than leaving the canopy intact.
BEHAVIOUR (0 STUDIES)
Assessment: likely to be beneficial (effectiveness 50%; certainty 40%; harms 0%).

https://www.conservationevidence.com/actions/989

Unknown effectiveness

Manage forest and woodland to encourage understorey growth

One study evaluated the effects of managing forest and woodland to encourage understorey growth on bat populations. The study was in Germany.
COMMUNITY RESPONSE (1 STUDY)
Richness/diversity (1 study): One paired sites study in Germany found more bat species and higher bat diversity in a forest managed to encourage understorey growth than in a managed forest without understorey growth.
POPULATION RESPONSE (1 STUDY)
Abundance (1 study): One paired sites study in Germany found higher overall bat activity (relative abundance) in a forest managed to encourage understorey growth than in a managed forest without understorey growth.
BEHAVIOUR (0 STUDIES)
Assessment: unknown effectiveness (effectiveness 60%; certainty 20%; harms 0%).

https://www.conservationevidence.com/actions/1986

Retain residual tree patches in logged areas

Three studies evaluated the effects of retaining residual tree patches in logged areas on bat populations. The three studies were in Canada.

COMMUNITY RESPONSE (0 STUDIES)

POPULATION RESPONSE (3 STUDIES)

Abundance (3 studies): Two replicated, site comparison studies in Canada found no difference in bat activity (relative abundance) along the edges of residual tree patches and the edges of clearcut blocks. One replicated, site comparison study in Canada found that the activity of smaller bat species was higher along the edge of residual tree patches than in the centre of clearcut blocks, but the activity of larger bat species did not differ. One replicated, controlled study in Canada found that residual tree patches had similar activity of little brown bats and northern long-eared bats and lower activity of silver-haired bats compared to clearcut forest patches.

BEHAVIOUR (0 STUDIES)

Assessment: unknown effectiveness (effectiveness 30%; certainty 35%; harms 0%).

https://www.conservationevidence.com/actions/995

Use shelterwood cutting instead of clearcutting

One study evaluated the effects of using shelterwood cutting instead of 'gap release' cutting on bat populations. The study was in Australia. We found no studies that evaluated the effects of shelterwood cutting instead of clearcutting.

COMMUNITY RESPONSE (0 STUDIES)

POPULATION RESPONSE (0 STUDIES)

BEHAVIOUR (1 STUDY)

Use (1 study): One site comparison study in Australia found more Gould's long-eared bats roosting in remnant trees within forests that had been shelterwood harvested than in forests harvested using gap release methods. Comparisons were not made with clearcutting.

Assessment: unknown effectiveness (effectiveness 15%; certainty 10%; harms 0%).

https://www.conservationevidence.com/actions/990

No evidence found (no assessment)

We have captured no evidence for the following interventions:

- Change timing of forestry operations

- Coppice woodland
- Encourage natural regeneration in former plantations
- Maintain forest and woodland edges for foraging bats
- Protect roost trees during forest operations
- Replant native trees in logged areas
- Retain buffers around roost trees in logged areas
- Retain riparian buffers in logged areas
- Strengthen cultural traditions such as sacred groves that prevent timber harvesting
- Train arborists and forestry operatives to identify potential bat roosts.

2.6 Threat: Human intrusions and disturbance

2.6.1 Caving and tourism

Based on the collated evidence, what is the current assessment of the effectiveness of interventions for caving and tourism?	
Likely to be beneficial	• Impose restrictions on cave visits
Trade-off between benefit and harms	• Install and maintain cave gates to restrict public access
Unknown effectiveness	• Install fencing around cave entrances to restrict public access • Minimize noise levels within caves • Restrict artificial lighting in caves and around cave entrances
No evidence found (no assessment)	• Inform the public of ways to reduce disturbance to bats in caves • Introduce guidelines for sustainable cave development and use • Minimize alterations to caves for tourism • Provide artificial subterranean bat roosts to replace roosts in disturbed caves • Restore and maintain microclimate in modified caves • Retain bat access points to caves • Train tourist guides to minimize disturbance and promote bat conservation

Likely to be beneficial

Impose restrictions on cave visits

Four studies evaluated the effects of imposing restrictions on cave visits on bat populations. One study was in each of the USA, Canada, Madagascar, and Turkey.

COMMUNITY RESPONSE (0 STUDIES)

POPULATION RESPONSE (2 STUDIES)

Abundance (2 studies): Two before-and-after studies in Canada and Turkey found that bat populations within caves increased after restrictions on cave visitors were imposed.

BEHAVIOUR (2 STUDIES)

Behaviour change (2 studies): One study in the USA found that reducing the number of people within cave tour groups did not have a significant effect on the number of take-offs, landings or overall activity (bat movements) of a cave myotis colony roosting within the cave. One study in Madagascar found that increasing visitor approach distances, along with avoiding direct illumination of bats, reduced the alertness and number of take-offs of Madagascan rousettes during experimental cave tours.

Assessment: likely to be beneficial (effectiveness 67%; certainty 50%; harms 0%).

https://www.conservationevidence.com/actions/1002

Trade-off between benefit and harms

Install and maintain cave gates to restrict public access

Eleven studies evaluated the effects of installing cave gates on bat populations. Six studies were in the USA and five studies were in Europe.

COMMUNITY RESPONSE (0 STUDIES)

POPULATION RESPONSE (7 STUDIES)

Abundance (7 studies): Three of four before-and-after studies (including one replicated study and one controlled study) in the Netherlands, the USA, Spain and Turkey found more or similar numbers of bats in caves and a bunker after gates were installed to restrict public access. The other study found fewer bats in caves after gates were installed. Two before-and-after studies in the USA and Spain found more bats within two caves after the size of the gated entrances were increased. One replicated, before-and-after study in the USA found that installing cave gates resulted in population increases or decreased rates of decline for 13 of 20 colonies of Indiana bat. One replicated,

site comparison study in Spain found no difference in the population growth rates of bats roosting in caves with and without cave gates.
Condition (1 study): One site comparison study in the USA found that bats hibernating in a cave with a wall and gate over the entrance lost more body mass than bats in a nearby unmodified cave.
BEHAVIOUR (5 STUDIES)
Use (1 study): One replicated, site comparison study in Spain found no difference in the occupancy rates of bats roosting in caves with and without cave gates.
Behaviour change (4 studies): One replicated, controlled, before-and-after and site comparison study in the USA found that bats at cave entrances circled more and entered caves less after gates were installed. One replicated study in the USA found that bats flew through gates with a funnel design more frequently than gates with a round bar or angle iron design. One randomized, controlled, before-and-after study in the UK found that fewer bats flew through cave gates when the spacing between horizontal bars was reduced. One before-and-after study in the USA found that significantly fewer bats emerged from a cave with a gate installed compared with a cave with a fence.
Assessment: trade-off between benefit and harms (effectiveness 70%; certainty 50%; harms 20%).

https://www.conservationevidence.com/actions/999

Unknown effectiveness

Install fencing around cave entrances to restrict public access

Two studies evaluated the effects of installing fencing around cave entrances on bat populations. One study was in the USA and one study was in Spain.
COMMUNITY RESPONSE (0 STUDIES)
POPULATION RESPONSE (1 STUDY)
Abundance (1 study): One replicated, site comparison study in Spain found no difference in the population growth rates of bats roosting in caves with and without fencing or gates installed.
BEHAVIOUR (2 STUDIES)
Use (1 study): One replicated, site comparison study in Spain found no difference in the occupancy rates of bats roosting in caves with and without fencing or gates installed.

Behaviour change (1 study): One controlled, before-and-after study in the USA found that significantly more southeastern myotis bats and gray myotis bats emerged from a cave after a steel gate was replaced with a fence.

Assessment: unknown effectiveness (effectiveness 40%; certainty 20%; harms 0%).

https://www.conservationevidence.com/actions/1991

Minimize noise levels within caves

One study evaluated the effects of minimizing noise levels within caves on bat populations. The study was in the USA.

COMMUNITY RESPONSE (0 STUDIES)

POPULATION RESPONSE (0 STUDIES)

BEHAVIOUR (1 STUDY)

Behaviour change (1 study): One controlled study in the USA found that experimental cave tours with groups that did not talk resulted in fewer bat flights than when groups did talk, but talking did not have an effect on the number of bat movements.

Assessment: unknown effectiveness (effectiveness 40%; certainty 21%; harms 0%).

https://www.conservationevidence.com/actions/1995

Restrict artificial lighting in caves and around cave entrances

One study evaluated the effects of restricting artificial lighting in caves on bat populations. The study was in the USA.

COMMUNITY RESPONSE (0 STUDIES)

POPULATION RESPONSE (0 STUDIES)

BEHAVIOUR (1 STUDY)

Behaviour change (1 study): One controlled study in the USA found that using low intensity white lights or red lights in caves resulted in fewer bat flights than with full white lighting, but the number of bat movements was similar between all three light treatments.

Assessment: unknown effectiveness (effectiveness 15%; certainty 12%; harms 0%).

https://www.conservationevidence.com/actions/1994

No evidence found (no assessment)

We have captured no evidence for the following interventions:

- Inform the public of ways to reduce disturbance to bats in caves
- Introduce guidelines for sustainable cave development and use
- Minimize alterations to caves for tourism
- Provide artificial subterranean bat roosts to replace roosts in disturbed caves
- Restore and maintain microclimate in modified caves
- Retain bat access points to caves
- Train tourist guides to minimize disturbance and promote bat conservation.

2.7 Threat: Natural system modifications

2.7.1 Fire or fire suppression

Based on the collated evidence, what is the current assessment of the effectiveness of interventions for fire or fire suppression?	
Trade-off between benefit and harms	• Use prescribed burning

Trade-off between benefit and harms

Use prescribed burning

Fifteen studies evaluated the effects of prescribed burning on bat populations. Thirteen studies were in the USA and two were in Australia.

COMMUNITY RESPONSE (2 STUDIES)

Community composition (2 studies): One of two replicated studies (one before-and-after with paired sites, one site comparison) in Australia found that the composition of bat species differed between burned and unburned woodland sites. The other study found that the composition of bat species was similar between unlogged forest blocks burned every two or four years and unburned blocks.

Richness/diversity (1 study): One replicated, randomized, site comparison study in Australia found more bat species in unlogged forest blocks burned every four years than in blocks burned every two years or unburned blocks.

POPULATION RESPONSE (9 STUDIES)

Abundance (9 studies): Two replicated, site comparison studies (including one controlled study) in the USA found that the activity (relative abundance) of open habitat bat species and evening bats increased with the number of prescribed fires, but there was no effect on other bat species, including cluttered habitat bat species. Four replicated, before-and-after or site comparison studies (including two controlled studies) in the USA and Australia found that prescribed burning, prescribed burning along with thinning or prescribed burning every four years resulted in higher overall bat activity or activity of Florida bonneted bats. One site comparison study in the USA found that two of seven sites that had been burned alongside other restoration practices had higher bat activity than unrestored sites. One replicated, randomized, site comparison study in the USA found that three of four burning and thinning treatments resulted in higher overall bat activity. One replicated, controlled, site comparison study in the USA found similar activity of three bat species in burned and unburned tree stands.

BEHAVIOUR (6 STUDIES)

Use (5 studies): One replicated, controlled before-and-after study in the USA found that more female northern myotis bats roosted in burned than unburned forest. Two replicated, controlled, site comparison studies in the USA found that fewer female northern myotis bats and male Indiana bats roosted in burned than unburned forest. One replicated study in the USA found that evening bats roosted in burned but not unburned forest. One replicated, paired sites study in the USA found that burned sites had a higher occupancy of five bat species/species groups than unburned sites, and burn severity had a negative effect on the occupancy of two bat species/species groups.

Behaviour change (4 studies): Two replicated, controlled, site comparison studies in the USA found no difference in roost switching frequency or the distance between roost trees for female northern myotis bats and male Indiana bats in burned and unburned forests. One replicated, controlled, before-and-after study in the USA found that female northern myotis home ranges and core areas did not differ in size between burned and unburned forests, but home ranges were closer to burned forest than unburned forest. One replicated, site comparison study in the USA found that home ranges of female Rafinesque's big-eared bats were located similar distances to burned and unburned forest, and male home ranges were closer to unburned forest.

Assessment: trade-off between benefit and harms (effectiveness 50%; certainty 50%; harms 20%).

https://www.conservationevidence.com/actions/1006

2.7.2 Dams and water management/use

Based on the collated evidence, what is the current assessment of the effectiveness of interventions for dams and water management/use?	
Unknown effectiveness	• Create or maintain small dams to provide foraging and drinking habitat for bats • Relocate bat colonies roosting inside dams

Unknown effectiveness

Create or maintain small dams to provide foraging and drinking habitat for bats

One study evaluated the effects of maintaining small dams as foraging and drinking habitat for bats on bat populations. The study was in Portugal.

COMMUNITY RESPONSE (0 STUDIES)

POPULATION RESPONSE (1 STUDY)

Abundance (1 study): One replicated, site comparison study in Portugal found that reservoirs created using small dams had greater activity (relative abundance) of four bat species than the streams feeding into them.

BEHAVIOUR (0 STUDIES)

Assessment: unknown effectiveness (effectiveness 51%; certainty 20%; harms 0%).

https://www.conservationevidence.com/actions/1997

Relocate bat colonies roosting inside dams

One study evaluated the effects of relocating bat colonies inside dams on bat populations. The study was in Argentina.

COMMUNITY RESPONSE (0 STUDIES)

POPULATION RESPONSE (1 STUDY)

Abundance (1 study): One study in Argentina found that almost two-thirds of a large colony of Brazilian free-tailed bats relocated to a different dam compartment five months after being displaced from six compartments where the colony originally roosted.

BEHAVIOUR (0 STUDIES)

Assessment: unknown effectiveness (effectiveness 5%; certainty 5%; harms 5%).

https://www.conservationevidence.com/actions/1998

2.8 Threat: Invasive or problematic species and disease

2.8.1 Invasive species

Based on the collated evidence, what is the current assessment of the effectiveness of interventions for invasive species?	
Likely to be beneficial	• Control invasive predators
Unknown effectiveness	• Control invasive plant species
No evidence found (no assessment)	• Control harmful invasive bat prey species • Control invasive non-predatory competitors • Exclude domestic and feral cats from bat roosts and roost entrances • Keep domestic cats indoors at night • Use collar-mounted devices on cats to reduce predation of bats

Likely to be beneficial

Control invasive predators

One study evaluated the effects of controlling invasive predators on bat populations. The study was in New Zealand.

COMMUNITY RESPONSE (0 STUDIES)

POPULATION RESPONSE (1 STUDY)

Survival (1 study): One replicated, before-and-after study in New Zealand found that controlling ship rats resulted in increased survival probabilities for female long-tailed bats.

BEHAVIOUR (0 STUDIES)

Assessment: likely to be beneficial (effectiveness 80%; certainty 40%; harms 0%).

https://www.conservationevidence.com/actions/1007

Unknown effectiveness

Control invasive plant species

One study evaluated the effects of controlling invasive plant species on bat populations. The study was in the USA.

COMMUNITY RESPONSE (0 STUDIES)

POPULATION RESPONSE (1 STUDY)

Abundance (1 study): One site comparison study in the USA found that two of seven forest fragments where invasive plant species had been removed alongside other restoration practices had significantly higher bat activity (relative abundance) than two unrestored forest fragments.

BEHAVIOUR (0 STUDIES)

Assessment: unknown effectiveness (effectiveness 20%; certainty 10%; harms 0%).

https://www.conservationevidence.com/actions/1008

No evidence found (no assessment)

We have captured no evidence for the following interventions:

- Control harmful invasive bat prey species
- Control invasive non-predatory competitors
- Exclude domestic and feral cats from bat roosts and roost entrances
- Keep domestic cats indoors at night
- Use collar-mounted devices on cats to reduce predation of bats.

2.8.2 White-nose syndrome

Based on the collated evidence, what is the current assessment of the effectiveness of interventions for white-nose syndrome?	
Unknown effectiveness	• Modify bat hibernacula environments to increase survival of bats infected with white-nose syndrome • Treat bats for infection with white-nose syndrome

Based on the collated evidence, what is the current assessment of the effectiveness of interventions for white-nose syndrome?	
No evidence found (no assessment)	• Breed bats in captivity to supplement wild populations affected by white-nose syndrome • Cull bats infected with white-nose syndrome • Decontaminate clothing and equipment after entering caves to reduce the spread of the white-nose syndrome pathogen • Restrict human access to bat caves to reduce the spread of the white-nose syndrome pathogen • Treat bat hibernacula environments to reduce the white-nose syndrome pathogen reservoir • Vaccinate bats against the white-nose syndrome pathogen

Unknown effectiveness

Modify bat hibernacula environments to increase survival of bats infected with white-nose syndrome

One study evaluated the effects of modifying hibernacula environments to increase the survival of bats infected with white-nose syndrome. The study was in the USA.

COMMUNITY RESPONSE (0 STUDIES)

POPULATION RESPONSE (1 STUDY)

Survival (1 study): One randomized, replicated, controlled study in the USA found that a greater number of little brown bats infected with the white-nose syndrome fungus survived in hibernation chambers at 4°C than at 10°

BEHAVIOUR (1 STUDY)

Behaviour change (1 study): One randomized, replicated, controlled study in the USA found that little brown bats infected with the white-nose syndrome fungus stayed in hibernation for longer in hibernation chambers at 4°C than at 10°

Assessment: unknown effectiveness (effectiveness 50%; certainty 30%; harms 10%).

https://www.conservationevidence.com/actions/1013

Treat bats for infection with white-nose syndrome

Two studies evaluated the effects of treating bats with a probiotic bacterium to reduce white-nose syndrome infection. One study was in Canada and one in the USA.

COMMUNITY RESPONSE (0 STUDIES)

POPULATION RESPONSE (2 STUDIES)

Survival (2 studies): One randomized, controlled study in Canada found that treating little brown bats with a probiotic bacterium at the time of infection with white-nose syndrome (but not 21 days prior) increased survival within cages in a laboratory. One randomized, controlled study in the USA found that treating little brown bats with a probiotic bacterium within a mine increased survival for free-flying bats, but not caged bats.

Condition (2 studies): One randomized, controlled study in Canada found that little brown bats caged in a laboratory and treated with a probiotic bacterium at the time of infection with white-nose syndrome had reduced symptoms of the disease, but bats treated 21 days prior to infection had worse symptoms. One randomized, controlled study in the USA found that little brown bats kept within cages in a mine and treated with a probiotic bacterium had a similar severity of white-nose syndrome to untreated bats.

BEHAVIOUR (0 STUDIES)

Assessment: unknown effectiveness (effectiveness 50%; certainty 35%; harms 15%).

https://www.conservationevidence.com/actions/2008

No evidence found (no assessment)

We have captured no evidence for the following interventions:

- Breed bats in captivity to supplement wild populations affected by white-nose syndrome
- Cull bats infected with white-nose syndrome
- Decontaminate clothing and equipment after entering caves to reduce the spread of the white-nose syndrome pathogen
- Restrict human access to bat caves to reduce the spread of the white-nose syndrome pathogen
- Treat bat hibernacula environments to reduce the white-nose syndrome pathogen reservoir
- Vaccinate bats against the white-nose syndrome pathogen.

2.8.3 Disease

Based on the collated evidence, what is the current assessment of the effectiveness of interventions for disease?	
No evidence found (no assessment)	• Carry out surveillance of bats to prevent the spread of disease/viruses to humans to reduce human-wildlife conflict

No evidence found (no assessment)

We have captured no evidence for the following interventions:

- Carry out surveillance of bats to prevent the spread of disease/viruses to humans to reduce human-wildlife conflict.

2.8.4 Problematic native species

Based on the collated evidence, what is the current assessment of the effectiveness of interventions for problematic native species?	
No evidence found (no assessment)	• Modify bats roosts to reduce negative impacts of one bat species on another • Protect bats within roosts from disturbance or predation by native species

No evidence found (no assessment)

We have captured no evidence for the following interventions:

- Modify bats roosts to reduce negative impacts of one bat species on another
- Protect bats within roosts from disturbance or predation by native species.

2.9 Threat: Pollution

2.9.1 Domestic and urban waste water

Based on the collated evidence, what is the current assessment of the effectiveness of interventions for domestic and urban waste water?	
Unknown effectiveness	• Change effluent treatments of domestic and urban waste water
No evidence found (no assessment)	• Prevent pollution from sewage treatment facilities from entering watercourses • Reduce or prevent the use of septic systems near caves

Unknown effectiveness

Change effluent treatments of domestic and urban waste water

One study evaluated the effects of different sewage treatments on the activity of foraging bats. The study was in the UK. We found no studies that evaluated the effects of changing effluent treatments of domestic and urban waste water discharged into rivers on bat populations.

COMMUNITY RESPONSE (0 STUDIES)

POPULATION RESPONSE (1 STUDY)

Abundance (1 study): One replicated, site comparison study in the UK found higher activity (relative abundance) of foraging bats over filter bed sewage treatment works than over active sludge systems.

BEHAVIOUR (0 STUDIES)

Assessment: unknown effectiveness (effectiveness 25%; certainty 20%; harms 0%).

https://www.conservationevidence.com/actions/1014

No evidence found (no assessment)

We have captured no evidence for the following interventions:

- Prevent pollution from sewage treatment facilities from entering watercourses
- Reduce or prevent the use of septic systems near caves.

2.9.2 Agricultural and forestry effluents

Based on the collated evidence, what is the current assessment of the effectiveness of interventions for agricultural and forestry effluents?	
Likely to be beneficial	• Reduce pesticide, herbicide or fertiliser use
No evidence found (no assessment)	• Change effluent treatments used in agriculture and forestry • Introduce legislation to control the use of hazardous substances • Plant riparian buffer strips • Prevent pollution from agricultural land or forestry from entering watercourses • Use organic pest control instead of synthetic pesticides

Likely to be beneficial

Reduce pesticide, herbicide or fertiliser use

Four studies evaluated the effects of reducing pesticide, herbicide and fertiliser use on bat populations. One study was in each of Mexico, Portugal, Germany and Columbia.

COMMUNITY RESPONSE (3 STUDIES)

Community composition (1 study): One replicated, site comparison study in Portugal found that farms using few or no chemicals had different compositions of bat species to farms using high chemical inputs.

Richness/diversity (2 studies): One site comparison study in Mexico found that coffee agroforestry plantations using few or no chemicals had a higher diversity of insect-eating bat species than plantations with high chemical inputs, but the diversity of fruit and nectar-eating bat species did not differ. One paired sites study in Germany recorded more bat species over grassland with moderate or no fertiliser applications than grassland with high fertiliser applications.

POPULATION RESPONSE (3 STUDIES)

Abundance (2 studies): Two site comparison or paired sites studies (one replicated) in Portugal and Germany found that farms or grasslands with few or no chemical inputs had higher overall bat activity (relative abundance) than those using high chemical inputs.

Condition (1 study): One replicated, site comparison study in Columbia found that great fruit-eating bats captured in 'silvopastoral' areas that used no chemicals, along with agroforestry, had higher body weights and body condition scores than those in conventional farming areas that used chemicals.

BEHAVIOUR (0 STUDIES)

Assessment: likely to be beneficial (effectiveness 77%; certainty 60%; harms 0%).

https://www.conservationevidence.com/actions/2013

No evidence found (no assessment)

We have captured no evidence for the following interventions:

- Change effluent treatments used in agriculture and forestry
- Introduce legislation to control the use of hazardous substances
- Plant riparian buffer strips
- Prevent pollution from agricultural land or forestry from entering watercourses
- Use organic pest control instead of synthetic pesticides.

2.9.3 Light pollution

Based on the collated evidence, what is the current assessment of the effectiveness of interventions for light pollution?	
Beneficial	• Leave bat roosts and roost entrances unlit
Likely to be beneficial	• Avoid illumination of bat commuting routes • Avoid illumination of bat foraging, drinking and swarming sites • Use low intensity lighting • Use red lighting rather than other lighting colours

Based on the collated evidence, what is the current assessment of the effectiveness of interventions for light pollution?	
Unknown effectiveness	• Restrict timing of lighting • Use ultraviolet filters on lights
No evidence found (no assessment)	• Direct lighting away from bat access points or habitats • Use 'warm white' rather than 'cool' LED lights • Use glazing treatments to reduce light spill from inside lit buildings

Beneficial

Leave bat roosts and roost entrances unlit

Five studies evaluated the effects of leaving bat roosts and roost entrances unlit on bat populations. Two studies were in the UK, and one study was in each of Canada, Hungary and Sweden.

COMMUNITY RESPONSE (0 STUDIES)

POPULATION RESPONSE (2 STUDIES)

Abundance (1 study): One replicated, controlled study in Canada found that numbers of big brown bats and little brown bats roosting in buildings increased when roosts were left unlit and decreased when roosts were illuminated with artificial lights.

Condition (1 study): One replicated, controlled study in Hungary found that juvenile bats had a higher body mass and greater forearm length at unlit roosts than at roosts with artificial lighting.

BEHAVIOUR (4 STUDIES)

Use (1 study): One replicated, before-and-after study in Sweden found that all of 13 unlit churches continued to be used by brown long-eared bat colonies over 25 years, but bat colonies abandoned their roosts at 14 of 23 churches that were either partly or fully lit with floodlights.

Behaviour change (3 studies): Three controlled studies (including two replicated studies) in the UK and Hungary found that more bats emerged, and bats emerged earlier and foraged for shorter periods, when roosts were left unlit than when they had artificial lighting.

Assessment: beneficial (effectiveness 90%; certainty 80%; harms 0%).

https://www.conservationevidence.com/actions/1017

Likely to be beneficial

Avoid illumination of bat commuting routes

Three studies evaluated the effects of avoiding the illumination of bat commuting routes on bat populations. Two studies were in the UK and one was in the Netherlands.

COMMUNITY RESPONSE (0 STUDIES)

POPULATION RESPONSE (3 STUDIES)

Abundance (3 studies): One replicated, before-and-after study in the Netherlands found similar numbers of pond bats flying along unlit canals and canals illuminated with lamps. Two replicated, controlled studies in the UK found greater activity (relative abundance) of lesser horseshoe bats and myotis bats along unlit hedges than along hedges illuminated with street lights, but activity was similar for common and soprano pipistrelles and Nyctalus/Eptesicus species along unlit and illuminated hedges.

BEHAVIOUR (2 STUDIES)

Behaviour change (2 studies): One replicated, before-and-after study in the Netherlands found that 28–96% of pond bats changed their flight paths along canals to avoid light spill from lamps. One replicated, controlled study in the UK found that lesser horseshoe bats were active earlier along unlit hedges than along those illuminated with street lights.

Assessment: likely to be beneficial (effectiveness 70%; certainty 50%; harms 0%).

https://www.conservationevidence.com/actions/2017

Avoid illumination of bat foraging, drinking and swarming sites

Two studies evaluated the effects of avoiding the illumination of bat drinking sites on bat populations. Both studies were in Italy and one was also in Israel.

COMMUNITY RESPONSE (0 STUDIES)

POPULATION RESPONSE (2 STUDIES)

Abundance (2 studies): Two replicated before-and-after studies (one randomized) in Italy found that unlit water troughs had greater activity (relative abundance) of five of six bat species/species groups and six of eight bat species/species groups than troughs illuminated with artificial light. One of the studies also found that unlit desert ponds in Israel had greater activity (relative abundance) of three bat species than illuminated ponds.

BEHAVIOUR (0 STUDIES)

Assessment: likely to be beneficial (effectiveness 80%; certainty 52%; harms 0%).

https://www.conservationevidence.com/actions/2018

Use low intensity lighting

Three studies evaluated the effects of using low intensity lighting on bat populations. The three studies were in the UK.

COMMUNITY RESPONSE (0 STUDIES)

POPULATION RESPONSE (2 STUDIES)

Abundance (2 studies): One replicated, randomized, controlled study in the UK found that activity (relative abundance) of lesser horseshoe bats, but not myotis bats, was higher along hedges with medium or low intensity lighting than hedges with high intensity lighting. One replicated, randomized, controlled study in the UK found that activity of myotis bats, but not common pipistrelles, was higher along treelined roads with street lights dimmed to an intensity of 25% than roads with streetlights dimmed to 50% or left undimmed.

BEHAVIOUR (1 STUDY)

Behaviour change (1 study): One replicated, controlled study in the UK found that more soprano pipistrelles emerged from two roosts when the intensity of red lights was reduced by placing filters over them.

Assessment: likely to be beneficial (effectiveness 65%; certainty 50%; harms 5%).

https://www.conservationevidence.com/actions/1018

Use red lighting rather than other lighting colours

Three studies evaluated the effects of red lighting on bat populations. Two studies were in the Netherlands and one was in the UK.

COMMUNITY RESPONSE (0 STUDIES)

POPULATION RESPONSE (2 STUDIES)

Abundance (2 studies): One replicated, controlled, site comparison study in the Netherlands found that red lighting resulted in higher activity (relative abundance) for one of three bat species groups than white or green lighting. One site comparison study in the Netherlands found that culverts illuminated with red light had similar activity of commuting Daubenton's bats as culverts illuminated with white or green light.

BEHAVIOUR (1 STUDY)

Behaviour (1 study): One replicated, controlled study in the UK found that more soprano pipistrelles emerged from a roost when lit with red light than when lit with white light, but no difference was found between red and blue lights.

Assessment: likely to be beneficial (effectiveness 50%; certainty 50%; harms 0%).

https://www.conservationevidence.com/actions/2021

Unknown effectiveness

Restrict timing of lighting

One study evaluated the effects of restricting the timing of lighting on bat populations. The study was in France.

COMMUNITY RESPONSE (0 STUDIES)

POPULATION RESPONSE (1 STUDY)

Abundance (1 study): One replicated, paired sites study in France found that turning off streetlights for part of the night resulted in mixed results for activity (relative abundance), depending on bat species, when compared with leaving streetlights switched on all night.

BEHAVIOUR (0 STUDIES)

Assessment: unknown effectiveness (effectiveness 38%; certainty 20%; harms 0%).

https://www.conservationevidence.com/actions/1019

Use ultraviolet filters on lights

One study evaluated the effects of using ultraviolet filters on lights on bat populations. The study was in the UK.

COMMUNITY RESPONSE (0 STUDIES)

POPULATION RESPONSE (1 STUDY)

Abundance (1 study): One randomized, replicated, controlled study in the UK found that hedges lit with ultraviolet filtered lights had higher soprano pipistrelle, but not common pipistrelle activity (relative abundance) than hedges lit with unfiltered light.

BEHAVIOUR (0 STUDIES)

Assessment: unknown effectiveness (effectiveness 45%; certainty 22%; harms 0%).

https://www.conservationevidence.com/actions/1020

No evidence found (no assessment)

We have captured no evidence for the following interventions:

- Direct lighting away from bat access points or habitats
- Use 'warm white' rather than 'cool' LED lights
- Use glazing treatments to reduce light spill from inside lit buildings.

2.9.4 Timber treatments

Based on the collated evidence, what is the current assessment of the effectiveness of interventions for timber treatments?	
Likely to be ineffective or harmful	• Restrict timing of timber treatment application
No evidence found (no assessment)	• Use mammal-safe timber treatments in roof spaces

Likely to be ineffective or harmful

Restrict timing of timber treatment application

One study evaluated the effects of restricting the timing of timber treatment application on bat populations. The study was in the UK.

COMMUNITY RESPONSE (0 STUDIES)

POPULATION RESPONSE (1 STUDY)

Survival (1 study): One replicated, controlled laboratory study in the UK found that treating timber with lindane and pentachlorophenol 14 months prior to exposure by bats increased survival but did not prevent death.

BEHAVIOUR (0 STUDIES)

Assessment: likely to be ineffective or harmful (effectiveness 5%; certainty 55%; harms 50%).

https://www.conservationevidence.com/actions/1023

No evidence found (no assessment)

We have captured no evidence for the following interventions:

- Use mammal-safe timber treatments in roof spaces.

2.9.5 Industrial effluents

Based on the collated evidence, what is the current assessment of the effectiveness of interventions for industrial effluents?	
No evidence found (no assessment)	• Introduce or enforce legislation to prevent ponds and streams from being contaminated by toxins

No evidence found (no assessment)

We have captured no evidence for the following interventions:

- Introduce or enforce legislation to prevent ponds and streams from being contaminated by toxins.

2.9.6 Noise pollution

Based on the collated evidence, what is the current assessment of the effectiveness of interventions for noise pollution?	
No evidence found (no assessment)	• Impose noise limits in proximity to bat roosts and habitats • Install sound barriers in proximity to bat roosts and habitats

No evidence found (no assessment)

We have captured no evidence for the following interventions:

- Impose noise limits in proximity to bat roosts and habitats
- Install sound barriers in proximity to bat roosts and habitats.

2.10 Threat: Climate change and severe weather

Based on the collated evidence, what is the current assessment of the effectiveness of interventions for climate change and severe weather?	
No evidence found (no assessment)	• Adapt bat roost structures to buffer against temperature extremes • Enhance natural habitat features to improve landscape connectivity to allow for range shifts of bats • Manage natural water bodies in arid areas to prevent desiccation • Provide suitable bat foraging and roosting habitat at expanding range fronts

No evidence found (no assessment)

We have captured no evidence for the following interventions:

- Adapt bat roost structures to buffer against temperature extremes
- Enhance natural habitat features to improve landscape connectivity to allow for range shifts of bats
- Manage natural water bodies in arid areas to prevent desiccation
- Provide suitable bat foraging and roosting habitat at expanding range fronts.

2.11 Habitat protection

Based on the collated evidence, what is the current assessment of the effectiveness of interventions for habitat protection?	
Likely to be beneficial	• Conserve roosting sites for bats in old structures or buildings • Legally protect bat habitats
No evidence found (no assessment)	• Retain buffer zones around core habitat • Retain connectivity between habitat patches • Retain existing bat commuting routes • Retain native forest and woodland • Retain remnant habitat patches • Retain veteran and standing dead trees as roosting sites for bats • Retain wetlands

Likely to be beneficial

Conserve roosting sites for bats in old structures or buildings

Three studies evaluated the effects of conserving roosting sites for bats in old structures or buildings on bat populations. Two studies were in the UK and one was in Germany.

COMMUNITY RESPONSE (0 STUDIES)

POPULATION RESPONSE (1 STUDY)

Abundance (1 study): One before-and-after study in the UK found that a greater number of bats hibernated in a railway tunnel after walls with access grilles were installed at the tunnel entrances and wood was attached to the tunnel walls.

BEHAVIOUR (2 STUDIES)

Uptake (1 study): One before-and-after study in Germany found that numbers of bats hibernating in a disused cellar after it was emptied of rubbish increased over 11 years.

Use (2 studies): One before-and-after study in Germany found that a disused cellar that was emptied of rubbish was used by hibernating bats of four species. One before-and-after study in the UK found that Natterer's bats used a roost that was 'boxed-in' within a church, but the number of bats using the roost was reduced by half.

Assessment: likely to be beneficial (effectiveness 72%; certainty 50%; harms 0%).

https://www.conservationevidence.com/actions/2046

Legally protect bat habitats

Five studies evaluated the effects of legally protecting bat habitats on bat populations. Four studies were in Europe and one was in India.

COMMUNITY RESPONSE (2 STUDIES)

Community composition (1 study): One replicated, site comparison study in India found that the composition of bat species was similar in protected forest and unprotected forest fragments.

Richness/diversity (2 studies): Two replicated, site comparison or paired sites studies in Europe and India found that the number of bat species did not differ between protected and unprotected forests or forest fragments. One replicated, site comparison study in France found that protected sites had a greater number of bat species than unprotected sites.

POPULATION RESPONSE (4 STUDIES)

Abundance (4 studies): One replicated, site comparison study in the UK found that the activity (relative abundance) of Daubenton's bats was higher over rivers on farms in protected areas than in unprotected areas. One replicated, paired sites study in Europe found that the activity of common noctule bats was higher in protected forests than unprotected forests, but bat activity overall did not differ. Two replicated, site comparison studies in France and India found higher overall bat activity, higher activity of three of six bat species/species groups and a greater number of bats in protected sites and forests than unprotected sites and forests.

BEHAVIOUR (1 STUDY)

Use (1 study): One study in Spain found that the distributions of 10 of 11 bat species overlapped with areas designated to protect them significantly more than by chance.

Assessment: likely to be beneficial (effectiveness 50%; certainty 41%; harms 0%).

https://www.conservationevidence.com/actions/2045

No evidence found (no assessment)

We have captured no evidence for the following interventions:

- Retain buffer zones around core habitat
- Retain connectivity between habitat patches
- Retain existing bat commuting routes
- Retain native forest and woodland
- Retain remnant habitat patches
- Retain veteran and standing dead trees as roosting sites for bats
- Retain wetlands.

2.12 Habitat restoration and creation

Based on the collated evidence, what is the current assessment of the effectiveness of interventions for habitat restoration and creation?	
Likely to be beneficial	• Create artificial caves or hibernacula for bats • Create artificial water sources • Restore or create wetlands
Unknown effectiveness	• Create artificial hollows and cracks in trees for roosting bats • Reinstate bat roosts in felled tree trunks • Restore or create forest or woodland • Restore or create grassland
No evidence found (no assessment)	• Create new unlit commuting routes using planting • Restore or create linear habitat features/green corridors

Likely to be beneficial

Create artificial caves or hibernacula for bats

Four studies evaluated the effects of creating artificial caves or hibernacula for bats on bat populations. Two studies were in the UK and two were in Germany.

COMMUNITY RESPONSE (0 STUDIES)
POPULATION RESPONSE (0 STUDIES)
BEHAVIOUR (4 STUDIES)

Uptake (1 study): One study in the UK found that the number of bats using an artificial hibernaculum increased in each of nine years after it was built.
Use (4 studies): One study in the UK found that an artificial cave was used by a small number of brown long-eared bats. Three studies in Germany and the UK found that artificial hibernacula were used by up to four bat species.
Assessment: likely to be beneficial (effectiveness 72%; certainty 60%; harms 0%).

https://www.conservationevidence.com/actions/2049

Create artificial water sources

Five studies evaluated the effects of creating artificial water sources for bats on bat populations. One study was in each of the USA, Germany, South Africa, Israel and Mexico.

COMMUNITY RESPONSE (1 STUDY)

Richness/diversity (1 study): One replicated, paired sites study in South Africa found a similar number of bat species over farm ponds and in grassland/crops, trees, vineyards or orchards.

POPULATION RESPONSE (5 STUDIES)

Abundance (5 studies): Five replicated studies (including four site comparisons and one paired sites study) in Israel, the USA, Germany, South Africa and Mexico found that bat activity (relative abundance) was similar or higher over reservoirs and waste water treatment pools, heliponds and drainage ditches, retention ponds and farm/cattle ponds compared to over natural wetlands, nearby vineyards, surrounding forest or grassland/crops, trees and orchards.

BEHAVIOUR (0 STUDIES)

Assessment: likely to be beneficial (effectiveness 70%; certainty 55%; harms 0%).

https://www.conservationevidence.com/actions/959

Restore or create wetlands

One study evaluated the effects of restoring wetlands on bat populations. The study was in the USA.

COMMUNITY RESPONSE (0 STUDIES)

POPULATION RESPONSE (1 STUDY)

Abundance (1 study): One replicated, controlled, before-and-after study in the USA found that restoring wetlands increased overall bat activity (relative abundance), and restored wetlands had similar bat activity to undisturbed wetlands.

BEHAVIOUR (0 STUDIES)

Assessment: likely to be beneficial (effectiveness 62%; certainty 40%; harms 0%).

https://www.conservationevidence.com/actions/2036

Unknown effectiveness

Create artificial hollows and cracks in trees for roosting bats

One study evaluated the effects of creating artificial hollows and cracks in trees for roosting bats. The study was in Australia.

COMMUNITY RESPONSE (0 STUDIES)

POPULATION RESPONSE (0 STUDIES)

BEHAVIOUR (1 STUDY)

Use (1 study): One replicated study in Australia found that eight of 16 artificial hollows cut into trees for bats, birds and marsupials with two different entrance designs were used by roosting long-eared bats.

Assessment: unknown effectiveness (effectiveness 50%; certainty 23%; harms 0%).

https://www.conservationevidence.com/actions/2047

Reinstate bat roosts in felled tree trunks

One study evaluated the effects of reinstating a bat roost within a felled tree trunk on bat populations. The study was in the UK.

COMMUNITY RESPONSE (0 STUDIES)

POPULATION RESPONSE (0 STUDIES)

BEHAVIOUR (1 STUDY)

Use (1 study): One before-and-after study in the UK found that a roost reinstated by attaching the felled tree trunk to a nearby tree continued to be used by common noctule bats as a maternity roost.

Assessment: unknown effectiveness (effectiveness 35%; certainty 10%; harms 0%).

https://www.conservationevidence.com/actions/2048

Restore or create forest or woodland

Two studies evaluated the effects of restoring forests on bat populations. One study was in Brazil and one in Australia.

COMMUNITY RESPONSE (1 STUDY)

Richness/diversity (1 study): One site comparison study in Brazil found that a reforested area had significantly lower bat diversity than a native forest fragment.

POPULATION RESPONSE (1 STUDY)

Abundance (1 study): One replicated, controlled, site comparison study in Australia found that forests restored after mining had significantly higher or similar bat activity (relative abundance) as unmined forests for five of seven bat species.

BEHAVIOUR (0 STUDIES)

Assessment: unknown effectiveness (effectiveness 40%; certainty 15%; harms 0%).

https://www.conservationevidence.com/actions/2050

Restore or create grassland

One study evaluated the effects of creating grassland on bat populations. The study was in the UK.

COMMUNITY RESPONSE (0 STUDIES)

POPULATION RESPONSE (1 STUDY)

Abundance (1 study): One replicated, paired sites study in the UK found that pipistrelle activity (relative abundance) did not differ between species-rich grassland created on agri-environment scheme farms and improved pasture or crop fields on conventional farms.

BEHAVIOUR (0 STUDIES)

Assessment: unknown effectiveness (effectiveness 2%; certainty 10%; harms 0%).

https://www.conservationevidence.com/actions/2051

No evidence found (no assessment)

We have captured no evidence for the following interventions:

- Create new unlit commuting routes using planting
- Restore or create linear habitat features/green corridors.

2.13 Species management

2.13.1 Species management

Based on the collated evidence, what is the current assessment of the effectiveness of interventions for species management?	
Likely to be beneficial	• Provide bat boxes for roosting bats
Unknown effectiveness	• Manage microclimate of artificial bat roosts
No evidence found (no assessment)	• Legally protect bat species • Regularly clean bat boxes to increase occupancy • Release captive-bred bats

Likely to be beneficial

Provide bat boxes for roosting bats

Forty-four studies evaluated the effects of providing bat boxes for roosting bats on bat populations. Twenty-seven studies were in Europe, nine studies were in North America, four studies were in Australia, two studies were in South America, and one study was a worldwide review.

COMMUNITY RESPONSE (0 STUDIES)

POPULATION RESPONSE (0 STUDIES)

BEHAVIOUR (44 STUDIES)

Uptake (9 studies): Nine replicated studies in Europe and the USA found that the number of bats using bat boxes increased by 2–10 times up to 10 years after installation.

Use (43 studies): Forty-one of 43 studies (including 34 replicated studies and two reviews) in Europe, the USA, South America, and Australia found that bats used bat boxes installed in forest or woodland, forestry plantations, farmland, pasture, wetlands, urban areas and buildings, bridges, underpasses or unknown habitats. The other two studies in the USA and UK found that bats displaced from buildings did not use any of 43 bat houses of four different designs or 12 heated bat boxes of one design. One review of 109 studies across Europe, North America and Asia found that 72 bat species used bat boxes, although only 18 species commonly used them, and 31 species used them as maternity roosts. Twenty-two studies (including 17 replicated studies, one before-and-after study and two reviews) found bats occupying less than half of bat boxes provided (0–49%). Nine replicated studies found bats occupying more than half of bat boxes provided (54–100%).

OTHER (23 STUDIES)

Bat box design (16 studies): Three studies in Germany, Portugal and Australia found that bats used black bat boxes more than grey, white or wooden boxes. One of two studies in Spain and the USA found higher occupancy rates in larger bat boxes. One study in the USA found that bats used both resin and wood cylindrical bat boxes, but another study in the USA found that resin bat boxes became occupied more quickly than wood boxes. One study in the UK found higher occupancy rates in concrete than wooden bat boxes. One study in the USA found that Indiana bats used rocket boxes more than wooden bat boxes or bark-mimic roosts. One study in Spain found that more bats occupied bat boxes that had two compartments than one compartment in the breeding season. One study in Lithuania found that bat breeding colonies occupied standard and four/five chamber bat boxes and individuals occupied flat bat boxes. Four studies in the USA, UK, Spain and Australia found bats selecting four of nine, three of five, three of four and one of five bat box designs. One study in the UK found that different bat box designs were used by different species. One study in Costa Rica found that bat boxes simulating tree trunks were used by 100% of bats and in group sizes similar to natural roosts.

Bat box position (11 studies): Three studies in Germany, Spain and the USA found that bat box orientation and/or the amount of exposure to sunlight affected bat occupancy, and one study in Spain found that orientation did not have a significant effect on occupancy. Two studies in the UK and Italy found that bat box height affected occupancy, and two studies in Spain and the USA found no effect of height. Two studies in the USA and Spain found higher occupancy of bat boxes on buildings than on trees. One study in Australia found that bat boxes were occupied more often in farm forestry sites than in native forest, one study in Poland found higher occupancy in pine relative to mixed deciduous stands, and one study in Costa Rica found higher occupancy in forest fragments than in pasture. One study in the USA found higher occupancy rates in areas where bats were known to roost prior to installing bat boxes. One review in the UK found that bat boxes were more likely to be occupied when a greater number of bat boxes were installed across a site.

Assessment: likely to be beneficial (effectiveness 70%; certainty 60%; harms 0%).

https://www.conservationevidence.com/actions/1024

Unknown effectiveness

Manage microclimate of artificial bat roosts

Three studies evaluated the effects of managing the microclimate of artificial bat roosts on bat populations. Two studies were in the UK and one in Spain.

COMMUNITY RESPONSE (0 STUDIES)

POPULATION RESPONSE (1 STUDY)

Abundance (1 study): One before-and-after study in Spain found more bats in two artificial roosts within buildings after they had been modified to reduce internal roost temperatures.

BEHAVIOUR (2 STUDIES)

Use (2 studies): One replicated, before-and-after study in the UK found that heated bat boxes were used by common pipistrelle bats at one of seven sites, but none were used by maternity colonies. One replicated study in the UK found that none of the 12 heated bat boxes installed within churches were used by displaced Natterer's bats.

Assessment: unknown effectiveness (effectiveness 40%; certainty 30%; harms 0%).

https://www.conservationevidence.com/actions/2052

No evidence found (no assessment)

We have captured no evidence for the following interventions:

- Legally protect bat species
- Regularly clean bat boxes to increase occupancy
- Release captive-bred bats.

2.13.2 Ex-situ conservation

Based on the collated evidence, what is the current assessment of the effectiveness of interventions for ex-situ conservation?	
Unknown effectiveness	• Breed bats in captivity • Rehabilitate injured/orphaned bats to maintain wild bat populations

Unknown effectiveness

Breed bats in captivity

Eight studies evaluated the effects of breeding bats in captivity on bat populations. Three studies were in the USA, two in the UK, and one in each of Italy, Brazil and New Zealand.

COMMUNITY RESPONSE (0 STUDIES)

POPULATION RESPONSE (8 STUDIES)

Reproductive success (6 studies): Six studies in the USA, UK, Italy and Brazil found that 6–100% of female bats captured in the wild successfully conceived, gave birth and reared young in captivity. Two studies in the UK and Brazil found that two of five and two of three bats born in captivity successfully gave birth to live young.

Survival (8 studies): Seven studies in the USA, UK, Italy and Brazil found that 20–100% of bat pups born in captivity survived from between 10 days to adulthood. One study in New Zealand found that two of five New Zealand lesser short-tailed bat pups born in captivity survived, both of which were hand-reared.

BEHAVIOUR (0 STUDIES)

Assessment: unknown effectiveness (effectiveness 50%; certainty 30%; harms 10%).

https://www.conservationevidence.com/actions/2053

Rehabilitate injured/orphaned bats to maintain wild bat populations

Four studies evaluated the effects of rehabilitating injured/orphaned bats on bat populations. Two studies were in the UK, one was in Italy and one in Brazil.

COMMUNITY RESPONSE (0 STUDIES)

POPULATION RESPONSE (4 STUDIES)
Survival (4 studies): One study in Brazil found that two hand-reared orphaned greater spear-nosed bats survived for over three months in captivity. Two studies in the UK and Italy found that 70–90% of hand-reared pipistrelle bats survived for at least 4–14 days after release into the wild, and six of 21 bats joined wild bat colonies. One study in the UK found that pipistrelle bats that flew in a large flight cage for long periods before release survived for longer and were more active than bats that flew for short periods or in a small enclosure. One study in the UK found that 13% of ringed hand-reared pipstrelle bats were found alive in bat boxes 38 days to almost four years after release into the wild.
Condition (1 study): One study in Brazil found that two orphaned greater spear-nosed bats increased in body weight and size after being hand-reared, and reached a normal size for the species after 60 days.
BEHAVIOUR (0 STUDIES)
Assessment: unknown effectiveness (effectiveness 47%; certainty 27%; harms 0%).

https://www.conservationevidence.com/actions/2054

2.13.3 Translocation

Based on the collated evidence, what is the current assessment of the effectiveness of interventions for translocation?	
Likely to be ineffective or harmful	• Translocate bats

Likely to be ineffective or harmful

Translocate bats

Two studies evaluated the effects of translocating bats on bat populations. One study was in New Zealand and one study was in Switzerland.
COMMUNITY RESPONSE (0 STUDIES)
POPULATION RESPONSE (2 STUDIES)
Reproductive success (1 study): One study in Switzerland found that a female greater horseshoe bat that settled at a release site after translocation had a failed pregnancy.

Survival (1 study): One study in Switzerland found that four of 18 bats died after translocation.

Condition (1 study): One study in New Zealand found that lesser short-tailed bats captured at release sites eight months after translocation were balding and had damaged, infected ears.

BEHAVIOUR (2 STUDIES)

Uptake (2 studies): Two studies in New Zealand and Switzerland found that low numbers of bats remained at release sites after translocation.

Behaviour change (1 study): One study in Switzerland found that bats homed after release at translocation sites less than 20 km from their original roosts.

Assessment: likely to be ineffective or harmful (effectiveness 5%; certainty 40%; harms 80%).

https://www.conservationevidence.com/actions/1009

2.14 Education and awareness raising

Based on the collated evidence, what is the current assessment of the effectiveness of interventions for education and awareness raising?	
No evidence found (no assessment)	• Educate farmers, land managers and local communities about the benefits of bats to improve management of bat habitats • Educate farmers, local communities and pest controllers to reduce indiscriminate culling of vampire bats • Educate pest controllers and homeowners/tenants to reduce the illegal use of pesticides in bat roosts • Educate the public to improve perception of bats to improve behaviour towards bats • Engage policymakers to make policy changes beneficial to bats • Promote careful bat-related eco-tourism to improve behaviour towards bats • Provide training to conservationists, land managers, and the building and development sector on bat ecology and conservation to reduce bat roost disturbance • Provide training to wildlife control operators on least harmful ways of removing bats from their roosts

No evidence found (no assessment)

We have captured no evidence for the following interventions:

- Educate farmers, land managers and local communities about the benefits of bats to improve management of bat habitats
- Educate farmers, local communities and pest controllers to reduce indiscriminate culling of vampire bats
- Educate pest controllers and homeowners/tenants to reduce the illegal use of pesticides in bat roosts
- Educate the public to improve perception of bats to improve behaviour towards bats
- Engage policymakers to make policy changes beneficial to bats
- Promote careful bat-related eco-tourism to improve behaviour towards bats
- Provide training to conservationists, land managers, and the building and development sector on bat ecology and conservation to reduce bat roost disturbance
- Provide training to wildlife control operators on least harmful ways of removing bats from their roosts.

3. BIRD CONSERVATION

David R. Williams, Matthew F. Child, Lynn V. Dicks, Nancy Ockendon, Robert G. Pople, David A. Showler, Jessica C. Walsh, Erasmus K. H. J. zu Ermgassen & William J. Sutherland

Expert assessors

Tatsuya Amano, University of Cambridge, UK
Andy Brown, Natural England, UK
Fiona Burns, Royal Society for the Protection of Birds, UK
Yohay Carmel, Israel Institute of Technology
Mick Clout, University of Auckland, New Zealand
Geoff Hilton, Wildfowl & Wetlands Trust, UK
Nancy Ockendon, University of Cambridge, UK
James Pearce-Higgins, British Trust for Ornithology, UK
Sugoto Roy, Food and Environment Research Agency, DEFRA, UK
Rebecca K. Smith, University of Cambridge, UK
William J. Sutherland, University of Cambridge, UK
Judit Szabo, Charles Darwin University, Australia
Bernie Tershy, University of California, USA
Des Thomson, Scottish Natural Heritage, UK
Stuart Warrington, National Trust, UK
David Williams, University of Cambridge, UK

Scope of assessment: for native wild bird species across the world.

Assessed: 2015.

Effectiveness measure is the median % score.

Certainty measure is the median % certainty of evidence, determined by the quantity and quality of the evidence in the synopsis.

Harm measure is the median % score for negative side-effects to the group of species of concern. This was not scored for section 3.11 on invasive species.

 https://doi.org/10.11647/OBP.0267.03

This book is meant as a guide to the evidence available for different conservation interventions and as a starting point in assessing their effectiveness. The assessments are based on the available evidence for the target group of species for each intervention. The assessment may therefore refer to different species or habitat to the one(s) you are considering. Before making any decisions about implementing interventions it is vital that you read the more detailed accounts of the evidence in order to assess their relevance for your study species or system.

Full details of the evidence are available at**www.conservationevidence.com**

There may also be significant negative side-effects on the target groups or other species or communities that have not been identified in this assessment.

A lack of evidence means that we have been unable to assess whether or not an intervention is effective or has any harmful impacts.

3.1 Habitat protection

Based on the collated evidence, what is the current assessment of the effectiveness of interventions for habitat protection?	
Likely to be beneficial	• Legally protect habitats
Trade-offs between benefit and harms	• Provide or retain un-harvested buffer strips
Unknown effectiveness (limited evidence)	• Ensure connectivity between habitat patches

Likely to be beneficial

Legally protect habitats for birds

Four studies from Europe found that populations increased after habitat protection and a review from China found high use of protected habitats by cranes. A replicated, randomised and controlled study from Argentina found that some, but not all bird groups had higher species richness or were at higher densities in protected habitats. *Assessment: likely to be beneficial (effectiveness 50%; certainty 52%; harms 0%).*

http://www.conservationevidence.com/actions/158

Trade-off between benefit and harms

Provide or retain un-harvested buffer strips

Three replicated studies from the USA found that species richness or abundances were higher in narrow (<100 m) strips of forest, but five replicated studies from North America found that wider strips retained a community more similar to that of uncut forest than narrow strips. Tw replicated studies from the USA found no differences in productivity between wide and narrow buffers, but that predation of artificial nests was higher in buffers than in continuous forest. *Assessment: trade-offs between benefits and harms (effectiveness 60%; certainty 55%; harms 20%).*

http://www.conservationevidence.com/actions/161

Unknown effectiveness (limited evidence)

Ensure connectivity between habitat patches

Two studies of a replicated, controlled experiment in Canadian forests found that some species (not forest specialists) were found at higher densities in forest patches connected to continuous forest, compared to isolated patches and that some species used corridors more than clearcuts between patches. *Assessment: unknown effectiveness — limited evidence (effectiveness 38%; certainty 38%; harms 0%).*

http://www.conservationevidence.com/actions/160

3.2 Education and awareness raising

Based on the collated evidence, what is the current assessment of the effectiveness of interventions for education and awareness raising?	
Likely to be beneficial	• Raise awareness amongst the general public through campaigns and public information
Unknown effectiveness (limited evidence)	• Provide bird feeding materials to families with young children
No evidence found (no assessment)	• Enhance bird taxonomy skills through higher education and training • Provide training to conservationists and land managers on bird ecology and conservation

Likely to be beneficial

Raise awareness amongst the general public through campaigns and public information

A literature review from North America found that education was not sufficient to change behaviour, but that it was necessary for the success of economic incentives and law enforcement. *Assessment: likely to be beneficial (effectiveness 45%; certainty 48%; harms 0%).*

http://www.conservationevidence.com/actions/162

Unknown effectiveness (limited evidence)

Provide bird feeding materials to families with young children

A single replicated, paired study from the USA found that most children involved in a programme providing families with bird food increased their knowledge of birds, but did not significantly change their environmental attitudes. *Assessment: unknown effectiveness — limited evidence (effectiveness 42%; certainty 20%; harms 0%).*

http://www.conservationevidence.com/actions/163

No evidence found (no assessment)

We have captured no evidence for the following interventions:

- Enhance bird taxonomy skills through higher education and training
- Provide training to conservationists and land managers on bird ecology and conservation

3.3 Threat: Residential and commercial development

Based on the collated evidence, what is the current assessment of the effectiveness of interventions for residential and commercial development?	
Unknown effectiveness (limited evidence)	• Angle windows to reduce bird collisions • Mark windows to reduce bird collisions

Unknown effectiveness (limited evidence)

Angle windows to reduce bird collisions

A single randomised, replicated and controlled experiment in the USA found that fewer birds collided with windows angled away from the vertical. *Assessment: unknown effectiveness — limited evidence (effectiveness 60%; certainty 20%; harms 0%).*

http://www.conservationevidence.com/actions/166

Mark windows to reduce bird collisions

Two randomised, replicated and controlled studies found that marking windows did not appear to reduce bird collisions. However, when windows were largely covered with white cloth, or tinted, fewer birds flew towards or collided with them. A third randomised, replicated and controlled study found that fewer birds collided with tinted windows than with un-tinted ones, although the authors noted that the poor reflective quality of the glass could have influenced the results. *Assessment: unknown effectiveness — limited evidence (effectiveness 20%; certainty 20%; harms 0%).*

http://www.conservationevidence.com/actions/167

3.4 Threat: Agriculture

3.4.1 All farming systems

Based on the collated evidence, what is the current assessment of the effectiveness of interventions for all farming systems?	
Beneficial	• Plant wild bird seed or cover mixture • Provide (or retain) set-aside areas in farmland
Likely to be beneficial	• Create uncultivated margins around intensive arable or pasture fields • Increase the proportion of natural/semi-natural habitat in the farmed landscape • Manage ditches to benefit wildlife • Pay farmers to cover the costs of conservation measures • Plant grass buffer strips/margins around arable or pasture fields • Plant nectar flower mixture/wildflower strips • Leave refuges in fields during harvest • Reduce conflict by deterring birds from taking crops: use bird scarers • Relocate nests at harvest time to reduce nestling mortality • Use mowing techniques to reduce mortality
Unknown effectiveness (limited evidence)	• Control scrub on farmland • Offer per clutch payment for farmland birds • Manage hedges to benefit wildlife • Plant new hedges • Reduce conflict by deterring birds from taking crops: use repellents • Take field corners out of management

Likely to be ineffective or harmful	• Mark bird nests during harvest or mowing
No evidence found (no assessment)	• Cross compliance standards for all subsidy payments • Food labelling schemes relating to biodiversity-friendly farming • Manage stone-faced hedge banks to benefit birds • Plant in-field trees • Protect in-field trees • Reduce field size (or maintain small fields) • Support or maintain low-intensity agricultural systems • Tree pollarding, tree surgery

Beneficial

Plant wild bird seed or cover mixture

Seven of 41 studies found that fields or farms with wild bird cover had higher diversity than other sites, or that wild bird cover held more species than other habitats. Thirty-two studies found that populations, or abundances of some or all species were higher on wild bird cover than other habitats, or that wild bird cover was used more than other habitats. Four of these studies investigated several interventions at once. Thirteen studies found that bird populations or densities were similar on wild bird cover and other habitats that some species were not associated with wild bird cover, or that birds rarely used wild bird cover. Three studies found higher productivities of birds on wild bird cover than other habitats. Two found no differences for some or all species studied. Two studies found that survival of grey partridge or artificial nests increased on wild bird cover; one found lower partridge survival in farms with wild bird cover than other farms. Five studies from the UK found that some wild bird cover crops were used more than others. A study and a review found that the arrangement of wild bird cover in the landscape affected its use by birds. *Assessment: beneficial (effectiveness 81%; certainty 81%; harms 0%).*

http://www.conservationevidence.com/actions/187

Provide (or retain) set-aside areas in farmland

Four out of 23 studies from Europe and North America found more species on set-aside than on crops. One study found fewer. Twenty-one studies found that some species were at higher densities on set-aside than other habitats, or that they used set-aside more often. Four found that some species were found at lower densities on set-aside than other habitats. Three studies found that waders and Eurasian skylarks had higher productivities on set-aside than other crops. One study found that skylarks on set-aside had lower similar or lower productivities than on crops. One study from the UK found that rotational set-aside was used more than non-rotational set-aside, another found no difference. A review from North America and Europe found that naturally regenerated set-aside held more birds and more species than sown set-aside. *Assessment: beneficial (effectiveness 70%; certainty 75%; harms 0%).*

http://www.conservationevidence.com/actions/175

Likely to be beneficial

Create uncultivated margins around intensive arable or pasture fields

One of eight studies found that three sparrow species found on uncultivated margins on a site in the USA were not found on mown field edges. A replicated study from Canada found fewer species in uncultivated margins than in hedges or trees. Three studies found that some bird species were associated with uncultivated margins, or that birds were more abundant on margins than other habitats. One study found that these effects were very weak and four studies of three experiments found that uncultivated margins contained similar numbers of birds as other habitats in winter, or that several species studied did not show associations with margins. A study from the UK found that yellowhammers used uncultivated margins more than crops in early summer. Use fell in uncut margins later in the year. A study from the UK found that grey partridge released on uncultivated margins had high survival. *Assessment: likely to be beneficial (effectiveness 45%; certainty 55%; harms 0%).*

http://www.conservationevidence.com/actions/190

Increase the proportion of natural/semi-natural habitat in the farmed landscape

Two studies from Switzerland and Australia, of the five we captured, found that areas with plantings of native species, or areas under a scheme designed to increase semi-natural habitats (the Swiss Ecological Compensation Areas scheme), held more bird species than other areas. One study from Switzerland found that populations of three bird species increased in areas under the Ecological Compensation Areas scheme. A third Swiss study found that some habitats near Ecological Compensation Areas held more birds than habitats further away, but the overall amount of Ecological Compensation Area had no effect on bird populations. A study from the UK found no effect of habitat-creation on grey partridge populations. *Assessment: likely to be beneficial (effectiveness 45%; certainty 44%; harms 0%).*

http://www.conservationevidence.com/actions/171

Manage ditches to benefit wildlife

One study of four from the UK found that bunded ditches were visited more often by birds than non-bunded ditches. Three studies found that some birds responded positively to ditches managed for wildlife, but that other species did not respond to management, or responded negatively. *Assessment: likely to be beneficial (effectiveness 40%; certainty 49%; harms 14%).*

http://www.conservationevidence.com/actions/180

Pay farmers to cover the costs of conservation measures

Three out of 31 studies found national population increases in three species after payment schemes targeted at their conservation. One found that many other species continued declining. Twenty-two studies found that at least some species were found at higher densities on sites with agri-environment schemes; some differences were present only in summer or only in winter. Fifteen studies found some species at similar densities on agri-environment schemes and non-agri-environment scheme sites or appeared to respond negatively to agri-environment schemes. One study found that grey partridge survival was higher in some years on agri-environment scheme sites. Two studies found higher productivity on agri-environment scheme sites for some species, one found no effect of agri-environment schemes. A review found that some agri-environment schemes options were not being used enough to benefit many species of bird. A study from the UK found that there was no difference in the densities of seed-eating birds in winter between two agri-environment scheme designations. *Assessment: likely to be beneficial (effectiveness 56%; certainty 80%; harms 0%).*

http://www.conservationevidence.com/actions/172

Plant grass buffer strips/margins around arable or pasture fields

One of 15 studies found more bird species in fields in the USA that were bordered by grass margins than in unbordered fields. Two studies from the UK found no effect of margins on species richness. One study found that more birds used grass strips in fields than used crops. Even more used grass margins. Nine studies from the USA and UK found that sites with grass margins had more positive population trends or higher populations for some birds, or that some species showed strong habitat associations with grass margins. Three studies found no such effect for some or all species. Two studies found that species used margins more than other habitats and one found that birds used cut margins more than uncut during winter, but less than other habitats during summer. A study from the UK found that grey partridge broods were smaller on grass margins than other habitat types. *Assessment: likely to be beneficial (effectiveness 47%; certainty 54%; harms 0%).*

http://www.conservationevidence.com/actions/191

Plant nectar flower mixture/wildflower strips

Three of seven studies found that birds used wildflower strips more than other habitats; two found strips were not used more than other habitats. A study from Switzerland found that Eurasian skylarks were more likely to nest in patches sown with annual weeds than in crops and were less likely to abandon nests. A study from the UK found that management of field margins affected their use more than the seed mix used. *Assessment: likely to be beneficial (effectiveness 55%; certainty 45%; harms 0%).*

http://www.conservationevidence.com/actions/189

Leave refuges in fields during harvest

One study found that fewer gamebirds came into contact with mowing machinery when refuges were left in fields. A study from the UK found that Eurasian skylarks did not nest at higher densities in uncut refuges than in the rest of the field. *Assessment: likely to be beneficial (effectiveness 50%; certainty 41%; harms 0%).*

http://www.conservationevidence.com/actions/193

Reduce conflict by deterring birds from taking crops (using bird scarers)

A controlled paired study in the USA found reduced levels of damage to almond orchards when American crow distress calls were broadcast. A study in Pakistan found that four pest species were less abundant when reflector ribbons were hung above crops compared to where ribbons were not used. *Assessment: likely to be beneficial (effectiveness 66%; certainty 44%; harms 0%).*

http://www.conservationevidence.com/actions/199

Relocate nests at harvest time to reduce nestling mortality

A study from Spain found that Montagu's harrier clutches had higher hatching and fledging rates when they were temporarily moved during harvest than control nests that were not moved. *Assessment: likely to be beneficial (effectiveness 55%; certainty 42%; harms 0%).*

http://www.conservationevidence.com/actions/195

Use mowing techniques to reduce mortality

One of three studies from the UK found a large increase in the national population of corncrakes after a scheme to delay mowing and promote corncrake-friendly mowing techniques. Two studies found lower levels of corncrake and Eurasian skylark mortality when wildlife-friendly mowing techniques were used. *Assessment: likely to be beneficial (effectiveness 85%; certainty 50%; harms 0%).*

http://www.conservationevidence.com/actions/192

Unknown effectiveness (limited evidence)

Control scrub on farmland

A study from the UK found farms with a combined intervention that included scrub control had lower numbers of young grey partridge per adult. *Assessment: unknown effectiveness — limited evidence (effectiveness 7%; certainty 9%; harms 1%).*

http://www.conservationevidence.com/actions/197

Offer per clutch payment for farmland birds

One of two studies from the Netherlands found slightly higher breeding densities of waders on farms with per clutch payment schemes but this and another study found no higher numbers overall. One study found higher hatching success on farms with payment schemes. *Assessment: unknown effectiveness — limited evidence (effectiveness 43%; certainty 35%; harms 0%).*

http://www.conservationevidence.com/actions/196

Manage hedges to benefit wildlife

One of seven studies found no differences in the number of species in a UK site with wildlife-friendly hedge management and sites without. Seven studies found that some species increased in managed hedges or were more likely to be found in them than other habitats. One investigated several interventions at the same time. Four studies found that some species responded negatively or not at all to hedge management or that effects varied across regions of the UK. *Assessment: unknown effectiveness — limited evidence (effectiveness 39%; certainty 38%; harms 3%).*

http://www.conservationevidence.com/actions/177

Plant new hedges

A study from the USA found that populations of northern bobwhites increased following several interventions including the planting of new hedges. *Assessment: unknown effectiveness — limited evidence (effectiveness 23%; certainty 19%; harms 0%).*

http://www.conservationevidence.com/actions/178

Reduce conflict by deterring birds from taking crops (using repellents)

A replicated, randomised and controlled *ex situ* study in the USA found that dickcissels consumed less rice if it was treated with two repellents compared to controls. *Assessment: unknown effectiveness — limited evidence (effectiveness 29%; certainty 27%; harms 0%).*

http://www.conservationevidence.com/actions/200

Take field corners out of management

A study from the UK found that overwinter survival of grey partridge was positively correlated with taking field corners out of management, but this relationship was only significant in one of three winters. There was no relationship with measures of productivity (brood size, young: adult). *Assessment: unknown effectiveness — limited evidence (effectiveness 30%; certainty 15%; harms 0%).*

http://www.conservationevidence.com/actions/198

Likely to be ineffective or harmful

Mark bird nests during harvest or mowing

A study from the Netherlands found that fewer northern lapwing nests were destroyed when they were marked with bamboo poles than when they were unmarked. *Assessment: likely to be ineffective or harmful (effectiveness 30%; certainty 45%; harms 20%).*

http://www.conservationevidence.com/actions/148

No evidence found (no assessment)

We have captured no evidence for the following interventions:

- Cross compliance standards for all subsidy payments
- Food labelling schemes relating to biodiversity-friendly farming
- Manage stone-faced hedge banks to benefit birds
- Plant in-field trees
- Protect in-field trees
- Reduce field size (or maintain small fields)
- Support or maintain low-intensity agricultural systems
- Tree pollarding, tree surgery

3.4.2 Arable farming

Based on the collated evidence, what is the current assessment of the effectiveness of interventions for arable farming systems?	
Likely to be beneficial	• Create 'skylark plots' • Leave overwinter stubbles • Leave uncropped cultivated margins or fallow land (includes lapwing and stone curlew plots) • Sow crops in spring rather than autumn • Undersow spring cereals, with clover for example
Trade-off between benefit and harms	• Reduce tillage
Unknown effectiveness (limited evidence)	• Implement mosaic management • Increase crop diversity to benefit birds • Plant more than one crop per field (intercropping)
Unlikely to be beneficial	• Create beetle banks
Likely to be ineffective or harmful	• Plant cereals in wide-spaced rows • Revert arable land to permanent grassland
No evidence found (no assessment)	• Add 1% barley into wheat crop for corn buntings • Create corn bunting plots • Leave unharvested cereal headlands within arable fields • Plant nettle strips

Likely to be beneficial

Create 'skylark plots' (undrilled patches in cereal fields)

One study of seven found that the Eurasian skylark population on a farm increased after skylark plots were provided. Another found higher skylark densities on fields with plots in. Two studies from the UK found that skylark productivity was higher for birds with skylark plots in their territories, a study from Switzerland found no differences. Two studies from Denmark and Switzerland found that skylarks used plots more than expected, but a study from the UK found that seed-eating songbirds did not. *Assessment: likely to be beneficial (effectiveness 65%; certainty 60%; harms 0%).*

http://www.conservationevidence.com/actions/214

Leave overwinter stubbles

Three of fourteen studies report positive population-level changes in two species after winter stubble provision. All investigated several interventions at once. Eight studies found that some farmland birds were found on stubbles or were positively associated with them, three investigated several interventions and one found no more positive associations than expected by chance. A study from the UK found that most species did not preferentially use stubble, compared to cover crops and another found that a greater area of stubble in a site meant lower grey partridge brood size. Five studies from the UK found that management of stubbles influenced their use by birds. One study found that only one species was more common on stubbles under agri-environment schemes. *Assessment: likely to be beneficial (effectiveness 40%; certainty 60%; harms 0%).*

http://www.conservationevidence.com/actions/203

Leave uncropped cultivated margins or fallow land (includes lapwing and stone curlew plots)

Three of nine studies report that the UK population of Eurasian thick-knees increased following a scheme to promote lapwing plots (and other interventions). A study from the UK found that plots did not appear to influence grey partridge populations. Four studies from the UK found that at least one species was associated with lapwing plots, or used them for foraging or nesting. One study found that 11 species were not associated with plots, another that fewer used plots than used crops in two regions of the UK. Two studies found that nesting success was higher on lapwing plots and fallow than in crops. A third found fewer grey partridge chicks per adult on sites with lots of lapwing plots. *Assessment: likely to be beneficial (effectiveness 59%; certainty 55%; harms 15%).*

http://www.conservationevidence.com/actions/213

Sow crops in spring rather than autumn

One study from Sweden, of three examining the effects of spring-sown crops, found that more birds were found on areas with spring, rather than autumn-sown crops. A study from the UK found that several species used the study site for the first time after spring-sowing was started. All three studies found that some populations increased after the start of spring sowing. A study from the UK found that some species declined as well. A study from Sweden found that hatching success of songbirds and northern lapwing was lower on spring-sown, compared with autumn-sown crops. *Assessment: likely to be beneficial (effectiveness 55%; certainty 67%; harms 10%).*

http://www.conservationevidence.com/actions/207

Undersow spring cereals, with clover for example

Four of five studies from the UK found that bird densities were higher on undersown fields or margins than other fields, or that use of fields increased if they were undersown. Two studies of the same experiment found that not all species nested at higher densities in undersown habitats. A study from the UK found that grey partridge populations were lower on sites with large amounts of undersown cereal. *Assessment: likely to be beneficial (effectiveness 60%; certainty 45%; harms 10%).*

http://www.conservationevidence.com/actions/208

Trade-off between benefit and harms

Reduce tillage

Six of ten studies found that some or all bird groups had higher species richness or diversity on reduced-tillage fields, compared to conventional fields in some areas. Two studies found that some groups had lower diversity on reduced-tillage sites, or that there was no difference between treatments. Nine studies found that some species were found at higher densities on reduced tillage fields, six found that some species were at similar or lower densities. Three studies found evidence for higher productivities on reduced-tillage fields. One found that not all measures of productivity were higher. *Assessment: trade-offs between benefits and harms (effectiveness 50%; certainty 48%; harms 51%).*

http://www.conservationevidence.com/actions/211

Unknown effectiveness (limited evidence)

Implement mosaic management

One of two studies from the Netherlands found that northern lapwing population trends, but not those of three other waders, became more positive following the introduction of mosaic management. The other found that black-tailed godwit productivity was higher under mosaic management than other management types. *Assessment: unknown effectiveness — limited evidence (effectiveness 20%; certainty 33%; harms 0%).*

http://www.conservationevidence.com/actions/130

Increase crop diversity to benefit birds

A study from the UK found that more barnacle geese used a site after the amount of land under cereals was decreased and several other interventions were used. *Assessment: unknown effectiveness — limited evidence (effectiveness 20%; certainty 19%; harms 0%).*

http://www.conservationevidence.com/actions/201

Plant more than one crop per field (intercropping)

A study from the USA found that 35 species of bird used fields with intercropping, with four nesting, but that productivity from the fields was very low. *Assessment: unknown effectiveness — limited evidence (effectiveness 30%; certainty 36%; harms 18%).*

http://www.conservationevidence.com/actions/209

Unlikely to be beneficial

Create beetle banks

Two of six studies from the UK found that some bird populations were higher on sites with beetle banks. Both investigated several interventions at once. Two studies found no relationships between bird species abundances or populations and beetle banks. Two studies (including a review) from the UK found that three bird species used beetle banks more than expected, one used them less than expected. *Assessment: unlikely to be beneficial (effectiveness 30%; certainty 41%; harms 0%).*

http://www.conservationevidence.com/actions/217

Likely to be ineffective or harmful

Plant cereals in wide-spaced rows

One of three studies from the UK found that fields with wide-spaced rows held more Eurasian skylark nests than control fields. One study found that fields with wide-spaced rows held fewer nests. Both found that fields with wide-spaced rows held fewer nests than fields with skylark plots. A study from the UK found that skylark chicks in fields with wide-spaced rows had similar diets to those in control fields. *Assessment: likely to be ineffective or harmful (effectiveness 20%; certainty 44%; harms 20%).*

http://www.conservationevidence.com/actions/216

Revert arable land to permanent grassland

All five studies looking at the effects of reverting arable land to grassland found no clear benefit to birds. The studies monitored birds in winter or grey partridges in the UK and wading birds in Denmark. They included three replicated controlled trials. *Assessment: likely to be ineffective or harmful (effectiveness 0%; certainty 64%; harms 10%).*

http://www.conservationevidence.com/actions/210

No evidence found (no assessment)

We have captured no evidence for the following interventions:

- Add 1% barley into wheat crop for corn buntings
- Create corn bunting plots
- Leave unharvested cereal headlands within arable fields
- Plant nettle strips

3.4.3 Livestock farming

Based on the collated evidence, what is the current assessment of the effectiveness of interventions for livestock farming systems?	
Likely to be beneficial	• Delay mowing date on grasslands • Leave uncut rye grass in silage fields • Maintain species-rich, semi-natural grassland • Maintain traditional water meadows • Mark fencing to avoid bird mortality • Plant cereals for whole crop silage • Reduce grazing intensity • Reduce management intensity of permanent grasslands
Trade-off between benefit and harms	• Exclude livestock from semi-natural habitat
Unknown effectiveness (limited evidence)	• Create open patches or strips in permanent grassland • Maintain upland heath/moor • Protect nests from livestock to reduce trampling • Provide short grass for waders • Raise mowing height on grasslands
Unlikely to be beneficial	• Use traditional breeds of livestock
No evidence found (no assessment)	• Maintain lowland heathland • Maintain rush pastures • Maintain wood pasture and parkland • Plant Brassica fodder crops • Use mixed stocking

Likely to be beneficial

Delay mowing date on grasslands

Two of five studies (both reviews) found that the UK corncrake populations increased following two schemes to encourage farmers to delay mowing. A study from the Netherlands found no evidence that waders and other birds were more abundant in fields with delayed mowing. Another study from the Netherlands found that fields with delayed mowing held more birds than other fields, but differences were present before the scheme began and population trends did not differ between treatments. A study from the USA found that fewer nests were destroyed by machinery in late-cut fields, compared with early-cut fields. *Assessment: likely to be beneficial (effectiveness 45%; certainty 52%; harms 0%).*

http://www.conservationevidence.com/actions/223

Leave uncut rye grass in silage fields

All four studies from the UK (including two reviews) found that seed-eating birds were benefited by leaving uncut (or once-cut) rye grass in fields, or that seed-eating species were more abundant on uncut plots. Three studies found that seed-eating birds were more abundant on uncut and ungrazed plots than on uncut and grazed plots. A study from the UK found that the responses of non-seed-eating birds were less certain than seed-eating species, with some species avoiding uncut rye grass. *Assessment: likely to be beneficial (effectiveness 67%; certainty 56%; harms 8%).*

http://www.conservationevidence.com/actions/224

Maintain species-rich, semi-natural grassland

One of two studies found that the populations of five species increased in an area of the UK after the start of management designed to maintain unimproved grasslands. A study from Switzerland found that wetland birds nested at greater densities on managed hay meadows than expected, but birds of open farmland used hay meadows less. *Assessment: likely to be beneficial (effectiveness 41%; certainty 44%; harms 0%).*

http://www.conservationevidence.com/actions/218

Maintain traditional water meadows

One of four studies (from the UK) found that the populations of two waders increased on reserves managed as water meadows. Two studies from the Netherlands found that there were more waders or birds overall on specially managed meadows or 12.5 ha plots, but one found that these differences were present before management began, the other found no differences between individual fields under different management. Two studies from the UK and Netherlands found that wader populations were no different between specially and conventionally managed meadows, or that wader populations decreased on specially-managed meadows. A study from the UK found that northern lapwing productivity was not high enough to maintain populations on three of four sites managed for waders. *Assessment: likely to be beneficial (effectiveness 50%; certainty 52%; harms 0%).*

http://www.conservationevidence.com/actions/229

Mark fencing to avoid bird mortality

A study from the UK found that fewer birds collided with marked sections of deer fences, compared to unmarked sections. *Assessment: likely to be beneficial (effectiveness 65%; certainty 46%; harms 0%).*

http://www.conservationevidence.com/actions/238

Plant cereals for whole crop silage

Three studies of one experiment found that seed-eating birds used cereal-based wholecrop silage crops more than other crops in summer and winter. Insect-eating species used other crops and grassland more often. *Assessment: likely to be beneficial (effectiveness 55%; certainty 43%; harms 0%).*

http://www.conservationevidence.com/actions/225

Reduce grazing intensity

Nine of eleven studies from the UK and USA found that the populations of some species were higher on fields with reduced grazing intensity, compared to conventionally-grazed fields, or found that birds used these fields more. Three studies investigated several interventions at once. Five studies from Europe found that some or all species were no more numerous, or were less abundant on fields with reduced grazing. A study from the UK found that black grouse populations increased at reduced grazing sites (whilst they declined elsewhere). However, large areas with reduced grazing had low female densities. A study from the USA found that the number of species on plots with reduced grazing increased over time. A study from four European countries found no differences in the number of species on sites with low- or high-intensity grazing. *Assessment: likely to be beneficial (effectiveness 46%; certainty 55%; harms 0%).*

http://www.conservationevidence.com/actions/220

Reduce management intensity of permanent grasslands

Seven of eight European studies found that some or all birds studied were more abundant on grasslands with reduced management intensity, or used them more than other habitats for foraging. Five studies of four experiments found that some or all species were found at lower or similar abundances on reduced-management grasslands, compared to intensively-managed grasslands. *Assessment: likely to be beneficial (effectiveness 65%; certainty 46%; harms 0%).*

http://www.conservationevidence.com/actions/219

Trade-off between benefit and harms

Exclude livestock from semi-natural habitat

Two studies from the USA, out of 11 overall, found higher species richness on sites with grazers excluded. A study from Argentina found lower species richness and one from the USA found no difference. Seven studies from the USA found that overall bird abundance, or the abundances of some species were higher in sites with grazers excluded. Seven studies from the USA and Argentina found that overall abundance or the abundance of some species were lower on sites without grazers, or did not differ. Three studies found that productivities were higher on sites with grazers excluded. In one, the difference was only found consistently in comparison with improved pastures, not unimproved. *Assessment: trade-offs between benefits and harms (effectiveness 50%; certainty 57%; harms 30%).*

http://www.conservationevidence.com/actions/236

Unknown effectiveness (limited evidence)

Create open patches or strips in permanent grassland

A study from the UK found that Eurasian skylarks used fields with open strips in, but that variations in skylark numbers were too great to draw conclusions from this finding. *Assessment: unknown effectiveness — limited evidence (effectiveness 20%; certainty 19%; harms 0%).*

http://www.conservationevidence.com/actions/239

Maintain upland heath/moor

A study from the UK found that bird populations in one region were increasing with agri-environment guidelines on moor management. There were some problems with overgrazing, burning and scrub encroachment. *Assessment: unknown effectiveness — limited evidence (effectiveness 30%; certainty 15%; harms 0%).*

http://www.conservationevidence.com/actions/230

Protect nests from livestock to reduce trampling

One of two studies found that a population of Chatham Island oystercatchers increased following several interventions including the erection of fencing around individual nests. A study from Sweden found that no southern dunlin nests were trampled when protected by cages; some unprotected nests were destroyed. *Assessment: unknown effectiveness — limited evidence (effectiveness 56%; certainty 19%; harms 0%).*

http://www.conservationevidence.com/actions/237

Provide short grass for waders

A study from the UK found that common starlings and northern lapwing spent more time foraging on areas with short swards, compared to longer swards. *Assessment: unknown effectiveness — limited evidence (effectiveness 41%; certainty 32%; harms 0%).*

http://www.conservationevidence.com/actions/221

Raise mowing height on grasslands

One of two studies from the UK found that no more foraging birds were attracted to plots with raised mowing heights, compared to plots with shorter grass. A review from the UK found that Eurasian skylarks had higher productivity on sites with raised mowing heights, but this increase was not enough to maintain local populations. *Assessment: unknown effectiveness — limited evidence (effectiveness 20%; certainty 36%; harms 0%).*

http://www.conservationevidence.com/actions/222

Unlikely to be beneficial

Use traditional breeds of livestock

A study from four countries in Europe found no differences in bird abundances in areas grazed with traditional or commercial breeds. *Assessment: unlikely to be beneficial (effectiveness 0%; certainty 44%; harms 0%).*

http://www.conservationevidence.com/actions/233

No evidence found (no assessment)

We have captured no evidence for the following interventions:

- Maintain lowland heathland
- Maintain rush pastures
- Maintain wood pasture and parkland
- Plant Brassica fodder crops
- Use mixed stocking

3.4.4 Perennial, non-timber crops

Based on the collated evidence, what is the current assessment of the effectiveness of interventions for perennial, non-timber crops?	
Unknown effectiveness (limited evidence)	• Maintain traditional orchards
No evidence found (no assessment)	• Manage perennial bioenergy crops to benefit wildlife

Unknown effectiveness (limited evidence)

Maintain traditional orchards

Two site comparison studies from the UK and Switzerland found that traditional orchards offer little benefit to birds. In Switzerland only one breeding bird species was associated with traditional orchards. In the UK, the population density of cirl bunting was negatively related to the presence of orchards. *Assessment: unknown effectiveness — limited evidence (effectiveness 10%; certainty 24%; harms 0%).*

http://www.conservationevidence.com/actions/240

No evidence found (no assessment)

We have captured no evidence for the following interventions:

- Manage perennial bioenergy crops to benefit wildlife

3.4.5 Aquaculture

Based on the collated evidence, what is the current assessment of the effectiveness of interventions for aquaculture?	
Likely to be beneficial	• Deter birds from landing on shellfish culture gear • Disturb birds at roosts • Provide refuges for fish within ponds • Use electric fencing to exclude fish-eating birds • Use 'mussel socks' to prevent birds from attacking shellfish • Use netting to exclude fish-eating birds
Unknown effectiveness (limited evidence)	• Increase water turbidity to reduce fish predation by birds • Translocate birds away from fish farms • Use in-water devices to reduce fish loss from ponds
Unlikely to be beneficial	• Disturb birds using foot patrols • Spray water to deter birds from ponds
Likely to be ineffective or harmful	• Scare birds from fish farms

Likely to be beneficial

Deter birds from landing on shellfish culture gear

A study from Canada found that fewer birds landed on oyster cages fitted with spikes than control cages. The same study found that fewer birds landed on oyster bags suspended 6 cm, but not 3 cm, underwater, compared to bags on the surface. *Assessment for using spikes on oyster cages: likely to be beneficial (effectiveness 60%; certainty 43%; harms 0%). Assessment for suspending oyster bags under water: likely to be beneficial (effectiveness 55%; certainty 43%; harms 0%).*

http://www.conservationevidence.com/actions/257 http://www.conservationevidence.com/actions/256

Disturb birds at roosts

One study from the USA found reduced fish predation after fish-eating birds were disturbed at roosts. Five studies from the USA and Israel found that birds foraged less near disturbed roosts, or left the area after being disturbed. One found the effects were only temporary. *Assessment: likely to be beneficial (effectiveness 67%; certainty 45%; harms 0%).*

http://www.conservationevidence.com/actions/245

Provide refuges for fish within ponds

A study from the UK found that cormorants caught fewer fish in a pond with fish refuges in, compared to a control pond. *Assessment: likely to be beneficial (effectiveness 65%; certainty 43%; harms 0%).*

http://www.conservationevidence.com/actions/253

Use electric fencing to exclude fish-eating birds

Two before-and-after trials from the USA found lower use of fish ponds by herons after electric fencing was installed. *Assessment: likely to be beneficial (effectiveness 60%; certainty 49%; harms 0%).*

http://www.conservationevidence.com/actions/247

Use 'mussel socks' to prevent birds from attacking shellfish

A study from Canada found that mussel socks with protective sleeves lost fewer medium-sized mussels (but not small or large mussels), compared to unprotected mussel socks. *Assessment: likely to be beneficial (effectiveness 50%; certainty 41%; harms 0%).*

http://www.conservationevidence.com/actions/250

Use netting to exclude fish-eating birds

Two studies from Germany and the USA, and a review, found that netting over ponds reduced the loss of fish to predatory birds. Two studies from the USA and the Netherlands found that birds still landed on ponds with netting, but that they altered their behaviour, compared to open ponds. Two studies from Germany and Israel found that some birds became entangled in netting over ponds. *Assessment: likely to be beneficial (effectiveness 60%; certainty 59%; harms 15%).*

http://www.conservationevidence.com/actions/248

Unknown effectiveness (limited evidence)

Increase water turbidity to reduce fish predation by birds

An *ex situ* study from France found that egret foraging efficiency was reduced in more turbid water. *Assessment: unknown effectiveness — limited evidence (effectiveness 50%; certainty 23%; harms 0%).*

http://www.conservationevidence.com/actions/252

Translocate birds away from fish farms

A study from the USA found that translocating birds appeared to reduce bird numbers at a fish farm. A study from Belgium found that it did not. *Assessment: unknown effectiveness — limited evidence (effectiveness 20%; certainty 33%; harms 0%).*

http://www.conservationevidence.com/actions/251

Use in-water devices to reduce fish loss from ponds

A study from the USA found that fewer cormorants used two ponds after underwater ropes were installed; a study from Australia found that no fewer cormorants used ponds with gill nets in. *Assessment: unknown effectiveness — limited evidence (effectiveness 34%; certainty 35%; harms 0%).*

http://www.conservationevidence.com/actions/254

Unlikely to be beneficial

Disturb birds using foot patrols

Two replicated studies from Belgium and Australia found that using foot patrols to disturb birds from fish farms did not reduce the number of birds present or fish consumption. *Assessment: unlikely to be beneficial (effectiveness 0%; certainty 45%; harms 0%).*

http://www.conservationevidence.com/actions/249

Spray water to deter birds from ponds

A study from Sweden found that spraying water deterred birds from fish ponds, but that some birds became habituated to the spray. *Assessment: unlikely to be beneficial (effectiveness 31%; certainty 43%; harms 0%).*

http://www.conservationevidence.com/actions/255

Likely to be ineffective or harmful

Scare birds from fish farms

One study from Israel found a population increase in fish-eating birds after efforts to scare them from fish farms, possibly due to lower persecution. One of two studies found evidence for reduced loss of fish when birds were scared from farms. Two studies from Australia and Belgium found that disturbing birds using foot patrols was not effective. Ten of 11 studies from across the world found some effects for acoustic deterrents, five of seven found that visual deterrents were effective. In both cases some studies found that results were temporary, birds became habituated or that some deterrents were effective, whilst others were not. One study found that trained raptors were effective, one found little evidence for the effectiveness of helicopters or light aircraft. *Assessment: likely to be ineffective or harmful (effectiveness 36%; certainty 64%; harms 0%).*

http://www.conservationevidence.com/actions/244

3.5 Threat: Energy production and mining

Based on the collated evidence, what is the current assessment of the effectiveness of interventions for energy production and mining?	
Unknown effectiveness (limited evidence)	• Paint wind turbines to increase their visibility

Unknown effectiveness (limited evidence)

Paint wind turbines to increase their visibility

A single, controlled *ex situ* experiment found that thick black stripes running across a wind turbine's blades made them more conspicuous to an American kestrel *Falco sparverius* than control (unpatterned) blades. Other designs were less visible or indistinguishable from controls. *Assessment: unknown effectiveness — limited evidence (effectiveness 16%; certainty 10%; harms 0%).*

http://www.conservationevidence.com/actions/258

3.6 Threat: Transportation and service corridors

3.6.1 Verges and airports

Based on the collated evidence, what is the current assessment of the effectiveness of interventions for verges and airports?	
Likely to be beneficial	• Scare or otherwise deter birds from airports
Unknown effectiveness (limited evidence)	• Mow roadside verges
No evidence found (no assessment)	• Sow roadside verges

Likely to be beneficial

Scare or otherwise deter birds from airports

Two replicated studies in the UK and USA found that fewer birds used areas of long grass at airports, but no data were provided on the effect of long grass on strike rates or bird mortality. *Assessment: likely to be beneficial (effectiveness 50%; certainty 44%; harms 0%).*

http://www.conservationevidence.com/actions/261

Unknown effectiveness (limited evidence)

Mow roadside verges

A single replicated, controlled trial in the USA found that mowed roadside verges were less attractive to ducks as nesting sites, but had higher nesting success after four years. *Assessment: unknown effectiveness — limited evidence (effectiveness 30%; certainty 30%; harms 9%).*

http://www.conservationevidence.com/actions/259

No evidence found (no assessment)

We have captured no evidence for the following intervention:

- Sow roadside verges

3.6.2 Power lines and electricity pylons

Based on the collated evidence, what is the current assessment of the effectiveness of interventions for power lines and electricity pylons?	
Beneficial	• Mark power lines
Likely to be beneficial	• Bury or isolate power lines • Insulate electricity pylons • Remove earth wires from power lines • Use perch-deterrents to stop raptors perching on pylons
Unknown effectiveness (limited evidence)	• Thicken earth wires
Unlikely to be beneficial	• Add perches to electricity pylons • Reduce electrocutions by using plastic, not metal, leg rings to mark birds • Use raptor models to deter birds from power lines

Beneficial

Mark power lines

A total of eight studies and two literature reviews from across the world found that marking power lines led to significant reductions in bird collision mortalities. Different markers had different impacts. *Assessment: beneficial (effectiveness 81%; certainty 85%; harms 0%).*

http://www.conservationevidence.com/actions/265

Likely to be beneficial

Bury or isolate power lines

A single before-and-after study in Spain found a dramatic increase in juvenile eagle survival following the burial or isolation of dangerous power lines. *Assessment: likely to be beneficial (effectiveness 60%; certainty 44%; harms 0%).*

http://www.conservationevidence.com/actions/262

Insulate electricity pylons

A single before-and-after study in the USA found that insulating power pylons significantly reduced the number of Harris's hawks electrocuted. *Assessment: likely to be beneficial (effectiveness 60%; certainty 45%; harms 0%).*

http://www.conservationevidence.com/actions/268

Remove earth wires from power lines

Two before-and-after studies from Norway and the USA describe significant reductions in bird collision mortalities after earth wires were removed from sections of power lines. *Assessment: likely to be beneficial (effectiveness 90%; certainty 60%; harms 0%).*

http://www.conservationevidence.com/actions/263

Use perch-deterrents to stop raptors perching on pylons

A single controlled study in the USA found that significantly fewer raptors were found near perch-deterrent lines, compared to controls, but no information on electrocutions was provided. *Assessment: likely to be beneficial (effectiveness 50%; certainty 45%; harms 0%).*

http://www.conservationevidence.com/actions/269

Unknown effectiveness (limited evidence)

Thicken earth wires

A single paired sites trial in the USA found no reduction in crane species collision rates in a wire span with an earth wire three times thicker than normal. *Assessment: unknown effectiveness — limited evidence (effectiveness 0%; certainty 25%; harms 0%).*

http://www.conservationevidence.com/actions/264

Unlikely to be beneficial

Add perches to electricity pylons

A single before-and-after study in Spain found that adding perches to electricity pylons did not reduce electrocutions of Spanish imperial eagles. *Assessment: unlikely to be beneficial (effectiveness 0%; certainty 42%; harms 0%).*

http://www.conservationevidence.com/actions/267

Reduce electrocutions by using plastic, not metal, leg rings to mark birds

A single replicated and controlled study in the USA found no evidence that using plastic leg rings resulted in fewer raptors being electrocuted. *Assessment: unlikely to be beneficial (effectiveness 0%; certainty 42%; harms 0%).*

http://www.conservationevidence.com/actions/270

Use raptor models to deter birds from power lines

A single paired sites trial in Spain found that installing raptor models near power lines had no impact on bird collision mortalities. *Assessment: unlikely to be beneficial (effectiveness 0%; certainty 43%; harms 0%)*

http://www.conservationevidence.com/actions/266

3.7 Threat: Biological resource use

3.7.1 Reducing exploitation and conflict

Based on the collated evidence, what is the current assessment of the effectiveness of interventions for reducing exploitation and conflict?	
Beneficial	• Use legislative regulation to protect wild populations
Likely to be beneficial	• Use wildlife refuges to reduce hunting disturbance
Unknown effectiveness (limited evidence)	• Employ local people as 'biomonitors' • Increase 'on-the-ground' protection to reduce unsustainable levels of exploitation • Introduce voluntary 'maximum shoot distances' • Mark eggs to reduce their appeal to collectors • Move fish-eating birds to reduce conflict with fishermen • Promote sustainable alternative livelihoods • Provide 'sacrificial grasslands' to reduce conflict with farmers • Relocate nestlings to reduce poaching • Use education programmes and local engagement to help reduce persecution or exploitation of species
No evidence found (no assessment)	• Use alerts during shoots to reduce mortality of non-target species

Scare fish-eating birds from areas to reduce conflict

Studies investigating scaring fish from fishing areas are discussed in 'Threat: Agriculture — Aquaculture'.

Beneficial

Use legislative regulation to protect wild populations

Five out of six studies from Europe, Asia, North America and across the world, found evidence that stricter legislative protection was correlated with increased survival, lower harvests or increased populations. The sixth, a before-and-after study from Australia, found that legislative protection did not reduce harvest rates. *Assessment: beneficial (effectiveness 65%; certainty 65%; harms 0%).*

http://www.conservationevidence.com/actions/271

Likely to be beneficial

Use wildlife refuges to reduce hunting disturbance

Three studies from the USA and Europe found that more birds used refuges where hunting was not allowed, compared to areas with hunting, and more used the refuges during the open season. However, no studies examined the population-level effects of refuges. *Assessment: likely to be beneficial (effectiveness 45%; certainty 45%; harms 0%).*

http://www.conservationevidence.com/actions/278

Unknown effectiveness (limited evidence)

Employ local people as 'biomonitors'

A single replicated study in Venezuela found that poaching of parrot nestlings was significantly lower in years following the employment of five local people as 'biomonitors'. *Assessment: unknown effectiveness — limited evidence (effectiveness 50%; certainty 19%; harms 0%).*

http://www.conservationevidence.com/actions/275

Increase 'on-the-ground' protection to reduce unsustainable levels of exploitation

Two before-and-after studies from Europe and Central America found increases in bird populations and recruitment following stricter anti-poaching methods or the stationing of a warden on the island in question. However, the increases in Central America were only short-term, and were lost when the intensive effort was reduced. *Assessment: unknown effectiveness — limited evidence (effectiveness 50%; certainty 25%; harms 0%).*

http://www.conservationevidence.com/actions/272

Introduce voluntary 'maximum shoot distances'

A single study from Denmark found a significant reduction in the injury rates of pink-footed geese following the implementation of a voluntary maximum shooting distance. *Assessment: unknown effectiveness — limited evidence (effectiveness 40%; certainty 20%; harms 0%).*

http://www.conservationevidence.com/actions/279

Mark eggs to reduce their appeal to collectors

A single before-and-after study in Australia found increased fledging success of raptor eggs in a year they were marked with a permanent pen. *Assessment: unknown effectiveness — limited evidence (effectiveness 50%; certainty 35%; harms 0%).*

http://www.conservationevidence.com/actions/276

Move fish-eating birds to reduce conflict with fishermen

A single before-and-after study in the USA found that Caspian tern chicks had a lower proportion of commercial fish in their diet following the movement of the colony away from an important fishery. *Assessment: unknown effectiveness — limited evidence (effectiveness 32%; certainty 24%; harms 0%).*

http://www.conservationevidence.com/actions/281

Promote sustainable alternative livelihoods

A single before-and-after study in Costa Rica found that a scarlet macaw population increased following several interventions including the promotion of sustainable, macaw-based livelihoods. *Assessment: unknown effectiveness — limited evidence (effectiveness 30%; certainty 19%; harms 0%).*

http://www.conservationevidence.com/actions/273

Provide 'sacrificial grasslands' to reduce conflict with farmers

Two UK studies found that more geese used areas of grassland managed for them, but that this did not appear to attract geese from outside the study site and therefore was unlikely to reduce conflict with farmers. *Assessment: unknown effectiveness — limited evidence (effectiveness 18%; certainty 20%; harms 0%).*

http://www.conservationevidence.com/actions/280

Relocate nestlings to reduce poaching

A single replicated study in Venezuela found a significant reduction in poaching rates and an increase in fledging rates of yellow-shouldered amazons when nestlings were moved into police premises overnight. *Assessment: unknown effectiveness — limited evidence (effectiveness 50%; certainty 30%; harms 0%).*

http://www.conservationevidence.com/actions/277

Use education programmes and local engagement to help reduce persecution or exploitation of species

Six out of seven studies from across the world found increases in bird populations or decreases in mortality following education programmes, whilst one study from Venezuela found no evidence that poaching decreased following an educational programme. In all but one study reporting successes, other interventions were also used, and a literature review from the USA and Canada argues that education was not sufficient to change behaviour, although a Canadian study found that there was a significant shift in local peoples' attitudes to conservation and exploited species following educational programmes. *Assessment: unknown effectiveness — limited evidence (effectiveness 50%; certainty 30%; harms 0%).*

http://www.conservationevidence.com/actions/274

No evidence found (no assessment)

We have captured no evidence for the following interventions:

- Use alerts during shoots to reduce mortality of non-target species

3.7.2 Reducing fisheries bycatch

Based on the collated evidence, what is the current assessment of the effectiveness of interventions for reducing fisheries bycatch?	
Beneficial	• Use streamer lines to reduce seabird bycatch on longlines
Likely to be beneficial	• Mark trawler warp cables to reduce seabird collisions • Reduce seabird bycatch by releasing offal overboard when setting longlines • Weight baits or lines to reduce longline bycatch of seabirds
Trade-off between benefit and harms	• Set lines underwater to reduce seabird bycatch • Set longlines at night to reduce seabird bycatch
Unknown effectiveness (limited evidence)	• Dye baits to reduce seabird bycatch • Thaw bait before setting lines to reduce seabird bycatch • Turn deck lights off during night-time setting of longlines to reduce bycatch • Use a sonic scarer when setting longlines to reduce seabird bycatch • Use acoustic alerts on gillnets to reduce seabird bycatch • Use bait throwers to reduce seabird bycatch • Use bird exclusion devices such as 'Brickle curtains' to reduce seabird mortality when hauling longlines • Use high visibility mesh on gillnets to reduce seabird bycatch • Use shark liver oil to deter birds when setting lines
Likely to be ineffective or harmful	• Use a line shooter to reduce seabird bycatch
No evidence found (no assessment)	• Reduce bycatch through seasonal or area closures • Reduce 'ghost fishing' by lost/discarded gear • Reduce gillnet deployment time to reduce seabird bycatch • Set longlines at the side of the boat to reduce seabird bycatch

	• Tow buoys behind longlining boats to reduce seabird bycatch • Use a water cannon when setting longlines to reduce seabird bycatch • Use high-visibility longlines to reduce seabird bycatch • Use larger hooks to reduce seabird bycatch on longlines

Beneficial

Use streamer lines to reduce seabird bycatch on longlines

Ten studies from coastal and pelagic fisheries across the globe found strong evidence for reductions in bycatch when streamer lines were used. Five studies from the South Atlantic, New Zealand and Australia were inconclusive, uncontrolled or had weak evidence for reductions. One study from the sub-Antarctic Indian Ocean found no evidence for reductions. Three studies from around the world found that bycatch rates were lower when two streamers were used compared to one, and one study found rates were lower still with three streamers. *Assessment: beneficial (effectiveness 65%; certainty 75%; harms 0%).*

http://www.conservationevidence.com/actions/285

Likely to be beneficial

Mark trawler warp cables to reduce seabird collisions

A single replicated and controlled study in Argentina found lower seabird mortality (from colliding with warp cables) when warp cables were marked with orange traffic cones. *Assessment: likely to be beneficial (effectiveness 54%; certainty 40%; harms 0%).*

http://www.conservationevidence.com/actions/305

Reduce seabird bycatch by releasing offal overboard when setting longlines

Two replicated and controlled studies in the South Atlantic and sub-Antarctic Indian Ocean found significantly lower seabird bycatch rates when offal was released overboard as lines were being set. *Assessment: likely to be beneficial (effectiveness 51%; certainty 50%; harms 0%).*

http://www.conservationevidence.com/actions/299

Weight baits or lines to reduce longline bycatch of seabirds

Three replicated and controlled studies from the Pacific found lower bycatch rates of some seabird species on weighted longlines. An uncontrolled study found low bycatch rates with weighted lines but that weights only increased sink rates in small sections of the line. Some species were found to attack weighted lines more than control lines. *Assessment: likely to be beneficial (effectiveness 46%; certainty 45%; harms 15%).*

http://www.conservationevidence.com/actions/296

Trade-off between benefit and harms

Set lines underwater to reduce seabird bycatch

Five studies in Norway, South Africa and the North Pacific found lower seabird bycatch rates on longlines set underwater. However, results were species-specific, with shearwaters and possibly albatrosses continuing to take baits set underwater. *Assessment: trade-offs between benefits and harms (effectiveness 61%; certainty 50%; harms 24%).*

http://www.conservationevidence.com/actions/288

Set longlines at night to reduce seabird bycatch

Six out of eight studies from around the world found lower bycatch rates when longlines were set at night, but the remaining two found higher bycatch rates (of northern fulmar in the North Pacific and white-chinned petrels in the South Atlantic, respectively). Knowing whether bycatch species are night- or day-feeding is therefore important in reducing bycatch rates. *Assessment: trade-offs between benefits and harms (effectiveness 60%; certainty 70%; harms 48%).*

http://www.conservationevidence.com/actions/283

Unknown effectiveness (limited evidence)

Dye baits to reduce seabird bycatch

A single randomised, replicated and controlled trial in Hawaii, USA, found that albatrosses attacked baits at significantly lower rates when baits were dyed blue. *Assessment: unknown effectiveness — limited evidence (effectiveness 50%; certainty 20%; harms 0%).*

http://www.conservationevidence.com/actions/293

Thaw bait before setting lines to reduce seabird bycatch

A study from Australia found that longlines set using thawed baits caught significantly fewer seabirds than controls. *Assessment: unknown effectiveness — limited evidence (effectiveness 50%; certainty 30%; harms 0%).*

http://www.conservationevidence.com/actions/298

Turn deck lights off during night-time setting of longlines to reduce bycatch

A single replicated and controlled study in the South Atlantic found lower seabird bycatch rates on night-set longlines when deck lights were turned off. *Assessment: unknown effectiveness — limited evidence (effectiveness 51%; certainty 21%; harms 0%).*

http://www.conservationevidence.com/actions/284

Use a sonic scarer when setting longlines to reduce seabird bycatch

A single study from the South Atlantic found that seabirds only temporarily changed behaviour when a sonic scarer was used, and seabird bycatch rates did not appear to be lower on lines set with a scarer. *Assessment: unknown effectiveness — limited evidence (effectiveness 2%; certainty 10%; harms 0%).*

http://www.conservationevidence.com/actions/295

Use acoustic alerts on gillnets to reduce seabird bycatch

A randomised, replicated and controlled trial in a coastal fishery in the USA found that fewer guillemots (common murres) but not rhinoceros auklets were caught in gillnets fitted with sonic alerts. *Assessment: unknown effectiveness — limited evidence (effectiveness 44%; certainty 21%; harms 0%).*

http://www.conservationevidence.com/actions/301

Use bait throwers to reduce seabird bycatch

A single analysis found significantly lower seabird bycatch on Australian longliners when a bait thrower was used to set lines. *Assessment: unknown effectiveness — limited evidence (effectiveness 46%; certainty 30%; harms 0%).*

http://www.conservationevidence.com/actions/291

Use bird exclusion devices such as 'Brickle curtains' to reduce seabird mortality when hauling longlines

A single replicated study found that Brickle curtains reduced the number of seabirds caught, when compared to an exclusion device using only a single boom. Using purse seine buoys as well as the curtain appeared to be even more effective, but sample sizes did not allow useful comparisons. *Assessment: unknown effectiveness — limited evidence (effectiveness 48%; certainty 30%; harms 0%).*

http://www.conservationevidence.com/actions/302

Use high visibility mesh on gillnets to reduce seabird bycatch

A single randomised, replicated and controlled trial in a coastal fishery in the USA found that fewer guillemots (common murres) and rhinoceros auklets were caught in gillnets with higher percentages of brightly coloured netting. However, such netting also reduced the catch of the target salmon. *Assessment: unknown effectiveness — limited evidence (effectiveness 60%; certainty 30%; harms 0%).*

http://www.conservationevidence.com/actions/303

Use shark liver oil to deter birds when setting lines

Two out of three replicated and controlled trials in New Zealand found that fewer birds followed boats or dived for baits when non-commercial shark oil was dripped off the back of the boat. *Assessment: unknown effectiveness — limited evidence (effectiveness 30%; certainty 25%; harms 0%).*

http://www.conservationevidence.com/actions/297

Likely to be ineffective or harmful

Use a line shooter to reduce seabird bycatch

Two randomised, replicated and controlled trials found that seabird bycatch rates were higher (in the North Pacific) or the same (in Norway) on longlines set using line shooters, compared to those set without a shooter. *Assessment: likely to be ineffective or harmful (effectiveness 0%; certainty 50%; harms 40%).*

http://www.conservationevidence.com/actions/290

No evidence found (no assessment)

We have captured no evidence for the following interventions:

- Reduce bycatch through seasonal or area closures
- Reduce 'ghost fishing' by lost/discarded gear
- Reduce gillnet deployment time to reduce seabird bycatch
- Set longlines at the side of the boat to reduce seabird bycatch
- Tow buoys behind longlining boats to reduce seabird bycatch
- Use a water cannon when setting longlines to reduce seabird bycatch
- Use high-visibility longlines to reduce seabird bycatch
- Use larger hooks to reduce seabird bycatch on longlines

3.8 Threat: Human intrusions and disturbance

Based on the collated evidence, what is the current assessment of the effectiveness of interventions for human intrusions and disturbance?	
Likely to be beneficial	• Provide paths to limit disturbance • Start educational programmes for personal watercraft owners • Use signs and access restrictions to reduce disturbance at nest sites • Use voluntary agreements with local people to reduce disturbance
Unknown effectiveness (limited evidence)	• Habituate birds to human visitors • Use nest covers to reduce the impact of research on predation of ground-nesting seabirds
No evidence found (no assessment)	• Reduce visitor group sizes • Set minimum distances for approaching birds (buffer zones)

Likely to be beneficial

Provide paths to limit disturbance

A study from the UK found that two waders nested closer to a path, or at higher densities near the path, following resurfacing, which resulted in far fewer people leaving the path. *Assessment: likely to be beneficial (effectiveness 50%; certainty 40%; harms 0%).*

http://www.conservationevidence.com/actions/311

Start educational programmes for personal watercraft owners

A before-and-after study in the USA found that common tern reproduction increased, and rates of disturbance decreased, following a series of educational programmes aimed at recreational boat users. *Assessment: likely to be beneficial (effectiveness 40%; certainty 40%; harms 0%).*

http://www.conservationevidence.com/actions/314

Use signs and access restrictions to reduce disturbance at nest sites

Six studies from across the world found increased numbers of breeders, higher reproductive success or lower levels of disturbance in waders and terns following the start of access restrictions or the erection of signs near nesting areas. Two studies from Europe and Antarctica found no effect of access restrictions on reproductive success in eagles and penguins, respectively. *Assessment: likely to be beneficial (effectiveness 59%; certainty 55%; harms 10%).*

http://www.conservationevidence.com/actions/309

Use voluntary agreements with local people to reduce disturbance

A before-and-after trial in the USA found significantly lower rates of waterfowl disturbance following the establishment of a voluntary waterfowl avoidance area, despite an overall increase in boat traffic. *Assessment: likely to be beneficial (effectiveness 50%; certainty 40%; harms 0%).*

http://www.conservationevidence.com/actions/313

Unknown effectiveness (limited evidence)

Habituate birds to human visitors

A study from Australia found that bridled terns from heavily disturbed sites had similar or higher reproductive success compared with less-disturbed sites, possibly suggesting that habituation had occurred. *Assessment: unknown effectiveness — limited evidence (effectiveness 20%; certainty 10%; harms 0%).*

http://www.conservationevidence.com/actions/315

Use nest covers to reduce the impact of research on predation of ground-nesting seabirds

A before-and-after study from Canada found that hatching success of Caspian terns was significantly higher when researchers protected nests after disturbing adults from them. *Assessment: unknown effectiveness — limited evidence (effectiveness 41%; certainty 35%; harms 19%).*

http://www.conservationevidence.com/actions/316

No evidence found (no assessment)

We have captured no evidence for the following interventions:

- Reduce visitor group sizes
- Set minimum distances for approaching birds (buffer zones)

3.9 Threat: Natural system modifications

Based on the collated evidence, what is the current assessment of the effectiveness of interventions for natural system modifications?	
Likely to be beneficial	• Create scrapes and pools in wetlands and wet grasslands • Provide deadwood/snags in forests: use ring-barking, cutting or silvicides • Use patch retention harvesting instead of clearcutting
Trade-off between benefit and harms	• Clear or open patches in forests • Employ grazing in artificial grassland/pastures • Employ grazing in natural grasslands • Employ grazing in non-grassland habitats • Manage water level in wetlands • Manually control or remove midstorey and ground-level vegetation (including mowing, chaining, cutting etc) in forests • Mow or cut natural grasslands • Mow or cut semi-natural grasslands/pastures • Manually control or remove midstorey and ground-level vegetation (including mowing, chaining, cutting etc) in shrubland • Raise water levels in ditches or grassland • Thin trees within forests

	• Use prescribed burning: grasslands • Use prescribed burning: pine forests • Use prescribed burning: savannahs • Use prescribed burning: shrublands • Use selective harvesting/logging instead of clearcutting
Unknown effectiveness (limited evidence)	• Clearcut and re-seed forests • Coppice trees • Fertilise grasslands • Manage woodland edges for birds • Manually control or remove midstorey and ground-level vegetation: reedbeds • Manually control or remove midstorey and ground-level vegetation: savannahs • Plant trees to act as windbreaks • Plough habitats • Provide deadwood/snags in forests: add woody debris to forests • Remove coarse woody debris from forests • Replace non-native species of tree/shrub • Re-seed grasslands • Use environmentally sensitive flood management • Use fire suppression/control • Use greentree reservoir management • Use prescribed burning: Australian sclerophyll forest • Use shelterwood cutting instead of clearcutting • Use variable retention management during forestry operations
Likely to be ineffective or harmful	• Apply herbicide to mid- and understorey vegetation • Treat wetlands with herbicides • Use prescribed burning: coastal habitats • Use prescribed burning: deciduous forests
No evidence found (no assessment)	• Protect nest trees before burning

Likely to be beneficial

Create scrapes and pools in wetlands and wet grasslands

Four out of six studies from the UK and North America found that more bird used sites, or breeding populations on sites increased, after ponds or scrapes were created. A study from the USA found that some duck species used newly created ponds and others used older ponds. A study from the UK found that northern lapwing chicks foraged in newly created features and that chick condition was higher in sites with a large number of footdrains. *Assessment: likely to be beneficial (effectiveness 75%; certainty 60%; harms 0%).*

http://www.conservationevidence.com/actions/359

Provide deadwood/snags in forests (use ring-barking, cutting or silvicides)

One of five studies found that forest plots provided with snags had higher bird diversity and abundance than plots without snags. Three of four studies from the USA and UK found that species used artificially-created snags for nesting and foraging. One study from the USA found that use increased with how long a snag had been dead. *Assessment: likely to be beneficial (effectiveness 45%; certainty 50%; harms 0%).*

http://www.conservationevidence.com/actions/343

Use patch retention harvesting instead of clearcutting

One of two studies (from the USA) found that areas under patch retention harvesting contained more birds of more species than clearcut areas, retaining similar numbers to unharvested areas. Two studies found that forest specialist species were found more frequently in patch retention plots than under other management. Habitat generalists declined on patch retention sites compared to other managements. *Assessment: likely to be beneficial (effectiveness 70%; certainty 46%; harms 0%).*

http://www.conservationevidence.com/actions/330

Trade-off between benefit and harms

Clear or open patches in forests

Seven out of nine studies from the UK and USA found that early-successional species increased in clearcut areas of forests, compared to other management. Two studies found that mature-forest species declined. One study found no differences in species richness between treatments, another found no consistent differences. A study from the USA found that a mosaic of cut and uncut areas supported a variety of species. *Assessment: trade-offs between benefits and harms (effectiveness 55%; certainty 60%; harms 30%).*

http://www.conservationevidence.com/actions/326

Employ grazing in artificial grasslands/pastures

Five studies from the UK and USA found use or nesting densities were higher in grazed compared to ungrazed areas. A study from Canada found an increase in duck populations following the start of grazing along with other interventions. Eight studies from the UK, Canada and the USA found species richness, community composition, abundances, use, nesting densities, nesting success or productivity were similar or lower on grazed compared with ungrazed areas. One found that several species were excluded by grazing. *Assessment: trade-offs between benefits and harms (effectiveness 43%; certainty 65%; harms 45%).*

http://www.conservationevidence.com/actions/349

Employ grazing in natural grasslands

Five of 12 studies from the USA and Canada found that densities of some species were higher on grazed than ungrazed sites. Eight studies from the USA, Canada and France found that some or all species studied were found at similar or lower densities on grazed compared to ungrazed sites or those under other management. Two controlled studies from the USA and Canada found that nesting success was higher on grazed than ungrazed sites. Five studies from the USA and Canada found that nesting success was similar or lower on grazed sites. *Assessment: trade-offs between benefits and harms (effectiveness 40%; certainty 60%; harms 50%).*

http://www.conservationevidence.com/actions/348

Employ grazing in non-grassland habitats

One of eight studies found more bird species on grazed than unmanaged sites, apart from in drought years. A study from the Netherlands found the number of species in a mixed habitat wetland site declined with increased grazing. Three studies in Sweden, the Netherlands and Kenya found that the overall abundance or densities of some species were higher in grazed than ungrazed sites. Four studies in Europe and Kenya found that some species were absent or at lower densities on grazed compared to ungrazed sites or those under different management. Five studies from across the world found no differences in abundances or densities of some or all species between grazed sites and those that were ungrazed or under different management. Two studies from the UK found that productivity was lower in grazed than ungrazed sites. A study from the UK found that songbirds and invertebrate-eating species, but not crows were more common on rough-grazed habitats than intensive pasture. *Assessment: trade-offs between benefits and harms (effectiveness 40%; certainty 67%; harms 40%).*

http://www.conservationevidence.com/actions/350

Manage water level in wetlands

Three studies (of six) from the USA, UK and Canada found that different species were more abundant at different water heights. One found that diversity levels also changed. One study found that great bitterns in the UK established territories earlier when deep water levels were maintained, but productivity did not vary. A study from Spain found that water management successfully retained water near a greater flamingo nesting area, but did not measure the effects on productivity or survival. *Assessment: trade-offs between benefits and harms (effectiveness 40%; certainty 41%; harms 35%).*

http://www.conservationevidence.com/actions/355

Manually control or remove midstorey and ground-level vegetation (including mowing, chaining, cutting etc.) in forests

Seven studies from Europe and the USA found that species richness, total density or densities of some species were higher in areas with mid- or understorey management compared to areas without management. Four studies also used other interventions. Seven studies from the USA and Canada found that species richness, densities, survival or competition for nest sites were similar or lower in areas with mid- or understorey control. Two studies investigated several interventions at once. Two studies from Canada found higher nest survival in forests with removal of deciduous trees compared to controls. One study found that chicks foraging success was higher in areas with cleared understorey vegetation compared to burned areas, but lower than under other managements. *Assessment: trade-offs between benefits and harms (effectiveness 40%; certainty 75%; harms 40%).*

http://www.conservationevidence.com/actions/335

Mow or cut natural grasslands

Two of six studies found higher densities of birds or nests on mown grasslands compared to unmanaged or burned areas. Two studies found lower densities or nests of some species and two found no differences in nesting densities or community composition on mown compared to unmown areas. One study from the USA found that grasshopper sparrow nesting success was higher on mown than grazed areas. One study from the USA found that duck nesting success was similar on cut and uncut areas. *Assessment: trade-offs between benefits and harms (effectiveness 40%; certainty 50%; harms 39%).*

http://www.conservationevidence.com/actions/338

Mow or cut semi-natural grasslands/pastures

One of four studies found that wader populations increased following annual cutting of semi-natural grasslands. One study from the UK found that ducks grazed at higher densities on cut areas. Another study in the UK found that goose grazing densities were unaffected by cutting frequency. One study from the USA found that Henslow's sparrows were more likely to be recaptured on unmown than mown grasslands. *Assessment: trade-offs between benefits and harms (effectiveness 40%; certainty 40%; harms 20%).*

http://www.conservationevidence.com/actions/339

Manually control or remove midstorey and ground-level vegetation (including mowing, chaining, cutting etc.) in shrublands

One of seven studies found that overall bird diversity and bird density was similar between chained areas, burned areas and controls. One found that overall diversity and abundance was lower on mown sites than controls, but that grassland-specialist species were present on managed sites. Five studies from the USA and Europe found than some species were at greater densities or abundances on sites with mechanical vegetation control than on sites with burning or no management. Three studies from the USA found that some species were less abundant on sites with mechanical vegetation removal. One study from the USA found no differences between areas cut in winter and summer. *Assessment: trade-offs between benefits and harms (effectiveness 43%; certainty 54%; harms 30%).*

http://www.conservationevidence.com/actions/337

Raise water levels in ditches or grassland

One of seven studies found that three waders were found to have recolonised a UK site or be found at very high densities after water levels were raised. Three studies from Europe found that raising water levels on grassland provided habitat for waders. A study from Denmark found that oystercatchers did not nest at higher densities on sites with raised water levels. A study from the UK found that birds visited sites with raised water levels more frequently than other fields, but another UK study found that feeding rates did not differ between sites with raised water levels and those without. A study from the USA found that predation rates on seaside sparrow nests increased as water levels were raised. *Assessment: trade-offs between benefits and harms (effectiveness 65%; certainty 55%; harms 25%).*

http://www.conservationevidence.com/actions/354

Thin trees within forests

One study of 14 (from the USA) found higher bird species richness in sites with tree thinning and several other interventions, compared to unmanaged sites. Three studies from the UK and USA found no such differences. Seven studies (four investigating multiple interventions) found that overall bird abundance or the abundance of some species was higher in thinned plots, compared to those under different management. Five studies found that found that abundances were similar, or that some species were less abundant in areas with thinning. Two studies from the USA found no effect of thinning on wood thrushes, a species thought to be sensitive to it. A study from the USA found that a higher proportion of nests were in nest boxes in a thinned site, compared to a control. A study from the USA found no differences in bird abundances between burned sites with high-retention thinning, compared to low-retention sites. *Assessment: trade-offs between benefits and harms (effectiveness 50%; certainty 60%; harms 30%).*

http://www.conservationevidence.com/actions/328

Use prescribed burning: grasslands

Four of 21 studies found that overall species richness and community composition did not vary between burned and unburned sites. Nine studies from across the world found that at least some species were more abundant or at higher densities in burned than unburned areas or areas under different management. Fourteen studies found that at least one species was at similar or lower abundances on burned areas. Responses varied depending on how soon after fires monitoring occurred. One study from the USA found that Florida grasshopper sparrow had significantly higher reproductive success soon after burns, whilst another found that dickcissel reproductive success was higher in patch-burned than burned and grazed areas. *Assessment: trade-offs between benefits and harms (effectiveness 45%; certainty 60%; harms 40%).*

http://www.conservationevidence.com/actions/322

Use prescribed burning: pine forests

Four of 28 studies in the USA found higher species richness, densities or abundances in sites with prescribed burning, tree thinning and in one case mid- or understorey control compared to controls. Fourteen studies found that some species were more abundant, or had higher productivities or survival in burned or burned and thinned areas than control areas. One study found that effects varied with geography and habitat. Fifteen studies found no differences in species richness or densities, community composition, productivity, behaviour or survival between sites with prescribed burning or burning and thinning, and controls or sites with other management. One study found that foraging success of chicks was lower in burned areas. Three studies found effects did not vary with burn season. *Assessment: trade-offs between benefits and harms (effectiveness 50%; certainty 77%; harms 35%).*

http://www.conservationevidence.com/actions/318

Use prescribed burning: savannahs

One of five studies found that burned areas of savannah tended to have more birds and species than control or grazed areas, although burned sites showed significant annual variation unlike grazed sites. A study from Australia found that effects on bird abundances depended on burn season and habitat type. Two studies in the USA found that some open country species were more common in burned areas than unburned. A study from the USA found that two eastern bluebirds successfully raised chicks after a local prescribed burn. *Assessment: trade-offs between benefits and harms (effectiveness 40%; certainty 50%; harms 35%).*

http://www.conservationevidence.com/actions/320

Use prescribed burning: shrublands

One of eight studies found that overall bird densities were similar between burned and unburned areas, whilst another found that species numbers and densities did not vary between areas burned in summer or winter. Three studies found that some species were more abundant on areas that were burned. Four found that species densities were similar or lower on burned compared to control areas or those under different management. One study found that sage sparrows chose different nest sites before and after burning. Another found no differences in greater sage grouse movement between burned and unburned areas. *Assessment: trade-offs between benefits and harms (effectiveness 43%; certainty 50%; harms 45%).*

http://www.conservationevidence.com/actions/321

Use selective harvesting/logging instead of clearcutting

Six of seven studies from the USA and Canada found that some species were more, and other less, abundant in selectively logged forests compared to unlogged stands, or those under other management. One study found that differences between treatments were not consistent. A study from the USA found that species richness of cavity-nesting birds was lower in selectively logged forests than in clearcuts. One study from the USA found that brood parasitism was higher in selectively logged forests for two species and lower for two others, compared to control stands. *Assessment: trade-offs between benefits and harms (effectiveness 65%; certainty 60%; harms 30%).*

http://www.conservationevidence.com/actions/331

Unknown effectiveness (limited evidence)

Clearcut and re-seed forests

One of two studies from the USA found that stands of pines replanted with native species held more species typical of scrub habitats than stands under different management. The other study found similar bird densities in clearcut and re-seeded sites and those under different management. *Assessment: unknown effectiveness — limited evidence (effectiveness 30%; certainty 35%; harms 0%).*

http://www.conservationevidence.com/actions/327

Coppice trees

One of three studies found a population increase in European nightjars on a UK site after the introduction of coppicing and other interventions. Two studies from the UK and USA found that the use of coppices by some bird species declined over time. A UK study found that species richness decreased with the age of a coppice, but that some species were more abundant in older stands. *Assessment: unknown effectiveness — limited evidence (effectiveness 34%; certainty 30%; harms 30%).*

http://www.conservationevidence.com/actions/329

Fertilise grasslands

All four studies captured (all from the UK) found that more geese grazed on fertilised areas of grass more than control areas. Two investigated cutting and fertilizing at the same time. One study found that fertilised areas were used less than re-seeded areas. One study found that fertilisation had an effect at applications of 50 kg N/ha, but not at 18 kg N/ha. Another found that the effects of fertilisation did not increase at applications over 80 kg N/ha. *Assessment: unknown effectiveness — limited evidence (effectiveness 60%; certainty 35%; harms 7%).*

http://www.conservationevidence.com/actions/353

Manage woodland edges for birds

One of three studies found that a local population of European nightjars increased at a UK site following the start of a management regime that included the management of woodland edges for birds. Two studies of an experiment in the USA found that bird abundance (but not species richness or nesting success) was higher in woodland edges managed for wildlife than unmanaged edges. *Assessment: unknown effectiveness — limited evidence (effectiveness 55%; certainty 39%; harms 30%).*

http://www.conservationevidence.com/actions/334

Manually control or remove midstorey and ground-level vegetation (including mowing, chaining, cutting etc.) (reedbeds)

One of three studies found that warblers nested at lower densities in cut areas of reeds. Productivity and success did not vary between treatments. A study from Denmark found that geese grazed at the highest densities on reedbeds cut 5–12 years previously. One study in the UK found that cutting reeds and changing water levels did not affect great bittern breeding productivity, but did delay territory establishment. *Assessment: unknown effectiveness — limited evidence (effectiveness 15%; certainty 36%; harms 14%).*

http://www.conservationevidence.com/actions/340

Manually control or remove midstorey and ground-level vegetation (including mowing, chaining, cutting etc.) (savannahs)

A study in Argentina found that in summer, but not overall, bird abundance and species richness was lower in an area where shrubs were removed compared to a control. Community composition also differed between treatments. *Assessment: unknown effectiveness — limited evidence (effectiveness 30%; certainty 10%; harms 30%).*

http://www.conservationevidence.com/actions/336

Plant trees to act as windbreaks

One of two studies found that a population of European nightjars increased at a UK site after multiple interventions including the planting of windbreak trees. A study from the USA found that such trees appeared to disrupt lekking behaviour in greater prairie chickens. *Assessment: unknown effectiveness — limited evidence (effectiveness 12%; certainty 25%; harms 20%).*

http://www.conservationevidence.com/actions/351

Plough habitats

One of four studies found that bird densities were higher on ploughed wetlands in the USA than unploughed ones. Three studies of one experiment in the UK found that few whimbrels nested on areas of heathland ploughed and re-seeded, but that they were used for foraging in early spring. There were no differences in chick survival between birds that used ploughed and re-seeded heathland and those that did not. *Assessment: unknown effectiveness — limited evidence (effectiveness 25%; certainty 36%; harms 10%).*

http://www.conservationevidence.com/actions/358

Provide deadwood/snags in forests (adding woody debris to forests)

One study from Australia found that brown treecreeper numbers were higher in plots with large amounts of dead wood added compared to plots with less or no debris added. *Assessment: unknown effectiveness — limited evidence (effectiveness 50%; certainty 29%; harms 0%).*

http://www.conservationevidence.com/actions/344

Remove coarse woody debris from forests

Two studies from the USA found that some species increased in sites with woody debris removal. One found that overall breeding bird abundance and diversity were lower in removal plots; the other that survival of black-chinned hummingbird nests was lower. *Assessment: unknown effectiveness — limited evidence (effectiveness 10%; certainty 33%; harms 60%).*

http://www.conservationevidence.com/actions/345

Replace non-native species of tree/shrub

A study from the USA found that the number of black-chinned hummingbird nests increased after fuel reduction and the planting of native species, but that the increase was smaller than at sites without planting. *Assessment: unknown effectiveness — limited evidence (effectiveness 5%; certainty 18%; harms 0%).*

http://www.conservationevidence.com/actions/341

Re-seed grasslands

One of two studies from the UK found that geese grazed at higher densities on re-seeded grasslands than on control or fertilised grasslands. Another study from the UK found that geese grazed at higher densities on areas sown with clover, rather than grass seed. *Assessment: unknown effectiveness — limited evidence (effectiveness 35%; certainty 19%; harms 0%).*

http://www.conservationevidence.com/actions/352

Use environmentally sensitive flood management

One of two studies found more bird territories on a stretch of river in the UK with flood beams, compared to a channelized river. The other found that 13 out of 20 species of bird increased at sites in the USA where a river's hydrological dynamics were restored. *Assessment: unknown effectiveness — limited evidence (effectiveness 41%; certainty 26%; harms 0%).*

http://www.conservationevidence.com/actions/356

Use fire suppression/control

All three studies we captured, from the USA, UK and Australia, found that some bird species increased after fire suppression, and in one case that woodland species appeared in a site. Two studies (from the UK and USA) found that some species declined following fire suppression. The USA study identified open country species as being negatively affected. *Assessment: unknown effectiveness — limited evidence (effectiveness 35%; certainty 34%; harms 30%).*

http://www.conservationevidence.com/actions/324

Use greentree reservoir management

A study from the USA found that fewer mid- and under-storey birds were found at a greentree reservoir site than at a control site. Canopy-nesting species were not affected. *Assessment: unknown effectiveness — limited evidence (effectiveness 0%; certainty 10%; harms 40%).*

http://www.conservationevidence.com/actions/357

Use prescribed burning (Australian sclerophyll forest)

Two of three studies from Australia found no differences in bird species richness in burned sites compared to unburned areas. All three found differences in species assemblages, with some species lost and others gained from areas after fire. *Assessment: unknown effectiveness — limited evidence (effectiveness 30%; certainty 31%; harms 30%).*

http://www.conservationevidence.com/actions/319

Use shelterwood cutting instead of clearcutting

A study from the USA found that bird community composition differed between shelterwood stands and those under other forestry practices: some species were more abundant, others less so. *Assessment: unknown effectiveness — limited evidence (effectiveness 40%; certainty 20%; harms 40%).*

http://www.conservationevidence.com/actions/333

Use variable retention management during forestry operations

A study from the USA found that nine species were more abundant and five less so in stands under variable retention management, compared to unmanaged stands. *Assessment: unknown effectiveness — limited evidence (effectiveness 45%; certainty 20%; harms 25%).*

http://www.conservationevidence.com/actions/332

Likely to be ineffective or harmful

Apply herbicide to mid- and understorey vegetation

One of seven studies from North America found that bird species richness in a forest declined after deciduous trees were treated with herbicide. Three studies found increases in total bird densities, or those of some species, after herbicide treatment, although one found no differences between treatment and control areas. One study found that densities of one species decreased and another remained steady after treatment. Three studies found that nest survival was lower in herbicide-treated areas and one found lower nesting densities. One study found that northern bobwhite chicks higher had foraging success in forest areas treated with herbicide compared to under other managements. *Assessment: likely to be ineffective or harmful (effectiveness 20%; certainty 50%; harms 60%).*

http://www.conservationevidence.com/actions/346

Treat wetlands with herbicides

All four studies from the USA found higher densities of birds in wetlands sprayed with herbicide, compared with unsprayed areas. Two found that some species were at lower densities compared to unsprayed areas or those under other management. *Assessment: likely to be ineffective or harmful (effectiveness 30%; certainty 42%; harms 40%).*

http://www.conservationevidence.com/actions/347

Use prescribed burning (coastal habitats)

One study from the USA found that breeding seaside sparrow numbers decreased the year a site was burned, but were higher than on an unburned site the following year. One study in Argentina found that tall-grass specialist species were lost from burned areas in the year of burning, but that some habitats recovered by the following year. One study from the USA found no differences in nest predation rates between burned and unburned areas for two years after burning. *Assessment: likely to be ineffective or harmful (effectiveness 20%; certainty 40%; harms 30%).*

http://www.conservationevidence.com/actions/323

Use prescribed burning (deciduous forests)

One of four studies found that bird species richness was similar in burned and unburned aspen forests, although relative abundances of some species changed. A study in the USA found no changes in community composition in oak and hickory forests following burning. One study in the USA found no differences in wood thrush nest survival in burned and unburned areas. Another study in the USA found a reduction in black-chinned hummingbird nests following fuel reduction treatments including burning. *Assessments: likely to be ineffective or harmful (effectiveness 32%; certainty 60%; harms 30%).*

http://www.conservationevidence.com/actions/317

No evidence found (no assessment)

We have captured no evidence for the following interventions:

- Protect nest trees before burning

3.10 Habitat restoration and creation

Based on the collated evidence, what is the current assessment of the effectiveness of interventions for habitat restoration and creation?	
Beneficial	• Restore or create forests • Restore or create wetlands and marine habitats: restore or create inland wetlands
Likely to be beneficial	• Restore or create grassland • Restore or create traditional water meadows • Restore or create wetlands and marine habitats: restore or create coastal and intertidal wetlands
Unknown effectiveness (limited evidence)	• Restore or create shrubland • Restore or create wetlands and marine habitats: restore or create kelp forests • Restore or create wetlands and marine habitats: restore or create lagoons
No evidence found (no assessment)	• Restore or create savannahs • Revegetate gravel pits

Beneficial

Restore or create forests

Thirteen of 15 studies from across the world found that restored forests were similar to in-tact forests, that species returned to restored sites, that species recovered significantly better at restored than unrestored sites or that bird species richness, diversity or abundances in restored forest sites increased over time. One study also found that restoration techniques themselves improved over time. Nine studies found that some species did not return to restored forests or were less common and a study found that territory densities decreased over time. A study from the USA found that no more birds were found in restored sites, compared with unrestored. One study investigated productivity and found it was similar between restored and intact forests. A study from the USA found that planting fast-growing species appeared to provide better habitat than slower-growing trees. *Assessment: beneficial (effectiveness 65%; certainty 76%; harms 0%).*

http://www.conservationevidence.com/actions/360

Restore or create wetlands and marine habitats (inland wetlands)

All eleven studies from the USA and Canada found that birds used restored or created wetlands. Two found that rates of use and species richness were similar or higher than on natural wetlands. One found that use was higher than on unrestored wetlands. Three studies from the USA and Puerto Rico found that restored wetlands held lower densities and fewer species or had similar productivity compared to natural wetlands. Two studies in the USA found that semi-permanent restored and larger wetlands were used more than temporary or seasonal or smaller ones. *Assessment: beneficial (effectiveness 70%; certainty 65%; harms 0%).*

http://www.conservationevidence.com/actions/366

Likely to be beneficial

Restore or create grassland

Three of 23 studies found that species richness on restored grasslands was higher than unrestored habitats, or similar to remnant grassland, and three found that target species used restored grassland. Two studies from the USA found that diversity or species richness fell after restoration or was lower than unrestored sites. Seven studies from the USA and UK found high use of restored sites, or that such sites held a disproportionate proportion of the local population of birds. Two studies found that densities or abundances were lower on restored than unrestored sites, potentially due to drought conditions in one case. Five studies found that at least some bird species had higher productivities in restored sites compared to unrestored; had similar or higher productivities than natural habitats; or had high enough productivities to sustain populations. Three studies found that productivities were lower in restored than unrestored areas, or that productivities on restored sites were too low to sustain populations. A study from the USA found that older restored fields held more nests, but fewer species than young fields. Three studies found no differences between restoration techniques; two found that sowing certain species increased the use of sites by birds. *Assessment: likely to be beneficial (effectiveness 45%; certainty 70%; harms 0%).*

http://www.conservationevidence.com/actions/361

Restore or create traditional water meadows

Four out of five studies found that the number of waders or wildfowl on UK sites increased after the restoration of traditional water meadows. One study from Sweden found an increase in northern lapwing population after an increase in meadow management. One study found that lapwing productivity was higher on meadows than some habitats, but not others. *Assessment: likely to be beneficial (effectiveness 65%; certainty 50%; harms 0%).*

http://www.conservationevidence.com/actions/363

Restore or create wetlands and marine habitats (coastal and intertidal wetlands)

All six studies from the USA and UK found that bird species used restored or created wetlands. Two found that numbers and/or diversity were similar to in natural wetlands and one that numbers were higher than in unrestored sites. Three found that bird numbers on wetlands increased over time. Two studies from the UK found that songbirds and waders decreased following wetland restoration, whilst a study from the USA found that songbirds were more common on unrestored sites than restored wetlands. *Assessment: likely to be beneficial (effectiveness 65%; certainty 55%; harms 3%).*

http://www.conservationevidence.com/actions/367

Unknown effectiveness (limited evidence)

Restore or create shrubland

Three studies from the UK, USA and the Azores found local bird population increases after shrubland restoration. Two studies investigated multiple interventions and one found an increase from no birds to one or two pairs.

One study from the UK found that several interventions, including shrubland restoration, were negatively related to the number of young grey partridges per adult bird on sites. *Assessment: unknown effectiveness — limited evidence (effectiveness 25%; certainty 20%; harms 3%).*

http://www.conservationevidence.com/actions/364

Restore or create wetlands and marine habitats (kelp forests)

One study in the USA found that the densities of five of the nine bird species increased following kelp forest restoration. *Assessment: unknown effectiveness — limited evidence (effectiveness 60%; certainty 15%; harms 0%).*

http://www.conservationevidence.com/actions/368

Restore or create wetlands and marine habitats (lagoons)

One study in the UK found that large numbers of bird species used and bred in a newly-created lagoon. *Assessment: unknown effectiveness — limited evidence (effectiveness 61%; certainty 20%; harms 0%).*

http://www.conservationevidence.com/actions/369

No evidence found (no assessment)

We have captured no evidence for the following interventions:

- Restore or create savannahs
- Revegetate gravel pits

3.11 Threat: Invasive alien and other problematic species

This assessment method for this chapter is described in Walsh, J. C., Dicks, L. V. & Sutherland, W. J. (2015) The effect of scientific evidence on conservation practitioners' management decisions. Conservation Biology, 29: 88–98. No harms were assessed for sections 3.11.1, 3.11.2, 3.11,3 and 3.11.4.

3.11.1 Reduce predation by other species

Based on the collated evidence, what is the current assessment of the effectiveness of interventions for reducing predation by other species?	
Beneficial	• Control mammalian predators on islands • Remove or control predators to enhance bird populations and communities
Likely to be beneficial	• Control avian predators on islands
Unknown effectiveness (limited evidence)	• Control invasive ants on islands • Reduce predation by translocating predators
Evidence not assessed	• Control predators not on islands

Beneficial

Control mammalian predators on islands

Of the 33 studies from across the world, 16 described population increases or recolonisations in at least some of the sites studied and 18 found higher reproductive success or lower mortality (on artificial nests in one case). Two studies that investigated population changes found only partial increases, in black oystercatchers *Haematopus bachmani* and two gamebird species, respectively. Eighteen of the studies investigated rodent control; 12 cat *Felis catus* control and 6 various other predators including pigs *Sus scrofa* and red foxes *Vulpes*. The two that found only partial increases examined cat, fox and other larger mammal removal. *Assessment: beneficial (effectiveness 81%; certainty 78%).*

http://www.conservationevidence.com/actions/373

Remove or control predators to enhance bird populations and communities

Both a meta-analysis and a systematic review (both global) found that bird reproductive success increased with predator control and that either post-breeding or breeding-season populations increased. The systematic review found that post-breeding success increased with predator control on mainland, but not islands. *Assessment: beneficial (effectiveness 66%; certainty 71%).*

http://www.conservationevidence.com/actions/371

Likely to be beneficial

Control avian predators on islands

Seven out of ten studies from North America, Australia and Europe found that controlling avian predators led to increased population sizes, reduced mortality, increased reproductive success or successful translocation of seabirds on islands. Two controlled studies on European islands found little effect of controlling crows on reproductive success in raptors or gamebirds. One study in the UK found that numbers of terns and small gulls on gravel islands declined despite the attempted control of large gulls. *Assessment: likely to be beneficial (effectiveness 50%; certainty 45%).*

http://www.conservationevidence.com/actions/372

Unknown effectiveness (limited evidence)

Control invasive ants on islands

A single study in the USA found that controlling the invasive tropical fire ant *Solenopsis geminata*, but not the big-headed ant *Pheidole megacephala*, led to lower rates of injuries and temporarily higher fledging success than on islands without ant control. The authors note that very few chicks were injured by *P. megacephala* on either experimental or control islands. *Assessment: unknown effectiveness — limited evidence (effectiveness 10%; certainty 15%).*

http://www.conservationevidence.com/actions/383

Reduce predation by translocating predators

Two studies from France and the USA found local population increases or reduced predation following the translocation of predators away from an area. *Assessment: unknown effectiveness — limited evidence (effectiveness 27%; certainty 20%).*

http://www.conservationevidence.com/actions/393

Evidence not assessed

Control predators not on islands

A study from the UK found higher bird community breeding densities and fledging success rates in plots with red fox *Vulpes vulpes* and carrion crow *Corvus corone* control. Of the 25 taxa-specific studies, only five found evidence for population increases with predator control, whilst one found a population decrease (with other interventions also used); one found lower or similar survival, probably because birds took bait. Nineteen studies found some evidence for increased reproductive success or decreased predation with predator control, with three studies (including a meta-analysis) finding no evidence for higher reproductive success or predation with predator control or translocation from the study site. One other study found evidence for increases in only three of six species studied. Most studies studied the removal of a number of different mammals, although several also removed bird predators, mostly carrion crows and gulls *Larus* spp. *Assessment: this intervention has not been assessed.*

http://www.conservationevidence.com/actions/384

3.11.2 Reduce incidental mortality during predator eradication or control

Based on the collated evidence, what is the current assessment of the effectiveness of interventions for reducing incidental mortality during predator eradication or control predation	
Unknown effectiveness (limited evidence)	• Distribute poison bait using dispensers • Use coloured baits to reduce accidental mortality during predator control • Use repellents on baits
Evidence not assessed	• Do birds take bait designed for pest control?

Unknown effectiveness (limited evidence)

Distribute poison bait using dispensers

A study from New Zealand found that South Island robin survival was higher when bait for rats and mice was dispensed from feeders, compared to being scattered. *Assessment: unknown effectiveness — limited evidence (effectiveness 40%; certainty 25%).*

http://www.conservationevidence.com/actions/157

Use coloured baits to reduce accidental mortality during predator control

Two out of three studies found that dyed baits were consumed at lower rates by songbirds and kestrels. An *ex situ* study from Australia found that dyeing food did not reduce its consumption by bush thick-knees. *Assessment: unknown effectiveness — limited evidence (effectiveness 20%; certainty 30%).*

http://www.conservationevidence.com/actions/182

Use repellents on baits

A study in New Zealand found that repellents reduced the rate of pecking at baits by North Island robins. A study from the USA found that treating bait with repellents did not reduce consumption by American kestrels. *Assessment: unknown effectiveness — limited evidence (effectiveness 10%; certainty 10%).*

http://www.conservationevidence.com/actions/159

Evidence not assessed

Do birds take bait designed for pest control?

Two studies from New Zealand and Australia, one *ex situ*, found no evidence that birds took bait meant for pest control. *Assessment: this intervention has not been assessed.*

http://www.conservationevidence.com/actions/395

3.11.3 Reduce nest predation by excluding predators from nests or nesting areas

Based on the collated evidence, what is the current assessment of the effectiveness of interventions for reducing nest predation by excluding predators from nests or nesting areas	
Likely to be beneficial	• Physically protect nests from predators using non-electric fencing • Physically protect nests with individual exclosures/barriers or provide shelters for chicks • Protect bird nests using electric fencing • Use artificial nests that discourage predation
Unknown effectiveness (limited evidence)	• Guard nests to prevent predation • Plant nesting cover to reduce nest predation • Protect nests from ants • Use multiple barriers to protect nests • Use naphthalene to deter mammalian predators • Use snakeskin to deter mammalian nest predators
No evidence found (no assessment)	• Play spoken-word radio programs to deter predators • Use 'cat curfews' to reduce predation • Use lion dung to deter domestic cats • Use mirrors to deter nest predators • Use ultrasonic devices to deter cats
Evidence not assessed	• Can nest protection increase nest abandonment? • Can nest protection increase predation of adults and chicks?

Likely to be beneficial

Physically protect nests from predators using non-electric fencing

Two of four studies from the UK and the USA found that fewer nests failed or were predated when predator exclusion fences were erected. Two studies found that nesting and fledging success was no higher when fences were used, one found that hatching success was higher. *Assessment: likely to be beneficial (effectiveness 45%; certainty 48%).*

http://www.conservationevidence.com/actions/183

Physically protect nests with individual exclosures/barriers or provide shelters for chicks

Nine of 23 studies found that fledging rates or productivity were higher for nests protected by individual barriers than for unprotected nests. Two found no higher productivity. Fourteen studies found that hatching rates or survival were higher, or that predation was lower for protected nests. Two found no differences between protected and unprotected nests and one found that adults were harassed by predators at protected nests. One study found that chick shelters were not used much and a review found that some exclosure designs were more effective than others. *Assessment: likely to be beneficial (effectiveness 50%; certainty 50%).*

http://www.conservationevidence.com/actions/397 http://www.conservationevidence.com/actions/398 http://www.conservationevidence.com/actions/399 http://www.conservationevidence.com/actions/400

Protect bird nests using electric fencing

Two of six studies found increased numbers of terns or tern nests following the erection of an electric fence around colonies. Five studies found higher survival or productivity of waders or seabirds when electric fences were used and one found lower predation by mammals inside electric fences. One study found that predation by birds was higher inside electric fences. *Assessment: likely to be beneficial (effectiveness 60%; certainty 59%).*

http://www.conservationevidence.com/actions/188

Use artificial nests that discourage predation

Three out of five studies from North America found lower predation rates or higher nesting success for wildfowl in artificial nests, compared with natural nests. An *ex situ* study found that some nest box designs prevented raccoons from entering. A study found that wood ducks avoided anti-predator nest boxes but only if given the choice of unaltered nest boxes. *Assessment: likely to be beneficial (effectiveness 59%; certainty 54%).*

http://www.conservationevidence.com/actions/402

Unknown effectiveness (limited evidence)

Guard nests to prevent predation

Nest guarding can be used as a response to a range of threats and is therefore discussed in 'General responses to small/declining populations – Guard nests'. *Assessment: unknown effectiveness — limited evidence (effectiveness 50%; certainty 30%).*

http://www.conservationevidence.com/actions/411

Plant nesting cover to reduce nest predation

Studies relevant to this intervention are discussed in 'Threat: Agriculture'. *Assessment: unknown effectiveness — limited evidence (effectiveness 28%; certainty 30%).*

http://www.conservationevidence.com/actions/405

Protect nests from ants

A study from the USA found that vireo nests protected from ants with a physical barrier and a chemical repellent had higher fledging success than unprotected nests. *Assessment: unknown effectiveness — limited evidence (effectiveness 45%; certainty 17%).*

http://www.conservationevidence.com/actions/410

Use multiple barriers to protect nests

One of two studies found that plover fledging success in the USA was no higher when an electric fence was erected around individual nest exclosures, compared to when just the exclosures were present. A study from the USA found that predation on chicks was lower when one of two barriers around nests was removed early, compared to when it was left for three more days.

Assessment: unknown effectiveness — limited evidence (effectiveness 7%; certainty 17%).

http://www.conservationevidence.com/actions/404

Use naphthalene to deter mammalian predators

A study from the USA found that predation rates on artificial nests did not differ when naphthalene moth balls were scattered around them. *Assessment: unknown effectiveness — limited evidence (effectiveness 0%; certainty 10%).*

http://www.conservationevidence.com/actions/408

Use snakeskin to deter mammalian nest predators

A study from the USA found that flycatcher nests were predated less frequently if they had a snakeskin wrapped around them. *Assessment: unknown effectiveness — limited evidence (effectiveness 33%; certainty 15%).*

http://www.conservationevidence.com/actions/406

No evidence found (no assessment)

We have captured no evidence for the following interventions:

- Play spoken-word radio programmes to deter predators
- Use 'cat curfews' to reduce predation
- Use lion dung to deter domestic cats
- Use mirrors to deter nest predators
- Use ultrasonic devices to deter cats

Evidence not assessed

Can nest protection increase nest abandonment?

One of four studies (from the USA) found an increase in abandonment after nest exclosures were used. Two studies from the USA and Sweden found no increases in abandonment when exclosures were used and a review from the USA found that some designs were more likely to cause abandonment than others. *Assessment: this intervention has not been assessed.*

http://www.conservationevidence.com/actions/401

Can nest protection increase predation of adults and chicks?

Four of five studies from the USA and Sweden found that predation on chicks and adults was higher when exclosures were used. One of these found that adults were harassed when exclosures were installed and the chicks rapidly predated when they were removed. One study from Sweden found that predation was no higher when exclosures were used. *Assessment: this intervention has not been assessed.*

http://www.conservationevidence.com/actions/403

3.11.4 Reduce mortality by reducing hunting ability or changing predator behaviour

Based on the collated evidence, what is the current assessment of the effectiveness of interventions for reducing mortality by reducing hunting ability or changing predator behaviour	
Unknown effectiveness (limited evidence)	• Reduce predation by translocating nest boxes • Use collar-mounted devices to reduce predation • Use supplementary feeding of predators to reduce predation
Unlikely to be beneficial	• Use aversive conditioning to reduce nest predation

Unknown effectiveness (limited evidence)

Reduce predation by translocating nest boxes

Two European studies found that predation rates were lower for translocated nest boxes than for controls. *Assessment: unknown effectiveness — limited evidence (effectiveness 48%; certainty 25%).*

http://www.conservationevidence.com/actions/420

Use collar-mounted devices to reduce predation

Two replicated randomised and controlled studies in the UK and Australia found that fewer birds were returned by cats wearing collars with anti-hunting devices, compared to cats with control collars. No differences were found between different devices. *Assessment: unknown effectiveness — limited evidence (effectiveness 48%; certainty 35%).*

http://www.conservationevidence.com/actions/416

Use supplementary feeding to reduce predation

One of three studies found that fewer grouse chicks were taken to harrier nests when supplementary food was provided to the harriers, but no effect on grouse adult survival or productivity was found. One study from the USA found reduced predation on artificial nests when supplementary food was provided. Another study from the USA found no such effect. *Assessment: unknown effectiveness — limited evidence (effectiveness 13%; certainty 20%).*

http://www.conservationevidence.com/actions/417

Unlikely to be beneficial

Use aversive conditioning to reduce nest predation

Nine out of 12 studies found no evidence for aversive conditioning or reduced nest predation after aversive conditioning treatment stopped. Ten studies found reduced consumption of food when it was treated with repellent chemicals, i.e. during the treatment. Three, all studying avian predators, found some evidence for reduced consumption after treatment but these were short-lived trials or the effect disappeared within a year. *Assessment: unlikely to be beneficial (effectiveness 9%; certainty 60%).*

http://www.conservationevidence.com/actions/418 http://www.conservationevidence.com/actions/419

3.11.5 Reduce competition with other species for food and nest sites

Based on the collated evidence, what is the current assessment of the effectiveness of interventions for reducing competition with other species for food and nest sites?	
Likely to be beneficial	• Reduce inter-specific competition for food by removing or controlling competitor species
Unknown effectiveness (limited evidence)	• Protect nest sites from competitors • Reduce competition between species by providing nest boxes • Reduce inter-specific competition for nest sites by modifying habitats to exclude competitor species • Reduce inter-specific competition for nest sites by removing competitor species: ground nesting seabirds • Reduce inter-specific competition for nest sites by removing competitor species: songbirds • Reduce inter-specific competition for nest sites by removing competitor species: woodpeckers

Likely to be beneficial

Reduce inter-specific competition for food by removing or controlling competitor species

Three out of four studies found that at least some of the target species increased following the removal or control of competitor species. Two studies found that some or all target species did not increase, or that there was no change in kleptoparasitic behaviour of competitor species after control efforts. *Assessment: likely to be beneficial (effectiveness 44%; certainty 40%; harms 0%).*

http://www.conservationevidence.com/actions/428

Unknown effectiveness (limited evidence)

Protect nest sites from competitors

Two studies from the USA found that red-cockaded woodpecker populations increased after the installation of 'restrictor plates' around nest holes to prevent larger woodpeckers for enlarging them. Several other interventions were used at the same time. A study from Puerto Rico found lower competition between species after nest boxes were altered. A study from the USA found weak evidence that exclusion devices prevented house sparrows from using nest boxes and another study from the USA found that fitting restrictor plates to red-cockaded woodpecker holes reduced the number that were enlarged by other woodpeckers. *Assessment: unknown effectiveness — limited evidence (effectiveness 39%; certainty 24%; harms 5%).*

http://www.conservationevidence.com/actions/426

Reduce competition between species by providing nest boxes

A study from the USA found that providing extra nest boxes did not reduce the rate at which common starlings usurped northern flickers from nests. *Assessment: unknown effectiveness — limited evidence (effectiveness 0%; certainty 16%; harms 0%).*

http://www.conservationevidence.com/actions/427

Reduce inter-specific competition for nest sites by modifying habitats to exclude competitor species

A study from the USA found that clearing midstorey vegetation did not reduce the occupancy of red-cockaded woodpecker nesting holes by southern flying squirrels. *Assessment: unknown effectiveness — limited evidence (effectiveness 0%; certainty 12%; harms 0%).*

http://www.conservationevidence.com/actions/425

Reduce inter-specific competition for nest sites by removing competitor species (ground nesting seabirds)

Four studies from Canada and the UK found increased tern populations following the control or exclusion of gulls, and in two cases with many additional interventions. Two studies from the UK and Canada found that controlling large gulls had no impact on smaller species. Two studies from the USA and UK found that exclusion devices successfully reduced the numbers of gulls at sites, although one found that they were only effective at small colonies and the other found that methods varied in their effectiveness and practicality. *Assessment: unknown effectiveness — limited evidence (effectiveness 41%; certainty 31%; harms 14%).*

http://www.conservationevidence.com/actions/422

Reduce inter-specific competition for nest sites by removing competitor species (songbirds)

Two studies from Australia found increases in bird populations and species richness after control of noisy miners. A study from Italy found that blue tits nested in more nest boxes when hazel dormice were excluded from boxes over winter. *Assessment: unknown effectiveness — limited evidence (effectiveness 50%; certainty 22%; harms 0%).*

http://www.conservationevidence.com/actions/424

Reduce inter-specific competition for nest sites by removing competitor species (woodpeckers)

Two studies in the USA found red-cockaded woodpecker populations increased following the removal of southern flying squirrels, in one case along with other interventions. A third found that red-cockaded woodpecker reintroductions were successful when squirrels were controlled. One study found fewer holes were occupied by squirrels following control efforts, but that occupancy by red-cockaded woodpeckers was no higher. *Assessment: unknown effectiveness — limited evidence (effectiveness 34%; certainty 28%; harms 0%).*

http://www.conservationevidence.com/actions/423

3.11.6 Reduce adverse habitat alteration by other species

Based on the collated evidence, what is the current assessment of the effectiveness of interventions for reducing adverse habitat alteration by other species?	
Likely to be beneficial	• Control or remove habitat-altering mammals • Reduce adverse habitat alterations by excluding problematic species (terrestrial species)
Unknown effectiveness (limited evidence)	• Reduce adverse habitat alterations by excluding problematic species (aquatic species) • Remove problematic vegetation • Use buffer zones to reduce the impact of invasive plant control

Likely to be beneficial

Control or remove habitat-altering mammals

Four out of five studies from islands in the Azores and Australia found that seabird populations increased after rabbits or other species were removed, although three studied several interventions at the same time. Two studies from Australia and Madeira found that seabird productivity increased after rabbit and house mouse eradication. *Assessment: likely to be beneficial (effectiveness 61%; certainty 41%; harms 0%).*

http://www.conservationevidence.com/actions/431

Reduce adverse habitat alterations by excluding problematic species (terrestrial species)

Three studies from the USA and the UK found higher numbers of certain songbird species and higher species richness in these groups when deer were excluded from forests. Intermediate canopy-nesting species in the USA and common nightingales in the UK were the species to benefit. A study from Hawaii found mixed effects of grazer exclusion. *Assessment: likely to be beneficial (effectiveness 48%; certainty 40%; harms 0%).*

http://www.conservationevidence.com/actions/429

Unknown effectiveness (limited evidence)

Reduce adverse habitat alterations by excluding problematic species (aquatic species)

A study in the USA found that waterbirds preferentially used wetland plots from which grass carp were excluded but moved as these became depleted over the winter. *Assessment: unknown effectiveness — limited evidence (effectiveness 30%; certainty 14%; harms 0%).*

http://www.conservationevidence.com/actions/430

Remove problematic vegetation

One of four studies (from Japan) found an increase in a bird population following the removal of an invasive plant. One study from the USA found lower bird densities in areas where a problematic native species was removed. One study from Australia found the Gould's petrel productivity was higher following the removal of native bird-lime trees, and a study from New Zealand found that Chatham Island oystercatchers could nest in preferable areas of beaches after invasive marram grass was removed. *Assessment: unknown effectiveness — limited evidence (effectiveness 43%; certainty 23%; harms 0%).*

http://www.conservationevidence.com/actions/432

Use buffer zones to reduce the impact of invasive plant control

A study from the USA found that no snail kite nests (built above water in cattail and bulrush) were lost during herbicide spraying when buffer zones were established around nests. *Assessment: unknown effectiveness — limited evidence (effectiveness 40%; certainty 10%; harms 0%).*

http://www.conservationevidence.com/actions/433

3.11.7 Reduce parasitism and disease

Based on the collated evidence, what is the current assessment of the effectiveness of interventions for reducing parasitism and disease?	
Likely to be beneficial	• Remove/control adult brood parasites
Trade-off between benefit and harms	• Remove/treat endoparasites and diseases
Unknown effectiveness (limited evidence)	• Alter artificial nest sites to discourage brood parasitism • Exclude or control 'reservoir species' to reduce parasite burdens • Remove brood parasite eggs from target species' nests • Remove/treat ectoparasites to increase survival or reproductive success: reduce nest ectoparasites by providing beneficial nesting material
	• Remove/treat ectoparasites to increase survival or reproductive success: remove ectoparasites from feathers • Use false brood parasite eggs to discourage brood parasitism
Unlikely to be beneficial	• Remove/treat ectoparasites to increase survival or reproductive success: remove ectoparasites from nests

Likely to be beneficial

Remove/control adult brood parasites

One of 12 studies, all from the Americas, found that a host species population increased after control of the parasitic cowbird, two studies found no effect. Five studies found higher productivities or success rates when cowbirds were removed, five found that some or all measures of productivity were no different. Eleven studies found that brood parasitism rates were lower after cowbird control. *Assessment: likely to be beneficial (effectiveness 48%; certainty 61%; harms 0%).*

http://www.conservationevidence.com/actions/441

Trade-off between benefit and harms

Remove/treat endoparasites and diseases

Two out of five studies found that removing endoparasites increased survival in birds and one study found higher productivity in treated birds. Two studies found no evidence, or uncertain evidence, for increases in survival with treatment and one study found lower parasite burdens, but also lower survival in birds treated with antihelmintic drugs. *Assessment: trade-offs between benefits and harms (effectiveness 48%; certainty 51%; harms 37%).*

http://www.conservationevidence.com/actions/434

Unknown effectiveness (limited evidence)

Alter artificial nest sites to discourage brood parasitism

A replicated trial from Puerto Rico found that brood parasitism levels were extremely high across all nest box designs tested. *Assessment: unknown effectiveness — limited evidence (effectiveness 0%; certainty 13%; harms 0%).*

http://www.conservationevidence.com/actions/446

Exclude or control 'reservoir species' to reduce parasite burdens

One of two studies found increased chick production in grouse when hares (carries of louping ill virus) were culled in the area, although a comment on the paper disputes this finding. A literature review found no compelling evidence for the effects of hare culling on grouse populations. *Assessment: unknown effectiveness — limited evidence (effectiveness 13%; certainty 20%; harms 0%).*

http://www.conservationevidence.com/actions/435

Remove brood parasite eggs from target species' nests

One of two studies found lower rates of parasitism when cowbird eggs were removed from host nests. One study found that nests from which cowbird eggs were removed had lower success than parasitised nests. *Assessment: unknown effectiveness — limited evidence (effectiveness 24%; certainty 20%; harms 21%).*

http://www.conservationevidence.com/actions/443

Remove/treat ectoparasites to increase survival or reproductive success (provide beneficial nesting material)

A study in Canada found lower numbers of some, but not all, parasites in nests provided with beneficial nesting material, but that there was no effect on fledging rates or chick condition. *Assessment: unknown effectiveness — limited evidence (effectiveness 15%; certainty 13%; harms 0%).*

http://www.conservationevidence.com/actions/439

Remove/treat ectoparasites to increase survival or reproductive success (remove ectoparasites from feathers)

A study in the UK found that red grouse treated with spot applications had lower tick and disease burdens and higher survival than controls, whilst birds with impregnated tags had lower tick burdens only. A study in Hawaii found that CO2 was the most effective way to remove lice from feathers, although lice were not killed. *Assessment: unknown effectiveness — limited evidence (effectiveness 42%; certainty 16%; harms 0%).*

http://www.conservationevidence.com/actions/437

Use false brood parasite eggs to discourage brood parasitism

A study from the USA found that parasitism rates were lower for red-winged blackbird nests with false or real cowbird eggs placed in them, than for control nests. *Assessment: unknown effectiveness — limited evidence (effectiveness 35%; certainty 19%; harms 0%).*

http://www.conservationevidence.com/actions/444

Unlikely to be beneficial

Remove/treat ectoparasites to increase survival or reproductive success (remove ectoparasites from nests)

Six of the seven studies found lower infestation rates in nests treated for ectoparasites, one (that used microwaves to treat nests) did not find fewer parasites. Two studies from the USA found higher survival or lower abandonment in nests treated for ectoparasites, whilst seven studies from across the world found no differences in survival, fledging rates or productivity between nests treated for ectoparasites and controls. Two of six studies found that chicks from nests treated for ectoparasites were in better condition than those from control nests. *Assessment: unlikely to be beneficial (effectiveness 25%; certainty 58%; harms 0%).*

http://www.conservationevidence.com/actions/438

3.11.8 Reduce detrimental impacts of other problematic species

Based on the collated evidence, what is the current assessment of the effectiveness of interventions for reducing detrimental impacts of other problematic species?	
Unknown effectiveness (limited evidence)	• Use copper strips to exclude snails from nests

Unknown effectiveness (limited evidence)

Use copper strips to exclude snails from nests

A study from Mauritius found no mortality from snails invading echo parakeet nests after the installation of copper strips around nest trees. Before installation, four chicks were killed by snails. *Assessment: unknown effectiveness — limited evidence (effectiveness 47%; certainty 15%; harms 0%).*

http://www.conservationevidence.com/actions/447

3.12 Threat: Pollution

3.12.1 Industrial pollution

Based on the collated evidence, what is the current assessment of the effectiveness of interventions for industrial pollution?	
Likely to be beneficial	• Use visual and acoustic 'scarers' to deter birds from landing on pools polluted by mining or sewage
Unknown effectiveness (limited evidence)	• Relocate birds following oil spills • Use repellents to deter birds from landing on pools polluted by mining
Unlikely to be beneficial	• Clean birds after oil spills

Likely to be beneficial

Use visual and acoustic 'scarers' to deter birds from landing on pools polluted by mining or sewage

Two studies from Australia and the USA found that deterrent systems reduced bird mortality on toxic pools. Four of five studies from the USA and Canada found that fewer birds landed on pools when deterrents were used, one found no effect. Two studies found that radar-activated systems were more effective than randomly-activated systems. One study found that loud noises were more effective than raptor models. *Assessment: likely to be beneficial (effectiveness 50%; certainty 46%; harms 0%).*

http://www.conservationevidence.com/actions/452

Unknown effectiveness (limited evidence)

Relocate birds following oil spills

A study from South Africa found that a high percentage of penguins relocated following an oil spill returned to and bred at their old colony. More relocated birds bred than oiled-and-cleaned birds. *Assessment: unknown effectiveness — limited evidence (effectiveness 39%; certainty 10%; harms 5%).*

http://www.conservationevidence.com/actions/449

Use repellents to deter birds from landing on pools polluted by mining

An *ex situ* study from the USA found that fewer common starlings consumed contaminated water laced with chemicals, compared to untreated water. *Assessment: unknown effectiveness — limited evidence (effectiveness 51%; certainty 10%; harms 0%).*

http://www.conservationevidence.com/actions/453

Unlikely to be beneficial

Clean birds after oil spills

Three studies from South Africa and Australia found high survival of oiled-and-cleaned penguins and plovers, but a large study from the USA found low survival of cleaned common guillemots. Two studies found that cleaned birds bred and had similar success to un-oiled birds. After a second spill, one study found that cleaned birds were less likely to breed. Two studies found that cleaned birds had lower breeding success than un-oiled birds. *Assessment: unlikely to be beneficial (effectiveness 30%; certainty 45%; harms 5%).*

http://www.conservationevidence.com/actions/448

3.12.2 Agricultural pollution

Based on the collated evidence, what is the current assessment of the effectiveness of interventions for agricultural pollution?	
Likely to be beneficial	• Leave headlands in fields unsprayed (conservation headlands) • Provide food for vultures to reduce mortality from diclofenac • Reduce pesticide, herbicide and fertiliser use generally
Unknown effectiveness (limited evidence)	• Reduce chemical inputs in permanent grassland management • Restrict certain pesticides or other agricultural chemicals
No evidence found (no assessment)	• Make selective use of spring herbicides • Provide buffer strips along rivers and streams • Provide unfertilised cereal headlands in arable fields • Use buffer strips around in-field ponds • Use organic rather than mineral fertilisers

Likely to be beneficial

Leave headlands in fields unsprayed (conservation headlands)

Three studies from Europe found that several species were strongly associated with conservation headlands; two of these found that other species were not associated with them. A review from the UK found larger grey partridge populations on sites with conservation headlands. Three studies found higher grey partridge adult or chick survival on sites with conservation headlands, one found survival did not differ. Four studies found higher grey partridge productivity on sites with conservation headlands, two found similar productivities and one found a negative relationship between conservation headlands and the number of chicks per adult partridge. *Assessment: likely to be beneficial (effectiveness 70%; certainty 50%; harms 0%).*

http://www.conservationevidence.com/actions/461

Provide food for vultures to reduce mortality from diclofenac

A before-and-after trial in Pakistan found that oriental white-backed vulture mortality rates were significantly lower when supplementary food was provided, compared to when it was not. *Assessment: likely to be beneficial (effectiveness 60%; certainty 40%; harms 0%).*

http://www.conservationevidence.com/actions/456

Reduce pesticide, herbicide and fertiliser use generally

One of nine studies found that the populations of some species increased when pesticide use was reduced and other interventions used. Three studies found that some or all species were found at higher densities on reduced-input sites. Five found that some of all species were not at higher densities. A study from the UK found that grey partridge chicks had higher survival on sites with reduced pesticide input. Another found that partridge broods were smaller on such sites and there was no relationship between reduced inputs and survival or the ratio of young to old birds. *Assessment: likely to be beneficial (effectiveness 55%; certainty 55%; harms 3%).*

http://www.conservationevidence.com/actions/454

Unknown effectiveness (limited evidence)

Reduce chemical inputs in permanent grassland management

A study from the UK found that no more foraging birds were attracted to pasture plots with no fertiliser, compared to control plots. *Assessment: unknown effectiveness — limited evidence (effectiveness 0%; certainty 10%; harms 0%).*

http://www.conservationevidence.com/actions/459

Restrict certain pesticides or other agricultural chemicals

A before-and-study from Spain found an increase in the regional griffon vulture population following the banning of strychnine, amongst several other interventions. *Assessment: unknown effectiveness — limited evidence (effectiveness 20%; certainty 10%; harms 0%).*

http://www.conservationevidence.com/actions/455

No evidence found (no assessment)

We have captured no evidence for the following interventions:

- Make selective use of spring herbicides
- Provide buffer strips along rivers and streams
- Provide unfertilised cereal headlands in arable fields
- Use buffer strips around in-field ponds
- Use organic rather than mineral fertilisers

3.12.3 Air-borne pollutants

Based on the collated evidence, what is the current assessment of the effectiveness of interventions for air-borne pollutants?	
Unknown effectiveness (limited evidence)	• Use lime to reduce acidification in lakes

Unknown effectiveness (limited evidence)

Use lime to reduce acidification in lakes

A study from Sweden found no difference in osprey productivity during a period of extensive liming of acidified lakes compared to two periods without liming. *Assessment: unknown effectiveness — limited evidence (effectiveness 0%; certainty 10%; harms 0%).*

http://www.conservationevidence.com/actions/465

3.12.4 Excess energy

Based on the collated evidence, what is the current assessment of the effectiveness of interventions for excess energy?	
Unknown effectiveness (limited evidence)	• Shield lights to reduce mortality from artificial lights • Turning off lights to reduce mortality from artificial lights • Use flashing lights to reduce mortality from artificial lights • Use lights low in spectral red to reduce mortality from artificial lights
No evidence found (no assessment)	• Reduce the intensity of lighthouse beams • Using volunteers to collect and rehabilitate downed birds

Unknown effectiveness (limited evidence)

Shield lights to reduce mortality from artificial lights

A study from the USA found that fewer shearwaters were downed when security lights were shielded, compared to nights with unshielded lights. *Assessment: unknown effectiveness — limited evidence (effectiveness 50%; certainty 15%; harms 0%).*

http://www.conservationevidence.com/actions/469

Turning off lights to reduce mortality from artificial lights

A study from the UK found that fewer seabirds were downed when artificial (indoor and outdoor) lighting was reduced at night, compared to nights with normal lighting. *Assessment: unknown effectiveness — limited evidence (effectiveness 49%; certainty 10%; harms 0%).*

http://www.conservationevidence.com/actions/467

Use flashing lights to reduce mortality from artificial lights

A study from the USA found that fewer dead birds were found beneath aviation control towers with only flashing lights, compared to those with both flashing and continuous lights. *Assessment: unknown effectiveness — limited evidence (effectiveness 54%; certainty 15%; harms 0%).*

http://www.conservationevidence.com/actions/470

Use lights low in spectral red to reduce mortality from artificial lights

Two studies from Europe found that fewer birds were attracted to low-red lights (including green and blue lights), compared with the number expected, or the number attracted to white or red lights. *Assessment: unknown effectiveness — limited evidence (effectiveness 56%; certainty 15%; harms 0%).*

http://www.conservationevidence.com/actions/471

No evidence found (no assessment)

We have captured no evidence for the following interventions:

- Reduce the intensity of lighthouse beams
- Using volunteers to collect and rehabilitate downed birds

3.13 Threat: Climate change, extreme weather and geological events

Based on the collated evidence, what is the current assessment of the effectiveness of interventions for climate change, extreme weather and geological events?	
Unknown effectiveness (limited evidence)	• Replace nesting habitats when they are washed away by storms • Water nesting mounds to increase incubation success in malleefowl

Unknown effectiveness (limited evidence)

Replace nesting habitats when they are washed away by storms

A before-and-after study found that a common tern colony increased following the replacement of nesting habitats, whilst a second found that a colony decreased. In both cases, several other interventions were used at the same time, making it hard to examine the effect of habitat provision. *Assessment: unknown effectiveness — limited evidence (effectiveness 8%; certainty 10%; harms 0%).*

http://www.conservationevidence.com/actions/474

Water nesting mounds to increase incubation success in malleefowl

A single small trial in Australia found that watering malleefowl nests increased their internal temperature but that a single application of water did not prevent the nests drying out and being abandoned during a drought. *Assessment: unknown effectiveness — limited evidence (effectiveness 9%; certainty 10%; harms 0%).*

http://www.conservationevidence.com/actions/473

3.14 General responses to small/declining populations

3.14.1 Inducing breeding, rehabilitation and egg removal

Based on the collated evidence, what is the current assessment of the effectiveness of interventions for inducing breeding, rehabilitation and egg removal?	
Unknown effectiveness (limited evidence)	• Rehabilitate injured birds • Remove eggs from wild nests to increase reproductive output • Use artificial visual and auditory stimuli to induce breeding in wild populations

Unknown effectiveness (limited evidence)

Rehabilitate injured birds

Two studies of four studies from the UK and USA found that 25–40% of injured birds taken in by centres were rehabilitated and released. Three studies from the USA found that rehabilitated birds appeared to have high survival. One found that mortality rates were higher for owls than raptors. *Assessment: unknown effectiveness — limited evidence (effectiveness 36%; certainty 30%; harms 0%).*

http://www.conservationevidence.com/actions/476

Remove eggs from wild nests to increase reproductive output

A study from Canada found that whooping crane reproductive success was higher for nests with one or two eggs removed than for controls. A study from the USA found that removing bald eagle eggs did not appear to affect the wild population and a replicated study from Mauritius found that removing entire Mauritius kestrel clutches appeared to increase productivity more than removing individual eggs as they were laid. *Assessment: unknown effectiveness — limited evidence (effectiveness 24%; certainty 25%; harms 5%).*

http://www.conservationevidence.com/actions/477

Use artificial visual and auditory stimuli to induce breeding in wild populations

A small study from the British Virgin Islands found an increase in breeding behaviour after the introduction of visual and auditory stimulants. *Assessment: unknown effectiveness — limited evidence (effectiveness 19%; certainty 11%; harms 0%).*

http://www.conservationevidence.com/actions/475

3.14.2 Provide artificial nesting sites

Based on the collated evidence, what is the current assessment of the effectiveness of interventions for providing artificial nesting sites?	
Beneficial	• Provide artificial nests: falcons • Provide artificial nests: owls • Provide artificial nests: songbirds • Provide artificial nests: wildfowl
Likely to be beneficial	• Clean artificial nests to increase occupancy or reproductive success • Provide artificial nests: burrow-nesting seabirds • Provide artificial nests: divers/loons • Provide artificial nests: ground- and tree-nesting seabirds • Provide artificial nests: oilbirds • Provide artificial nests: raptors • Provide artificial nests: wildfowl — artificial/ floating islands

Unknown effectiveness (limited evidence)	• Artificially incubate eggs or warm nests • Guard nests • Provide artificial nests: gamebirds • Provide artificial nests: grebes • Provide artificial nests: ibises and flamingos • Provide artificial nests: parrots • Provide artificial nests: pigeons • Provide artificial nests: rails • Provide artificial nests: rollers • Provide artificial nests: swifts • Provide artificial nests: trogons • Provide artificial nests: waders • Provide artificial nests: woodpeckers • Provide nesting habitat for birds that is safe from extreme weather • Provide nesting material for wild birds • Remove vegetation to create nesting areas • Repair/support nests to support breeding • Use differently-coloured artificial nests

Beneficial

Provide artificial nests (falcons)

Four studies from the USA and Europe found that local populations of falcons increased following the installation of artificial nesting sites. However, a study from Canada found no increase in the local population of falcons following the erection of nest boxes. Eight studies from across the world found that the success and productivity of falcons in nest boxes was higher than or equal to those in natural nests. Four studies from across the world found that productivity in nest boxes was lower than in natural nests, or that some falcons were evicted from their nests by owls. Four studies from across the world found no differences in productivity between nest box designs or positions, whilst two from Spain and Israel found that productivity in boxes varied between designs and habitats. Twenty-one studies from across the world found nest boxes were used by falcons, with one in the UK finding that nest boxes were not used at all. Seven studies found that position or design affected use, whilst three found no differences between design or positioning. *Assessment: beneficial (effectiveness 65%; certainty 65%; harms 0%).*

http://www.conservationevidence.com/actions/489

Provide artificial nests (owls)

Three studies from the UK appeared to show increases in local populations of owls following the installation of artificial nests. Another UK study found that providing nesting sites when renovating buildings maintained owl populations, whilst they declined at sites without nests. Four studies from the USA and the UK found high levels of breeding success in artificial nests. Two studies from the USA and Hungary found lower productivity or fledgling survival from breeding attempts in artificial nests, whilst a study from Finland found that artificial nests were only successful in the absence of larger owls. Four studies from the USA and Europe found that artificial nests were used as frequently as natural sites. Five studies from across the world found that owls used artificial nests. Seven studies found that nest position or design affected occupancy or productivity. However four studies found occupancy and/or productivity did not differ between different designs of nest box. *Assessment: beneficial (effectiveness 65%; certainty 66%; harms 5%).*

http://www.conservationevidence.com/actions/490

Provide artificial nests (songbirds)

Only three out of 66 studies from across the world found low rates of nest box occupancy in songbirds. Low rates of use were seen in thrushes, crows, swallows and New World warblers. Thrushes, crows, finches, swallows, wrens, tits, Old World and tyrant flycatchers, New World blackbirds, sparrows, waxbills, starlings and ovenbirds all used nest boxes. Five studies from across the world found higher population densities or growth rates, and one study from the USA found higher species richness, in areas with nest boxes. Twelve studies from across the world found that productivity in nest boxes was higher than or similar to natural nests. One study found there were more nesting attempts in areas with more nest boxes, although a study from Canada found no differences in productivity between areas with different nest box densities. Two studies from Europe found lower predation of species using nest boxes but three studies from the USA found low production in nest boxes. Thirteen studies from across the world found that use, productivity or usurpation rate varied with nest box design, whilst seven found no difference in occupation rates or success between different designs. Similarly, fourteen studies found different occupation or success rates depending on the position of artificial nest sites but two studies found no such differences. *Assessment: beneficial (effectiveness 67%; certainty 85%; harms 0%).*

http://www.conservationevidence.com/actions/498

Provide artificial nests (wildfowl)

Six studies from North America and Europe found that wildfowl populations increased with the provision of artificial nests, although one study from Finland found no increase in productivity in areas with nest boxes. Nine out of twelve studies from North America found that productivity was high in artificial nests. Two studies found that success for some species in nest boxes was lower than for natural nests. Nineteen studies from across the world found that occupancy rates varied from no use to 100% occupancy. Two studies found that occupancy rates were affected by design or positioning. Three studies from North America found that nest boxes could have other impacts on reproduction and behaviour. *Assessment: beneficial (effectiveness 62%; certainty 76%; harms 0%).*

http://www.conservationevidence.com/actions/482

Likely to be beneficial

Clean artificial nests to increase occupancy or reproductive success

Five out of ten studies from North America and Europe found that songbirds preferentially nested in cleaned nest boxes or those sterilised using microwaves, compared to used nest boxes. One study found that the preference was not strong enough for birds to switch nest boxes after they were settled. One study found that birds avoided heavily-soiled nest boxes. Two studies birds had a preference for used nest boxes and one found no preference for cleaned or uncleaned boxes. None of the five studies that examined it found any effect of nest box cleanliness on nesting success or parasitism levels. *Assessment: likely to be beneficial (effectiveness 40%; certainty 40%; harms 15%).*

http://www.conservationevidence.com/actions/499

Provide artificial nests (burrow-nesting seabirds)

Four studies from across the world found population increases or population establishment following the provision of nest boxes. In two cases this was combined with other interventions. Six studies from across the world found high occupancy rates for artificial burros by seabirds but three studies from across the world found very low occupancy rates for artificial burrows used by petrels. Eight studies from across the world found that the productivity of birds in artificial burrows was high although two studies from the USA and the Galapagos found low productivity in petrels. *Assessment: likely to be beneficial (effectiveness 60%; certainty 71%; harms 5%).*

http://www.conservationevidence.com/actions/481

Provide artificial nests (divers/loons)

Three studies from the UK and the USA found increases in loon productivity on lakes provided with nesting rafts. A study in the UK found that usage of nesting rafts varied between sites. *Assessment: likely to be beneficial (effectiveness 50%; certainty 50%; harms 0%).*

http://www.conservationevidence.com/actions/478

Provide artificial nests (ground- and tree-nesting seabirds)

Three studies from the UK and the Azores found increases in gull and tern populations following the provision of rafts/islands or nest boxes alongside other interventions. Five studies from Canada and Europe found that terns used artificial nesting sites. A study from the USA found that terns had higher nesting success on artificial rafts in some years and a study from Japan found increased nesting success after provision of nesting substrate. Design of nesting structure should be considered. *Assessment: likely to be beneficial (effectiveness 60%; certainty 49%; harms 0%).*

http://www.conservationevidence.com/actions/480

Provide artificial nests (oilbirds)

A study in Trinidad and Tobago found an increase in the size of an oilbird colony after the creation of artificial nesting lodges. *Assessment: likely to be beneficial (effectiveness 50%; certainty 45%; harms 0%).*

http://www.conservationevidence.com/actions/491

Provide artificial nests (raptors)

Nine studies from North America and Spain found that raptors used artificial nesting platforms. Two studies from the USA found increases in populations or densities following the installation of platforms. Three studies describe successful use of platforms but three found lower productivity or failed nesting attempts, although these studies only describe a single nesting attempt. *Assessment: likely to be beneficial (effectiveness 55%; certainty 55%; harms 0%).*

http://www.conservationevidence.com/actions/488

Provide artificial nests (wildfowl — artificial/floating islands)

Two studies from North America found that wildfowl used artificial islands and floating rafts and had high nesting success. A study in the UK found that wildfowl preferentially nested on vegetated rather than bare islands. *Assessment: likely to be beneficial (effectiveness 45%; certainty 45%; harms 0%).*

http://www.conservationevidence.com/actions/483

Unknown effectiveness (limited evidence)

Artificially incubate eggs or warm nests

One of two studies found that no kakapo chicks or eggs died of cold when they were artificially warmed when females left the nest. A study from the UK found that great tits were less likely to interrupt their laying sequence if their nest boxes were warmed, but there was no effect on egg or clutch size. *Assessment: unknown effectiveness — limited evidence (effectiveness 26%; certainty 16%; harms 0%).*

http://www.conservationevidence.com/actions/503

Guard nests

We captured four studies describing the effects of guarding nests. One, from Costa Rica, found an increase in scarlet macaw population after nest monitoring and several other interventions. Two studies from Puerto Rico and New Zealand found that nest success was higher, or mortality lower, when nests were monitored. A study from New Zealand found that nest success was high overall when nests were monitored. *Assessment: unknown effectiveness — limited evidence (effectiveness 41%; certainty 24%; harms 0%).*

http://www.conservationevidence.com/actions/506

Provide artificial nests (gamebirds)

A study in China found that approximately 40% of the local population of Cabot's tragopans used nesting platforms. *Assessment: unknown effectiveness — limited evidence (effectiveness 40%; certainty 13%; harms 0%).*

http://www.conservationevidence.com/actions/484

Provide artificial nests (grebes)

A study from the UK found that grebes used nesting rafts in some areas but not others. *Assessment: unknown effectiveness — limited evidence (effectiveness 10%; certainty 9%; harms 0%).*

http://www.conservationevidence.com/actions/479

Provide artificial nests (ibises and flamingos)

A study from Turkey found that ibises moved to a site with artificial breeding ledges. A study from Spain and France found that large numbers of flamingos used artificial nesting islands. *Assessment: unknown effectiveness — limited evidence (effectiveness 42%; certainty 31%; harms 0%).*

http://www.conservationevidence.com/actions/487

Provide artificial nests (parrots)

A study from Costa Rica found that the local population of scarlet macaws increased following the installation of nest boxes along with several other interventions. Five studies from South and Central America and Mauritius found that nest boxes were used by several species of parrots. One study from Peru found that blue-and-yellow macaws only used modified palms, not 'boxes', whilst another study found that scarlet macaws used both PVC and wooden boxes. Four studies from Venezuela and Columbia found that several species rarely, if ever, used nest boxes. Six studies from Central and South America found that parrots nested successfully in nest boxes, with two species showing higher levels of recruitment into the population following nest box erection and another finding that success rates for artificial nests were similar to natural nests. Three studies from South America found that artificial nests had low success rates, in two cases due to poaching. *Assessment: unknown effectiveness — limited evidence (effectiveness 25%; certainty 38%; harms 11%).*

http://www.conservationevidence.com/actions/497

Provide artificial nests (pigeons)

Two studies from the USA and the Netherlands found high use rates and high nesting success of pigeons and doves using artificial nests. *Assessment: unknown effectiveness — limited evidence (effectiveness 30%; certainty 16%; harms 0%).*

http://www.conservationevidence.com/actions/492

Provide artificial nests (rails)

A study from the UK found that common moorhens and common coot readily used artificial nesting islands. *Assessment: unknown effectiveness — limited evidence (effectiveness 30%; certainty 11%; harms 0%).*

http://www.conservationevidence.com/actions/485

Provide artificial nests (rollers)

A study from Spain found that the use of nest boxes by rollers increased over time and varied between habitats. Another study from Spain found no difference in success rates between new and old nest boxes. *Assessment: unknown effectiveness — limited evidence (effectiveness 20%; certainty 20%; harms 0%).*

http://www.conservationevidence.com/actions/494

Provide artificial nests (swifts)

A study from the USA found that Vaux's swifts successfully used nest boxes provided. *Assessment: unknown effectiveness — limited evidence (effectiveness 25%; certainty 16%; harms 0%).*

http://www.conservationevidence.com/actions/495

Provide artificial nests (trogons)

A small study from Guatemala found that at least one resplendent quetzal nested in nest boxes provided. *Assessment: unknown effectiveness — limited evidence (effectiveness 19%; certainty 11%; harms 0%).*

http://www.conservationevidence.com/actions/493

Provide artificial nests (waders)

Two studies from the USA and the UK found that waders used artificial islands and nesting sites. *Assessment: unknown effectiveness — limited evidence (effectiveness 25%; certainty 20%; harms 0%).*

http://www.conservationevidence.com/actions/486

Provide artificial nests (woodpeckers)

Four studies from the USA found local increases in red-cockaded woodpecker populations or the successful colonisation of new areas following the installation of 'cavity inserts'. One study also found that the productivity of birds using the inserts was higher than the regional average. Two studies from the USA found that red-cockaded woodpeckers used cavity inserts, in one case more frequently than making their own holes or using natural cavities. One study from the USA found that woodpeckers roosted, but did not nest, in nest boxes. Five studies from the USA found that some woodpeckers excavated holes in artificial snags but only roosted in excavated holes or nest boxes. *Assessment: unknown effectiveness — limited evidence (effectiveness 35%; certainty 39%; harms 0%).*

http://www.conservationevidence.com/actions/496

Provide nesting habitat for birds that is safe from extreme weather

Two of three studies found that nesting success of waders and terns was no higher on raised areas of nesting substrate, with one finding that similar numbers were lost to flooding. The third study found that Chatham Island oystercatchers used raised nest platforms, but did not report on nesting success. *Assessment: unknown effectiveness — limited evidence (effectiveness 28%; certainty 23%; harms 0%).*

http://www.conservationevidence.com/actions/504

Provide nesting material for wild birds

One of two studies found that wild birds took nesting material provided; the other found only very low rates of use. *Assessment: unknown effectiveness — limited evidence (effectiveness 11%; certainty 9%; harms 0%).*

http://www.conservationevidence.com/actions/501

Remove vegetation to create nesting areas

Two out of six studies found increases in population sizes at seabird and wader colonies after vegetation was cleared and a third found that an entire colony moved to a new site that was cleared of vegetation. Two of these studies found that several interventions were used at once. Two studies found that gulls and terns used plots cleared of vegetation, one of these found that nesting densities were higher on partially-cleared plots than totally cleared, or uncleared, plots. One study found that tern nesting success was higher on plots after they were cleared of vegetation and other interventions were used. *Assessment: unknown effectiveness — limited evidence (effectiveness 45%; certainty 28%; harms 10%).*

http://www.conservationevidence.com/actions/505

Repair/support nests to support breeding

A study from Puerto Rico found that no chicks died from chilling after nine nests were repaired to prevent water getting in. *Assessment: unknown effectiveness — limited evidence (effectiveness 20%; certainty 10%; harms 0%).*

http://www.conservationevidence.com/actions/502

Use differently-coloured artificial nests

A study from the USA found that two bird species (a thrush and a pigeon) both showed colour preferences for artificial nests, but that these preferences differed between species. In each case, clutches in the preferred colour nest were less successful than those in the other colour. *Assessment: unknown effectiveness — limited evidence (effectiveness 3%; certainty 9%; harms 0%).*

http://www.conservationevidence.com/actions/500

3.14.3 Foster chicks in the wild

Based on the collated evidence, what is the current assessment of the effectiveness of interventions for fostering chicks in the wild?	
Likely to be beneficial	• Foster eggs or chicks with wild conspecifics: raptors • Foster eggs or chicks with wild non-conspecifics (cross-fostering): songbirds
Unknown effectiveness (limited evidence)	• Foster eggs or chicks with wild conspecifics: bustards • Foster eggs or chicks with wild conspecifics: cranes • Foster eggs or chicks with wild conspecifics: gannets and boobies • Foster eggs or chicks with wild conspecifics: owls • Foster eggs or chicks with wild conspecifics: parrots • Foster eggs or chicks with wild conspecifics: vultures • Foster eggs or chicks with wild conspecifics: waders

	• Foster eggs or chicks with wild conspecifics: woodpeckers • Foster eggs or chicks with wild non-conspecifics (cross-fostering): cranes • Foster eggs or chicks with wild non-conspecifics (cross-fostering): ibises • Foster eggs or chicks with wild non-conspecifics (cross-fostering): petrels and shearwaters • Foster eggs or chicks with wild non-conspecifics (cross-fostering): waders

Likely to be beneficial

Foster eggs or chicks with wild conspecifics (raptors)

Ten out of 11 studies from across the world found that fostering raptor chicks to wild conspecifics had high success rates. A single study from the USA found that only one of six eggs fostered to wild eagle nests hatched and was raised. A study from Spain found that Spanish imperial eagle chicks were no more likely to survive to fledging if they were transferred to foster nests from three chick broods (at high risk from siblicide). A study from Spain found that young (15–20 day old) Montagu's harrier chicks were successfully adopted, but three older (27–29 day old) chicks were rejected. *Assessment: likely to be beneficial (effectiveness 70%; certainty 60%; harms 10%).*

http://www.conservationevidence.com/actions/510

Foster eggs or chicks with wild non-conspecifics (cross-fostering) (songbirds)

A study from the USA found that the survival of cross-fostered yellow warbler chicks was lower than previously-published rates for the species. A study from Norway found that the success of cross-fostering small songbirds varied depending on the species of chick and foster birds but recruitment was the same or higher than control chicks. The pairing success of cross-fostered chicks varied depending on species of chick and foster birds. *Assessment: likely to be beneficial (effectiveness 45%; certainty 45%; harms 10%).*

http://www.conservationevidence.com/actions/520

Unknown effectiveness (limited evidence)

Foster eggs or chicks with wild conspecifics (bustards)

A small study in Saudi Arabia found that a captive-bred egg was successfully fostered to a female in the wild. *Assessment: unknown effectiveness — limited evidence (effectiveness 20%; certainty 5%; harms 0%).*

http://www.conservationevidence.com/actions/513

Foster eggs or chicks with wild conspecifics (cranes)

A small study in Canada found high rates of fledging for whooping crane eggs fostered to first time breeders. *Assessment: unknown effectiveness — limited evidence (effectiveness 26%; certainty 11%; harms 0%).*

http://www.conservationevidence.com/actions/512

Foster eggs or chicks with wild conspecifics (gannets and boobies)

A small study in Australia found that gannet chicks were lighter, and hatching and fledging success lower in nests which had an extra egg or chick added. However, overall productivity was non-significantly higher in experimental nests. *Assessment: unknown effectiveness — limited evidence (effectiveness 9%; certainty 11%; harms 0%).*

http://www.conservationevidence.com/actions/507

Foster eggs or chicks with wild conspecifics (owls)

A study in the USA found high fledging rates for barn owl chicks fostered to wild pairs. A study from Canada found that captive-reared burrowing owl chicks fostered to wild nests did not have lower survival or growth rates than wild chicks. *Assessment: unknown effectiveness — limited evidence (effectiveness 35%; certainty 21%; harms 0%).*

http://www.conservationevidence.com/actions/511

Foster eggs or chicks with wild conspecifics (parrots)

A study from Venezuela found that yellow-shouldered Amazon chicks had high fledging rates when fostered to conspecific nests in the wild. A second study from Venezuela found lower poaching rates of yellow-shouldered Amazons when chicks were moved to foster nests closer to a field base. *Assessment: unknown effectiveness — limited evidence (effectiveness 30%; certainty 14%; harms 0%).*

http://www.conservationevidence.com/actions/515

Foster eggs or chicks with wild conspecifics (vultures)

Two small studies in Italy and the USA found that single chicks were successfully adopted by foster conspecifics, although in one case this led to the death of one of the foster parents' chicks. *Assessment: unknown effectiveness — limited evidence (effectiveness 30%; certainty 15%; harms 41%).*

http://www.conservationevidence.com/actions/509

Foster eggs or chicks with wild conspecifics (waders)

Two small trials in North America found that piping plovers accepted chicks introduced into their broods, although in one case the chick died. A study from New Zealand found that survival of fostered black stilts was higher for birds fostered to conspecifics rather than a closely related species. *Assessment: unknown effectiveness — limited evidence (effectiveness 29%; certainty 9%; harms 0%).*

http://www.conservationevidence.com/actions/508

Foster eggs or chicks with wild conspecifics (woodpeckers)

Three studies from the USA found that red-cockaded woodpecker chicks fostered to conspecifics had high fledging rates. One small study found that fostered chicks survived better than chicks translocated with their parents. *Assessment: unknown effectiveness — limited evidence (effectiveness 41%; certainty 29%; harms 0%).*

http://www.conservationevidence.com/actions/514

Foster eggs or chicks with wild non-conspecifics (cross-fostering) (cranes)

Two studies from the USA found low fledging success for cranes fostered to non-conspecifics' nests. *Assessment: unknown effectiveness — limited evidence (effectiveness 14%; certainty 35%; harms 10%).*

http://www.conservationevidence.com/actions/519

Foster eggs or chicks with wild non-conspecifics (cross-fostering) (ibises)

A 2007 literature review describes attempting to foster northern bald ibis chicks with cattle egrets as unsuccessful. *Assessment: unknown effectiveness — limited evidence (effectiveness 0%; certainty 10%; harms 0%).*

http://www.conservationevidence.com/actions/518

Foster eggs or chicks with wild non-conspecifics (cross-fostering) (petrels and shearwaters)

A study from Hawaii found that Newell's shearwater eggs fostered to wedge-tailed shearwater nests had high fledging rates. *Assessment: unknown effectiveness — limited evidence (effectiveness 45%; certainty 6%; harms 0%).*

http://www.conservationevidence.com/actions/516

Foster eggs or chicks with wild non-conspecifics (cross-fostering) (waders)

A study from the USA found that killdeer eggs incubated and raised by spotted sandpipers had similar fledging rates to parent-reared birds. A study from New Zealand found that cross-fostering black stilt chicks to black-winged stilt nests increased nest success, but cross-fostered chicks had lower success than chicks fostered to conspecifics' nests. *Assessment: unknown effectiveness — limited evidence (effectiveness 35%; certainty 30%; harms 0%).*

http://www.conservationevidence.com/actions/517

3.14.4 Provide supplementary food

Based on the collated evidence, what is the current assessment of the effectiveness of interventions for providing supplementary food?	
Beneficial	• Provide supplementary food to increase adult survival: songbirds
Likely to be beneficial	• Place feeders close to windows to reduce collisions • Provide calcium supplements to increase survival or reproductive success • Provide supplementary food to increase adult survival: cranes • Provide supplementary food to increase reproductive success: gulls, terns and skuas • Provide supplementary food to increase reproductive success: owls • Provide supplementary food to increase reproductive success: raptors • Provide supplementary food to increase reproductive success: songbirds

Unknown effectiveness (limited evidence)	• Provide perches to improve foraging success • Provide supplementary food through the establishment of food populations • Provide supplementary food to allow the rescue of a second chick • Provide supplementary food to increase adult survival: gamebirds • Provide supplementary food to increase adult survival: gulls, terns and skuas • Provide supplementary food to increase adult survival: hummingbirds • Provide supplementary food to increase adult survival: nectar-feeding songbirds • Provide supplementary food to increase adult survival: pigeons • Provide supplementary food to increase adult survival: raptors • Provide supplementary food to increase adult survival: vultures • Provide supplementary food to increase adult survival: waders • Provide supplementary food to increase adult survival: wildfowl • Provide supplementary food to increase adult survival: woodpeckers • Provide supplementary food to increase reproductive success: auks • Provide supplementary food to increase reproductive success: gamebirds • Provide supplementary food to increase reproductive success: gannets and boobies • Provide supplementary food to increase reproductive success: ibises • Provide supplementary food to increase reproductive success: kingfishers • Provide supplementary food to increase reproductive success: parrots • Provide supplementary food to increase reproductive success: petrels

	• Provide supplementary food to increase reproductive success: pigeons • Provide supplementary food to increase reproductive success: rails and coots • Provide supplementary food to increase reproductive success: vultures • Provide supplementary food to increase reproductive success: waders • Provide supplementary food to increase reproductive success: wildfowl • Provide supplementary water to increase survival or reproductive success

Beneficial

Provide supplementary food to increase adult survival (songbirds)

Seven studies from Europe and the USA found higher densities or larger populations of songbird species in areas close to supplementary food. Six studies from Europe, Canada and Japan found that population trends or densities were no different between fed and unfed areas. Four studies from around the world found that birds had higher survival when supplied with supplementary food. However, in two studies this was only apparent in some individuals or species and one study from the USA found that birds with feeding stations in their territories had lower survival. Six studies from Europe and the USA found that birds supplied with supplementary food were in better physical condition than unfed birds. However, in four studies this was only true for some individuals, species or seasons. Two studies investigated the effect of feeding on behaviours: one in the USA found that male birds spent more time singing when supplied with food and one in Sweden found no behavioural differences between fed and unfed birds. Thirteen studies from the UK, Canada and the USA investigated use of feeders. Four studies from the USA and the UK found high use of supplementary food, with up to 21% of birds' daily energy needs coming from feeders. However, another UK study found very low use of food. The timing of peak feeder use varied. Two trials from the UK found that the use of feeders increased with distance to houses and decreased with distance to cover. Two studies in Canada and the

UK, found that preferences for feeder locations and positions varies between species. *Assessment: beneficial (effectiveness 65%; certainty 75%; harms 0%).*

http://www.conservationevidence.com/actions/552

Likely to be beneficial

Place feeders close to windows to reduce collisions

A randomised, replicated and controlled study in the USA found that fewer birds hit windows, and fewer were killed, when feeders were placed close to windows, compared to when they were placed further away. *Assessment: likely to be beneficial (effectiveness 44%; certainty 43%; harms 0%).*

http://www.conservationevidence.com/actions/557

Provide calcium supplements to increase survival or reproductive success

Eight of 13 studies (including a literature review) from across the world found some positive effects of calcium provisioning on birds' productivites (six studies) or health (two studies). Six studies (including the review) found no evidence for positive effects on some of the species studied. One study from Europe found that birds at polluted sites took more calcium supplement than those at cleaner sites. *Assessment: likely to be beneficial (effectiveness 55%; certainty 50%; harms 0%).*

http://www.conservationevidence.com/actions/559

Provide supplementary food to increase adult survival (cranes)

A study from Japan and a global literature review found that local crane populations increased after the provision of supplementary food. *Assessment: likely to be beneficial (effectiveness 40%; certainty 40%; harms 15%).*

http://www.conservationevidence.com/actions/547

Provide supplementary food to increase reproductive success (gulls, terns and skuas)

Four studies of three experiments from Europe and Alaska found that providing supplementary food increased fledging success or chick survival in two gull species, although a study from the UK found that this was only true for one of two islands. One study from the Antarctic found no effect of feeding parent skuas on productivity. One study from Alaska found increased chick growth when parents were fed but a study from the Antarctic found no such increase. *Assessment: likely to be beneficial (effectiveness 42%; certainty 41%; harms 0%).*

http://www.conservationevidence.com/actions/525

Provide supplementary food to increase reproductive success (owls)

Two replicated, controlled trials from Europe and the USA found that owls supplied with supplementary food had higher hatching and fledging rates. The European study, but not the American, also found that fed pairs laid earlier and had larger clutches. The study in the USA also found that owls were no more likely to colonise nest boxes provided with supplementary food. *Assessment: likely to be beneficial (effectiveness 50%; certainty 42%; harms 0%).*

http://www.conservationevidence.com/actions/533

Provide supplementary food to increase reproductive success (raptors)

A small study in Italy described a small increase in local kite populations following the installation of a feeding station. Four European studies found that kestrels and Eurasian sparrowhawks laid earlier than control birds when supplied with supplementary food. Three studies from the USA and Europe found higher chick survival or condition when parents were supplied with food, whilst three from Europe found fed birds laid larger clutches and another found that fed male hen harriers bred with more females than control birds. Four studies from across the world found no evidence that feeding increased breeding frequency, clutch size, laying date, eggs size or hatching or fledging success. A study from Mauritius found uncertain effects of feeding on Mauritius kestrel reproduction. There was some evidence that the impact of feeding was lower in years with peak numbers of prey species. *Assessment: likely to be beneficial (effectiveness 55%; certainty 52%; harms 0%).*

http://www.conservationevidence.com/actions/532

Provide supplementary food to increase reproductive success (songbirds)

Two studies from the USA found evidence for higher population densities of magpies and American blackbirds in areas provided with supplementary food, whilst two studies from the UK and Canada found that population densities were not affected by feeding. Twelve studies from across the world found that productivity was higher for fed birds than controls. Eleven studies from Europe and the USA found that fed birds had the same, or even lower, productivity or chick survival than control birds. Nine studies from Europe and North America found that the eggs of fed birds were larger or heavier, or that the chicks of fed birds were in better physical condition. However, eight studies from across the world found no evidence for better condition or increased size in the eggs or chicks of fed birds. Six studies from across the world found that food-supplemented pairs laid larger clutches, whilst 14 studies from Europe and North America found that fed birds did not lay larger clutches. Fifteen studies from across the world found that birds supplied with supplementary food began nesting earlier than controls, although in two cases only certain individuals, or those in particular habitats, laid earlier. One study found that fed birds had shorter incubations than controls whilst another found that fed birds re-nested quicker and had shorter second incubations. Four studies from the USA and Europe found that fed birds did not lay any earlier than controls. Seven studies from across the world found that fed parent birds showed positive behavioural responses to feeding. However, three studies from across the world found neutral or negative responses to feeding. *Assessment: likely to be beneficial (effectiveness 51%; certainty 85%; harms 6%).*

http://www.conservationevidence.com/actions/537

Unknown effectiveness (limited evidence)

Provide perches to improve foraging success

One of four studies, from Sweden, found that raptors used clearcuts provided with perches more than clearcuts without perches. Two studies found that birds used perches provided, but a controlled study from the USA found that shrikes did not alter foraging behaviour when perches were present. *Assessment: unknown effectiveness — limited evidence (effectiveness 45%; certainty 30%; harms 0%).*

http://www.conservationevidence.com/actions/556

Provide supplementary food through the establishment of food populations

One of four studies that established prey populations found that wildfowl fed on specially-planted rye grass. Two studies found that cranes in the USA and owls in Canada did not respond to established prey populations. A study from Sweden found that attempts to increase macroinvertebrate numbers for wildfowl did not succeed. *Assessment: unknown effectiveness — limited evidence (effectiveness 9%; certainty 26%; harms 0%)..*

http://www.conservationevidence.com/actions/555

Provide supplementary food to allow the rescue of a second chick

A study from Spain found that second chicks from lammergeier nests survived longer if nests were provided with food, in one case allowing a chick to be rescued. *Assessment: unknown effectiveness — limited evidence (effectiveness 15%; certainty 14%; harms 0%).*

http://www.conservationevidence.com/actions/541

Provide supplementary food to increase adult survival (gamebirds)

Two European studies found increased numbers of birds in fed areas, compared to unfed areas. There was only an increase in the overall population in the study area in one of these studies. Of four studies in the USA on northern bobwhites, one found that birds had higher overwinter survival in fed areas, one found lower survival, one found fed birds had higher body fat percentages and a literature review found no overall effect of feeding. *Assessment: unknown effectiveness — limited evidence (effectiveness 49%; certainty 38%; harms 0%).*

http://www.conservationevidence.com/actions/544

Provide supplementary food to increase adult survival (gulls, terns and skuas)

A study in the Antarctic found that fed female south polar skuas lost more weight whilst feeding two chicks than unfed birds. There was no difference for birds with single chicks, or male birds. *Assessment: unknown effectiveness — limited evidence (effectiveness 0%; certainty 20%; harms 10%).*

http://www.conservationevidence.com/actions/548

Provide supplementary food to increase adult survival (hummingbirds)

Four studies from the USA found that three species of hummingbird preferred higher concentrations of sucrose, consuming more and visiting feeders more frequently. A study from the USA found that hummingbirds preferentially fed on sugar solutions over artificial sweeteners, and that the viscosity of these solutions did not affect their consumption. Two studies from Mexico and Argentina found that four species showed preferences for sucrose over fructose or glucose and sucrose over a sucrose-glucose mix, but no preference for sucrose over a glucose-fructose mix. A study from the USA found that birds showed a preference for red-dyed sugar solutions over five other colours. A study from the USA found that rufous hummingbirds preferentially fed on feeders that were placed higher. *Assessment: unknown effectiveness — limited evidence (effectiveness 10%; certainty 24%; harms 0%).*

http://www.conservationevidence.com/actions/550

Provide supplementary food to increase adult survival (nectar-feeding songbirds)

Two studies from Australia and New Zealand found that ten species of honeyeaters and stitchbirds readily used feeders supplying sugar solutions, with seasonal variations varying between species. A series of *ex situ* trials using southern African birds found that most species preferred sucrose solutions over glucose or fructose. One study found that sunbirds and sugarbirds only showed such a preference at low concentrations. Two studies found that two species showed preferences for sucrose when comparing 20% solutions, although a third species did not show this preference. All species rejected solutions with xylose added. A final study found that sucrose preferences were only apparent at equicalorific concentrations high enough for birds to subsist on. *Assessment: unknown effectiveness — limited evidence (effectiveness 10%; certainty 23%; harms 0%).*

http://www.conservationevidence.com/actions/553

Provide supplementary food to increase adult survival (pigeons)

The first of two studies of a recently-released pink pigeon population on Mauritius found that fewer than half the birds took supplementary food. However, the later study found that almost all birds used supplementary feeders. *Assessment: unknown effectiveness — limited evidence (effectiveness 10%; certainty 19%; harms 0%).*

http://www.conservationevidence.com/actions/549

Provide supplementary food to increase adult survival (raptors)

Two studies in the USA found that nesting northern goshawks were significantly heavier in territories supplied with supplementary food, compared with those from unfed territories. *Assessment: unknown effectiveness— limited evidence (effectiveness 30%; certainty 23%; harms 0%).*

http://www.conservationevidence.com/actions/546

Provide supplementary food to increase adult survival (vultures)

A study from Spain found a large increase in griffon vulture population in the study area following multiple interventions including supplementary feeding. Two studies from the USA and Israel found that vultures fed on the carcasses provided for them. In the study in Israel vultures were sometimes dominated by larger species at a feeding station supplied twice a month, but not at one supplied every day. *Assessment: unknown effectiveness — limited evidence (effectiveness 18%; certainty 18%; harms 0%).*

http://www.conservationevidence.com/actions/545

Provide supplementary food to increase adult survival (waders)

A study in Northern Ireland found that waders fed on millet seed when provided, but were dominated by other ducks when larger seeds were provided. *Assessment: unknown effectiveness — limited evidence (effectiveness 22%; certainty 9%; harms 0%).*

http://www.conservationevidence.com/actions/543

Provide supplementary food to increase adult survival (wildfowl)

Two studies from Canada and Northern Ireland found that five species of wildfowl readily consumed supplementary grains and seeds. The Canadian study found that fed birds were heavier and had larger hearts or flight muscles or more body fat than controls. *Assessment: unknown effectiveness — limited evidence (effectiveness 14%; certainty 15%; harms 0%).*

http://www.conservationevidence.com/actions/542

Provide supplementary food to increase adult survival (woodpeckers)

One replicated, controlled study from the USA found that 12 downy woodpeckers supplied with supplementary food had higher nutritional statuses than unfed birds. However, two analyses of a replicated, controlled study of 378 downy woodpeckers from the USA found that they did not have higher survival rates or nutritional statuses than unfed birds. *Assessment: unknown effectiveness — limited evidence (effectiveness 10%; certainty 30%; harms 0%).*

http://www.conservationevidence.com/actions/551

Provide supplementary food to increase reproductive success (auks)

Two replicated studies from the UK found that Atlantic puffin chicks provided with supplementary food were significantly heavier than control chicks, but fed chicks fledged at the same time as controls. A randomised, replicated and controlled study from Canada found that tufted puffin chicks supplied with supplementary food fledged later than controls and that fed chicks had faster growth by some, but not all, metrics. *Assessment: unknown effectiveness — limited evidence (effectiveness 30%; certainty 38%; harms 0%).*

http://www.conservationevidence.com/actions/524

Provide supplementary food to increase reproductive success (gamebirds)

A controlled study in Tibet found that Tibetan eared pheasants fed supplementary food laid significantly larger eggs and clutches than control birds. Nesting success and laying dates were not affected. *Assessment: unknown effectiveness — limited evidence (effectiveness 23%; certainty 10%; harms 0%).*

http://www.conservationevidence.com/actions/527

Provide supplementary food to increase reproductive success (gannets and boobies)

A small controlled study in Australia found that Australasian gannet chicks were significantly heavier if they were supplied with supplementary food, but only in one of two years. Fledging success of fed nests was also higher, but not significantly so. A randomised replicated and controlled study in the

Galapagos Islands found that fed female Nazca boobies were more likely to produce two-egg clutches, and that second eggs were significantly heavier. *Assessment: unknown effectiveness — limited evidence (effectiveness 33%; certainty 25%; harms 0%).*

http://www.conservationevidence.com/actions/523

Provide supplementary food to increase reproductive success (ibises)

A study from China found that breeding success of crested ibis was correlated with the amount of supplementary food provided, although no comparison was made with unfed nests. *Assessment: unknown effectiveness — limited evidence (effectiveness 25%; certainty 11%; harms 0%).*

http://www.conservationevidence.com/actions/530

Provide supplementary food to increase reproductive success (kingfishers)

A controlled study in the USA found that belted kingfishers supplied with food had heavier nestlings and were more likely to renest. There was mixed evidence for the effect of feeding on laying date. *Assessment: unknown effectiveness — limited evidence (effectiveness 33%; certainty 13%; harms 0%).*

http://www.conservationevidence.com/actions/534

Provide supplementary food to increase reproductive success (parrots)

Two studies from New Zealand found evidence that providing supplementary food for kakapos increased the number of breeding attempts made, whilst a third study found that birds provided with specially-formulated pellets appeared to have larger clutches than those fed on nuts. One study found no evidence that providing food increased the number of nesting attempts. *Assessment: unknown effectiveness — limited evidence (effectiveness 33%; certainty 11%; harms 0%).*

http://www.conservationevidence.com/actions/536

Provide supplementary food to increase reproductive success (petrels)

A replicated controlled study in Australia found that Gould's petrel chicks provided with supplementary food had similar fledging rates to both control and hand-reared birds, but were significantly heavier than other birds. *Assessment: unknown effectiveness — limited evidence (effectiveness 19%; certainty 14%; harms 0%).*

http://www.conservationevidence.com/actions/522

Provide supplementary food to increase reproductive success (pigeons)

A study in the UK found no differences in reproductive parameters of European turtle doves between years when food was supplied and those when it was not. *Assessment: unknown effectiveness — limited evidence (effectiveness 0%; certainty 21%; harms 0%).*

http://www.conservationevidence.com/actions/535

Provide supplementary food to increase reproductive success (rails and coots)

A small trial in the USA found that fed American coots laid heavier eggs, but not larger clutches, than controls. However, a randomised, replicated and controlled study in Canada found that clutch size, but not egg size, was larger in fed American coot territories. The Canadian study also found that coots laid earlier when fed, whilst a replicated trial from the UK found there was a shorter interval between common moorhens clutches in fed territories, but that fed birds were no more likely to produce second broods. *Assessment: unknown effectiveness — limited evidence (effectiveness 33%; certainty 26%; harms 0%).*

http://www.conservationevidence.com/actions/528

Provide supplementary food to increase reproductive success (vultures)

Two studies from the USA and Greece found that there were local increases in two vulture populations following the provision of food in the area. A study from Israel found that a small, regularly supplied feeding station could provide sufficient food for breeding Egyptian vultures. A study from Italy found that a small population of Egyptian vultures declined following the provision of food, and only a single vulture was seen at the feeding station. *Assessment: unknown effectiveness — limited evidence (effectiveness 30%; certainty 24%; harms 0%).*

http://www.conservationevidence.com/actions/531

Provide supplementary food to increase reproductive success (waders)

A small controlled trial from the Netherlands found that Eurasian oystercatchers did not produce larger replacement eggs if provided with supplementary food. Instead their eggs were smaller than the first clutch, whereas control females laid larger replacement eggs. *Assessment: unknown effectiveness — limited evidence (effectiveness 0%; certainty 10%; harms 0%).*

http://www.conservationevidence.com/actions/529

Provide supplementary food to increase reproductive success (wildfowl)

A small randomised controlled *ex situ* study from Canada found faster growth and higher weights for fed greater snow goose chicks than unfed ones, but no differences in mortality rates. *Assessment: unknown effectiveness — limited evidence (effectiveness 30%; certainty 10%; harms 0%).*

http://www.conservationevidence.com/actions/526

Provide supplementary water to increase survival or reproductive success

A controlled study from Morocco found that northern bald ibises provided with supplementary water had higher reproductive success than those a long way from water sources. *Assessment: unknown effectiveness — limited evidence (effectiveness 43%; certainty 14%; harms 0%).*

http://www.conservationevidence.com/actions/558

3.14.5 Translocations

Based on the collated evidence, what is the current assessment of the effectiveness of interventions for translocations?	
Beneficial	• Translocate birds to re-establish populations or increase genetic variation (birds in general) • Translocate birds to re-establish populations or increase genetic variation: raptors

Likely to be beneficial	• Translocate birds to re-establish populations or increase genetic variation: parrots • Translocate birds to re-establish populations or increase genetic variation: pelicans • Translocate birds to re-establish populations or increase genetic variation: petrels and shearwaters • Translocate birds to re-establish populations or increase genetic variation: rails • Translocate birds to re-establish populations or increase genetic variation: songbirds • Translocate birds to re-establish populations or increase genetic variation: wildfowl • Translocate birds to re-establish populations or increase genetic variation: woodpeckers • Use decoys to attract birds to new sites • Use techniques to increase the survival of species after capture • Use vocalisations to attract birds to new sites
Trade-off between benefit and harms	• Translocate birds to re-establish populations or increase genetic variation: gamebirds
Unknown effectiveness (limited evidence)	• Alter habitats to encourage birds to leave • Ensure translocated birds are familiar with each other before release • Translocate birds to re-establish populations or increase genetic variation: auks • Translocate birds to re-establish populations or increase genetic variation: herons, storks and ibises • Translocate birds to re-establish populations or increase genetic variation: megapodes • Translocate birds to re-establish populations or increase genetic variation: owls • Translocate nests to avoid disturbance
No evidence found (no assessment)	• Ensure genetic variation to increase translocation success

Beneficial

Translocate birds to re-establish populations or increase genetic variation (birds in general)

A review of 239 bird translocation programmes found 63–67% resulted in establishment of a self-sustaining population. *Assessment: beneficial (effectiveness 64%; certainty 65%; harms 0%).*

http://www.conservationevidence.com/actions/566

Translocate birds to re-establish populations or increase genetic variation (raptors)

Six studies of three translocation programmes in the UK and the USA found that all successfully established populations of white-tailed eagles, red kites and ospreys. A study in Spain found high survival of translocated Montagu's harrier fledglings. *Assessment: beneficial (effectiveness 65%; certainty 66%; harms 0%).*

http://www.conservationevidence.com/actions/574

Likely to be beneficial

Translocate birds to re-establish populations or increase genetic variation (parrots)

Three studies of two translocation programmes from the Pacific and New Zealand found that populations of parrots successfully established on islands after translocation. Survival of translocated birds ranged from 41% to 98% globally. Despite high survival, translocated kakapos in New Zealand had very low reproductive output. *Assessment: likely to be beneficial (effectiveness 50%; certainty 60%; harms 10%).*

http://www.conservationevidence.com/actions/578

Translocate birds to re-establish populations or increase genetic variation (pelicans)

Two reviews of a pelican translocation programme in the USA found high survival of translocated nestlings and rapid target population growth. Some growth may have been due to additional immigration from the source populations. *Assessment: likely to be beneficial (effectiveness 55%; certainty 49%; harms 0%).*

http://www.conservationevidence.com/actions/569

Translocate birds to re-establish populations or increase genetic variation (petrels and shearwaters)

Three studies from Australia and New Zealand found that colonies of burrow-nesting petrels and shearwaters were successfully established following the translocation and hand-rearing of chicks. *Assessment: likely to be beneficial (effectiveness 60%; certainty 50%; harms 0%).*

http://www.conservationevidence.com/actions/568

Translocate birds to re-establish populations or increase genetic variation (rails)

Three studies of two translocation programmes in the Seychelles and New Zealand found high survival rates among translocated rail. All three studies round that the birds bred successfully. *Assessment: likely to be beneficial (effectiveness 54%; certainty 44%; harms 14%).*

http://www.conservationevidence.com/actions/573

Translocate birds to re-establish populations or increase genetic variation (songbirds)

Nine studies from across the world, including a review of 31 translocation attempts, found that translocations led to the establishment of songbird populations. Eight studies were on islands. Three studies reported on translocations that failed to establish populations. One study found nesting success decreased as the latitudinal difference between source area and release site increased. *Assessment: likely to be beneficial (effectiveness 50%; certainty 68%; harms 0%).*

http://www.conservationevidence.com/actions/580

Translocate birds to re-establish populations or increase genetic variation (wildfowl)

Three studies of two duck translocation programmes in New Zealand and Hawaii found high survival, breeding and successful establishment of new populations. However a study in the USA found that no ducks stayed at the release site and there was high mortality after release. A study in the USA found wing-clipping prevented female ducks from abandoning their ducklings. *Assessment: likely to be beneficial (effectiveness 42%; certainty 50%; harms 19%).*

http://www.conservationevidence.com/actions/571

Translocate birds to re-establish populations or increase genetic variation (woodpeckers)

Six studies of four programmes found that >50% translocated birds remained at their new sites, and two studies reported large population increases. Birds from four programmes were reported as forming pairs or breeding and one study round translocated nestlings fledged at similar rates to native chicks. All studies were of red-cockaded woodpeckers. *Assessment: likely to be beneficial (effectiveness 51%; certainty 42%; harms 0%).*

http://www.conservationevidence.com/actions/577

Use decoys to attract birds to new sites

Ten studies found that birds nested in areas where decoys were placed or that more birds landed in areas with decoys than control areas. Six studies used multiple interventions at once. One study found that three-dimensional models appeared more effective than two-dimensional ones, and that plastic models were more effective than rag decoys. *Assessment: likely to be beneficial (effectiveness 51%; certainty 45%; harms 0%).*

http://www.conservationevidence.com/actions/586

Use techniques to increase the survival of species after capture

A study from the USA found that providing dark, quiet environments with readily-available food and water increased the survival of small songbirds after capture and the probability that they would adapt to captivity. A study from the USA found that keeping birds warm during transit increased survival. *Assessment: likely to be beneficial (effectiveness 49%; certainty 41%; harms 0%).*

http://www.conservationevidence.com/actions/581

Use vocalisations to attract birds to new sites

Seven out of ten studies from around the world found that seabirds were more likely to nest or land to areas where vocalisations were played, or moved to new nesting areas after vocalisations were played. Four of these studied multiple interventions at once. Three studies found that birds were no more likely to nest or land in areas where vocalisations were played. *Assessment: likely to be beneficial (effectiveness 45%; certainty 50%; harms 0%).*

http://www.conservationevidence.com/actions/585

Trade-off between benefit and harms

Translocate birds to re-establish populations or increase genetic variation (gamebirds)

Three studies from the USA found that translocation of gamebirds led to population establishment or growth or an increase in lekking sites. Four studies from the USA found that translocated birds had high survival, but two found high mortality in translocated birds. Four studies from the USA found breeding rates among translocated birds were high or similar to resident birds. *Assessment: trade-offs between benefits and harms (effectiveness 50%; certainty 47%; harms 35%).*

http://www.conservationevidence.com/actions/572

Unknown effectiveness (limited evidence)

Alter habitats to encourage birds to leave

A study from Canada found that an entire Caspian tern population moved after habitat was altered at the old colony site, alongside several other interventions. *Assessment: unknown effectiveness — limited evidence (effectiveness 20%; certainty 10%; harms 0%).*

http://www.conservationevidence.com/actions/587

Ensure translocated birds are familiar with each other before release

Two studies from New Zealand found no evidence that ensuring birds were familiar with each other increased translocation success. *Assessment: unknown effectiveness — limited evidence (effectiveness 0%; certainty 33%; harms 0%).*

http://www.conservationevidence.com/actions/582

Translocate birds to re-establish populations or increase genetic variation (auks)

A study in the USA and Canada found that 20% of translocated Atlantic puffins remained in or near the release site, with up to 7% breeding. *Assessment: unknown effectiveness — limited evidence (effectiveness 36%; certainty 38%; harms 0%).*

http://www.conservationevidence.com/actions/570

Translocate birds to re-establish populations or increase genetic variation (herons, storks and ibises)

A study in the USA found that a colony of black-crowned night herons was successfully translocated and bred the year after translocation. *Assessment: unknown effectiveness — limited evidence (effectiveness 44%; certainty 3%; harms 0%).*

http://www.conservationevidence.com/actions/575

Translocate birds to re-establish populations or increase genetic variation (megapodes)

A study from Indonesia found that up to 78% maleo eggs hatched after translocation. *Assessment: unknown effectiveness — limited evidence (effectiveness 49%; certainty 29%; harms 0%).*

http://www.conservationevidence.com/actions/567

Translocate birds to re-establish populations or increase genetic variation (owls)

A small study from New Zealand found that translocating two male boobooks allowed the establishment of a population when they interbred with a Norfolk Island boobook. A study in the USA found high survival amongst burrowing owls translocated as juveniles, although birds were not seen after release. *Assessment: unknown effectiveness — limited evidence (effectiveness 20%; certainty 19%; harms 0%).*

http://www.conservationevidence.com/actions/576

Translocate nests to avoid disturbance

All five studies captured found some success in relocating nests while they were in use, but one found that fewer than half of the burrowing owls studied were moved successfully; a study found that repeated disturbance caused American kestrels to abandon their nest and a study found that one barn swallow abandoned its nest after it was moved. *Assessment: unknown effectiveness — limited evidence (effectiveness 24%; certainty 39%; harms 30%).*

http://www.conservationevidence.com/actions/584

No evidence found (no assessment)

We have captured no evidence for the following interventions:

- Ensure genetic variation to increase translocation success.

3.15 Captive breeding, rearing and releases (*ex situ* conservation)

3.15.1 Captive breeding

Based on the collated evidence, what is the current assessment of the effectiveness of interventions for captive breeding?	
Likely to be beneficial	• Artificially incubate and hand-rear birds in captivity: raptors • Artificially incubate and hand-rear birds in captivity: seabirds • Artificially incubate and hand-rear birds in captivity: songbirds • Artificially incubate and hand-rear birds in captivity: waders • Use captive breeding to increase or maintain populations: raptors
Unknown effectiveness (limited evidence)	• Artificially incubate and hand-rear birds in captivity: bustards • Artificially incubate and hand-rear birds in captivity: cranes • Artificially incubate and hand-rear birds in captivity: gamebirds • Artificially incubate and hand-rear birds in captivity: parrots • Artificially incubate and hand-rear birds in captivity: penguins • Artificially incubate and hand-rear birds in captivity: rails • Artificially incubate and hand-rear birds in captivity: storks and ibises

	• Artificially incubate and hand-rear birds in captivity: vultures • Artificially incubate and hand-rear birds in captivity: wildfowl • Freeze semen for artificial insemination • Use artificial insemination in captive breeding • Use captive breeding to increase or maintain populations: bustards • Use captive breeding to increase or maintain populations: cranes • Use captive breeding to increase or maintain populations: pigeons • Use captive breeding to increase or maintain populations: rails • Use captive breeding to increase or maintain populations: seabirds • Use captive breeding to increase or maintain populations: songbirds • Use captive breeding to increase or maintain populations: storks and ibises • Use captive breeding to increase or maintain populations: tinamous • Use puppets to increase the success of hand-rearing • Wash contaminated semen and use it for artificial insemination
Evidence not assessed	• Can captive breeding have deleterious effects on individual fitness?

Likely to be beneficial

Artificially incubate and hand-rear birds in captivity (raptors)

Six studies from across the world found high success rates for artificial incubation and hand-rearing of raptors. A replicated and controlled study from France found that artificially incubated raptor eggs had lower hatching success than parent-incubated eggs but fledging success for hand-reared chicks was similar to wild chicks. A study from Canada found that hand-reared chicks had slower growth and attained a lower weight than parent-reared birds. A replicated study from Mauritius found that hand-rearing of wild eggs had higher success than hand-rearing captive-bred chicks. Three studies that provided methodological comparisons reported that incubation temperature affected hatching success and adding saline to the diet of falcon chicks increased their weight gain. *Assessment: likely to be beneficial (effectiveness 60%; certainty 52%; harms 5%).*

http://www.conservationevidence.com/actions/614

Artificially incubate and hand-rear birds in captivity (seabirds)

Five studies from across the world found evidence for the success of hand-rearing seabirds. One small study in Spain found that one of five hand-reared Audouin's gulls successfully bred in the wild. Four studies found that various petrel species successfully fledged after hand-rearing. One controlled study found that fledging rates of hand-reared birds was similar to parent-reared birds, although a study on a single bird found that the chick fledged at a lower weight and later than parent-reared chicks. *Assessment: likely to be beneficial (effectiveness 67%; certainty 45%; harms 2%).*

http://www.conservationevidence.com/actions/604

Artificially incubate and hand-rear birds in captivity (songbirds)

Four studies from the USA found high rates of success for artificial incubation and hand-rearing of songbirds. One study found that crow chicks fed more food had higher growth rates, but these rates never matched those of wild birds. *Assessment: likely to be beneficial (effectiveness 51%; certainty 44%; harms 1%).*

http://www.conservationevidence.com/actions/616

Artificially incubate and hand-rear birds in captivity (waders)

Three out of four replicated and controlled studies from the USA and New Zealand found that artificially incubated and/or hand-reared waders had higher hatching and fledging success than controls. One study from New Zealand found that hatching success of black stilt was lower for artificially-incubated eggs. *Assessment: likely to be beneficial (effectiveness 64%; certainty 41%; harms 4%).*

http://www.conservationevidence.com/actions/611

Use captive breeding to increase or maintain populations (raptors)

Three small studies and a review from around the world found that raptors bred successfully in captivity. Two of these studies found that wild-caught birds bred in captivity after a few years, with one pair of brown goshawks producing 15 young over four years, whilst a study on bald eagle captive breeding found low fertility in captive-bred eggs, but that birds still produced chicks after a year. A review of Mauritius kestrel captive breeding found that 139 independent young were raised over 12 years from 30 eggs and chicks taken from the wild. An update of the same programme found that hand-reared Mauritius kestrels were less successful if they came from captive-bred eggs compared to wild 'harvested' eggs. *Assessment: likely to be beneficial (effectiveness 50%; certainty 41%; harms 10%).*

http://www.conservationevidence.com/actions/596

Unknown effectiveness (limited evidence)

Artificially incubate and hand-rear birds in captivity (bustards)

Two reviews of a houbara bustard captive breeding programme in Saudi Arabia found no difference in survival between artificially and parentally incubated eggs, and that removing eggs from clutches as they were laid increased the number laid by females. *Assessment: unknown effectiveness — limited evidence (effectiveness 31%; certainty 10%; harms 0%).*

http://www.conservationevidence.com/actions/610

Artificially incubate and hand-rear birds in captivity (cranes)

Two studies from the USA found that hand-reared birds showed normal reproductive behaviour and higher survival than parent-reared birds. *Assessment: unknown effectiveness — limited evidence (effectiveness 76%; certainty 31%; harms 0%).*

http://www.conservationevidence.com/actions/609

Artificially incubate and hand-rear birds in captivity (gamebirds)

A study in Finland found that hand-reared grey partridges did not take off to fly as effectively as wild-caught birds, potentially making them more vulnerable to predation. *Assessment: unknown effectiveness — limited evidence (effectiveness 11%; certainty 10%; harms 50%).*

http://www.conservationevidence.com/actions/607

Artificially incubate and hand-rear birds in captivity (parrots)

Two studies from South America describe the successful hand-rearing of parrot chicks. A review of the kakapo management programme found that chicks could be successfully raised and released, but that eggs incubated from a young age had low success. A study from the USA found that all hand-reared thick-billed parrots died within a month of release: significantly lower survival than for wild-caught birds translocated to the release site. *Assessment: unknown effectiveness — limited evidence (effectiveness 19%; certainty 30%; harms 11%).*

http://www.conservationevidence.com/actions/615

Artificially incubate and hand-rear birds in captivity (penguins)

Two replicated and controlled studies from South Africa found that hand-reared and released African penguins had similar survival and breeding success as birds which were not hand-reared. *Assessment: unknown effectiveness — limited evidence (effectiveness 41%; certainty 15%; harms 0%).*

http://www.conservationevidence.com/actions/605

Artificially incubate and hand-rear birds in captivity (rails)

A controlled study from New Zealand found that post-release survival of hand-reared takahe was as high as wild-reared birds and that six of ten released females raised chicks. *Assessment: unknown effectiveness — limited evidence (effectiveness 64%; certainty 13%; harms 0%).*

http://www.conservationevidence.com/actions/608

Artificially incubate and hand-rear birds in captivity (storks and ibises)

A small study in the USA describes the successful artificial incubation and hand-rearing of two Abdim's stork chicks, whilst a review of northern bald ibis conservation found that only very intensive rearing of a small number of chicks appeared to allow strong bonds, thought to be important for the successful release of birds into the wild, to form between chicks. *Assessment: unknown effectiveness — limited evidence (effectiveness 18%; certainty 10%; harms 0%).*

http://www.conservationevidence.com/actions/612

Artificially incubate and hand-rear birds in captivity (vultures)

A study in Peru found that hand-reared Andean condors had similar survival to parent-reared birds after release into the wild. *Assessment: unknown effectiveness — limited evidence (effectiveness 30%; certainty 10%; harms 0%).*

http://www.conservationevidence.com/actions/613

Artificially incubate and hand-rear birds in captivity (wildfowl)

Two studies in Canada and India found high success rates for hand-rearing buffleheads and bar-headed geese in captivity. Eggs were artificially incubated or incubated under foster parents. A replicated, controlled study in England found that Hawaiian geese (nene) chicks showed less well-adapted behaviours if they were raised without parental contact. *Assessment: unknown effectiveness — limited evidence (effectiveness 50%; certainty 20%; harms 10%).*

http://www.conservationevidence.com/actions/606

Freeze semen for artificial insemination

Two small trials from the USA found that using thawed frozen semen for artificial insemination resulted in low fertility rates. A small trial from the USA found that a cryprotectant increased fertility rates achieved using frozen semen. *Assessment: unknown effectiveness — limited evidence (effectiveness 10%; certainty 10%; harms 45%).*

http://www.conservationevidence.com/actions/602

Use artificial insemination in captive breeding

A replicated study from Saudi Arabia found that artificial insemination could increase fertility in houbara bustards. A study of the same programme and a review found that repeated inseminations increased fertility, with the review arguing that artificial insemination had the potential to be a useful technique. Two studies from the USA found that artificially-inseminated raptors had either zero fertility, or approximately 50%. *Assessment: unknown effectiveness — limited evidence (effectiveness 33%; certainty 21%; harms 0%).*

http://www.conservationevidence.com/actions/601

Use captive breeding to increase or maintain populations (bustards)

Four studies of a captive breeding programme in Saudi Arabia reported that the houbara bustard chicks were successfully raised in captivity, with 285 chicks hatched in the 7th year of the project after 232 birds were used to start the captive population. Captive birds bred earlier and appeared to lay more eggs than wild birds. Forty-six percent of captive eggs hatched and 43% of chicks survived to ten years old. *Assessment: unknown effectiveness — limited evidence (effectiveness 41%; certainty 16%; harms 5%).*

http://www.conservationevidence.com/actions/592

Use captive breeding to increase or maintain populations (cranes)

A study from Canada over 32 years found that whooping cranes successfully bred in captivity eight years after the first eggs were removed from the wild. *Assessment: unknown effectiveness — limited evidence (effectiveness 51%; certainty 17%; harms 6%).*

http://www.conservationevidence.com/actions/591

Use captive breeding to increase or maintain populations (pigeons)

A review of a captive-breeding programme on Mauritius and in the UK found that 42 pink pigeons were successfully bred in captivity. *Assessment: unknown effectiveness — limited evidence (effectiveness 69%; certainty 21%; harms 0%).*

http://www.conservationevidence.com/actions/597

Use captive breeding to increase or maintain populations (rails)

A study from Australia found that three pairs of Lord Howe Islandwoodhens successfully bred in captivity, with 66 chicks being produced over four years. *Assessment: unknown effectiveness — limited evidence (effectiveness 26%; certainty 11%; harms 5%).*

http://www.conservationevidence.com/actions/590

Use captive breeding to increase or maintain populations (seabirds)

A study from Spain found that a single pair of Audouin's gulls successfully bred in captivity. *Assessment: unknown effectiveness — limited evidence (effectiveness 20%; certainty 4%; harms 5%).*

http://www.conservationevidence.com/actions/589

Use captive breeding to increase or maintain populations (songbirds)

Three studies from Australia and the USA found that three species of songbird bred successfully in captivity. Four out of five pairs of wild-bred, hand-reared puaiohi formed pairs and laid a total of 39 eggs and a breeding population of helmeted honeyeaters was successfully established through a breeding programme. Only one pair of loggerhead shrikes formed pairs from eight wild birds caught and their first clutch died. *Assessment: unknown effectiveness — limited evidence (effectiveness 77%; certainty 31%; harms 5%).*

http://www.conservationevidence.com/actions/598

Use captive breeding to increase or maintain populations (storks and ibises)

We captured a small study and a review both from the USA describing the captive breeding of storks. The study found that a pair bred; the review found that only seven of 19 species had been successfully bred in captivity. A review of bald ibis conservation found that 1,150 birds had been produced in captivity from 150 founders over 20 years. However, some projects had failed, and a study from Turkey found that captive birds had lower productivity than wild birds. *Assessment: unknown effectiveness — limited evidence (effectiveness 31%; certainty 30%; harms 8%).*

http://www.conservationevidence.com/actions/595

Use captive breeding to increase or maintain populations (tinamous)

A replicated study from Costa Rica found that great tinamous successfully bred in captivity, with similar reproductive success to wild birds. *Assessment: unknown effectiveness — limited evidence (effectiveness 51%; certainty 15%; harms 5%).*

http://www.conservationevidence.com/actions/588

Use puppets to increase the success of hand-rearing

Three studies from the USA and Saudi Arabia found that crows and bustards raised using puppets did not have higher survival, dispersal or growth than chicks hand-reared conventionally. *Assessment: unknown effectiveness — limited evidence (effectiveness 4%; certainty 20%; harms 0%).*

http://www.conservationevidence.com/actions/617

Wash contaminated semen and use it for artificial insemination

A replicated, controlled study from Spain found that washed, contaminated semen could be used to successfully inseminate raptors. *Assessment: unknown effectiveness — limited evidence (effectiveness 31%; certainty 15%; harms 0%).*

http://www.conservationevidence.com/actions/603

Evidence not assessed

Can captive breeding have deleterious effects?

We captured no studies investigating the effects of captive-breeding on fitness. Three studies using wild and captive populations or museum specimens found physiological or genetic changes in populations that had been bred in captivity. One found that changes were more likely to be caused by extremely low population levels than by captivity.

http://www.conservationevidence.com/actions/599

3.15.2 Release captive-bred individuals

Based on the collated evidence, what is the current assessment of the effectiveness of interventions for captive breeding?	
Likely to be beneficial	• Provide supplementary food after release • Release captive-bred individuals into the wild to restore or augment wild populations: cranes • Release captive-bred individuals into the wild to restore or augment wild populations: raptors • Release captive-bred individuals into the wild to restore or augment wild populations: songbirds • Release captive-bred individuals into the wild to restore or augment wild populations: vultures
Unknown effectiveness (limited evidence)	• Clip birds' wings on release • Release birds as adults or sub-adults not juveniles • Release birds in groups • Release captive-bred individuals into the wild to restore or augment wild populations: bustards • Release captive-bred individuals into the wild to restore or augment wild populations: gamebirds • Release captive-bred individuals into the wild to restore or augment wild populations: owls • Release captive-bred individuals into the wild restore or augment wild populations: parrots • Release captive-bred individuals into the wild to restore or augment wild populations: pigeons • Release captive-bred individuals into the wild to restore or augment wild populations: rails • Release captive-bred individuals into the wild to restore or augment wild populations: storks and ibises • Release captive-bred individuals into the wild to restore or augment wild populations: waders • Release captive-bred individuals into the wild to restore or augment wild populations: wildfowl • Release chicks and adults in 'coveys' • Use 'anti-predator training' to improve survival after release

	• Use appropriate populations to source released populations • Use 'flying training' before release • Use holding pens at release sites • Use microlites to help birds migrate

Likely to be beneficial

Provide supplementary food after release

All three studies captured found that released birds used supplementary food provided. One study from Australia found that malleefowl had higher survival when provided with food and a study from Peru found that supplementary food could be used to increase the foraging ranges of Andean condors after release. *Assessment: likely to be beneficial (effectiveness 45%; certainty 48%; harms 0%).*

http://www.conservationevidence.com/actions/639

Release captive-bred individuals into the wild to restore or augment wild populations (cranes)

Four studies of five release programmes from the USA and Russia found that released cranes had high survival or bred in the wild. Two studies from two release programmes in the USA found low survival of captive-bred eggs fostered to wild birds compared with wild eggs, or a failure to increase the wild flock size. A worldwide review found that releases of migratory species were more successful if birds were released into existing flocks, and for non-migratory populations. One study from the USA found that birds released as sub-adults had higher survival than birds cross-fostered to wild birds. *Assessment: likely to be beneficial (effectiveness 55%; certainty 50%; harms 5%).*

http://www.conservationevidence.com/actions/621

Release captive-bred individuals into the wild to restore or augment wild populations (raptors)

Five studies of three release programmes from across the world found the establishment or increase of wild populations of falcons. Five studies from the USA found high survival of released raptors although one study from Australia found that a wedge-tailed eagle had to be taken back into captivity after acting aggressively towards humans, and another Australian study found that only one of 15 brown goshawks released was recovered. *Assessment: likely to be beneficial (effectiveness 69%; certainty 56%; harms 10%).*

http://www.conservationevidence.com/actions/626

Release captive-bred individuals into the wild to restore or augment wild populations (songbirds)

A study in Mauritius describes the establishment of a population of Mauritius fody following the release of captive-bred individuals. Four studies of three release programmes on Hawaii found high survival of all three species released, with two thrush species successfully breeding. A replicated, controlled study from the USA found that shrike pairs with captive-bred females had lower reproductive success than pairs where both parents were wild-bred. *Assessment: likely to be beneficial (effectiveness 42%; certainty 40%; harms 5%).*

http://www.conservationevidence.com/actions/630

Release captive-bred individuals into the wild to restore or augment wild populations (vultures)

Four studies of two release programmes found that release programmes led to large population increases in Andean condors in Colombia and griffon vultures in France. A small study in Peru found high survival of released Andean condors over 18 months, with all fatalities occurring in the first six months after release. *Assessment: likely to be beneficial (effectiveness 73%; certainty 54%; harms 0%).*

http://www.conservationevidence.com/actions/625

Unknown effectiveness (limited evidence)

Clip birds' wings on release

Two of four studies found that bustards and geese had lower survival when released into holding pens with clipped wings compared to birds released without clipped wings. One study found no differences in survival for clipped or unclipped northern bald ibis. One study found that adult geese released with clipped wings survived better than geese released before they were able to fly. *Assessment: unknown effectiveness — limited evidence (effectiveness 10%; certainty 30%; harms 5%).*

http://www.conservationevidence.com/actions/633

Release birds as adults or sub-adults not juveniles

Three out of nine studies from across the world found that birds released as sub-adults had higher survival than those released as juveniles. Two studies found lower survival of wing-clipped sub-adult geese and bustards, compared with juveniles and one study found lower survival of all birds released as sub-adults, compared to those released as juveniles. Three studies found no differences in survival for birds released at different ages, although one found higher reproduction in birds released at greater ages. *Assessment: unknown effectiveness — limited evidence (effectiveness 35%; certainty 15%; harms 19%).*

http://www.conservationevidence.com/actions/636

Release birds in groups

A study from New Zealand found that released stilts were more likely to move long distances after release if they were released in larger groups. *Assessment: unknown effectiveness — limited evidence (effectiveness 32%; certainty 26%; harms 2%).*

http://www.conservationevidence.com/actions/634

Release captive-bred individuals into the wild to restore or augment wild populations (bustards)

Three reviews of a release programme for houbara bustard in Saudi Arabia found low initial survival of released birds, but the establishment of a breeding population and an overall success rate of 41%. The programme tested many different release techniques, the most successful of which was release of sub-adults, which were able to fly, into a large exclosure. *Assessment: unknown effectiveness — limited evidence (effectiveness 34%; certainty 26%; harms 5%).*

http://www.conservationevidence.com/actions/622

Release captive-bred individuals into the wild to restore or augment wild populations (gamebirds)

One of five studies from across the world found that releasing gamebirds established a population or bolstered an existing population. A review of a reintroduction programme in Pakistan found some breeding success in released cheer pheasants, but habitat change at the release site then excluded released birds. Three studies from Europe and the USA found that released birds had low survival, low reproductive success and no impact on the wild population. *Assessment: unknown effectiveness — limited evidence (effectiveness 5%; certainty 35%; harms 1%).*

http://www.conservationevidence.com/actions/619

Release captive-bred individuals into the wild to restore or augment wild populations (owls)

A study in the USA found that a barn owl population was established following the release of 157 birds in the area over three years. A replicated, controlled study in Canada found that released burrowing owls had similar reproductive output but higher mortality than wild birds. *Assessment: unknown effectiveness — limited evidence (effectiveness 24%; certainty 15%; harms 0%).*

http://www.conservationevidence.com/actions/627

Release captive-bred individuals into the wild to restore or augment wild populations (parrots)

A study from Venezuela found that the population of yellow-shouldered amazons increased significantly following the release of captive-bred birds along with other interventions. A study in Costa Rica and Peru found high survival and some breeding of scarlet macaw after release. Three replicated studies in the USA, Dominican Republic and Puerto Rico found low survival in released birds, although the Puerto Rican study also found that released birds bred successfully. *Assessment: unknown effectiveness — limited evidence (effectiveness 50%; certainty 30%; harms 3%).*

http://www.conservationevidence.com/actions/629

Release captive-bred individuals into the wild to restore or augment wild populations (pigeons)

A single review of a captive-release programme in Mauritius found that that released pink pigeons had a first year survival of 36%. *Assessment: unknown effectiveness — limited evidence (effectiveness 20%; certainty 5%; harms 1%).*

http://www.conservationevidence.com/actions/628

Release captive-bred individuals into the wild to restore or augment wild populations (rails)

One study from Australia found that released Lord Howe Island woodhens successfully bred in the wild, re-establishing a wild population and a study from the UK found high survival of released corncrake in the first summer after release. A replicated study in New Zealand found very low survival of North Island weka following release, mainly due to predation. *Assessment: unknown effectiveness — limited evidence (effectiveness 26%; certainty 16%; harms 0%).*

http://www.conservationevidence.com/actions/620

Release captive-bred individuals into the wild to restore or augment wild populations (storks and ibises)

A replicated study and a review of northern bald ibis release programmes in Europe and the Middle East found that only one of four resulted in a wild population being established or supported, with many birds dying or dispersing, rather than forming stable colonies. *Assessment: unknown effectiveness — limited evidence (effectiveness 20%; certainty 20%; harms 2%).*

http://www.conservationevidence.com/actions/624

Release captive-bred individuals into the wild to restore or augment wild populations (waders)

A review of black stilt releases in New Zealand found that birds had low survival (13–20%) and many moved away from their release sites. *Assessment: unknown effectiveness — limited evidence (effectiveness 10%; certainty 5%; harms 15%).*

http://www.conservationevidence.com/actions/623

Release captive-bred individuals into the wild to restore or augment wild populations (wildfowl)

Two studies of reintroduction programmes of ducks in New Zealand found high survival of released birds and population establishment. A study from Alaska found low survival of released cackling geese, but the population recovered from 1,000 to 6,000 birds after releases and the control of mammalian predators. A review of a reintroduction programme from Hawaii found that the release of Hawaiian geese (nene) did not result in the establishment of a self-sustaining population. Two studies from Canada found very low return rates for released ducks with one finding no evidence for survival of released birds over two years, although there was some evidence that breeding success was higher for released birds than wild ones. *Assessment: unknown effectiveness — limited evidence (effectiveness 30%; certainty 24%; harms 0%).*

http://www.conservationevidence.com/actions/618

Release chicks and adults in 'coveys'

Two out of three studies found that geese and partridges released in coveys had higher survival than young birds released on their own or adults released in pairs. A study from Saudi Arabia found that bustard chicks had low survival when released in coveys with flightless females. *Assessment: unknown effectiveness — limited evidence (effectiveness 40%; certainty 36%; harms 6%).*

http://www.conservationevidence.com/actions/635

Use 'anti-predator training' to improve survival after release

Both studies captured found higher survival for birds given predator training before release, compared with un-trained birds. One found that using a live fox, but not a model, for training increased survival in bustards, but that several birds were injured during training. *Assessment: unknown effectiveness — limited evidence (effectiveness 50%; certainty 20%; harms 9%).*

http://www.conservationevidence.com/actions/637

Use appropriate populations to source released populations

Two studies from Europe found that birds from populations near release sites adapted better and in one case had higher reproductive productivity than those from more distant populations. *Assessment: unknown effectiveness — limited evidence (effectiveness 53%; certainty 31%; harms 0%).*

http://www.conservationevidence.com/actions/631

Use 'flying training' before release

A study from the Dominican Republic found that parrots had higher first-year survival if they were given pre-release flying training. *Assessment: unknown effectiveness — limited evidence (effectiveness 30%; certainty 10%; harms 0%).*

http://www.conservationevidence.com/actions/638

Use holding pens at release sites

Three of four studies from North America and Saudi Arabia found that birds released into holding pens were more likely to form pairs or had higher survival than birds released into the open. One study found that parrots released into pens had lower survival than those released without preparation. A review of northern bald ibis releases found that holding pens could be used to prevent birds from migrating from the release site and so increase survival. *Assessment: unknown effectiveness — limited evidence (effectiveness 51%; certainty 36%; harms 2%).*

http://www.conservationevidence.com/actions/632

Use microlites to help birds migrate

A study from Europe found that northern bald ibises followed a microlite south in the winter but failed to make the return journey the next year. *Assessment: unknown effectiveness — limited evidence (effectiveness 3%; certainty 5%; harms 5%).*

http://www.conservationevidence.com/actions/640

4. FARMLAND CONSERVATION

Lynn V. Dicks, Joscelyne E. Ashpole, Juliana Dänhardt, Katy James, Annelie Jönsson, Nicola Randall, David A. Showler, Rebecca K. Smith, Susan Turpie, David R. Williams & William J. Sutherland

Expert assessors
Lynn V. Dicks, University of Cambridge, UK
Ian Hodge, University of Cambridge, UK
Clunie Keenleyside, Institute for European Environmental Policy, UK
Will Peach, Royal Society for the Protection of Birds, UK
Nicola Randall, Harper Adams University, UK
Jörn Scharlemann, United Nations Environment Programme — World Conservation Monitoring Centre, UK
Gavin Siriwardena, British Trust for Ornithology, UK
Henrik Smith, Lund University, Sweden
Rebecca K. Smith, University of Cambridge, UK
William J. Sutherland, University of Cambridge, UK

Scope of assessment: for native farmland wildlife in northern and western Europe (European countries west of Russia, but not south of France, Switzerland, Austria, Hungary and Romania).

Assessed: 2014.

Effectiveness measure is the % of experts that answered yes to the question: based on the evidence presented does this intervention benefit wildlife? (Yes, no or don't know).

Certainty measure is the median % score for the question: how much do we understand the extent to which this intervention benefits wildlife on farmland? (0 = no evidence, 100% = certainty).

Harm measure was not scored for this synopsis.

 https://doi.org/10.11647/OBP.0267.04

This book is meant as a guide to the evidence available for different conservation interventions and as a starting point in assessing their effectiveness. The assessments are based on the available evidence for the target group of species for each intervention. The assessment may therefore refer to different species or habitat to the one(s) you are considering. Before making any decisions about implementing interventions it is vital that you read the more detailed accounts of the evidence in order to assess their relevance for your study species or system.

Full details of the evidence are available at**www.conservationevidence.com**

There may also be significant negative side-effects on the target groups or other species or communities that have not been identified in this assessment.

A lack of evidence means that we have been unable to assess whether or not an intervention is effective or has any harmful impacts.

4.1 All farming systems

Based on the collated evidence, what is the current assessment of the effectiveness of interventions for all farming systems?	
Beneficial	• Create uncultivated margins around intensive arable or pasture fields • Plant grass buffer strips/margins around arable or pasture fields • Plant nectar flower mixture/wildflower strips • Plant wild bird seed or cover mixture • Provide or retain set-aside areas in farmland
Likely to be beneficial	• Manage ditches to benefit wildlife • Manage hedgerows to benefit wildlife (includes no spray, gap-filling and laying) • Pay farmers to cover the costs of conservation measures • Provide supplementary food for birds or mammals
Unknown effectiveness (limited evidence)	• Connect areas of natural or semi-natural habitat • Increase the proportion of natural or semi-natural habitat in the farmed landscape • Make direct payments per clutch for farmland birds • Manage the agricultural landscape to enhance floral resources • Mark bird nests during harvest or mowing • Plant new hedges • Provide nest boxes for bees (solitary bees or bumblebees) • Provide nest boxes for birds • Provide other resources for birds (water, sand for bathing) • Provide refuges during harvest or mowing

No evidence found (no assessment)	• Apply 'cross compliance' environmental standards linked to all subsidy payments • Implement food labelling schemes relating to biodiversity-friendly farming (organic, LEAF marque) • Introduce nest boxes stocked with solitary bees • Maintain in-field elements such as field islands and rockpiles • Manage stone-faced hedge banks to benefit wildlife • Manage woodland edges to benefit wildlife • Plant in-field trees (not farm woodland) • Protect in-field trees (includes management such as pollarding and surgery) • Provide badger gates • Provide foraging perches (e.g. for shrikes) • Provide otter holts • Provide red squirrel feeders • Reduce field size (or maintain small fields) • Restore or maintain dry stone walls • Support or maintain low-intensity agricultural systems

Beneficial

Create uncultivated margins around intensive arable or pasture fields

Twenty studies (including one randomized, replicated, controlled trial) from seven countries found uncultivated margins support more invertebrates, small mammal species or higher plant diversity than other habitats. Four studies (including two replicated studies from the UK) found positive associations between birds and uncultivated margins. Fifteen studies (including one randomized, replicated, controlled trial) from four countries found naturally regenerated margins had lower invertebrate or plant abundance or diversity than conventional fields or sown margins. Six studies (one randomized, replicated, controlled) from three countries found uncultivated margins did not have higher plant or invertebrate abundance or diversity than cropped or sown margins. *Assessment: beneficial (effectiveness 100%; certainty 63%).*

http://www.conservationevidence.com/actions/63

Plant grass buffer strips/margins around arable or pasture fields

Twenty studies (including two randomized, replicated, controlled studies) from four countries found grass margins benefited invertebrates, including increases in abundance or diversity. Nine studies (including two replicated, controlled trials) from the UK found grass buffer strips benefit birds, with increased numbers, diversity or use. Seven replicated studies (four controlled, two randomized) from two countries found grass buffer strips increased plant cover and species richness, a review found benefits to plants. Five studies (two replicated, controlled) from two countries found benefits to small mammals. Six (including three replicated, controlled trials) from two countries found no clear effect on invertebrate or bird numbers. *Assessment: beneficial (effectiveness 90%; certainty 65%).*

http://www.conservationevidence.com/actions/246

Plant nectar flower mixture/wildflower strips

Forty-one studies (including one randomized, replicated, controlled trial) from eight countries found flower strips increased invertebrate numbers or diversity. Ten studies (two replicated, controlled) found invertebrates visited flower strips. Fifteen studies (two randomized, replicated, controlled) found mixed or negative effects on invertebrates. Seventeen studies (one randomized, replicated, controlled) from seven countries found more plants or plant species on flower strips, four did not. Five studies (two randomized, replicated, controlled) from two countries found bird numbers, diversity or use increased in flower strips, two studies did not. Five studies (four replicated) found increases in small mammal abundance or diversity in flower strips. *Assessment: beneficial (effectiveness 100%; certainty 75%).*

http://www.conservationevidence.com/actions/442

Plant wild bird seed or cover mixture

Fifteen studies (including a systematic review) from the UK found fields sown with wild bird cover mix had more birds or bird species than other farmland habitats. Six studies (including two replicated trials) from the UK found birds used wild bird cover more than other habitats. Nine replicated studies from France and the UK found mixed or negative effects on birds. Eight studies (including two randomized, replicated, controlled studies) from the UK found wild bird cover had more invertebrates, four (including two replicated trials) found mixed or negative effects on invertebrate numbers. Six studies (including two replicated, controlled trials) from the UK found wild bird cover mix benefited plants, two replicated studies did not. *Assessment: beneficial (effectiveness 100%; certainty 65%).*

http://www.conservationevidence.com/actions/594

Provide or retain set-aside areas in farmland

Thirty-seven studies (one systematic review, no randomized, replicated, controlled trials) compared use of set-aside areas with control farmed fields. Twenty-one (including the systematic review) showed benefits to, or higher use by, all wildlife groups considered. Thirteen studies found some species or groups used set-aside more than crops; others did not. Two found higher Eurasian skylark reproductive success and one study found lower success on set-aside than control fields. Four studies found set-aside had no effect on wildlife, one found an adverse effect. Two studies found neither insects nor small mammals preferred set-aside. *Assessment: beneficial (effectiveness 90%; certainty 70%).*

http://www.conservationevidence.com/actions/156

Likely to be beneficial

Manage ditches to benefit wildlife

Five studies (including one replicated, controlled study) from the UK and the Netherlands found ditch management had positive effects on numbers, diversity or biomass of some or all invertebrates, amphibians, birds or plants studied. Three studies from the Netherlands and the UK (including two replicated site comparisons) found negative or no clear effects on plants or some birds. *Assessment: likely to be beneficial (effectiveness 40%; certainty 45%).*

http://www.conservationevidence.com/actions/135

Manage hedgerows to benefit wildlife (includes no spray, gap-filling and laying)

Ten studies from the UK and Switzerland (including one randomized, replicated, controlled trial) found managing hedges for wildlife increased berry yields, diversity or abundance of plants, invertebrates or birds. Five

UK studies (including one randomized, replicated, controlled trial) found plants, bees and farmland birds were unaffected by hedge management. Two replicated studies found hedge management had mixed effects on invertebrates or reduced hawthorn berry yield. *Assessment: likely to be beneficial (effectiveness 70%; certainty 50%).*

http://www.conservationevidence.com/actions/116

Pay farmers to cover the cost of conservation measures (as in agri-environment schemes)

For birds, twenty-four studies (including one systematic review) found increases or more favourable trends in bird populations, while eleven studies (including one systematic review) found negative or no effects of agri-environment schemes. For plants, three studies found more plant species, two found fewer plant species and seven found little or no effect of agri-environment schemes. For invertebrates, five studies found increases in abundance or species richness, while six studies found little or no effect of agri-environment schemes. For mammals, one replicated study found positive effects of agri-environment schemes and three studies found mixed effects in different regions or for different species. *Assessment: likely to be beneficial (effectiveness 60%; certainty 50%).*

http://www.conservationevidence.com/actions/700

Provide supplementary food for birds or mammals

Nine studies (two randomized, replicated, controlled) from France, Sweden and the UK found providing supplementary food increased abundance, overwinter survival or productivity of some birds. Two of the studies did not separate the effects of several interventions. Four studies (one replicated, controlled and one randomized, replicated) from Finland and the UK found some birds or mammals used supplementary food. Six replicated studies (three controlled) from Sweden and the UK found no clear effect on some birds or plants. *Assessment: likely to be beneficial (effectiveness 90%; certainty 50%).*

http://www.conservationevidence.com/actions/648

Unknown effectiveness (limited evidence)

Connect areas of natural or semi-natural habitat

All four studies (including two replicated trials) from the Czech Republic, Germany and the Netherlands investigating the effects of linking patches of natural or semi-natural habitat found some colonization by invertebrates or mammals. Colonization by invertebrates was slow or its extent varied between taxa. *Assessment: unknown effectiveness — limited evidence (effectiveness 0%; certainty 15%).*

http://www.conservationevidence.com/actions/579

Increase the proportion of semi-natural habitat in the farmed landscape

Of five studies monitoring the effects of the Swiss Ecological Compensation Areas scheme at a landscape scale (including three replicated site comparisons), one found an increase in numbers of birds of some species, two found no effect on birds and three found some species or groups increasing and others decreasing. *Assessment: unknown effectiveness — limited evidence (effectiveness 20%; certainty 20%).*

http://www.conservationevidence.com/actions/145

Make direct payments per clutch for farmland birds

Two replicated, controlled studies from the Netherlands found per clutch payments did not increase overall bird numbers. A replicated site comparison from the Netherlands found more birds bred on 12.5 ha plots under management including per-clutch payments but there were no differences at the field-scale. *Assessment: unknown effectiveness — limited evidence (effectiveness 0%; certainty 20%).*

http://www.conservationevidence.com/actions/146

Manage the agricultural landscape to enhance floral resources

A large replicated, controlled study from the UK found the number of long-tongued bumblebees on field margins was positively correlated with the number of 'pollen and nectar' agri-environment agreements in a 10 km square. *Assessment: unknown effectiveness — limited evidence (effectiveness 40%; certainty 10%).*

http://www.conservationevidence.com/actions/362

Mark bird nests during harvest or mowing

A replicated study from the Netherlands found that marked northern lapwing nests were less likely to fail as a result of farming operations than unmarked nests. *Assessment: unknown effectiveness — limited evidence (effectiveness 20%; certainty 15%).*

http://www.conservationevidence.com/actions/148

Plant new hedges

Two studies (including one replicated trial) from France and the UK found new hedges had more invertebrates or plant species than fields or field margins. A review found new hedges had more ground beetles than older hedges. However, an unreplicated site comparison from Germany found only two out of 85 ground beetle species dispersed along new hedges. A review found lower pest outbreaks in areas with new hedges. *Assessment: unknown effectiveness — limited evidence (effectiveness 60%; certainty 25%).*

http://www.conservationevidence.com/actions/538

Provide nest boxes for bees (solitary bees or bumblebees)

Ten studies (nine replicated) from Germany, Poland and the UK found solitary bee nest boxes were used by bees. Two replicated trials from the UK found bumblebee nest boxes had very low uptake. Two replicated studies found the local population size or number of emerging red mason bees increased when nest boxes were provided. A replicated trial in Germany found the number of occupied solitary bee nests almost doubled over three years with repeated nest box provision. *Assessment: unknown effectiveness — limited evidence (effectiveness 90%; certainty 38%).*

http://www.conservationevidence.com/actions/80

Provide nest boxes for birds

Two studies (including one before-and-after trial) from the Netherlands and the UK found providing nest boxes increased the number of clutches or breeding adults of two bird species. A replicated study from Switzerland found nest boxes had mixed effects on the number of broods produced by two species. Eight studies (six replicated) from five countries found nest boxes were used by birds. A controlled study from the UK found one species did not use artificial nest sites. Three replicated studies (one paired) from the UK and Sweden found box location influenced use or nesting success. *Assessment: unknown effectiveness — limited evidence (effectiveness 30%; certainty 23%).*

http://www.conservationevidence.com/actions/155

Provide other resources for birds (water, sand for bathing)

A small study in France found grey partridge density was higher in areas where water, shelter, sand and food were provided. *Assessment: unknown effectiveness — limited evidence (effectiveness 0%; certainty 1%).*

http://www.conservationevidence.com/actions/117

Provide refuges during harvest or mowing

A replicated study from France found mowing refuges reduced contact between mowing machinery and unfledged quails and corncrakes. A replicated controlled study and a review from the UK found Eurasian skylark did not use nesting refuges more than other areas. *Assessment: unknown effectiveness — limited evidence (effectiveness 20%; certainty 11%).*

http://www.conservationevidence.com/actions/147

No evidence found (no assessment)

We have captured no evidence for the following interventions:

- Apply 'cross compliance' environmental standards linked to all subsidy payments
- Implement food labelling schemes relating to biodiversity-friendly farming (organic, LEAF marque)
- Introduce nest boxes stocked with solitary bees
- Maintain in-field elements such as field islands and rockpiles
- Manage stone-faced hedge banks to benefit wildlife
- Manage woodland edges to benefit wildlife
- Plant in-field trees (not farm woodland)
- Protect in-field trees (includes management such as pollarding and surgery)
- Provide badger gates
- Provide foraging perches (e.g. for shrikes)
- Provide otter holts
- Provide red squirrel feeders
- Reduce field size (or maintain small fields)
- Restore or maintain dry stone walls
- Support or maintain low intensity agricultural systems

4.2 Arable farming

Based on the collated evidence, what is the current assessment of the effectiveness of interventions for arable farming systems?	
Beneficial	• Create skylark plots • Leave cultivated, uncropped margins or plots (includes 'lapwing plots')
Likely to be beneficial	• Create beetle banks • Leave overwinter stubbles • Reduce tillage • Undersow spring cereals, with clover for example
Unknown effectiveness (limited evidence)	• Convert or revert arable land to permanent grassland • Create rotational grass or clover leys • Increase crop diversity • Plant cereals in wide-spaced rows • Plant crops in spring rather than autumn • Plant nettle strips • Sow rare or declining arable weeds
No evidence found (no assessment)	• Add 1% barley into wheat crop for corn buntings • Create corn bunting plots • Leave unharvested cereal headlands within arable fields • Use new crop types to benefit wildlife (such as perennial cereal crops)
Evidence not assessed	• Implement 'mosaic management', a Dutch agri-environment option • Plant more than one crop per field (intercropping) • Take field corners out of management

Beneficial

Create skylark plots

All four studies (two replicated, controlled trials) from Switzerland and the UK investigating the effect of skylark plots on Eurasian skylarks found positive effects, including increases in population size. A replicated study from Denmark found skylarks used undrilled patches in cereal fields. Three studies (one replicated, controlled) from the UK found benefits to plants and invertebrates. Two replicated studies (one controlled) from the UK found no significant differences in numbers of invertebrates or seed-eating songbirds. *Assessment: beneficial (effectiveness 100%; certainty 80%).*

http://www.conservationevidence.com/actions/540

Leave cultivated, uncropped margins or plots (includes 'lapwing plots')

Seventeen of nineteen individual studies looking at uncropped, cultivated margins or plots (including one replicated, randomized, controlled trial) primarily from the UK found benefits to some or all target farmland bird species, plants, invertebrates or mammals. Two studies (one replicated) from the UK found no effect on ground beetles or most farmland birds. Two replicated site comparisons from the UK found cultivated, uncropped margins were associated with lower numbers of some bird species or age groups in some areas. *Assessment: beneficial (effectiveness 100%; certainty 65%).*

http://www.conservationevidence.com/actions/562

Likely to be beneficial

Create beetle banks

Five reports from two replicated studies (one controlled) and a review from Denmark and the UK found beetle banks had positive effects on invertebrate numbers, diversity or distributions. Five replicated studies (two controlled) found lower or no difference in invertebrate numbers. Three studies (including a replicated, controlled trial) from the UK found beetle banks, alongside other management, had positive effects on bird numbers or usage. Three studies (one replicated site comparison) from the UK found mixed or no effects on birds, two found negative on no clear effects on plants. Two studies (one controlled) from the UK found harvest mice nested on beetle banks. *Assessment: likely to be beneficial (effectiveness 80%; certainty 60%).*

http://www.conservationevidence.com/actions/651

Leave overwinter stubbles

Eighteen studies investigated the effects of overwinter stubbles. Thirteen studies (including two replicated site comparisons and a systematic review) from Finland, Switzerland and the UK found leaving overwinter stubbles benefits some plants, invertebrates, mammals or birds. Three UK studies (one randomized, replicated, controlled) found only certain birds were positively associated with overwinter stubbles. *Assessment: likely to be beneficial (effectiveness 90%; certainty 50%).*

http://www.conservationevidence.com/actions/695

Reduce tillage

Thirty-four studies (including seven randomized, replicated, controlled trials) from nine countries found reducing tillage had some positive effects on invertebrates, weeds or birds. Twenty-seven studies (including three randomized, replicated, controlled trials) from nine countries found reducing tillage had negative or no clear effects on some invertebrates, plants, mammals or birds. Three of the studies did not distinguish between the effects of reducing tillage and reducing chemical inputs. *Assessment: likely to be beneficial (effectiveness 60%; certainty 60%).*

http://www.conservationevidence.com/actions/126

Undersow spring cereals, with clover for example

Eleven studies (including three randomized, replicated, controlled trials) from Denmark, Finland, Switzerland and the UK found undersowing spring cereals benefited some birds, plants or invertebrates, including increases in numbers or species richness. Five studies (including one replicated, randomized, controlled trial) from Austria, Finland and the UK found no benefits to invertebrates, plants or some birds. *Assessment: likely to be beneficial (effectiveness 60%; certainty 43%).*

http://www.conservationevidence.com/actions/136

Unknown effectiveness (limited evidence)

Convert or revert arable land to permanent grassland

All seven individual studies (including two replicated, controlled trials) from the Czech Republic, Denmark and the UK looking at the effects of reverting arable land to grassland found no clear benefits to birds, mammals or plants. *Assessment: unknown effectiveness — limited evidence (effectiveness 0%; certainty 20%).*

http://www.conservationevidence.com/actions/561

Create rotational grass or clover leys

A controlled study from Finland found more spiders and fewer pest insects in clover leys than the crop. A replicated study from the UK found grass leys had fewer plant species than other conservation habitats. A UK study found newer leys had lower earthworm abundance and species richness than older leys. *Assessment: unknown effectiveness — limited evidence (effectiveness 0%; certainty 10%).*

http://www.conservationevidence.com/actions/643

Increase crop diversity

Four studies (including one replicated, controlled trial) from Belgium, Germany and Hungary found more ground beetle or plant species or individuals in fields with crop rotations or on farms with more crops in rotation than monoculture fields. *Assessment: unknown effectiveness — limited evidence (effectiveness 0%; certainty 9%).*

http://www.conservationevidence.com/actions/560

Plant cereals in wide-spaced rows

Two studies (one randomized, replicated, controlled) from the UK found planting cereals in wide-spaced rows had inconsistent, negative or no effects on plant and invertebrate abundance or species richness. *Assessment: unknown effectiveness — limited evidence (effectiveness 0%; certainty 18%).*

http://www.conservationevidence.com/actions/564

Plant crops in spring rather than autumn

Seven studies (including two replicated, controlled trials) from Denmark, Sweden and the UK found sowing crops in spring had positive effects on farmland bird numbers or nesting rates, invertebrate numbers or weed diversity or density. Three of the studies found the effects were seasonal. A review of European studies found fewer invertebrates in spring wheat than winter wheat. *Assessment: unknown effectiveness — limited evidence (effectiveness 40%; certainty 35%).*

http://www.conservationevidence.com/actions/137

Plant nettle strips

A small study from Belgium found nettle strips in field margins had more predatory invertebrate species than the crop, but fewer individuals than the crop or natural nettle stands. *Assessment: unknown effectiveness — limited evidence (effectiveness 50%; certainty 10%).*

http://www.conservationevidence.com/actions/118

Sow rare or declining arable weeds

Two randomized, replicated, controlled studies from the UK identified factors important in establishing rare or declining arable weeds, including type of cover crop, cultivation and herbicide treatment. *Assessment: unknown effectiveness — limited evidence (effectiveness 40%; certainty 15%).*

http://www.conservationevidence.com/actions/642

No evidence found (no assessment)

We have captured no evidence for the following interventions:

- Add 1% barley into wheat crop for corn buntings
- Create corn bunting plots
- Leave unharvested cereal headlands in arable fields
- Use new crop types to benefit wildlife (such as perennial cereal crops)

Evidence not assessed

Implement 'mosaic management', a Dutch agri-environment option

A replicated, controlled before-and-after study from the Netherlands found mosaic management had mixed effects on population trends of wading bird species. A replicated, paired sites study from the Netherlands found one bird species had higher productivity under mosaic management. *Assessment: this intervention has not been assessed.*

http://www.conservationevidence.com/actions/130

Plant more than one crop per field (intercropping)

All five studies (including three randomized, replicated, controlled trials) from the Netherlands, Poland, Switzerland and the UK looking at the effects of planting more than one crop per field found increases in the number of earthworms or ground beetles. *Assessment: this intervention has not been assessed.*

http://www.conservationevidence.com/actions/124

Take field corners out of management

A replicated site comparison from the UK found a positive correlation between grey partridge overwinter survival and taking field corners out of management. Brood size, ratio of young to old birds and density changes were unaffected. *Assessment: this intervention has not been assessed.*

http://www.conservationevidence.com/actions/128

4.3 Perennial (non-timber) crops

Based on the collated evidence, what is the current assessment of the effectiveness of interventions for perennial (non-timber) crops?	
Unknown effectiveness (limited evidence)	• Maintain traditional orchards
No evidence found (no assessment)	• Manage short-rotation coppice to benefit wildlife (includes 8m rides) • Restore or create traditional orchards

Unknown effectiveness (limited evidence)

Maintain traditional orchards

A replicated, controlled site comparison from Germany found more plant species in mown orchards than grazed or abandoned ones, but found no effects on wasps or bees. Two replicated site comparisons from Germany and Switzerland found traditional orchards managed under agri-environment schemes either did not have more plant species than controls or offered no clear benefits to birds. *Assessment: unknown effectiveness — limited evidence (effectiveness 10%; certainty 15%).*

http://www.conservationevidence.com/actions/703

No evidence found (no assessment)

We have captured no evidence for the following interventions:

- Manage short-rotation coppice to benefit wildlife (includes 8 m rides)
- Restore or create traditional orchards

4.4 Livestock farming

Based on the collated evidence, what is the current assessment of the effectiveness of interventions for livestock farming?	
Beneficial	• Restore or create species-rich semi-natural grassland • Use mowing techniques to reduce mortality
Likely to be beneficial	• Delay mowing or first grazing date on grasslands • Leave uncut strips of rye grass on silage fields • Maintain species-rich, semi-natural grassland • Maintain traditional water meadows (includes management for breeding and/or wintering waders/waterfowl) • Maintain upland heath/moorland • Reduce management intensity on permanent grasslands (several interventions at once) • Restore or create traditional water meadows
Unknown effectiveness (limited evidence)	• Add yellow rattle seed Rhi*nanthus minor to* hay meadows • Employ areas of semi-natural habitat for rough grazing (includes salt marsh, lowland heath, bog, fen) • Exclude livestock from semi-natural habitat (including woodland) • Maintain wood pasture and parkland • Plant cereals for whole crop silage • Raise mowing height on grasslands • Restore or create upland heath/moorland • Restore or create wood pasture • Use traditional breeds of livestock

Likely to be ineffective or harmful	• Reduce grazing intensity on grassland (including seasonal removal of livestock)
No evidence found (no assessment)	• Maintain rush pastures • Mark fencing to avoid bird mortality • Plant Brassica fodder crops (grazed in *situ)*
Evidence not assessed	• Create open patches or strips in permanent grassland • Provide short grass for birds • Use mixed stocking

Beneficial

Restore or create species-rich, semi-natural grassland

Twenty studies (including three randomized, replicated, controlled trials) from six countries found restored species-rich, semi-natural grasslands had similar invertebrate, plant or bird diversity or abundance to other grasslands. Seven studies (two randomized, replicated, controlled trials) from five countries found no clear effect on plant or invertebrate numbers, three replicated studies (of which two site comparisons) from two countries found negative effects. Forty studies (including six randomized, replicated, controlled trials) from nine countries identified effective techniques for restoring species-rich grassland. *Assessment: beneficial (effectiveness 100%; certainty 73%).*

http://www.conservationevidence.com/actions/133

Use mowing techniques to reduce mortality

Seven studies (including two replicated trials, one controlled and one randomized) from Germany, Ireland, Switzerland and the UK found mowing techniques that reduced mortality or injury in amphibians, birds, invertebrates or mammals. A review found the UK corncrake population increased around the same time that Corncrake Friendly Mowing was introduced and a replicated trial found mowing from the field centre outwards reduced corncrake chick mortality. *Assessment: beneficial (effectiveness 100%; certainty 78%).*

http://www.conservationevidence.com/actions/698

Likely to be beneficial

Delay mowing or first grazing date on grasslands

Eight studies (including a European systematic review) from the Netherlands, Sweden and the UK found delaying mowing or grazing benefited some or all plants, invertebrates or birds, including increases in numbers or productivity. Three reviews found the UK corncrake population increased following management that included delayed mowing. Six studies (including a European systematic review) from five countries found no clear effect on some plants, invertebrates or birds. *Assessment: likely to be beneficial (effectiveness 60%; certainty 45%).*

http://www.conservationevidence.com/actions/131

Leave uncut strips of rye grass on silage fields

Four studies (including two replicated, controlled trials) from the UK found uncut strips of rye grass benefited some birds, with increased numbers. A randomized, replicated, controlled study from the UK found higher ground beetle diversity on uncut silage plots, but only in the third study year. *Assessment: likely to be beneficial (effectiveness 80%; certainty 49%).*

http://www.conservationevidence.com/actions/132

Maintain species-rich, semi-natural grassland

Nine studies (including two randomized, replicated before-and-after trials) from Switzerland and the UK looked at the effectiveness of agri-environment schemes in maintaining species-rich grassland and all except one found mixed results. All twelve studies (including a systematic review) from six countries looking at grassland management options found techniques that improved or maintained vegetation quality. A site comparison from Finland and Russia found butterfly communities were more affected by grassland age and origin than present management. *Assessment: likely to be beneficial (effectiveness 80%; certainty 60%).*

http://www.conservationevidence.com/actions/702

Maintain traditional water meadows (includes management for breeding and/or wintering waders/waterfowl)

Four studies (including a replicated site comparison) from Belgium, Germany, the Netherlands and the UK found maintaining traditional water meadows increased numbers of some birds or plant diversity. One bird species declined.

Two studies (including a replicated site comparison from the Netherlands) found mixed or inconclusive effects on birds, plants or wildlife generally. A replicated study from the UK found productivity of one wading bird was too low to sustain populations in some areas of wet grassland managed for wildlife. *Assessment: likely to be beneficial (effectiveness 56%; certainty 50%).*

http://www.conservationevidence.com/actions/696

Maintain upland heath/moorland

Eight studies (including one randomized, replicated, controlled trial) from the UK found management, including reducing grazing, can help to maintain the conservation value of upland heath or moorland. Benefits included increased numbers of plants or invertebrates. Three studies (including a before-and-after trial) from the UK found management to maintain upland heath or moorland had mixed effects on some wildlife groups. Four studies (including a controlled site comparison) from the UK found reducing grazing had negative impacts on soil organisms, but a randomized, replicated before-and-after study found heather cover declined where grazing intensity had increased. *Assessment: likely to be beneficial (effectiveness 90%; certainty 50%).*

http://www.conservationevidence.com/actions/647

Reduce management intensity on permanent grasslands (several interventions at once)

Eleven studies (including four replicated site comparisons) from three countries found reducing management intensity benefited plants. Sixteen studies (including four paired site comparisons) from four countries found benefits to some or all invertebrates. Five studies (including one paired, replicated site comparison) from four countries found positive effects on some or all birds. Twenty-one studies (including two randomized, replicated, controlled trials) from six countries found no clear effects of reducing management intensity on some or all plants, invertebrates or birds. Five studies (including two paired site comparisons) from four countries found negative effects on plants, invertebrates or birds. *Assessment: likely to be beneficial (effectiveness 100%; certainty 60%).*

http://www.conservationevidence.com/actions/69

Restore or create traditional water meadows

Three studies (two before-and-after trials) from Sweden and the UK looked at bird numbers following water meadow restoration, one found increases, one found increases and decreases, one found no increases. Seventeen studies (two randomized, replicated, controlled) from six countries found successful techniques for restoring wet meadow plant communities. Three studies (one replicated, controlled) from four countries found restoration of wet meadow plant communities had reduced or limited success. *Assessment: likely to be beneficial (effectiveness 100%; certainty 50%).*

http://www.conservationevidence.com/actions/119

Unknown effectiveness (limited evidence)

Add yellow rattle seed *Rhinanthus minor* to hay meadows

A review from the UK reported that hay meadows had more plant species when yellow rattle was present. A randomized, replicated controlled trial in the UK found yellow rattle could be established by 'slot seeding'. *Assessment: unknown effectiveness — limited evidence (effectiveness 70%; certainty 20%).*

http://www.conservationevidence.com/actions/129

Employ areas of semi-natural habitat for rough grazing (includes salt marsh, lowland heath, bog, fen)

Three studies (two replicated) from the UK and unspecified European countries found grazing had positive effects on birds, butterflies or biodiversity generally. A series of site comparisons from the UK found one bird species used heathland managed for grazing as feeding but not nesting sites. Two studies (one replicated site comparison) from the UK found grazing had negative effects on two bird species. *Assessment: unknown effectiveness — limited evidence (effectiveness 20%; certainty 10%).*

http://www.conservationevidence.com/actions/697

Exclude livestock from semi-natural habitat (including woodland)

Three studies (including one randomized, replicated, controlled trial) from Ireland and the UK found excluding livestock from semi-natural habitats benefited plants and invertebrates. Three studies (one replicated, controlled and one replicated paired sites comparison) from Ireland and the UK did not find benefits to plants or birds. Two studies (one replicated, controlled and a review) from Poland and the UK found limited or mixed effects. *Assessment: unknown effectiveness — limited evidence (effectiveness 20%; certainty 15%).*

http://www.conservationevidence.com/actions/150

Maintain wood pasture and parkland

A randomized, replicated, controlled trial in Sweden found annual mowing on wood pasture maintained the highest number of plant species. *Assessment: unknown effectiveness — limited evidence (effectiveness 40%; certainty 10%).*

http://www.conservationevidence.com/actions/649

Plant cereals for whole crop silage

A replicated study from the UK found cereal-based whole crop silage had higher numbers of some birds than other crops. A review from the UK reported that seed-eating birds avoided cereal-based whole crop silage in winter, but used it as much as spring barley in summer. *Assessment: unknown effectiveness — limited evidence (effectiveness 80%; certainty 28%).*

http://www.conservationevidence.com/actions/149

Raise mowing height on grasslands

Three studies (including one replicated, controlled trial) from the UK or unspecified European countries found raised mowing heights caused less damage to amphibians and invertebrates or increased Eurasian skylark productivity. Two studies (one randomized, replicated, controlled) from the UK found no effect on bird or invertebrate numbers and a replicated study from the UK found young birds had greater foraging success in shorter grass. *Assessment: unknown effectiveness — limited evidence (effectiveness 0%; certainty 35%).*

http://www.conservationevidence.com/actions/138

Restore or create upland heath/moorland

A small trial in northern England found moorland restoration increased the number of breeding northern lapwing. A UK review concluded that vegetation changes were slow during the restoration of heather moorland from upland grassland. *Assessment: unknown effectiveness — limited evidence (effectiveness 78%; certainty 20%).*

http://www.conservationevidence.com/actions/650

Restore or create wood pasture

A replicated, controlled trial in Belgium found survival and growth of tree seedlings planted in pasture was enhanced when they were protected from grazing. A replicated study in Switzerland found cattle browsing had negative effects on tree saplings. *Assessment: unknown effectiveness — limited evidence (effectiveness 40%; certainty 5%).*

http://www.conservationevidence.com/actions/644

Use traditional breeds of livestock

Three studies (one replicated) from the UK found the breed of livestock affected vegetation structure, invertebrate communities and the amount of plants grazed. A replicated trial from France, Germany and the UK found no difference in the number of plant species or the abundance of birds, invertebrates or mammals between areas grazed by traditional or commercial livestock. *Assessment: unknown effectiveness — limited evidence (effectiveness 0%; certainty 20%).*

http://www.conservationevidence.com/actions/539

Likely to be ineffective or harmful

Reduce grazing intensity on grassland (including seasonal removal of livestock)

Fifteen studies (including three randomized, replicated, controlled trials) from four countries found reducing grazing intensity benefited birds, invertebrates or plants. Three studies (including one randomized, replicated, controlled trial) from the Netherlands and the UK found no benefit to plants or invertebrates. Nine studies (including a systematic review) from France, Germany and the UK found mixed effects for some or all wildlife groups. The systematic review concluded that intermediate grazing levels are usually optimal but different wildlife groups are likely to have different grazing requirements. *Assessment: likely to be ineffective (effectiveness 30%; certainty 70%).*

http://www.conservationevidence.com/actions/704

No evidence found (no assessment)

We have captured no evidence for the following interventions:

- Maintain rush pastures
- Mark fencing to avoid bird mortality
- Plant brassica fodder crops (grazed *in situ*)

Evidence not assessed

Create open patches or strips in permanent grassland

A randomized, replicated, controlled study from the UK found more Eurasian skylarks used fields containing open strips, but numbers varied. A randomized, replicated, controlled study from the UK found insect numbers on grassy headlands initially dropped when strips were cleared. *Assessment:this intervention has not been assessed.*

http://www.conservationevidence.com/actions/563

Provide short grass for birds

A replicated UK study found two bird species spent more time foraging on short grass than longer grass. *Assessment: this intervention has not been assessed.*

http://www.conservationevidence.com/actions/115

Use mixed stocking

A replicated, controlled study in the UK found more spiders, harvestmen and pseudoscorpions in grassland grazed by sheep-only than grassland grazed by sheep and cattle. Differences were only found when suction sampling not pitfall-trapping. *Assessment: this intervention has not been assessed.*

http://www.conservationevidence.com/actions/93

4.5 Threat: Residential and commercial development

Based on the collated evidence, what is the current assessment of the effectiveness of interventions for residential and commercial development?	
Unknown effectiveness (limited evidence)	• Provide owl nest boxes (tawny owl, barn owl)
No evidence found (no assessment)	• Maintain traditional farm buildings • Provide bat boxes, bat grilles, improvements to roosts

Unknown effectiveness (limited evidence)

Provide owl nest boxes (tawny owl, barn owl)

Two studies (one before-and-after study) from the Netherlands and the UK found providing nest boxes increased barn owl populations. A replicated study from the UK found a decrease in the proportion of breeding barn owls was not associated with the number of nest boxes. *Assessment: unknown effectiveness — limited evidence (effectiveness 100%; certainty 33%).*

http://www.conservationevidence.com/actions/154

No evidence found (no assessment)

We have captured no evidence for the following interventions:

- Maintain traditional farm buildings
- Provide bat boxes, bat grilles, improvements to roosts

4.6 Threat: Agri-chemicals

Based on the collated evidence, what is the current assessment of the effectiveness of interventions for agri-chemicals?	
Beneficial	• Leave headlands in fields unsprayed (conservation headlands) • Reduce fertilizer, pesticide or herbicide use generally • Use organic rather than mineral fertilisers
Likely to be beneficial	• Reduce chemical inputs in grassland management
Unknown effectiveness (limited evidence)	• Provide buffer strips alongside water courses (rivers and streams) • Restrict certain pesticides
No evidence found (no assessment)	• Buffer in-field ponds
Evidence not assessed	• Make selective use of spring herbicides

Beneficial

Leave headlands in fields unsprayed (conservation headlands)

Twenty-two studies from 14 experiments (including two randomized, replicated, controlled) from five countries found conservation headlands had higher invertebrate or plant diversity than other habitats, twelve studies from ten experiments (three randomized, replicated, controlled) did not.

Twenty-seven studies from 15 experiments (of which 13 replicated, controlled) from five countries found positive effects on abundance or behaviour of some wildlife groups. Nineteen studies from 13 experiments (12 replicated, controlled) from four countries found similar, or lower, numbers of birds, invertebrates or plants on conservation headlands than other habitats. *Assessment: beneficial (effectiveness 90%; certainty 75%).*

http://www.conservationevidence.com/actions/652

Reduce fertilizer, pesticide or herbicide use generally

Thirty-four studies (including a systematic review) from 10 countries found reducing fertilizer, pesticide or herbicide inputs benefited some invertebrates, plants or birds. Twenty-five studies (including seven randomized, replicated, controlled trials) from eight countries found negative or no clear effects on some invertebrates, plants or birds. *Assessment: beneficial (effectiveness 100%; certainty 70%).*

http://www.conservationevidence.com/actions/139

Use organic rather than mineral fertilizers

Fourteen studies (including four randomized, replicated, controlled trials) from six countries found areas treated with organic rather than mineral fertilizers had more plants or invertebrates or higher diversity. A randomized, replicated, controlled trial from the UK found no effect on weed numbers. Two studies (including a small trial from Belgium) found organic fertilizers benefited invertebrates, a UK review found that in large quantities they did not. *Assessment: beneficial (effectiveness 100%; certainty 70%).*

http://www.conservationevidence.com/actions/134

Likely to be beneficial

Reduce chemical inputs in grassland management

Six studies (including a randomized, replicated, controlled before-and-after trial) from three countries found stopping fertilizer inputs on grassland improved plant or invertebrate species richness or abundance. Two reviews from the Netherlands and the UK found no or low fertilizer input grasslands favour some birds and invertebrates. Five studies (two replicated trials of which one randomized and one replicated) from three countries found no clear effects on invertebrates or plants. *Assessment: likely to be beneficial (effectiveness 90%; certainty 60%).*

http://www.conservationevidence.com/actions/694

Unknown effectiveness (limited evidence)

Provide buffer strips alongside water courses (rivers and streams)

Three studies (including one replicated site comparison) from the Netherlands and the UK found riparian buffer strips increased diversity or abundance of plants, invertebrates or birds and supported vegetation associated with water vole habitats. Two replicated site comparisons from France and Ireland found farms with buffer strips did not have more plant species than farms without strips. *Assessment: unknown effectiveness — limited evidence (effectiveness 10%; certainty 15%).*

http://www.conservationevidence.com/actions/120

Restrict certain pesticides

A small UK study found two fungicides that reduced insect abundance less than an alternative. A replicated, controlled trial in Switzerland found applying slug pellets in a band at the field edge was as effective as spreading the pellets across the field. *Assessment: unknown effectiveness — limited evidence (effectiveness 50%; certainty 5%).*

http://www.conservationevidence.com/actions/565

No evidence found (no assessment)

We have captured no evidence for the following intervention:

- Buffer in-field ponds

Evidence not assessed

Make selective use of spring herbicides

A randomized, replicated, controlled study from the UK found spring herbicides had some benefits for beneficial weeds and arthropods. *Assessment: this intervention has not been assessed.*

http://www.conservationevidence.com/actions/98

4.7 Threat: Transport and service corridors

Based on the collated evidence, what is the current assessment of the effectiveness of interventions for transport and service corridors?	
No evidence found (no assessment)	• Manage land under power lines to benefit wildlife

No evidence found (no assessment)

We have captured no evidence for the following intervention:

- Manage land under power lines to benefit wildlife

4.8 Threat: Hunting and trapping (for pest control, food or sport)

Based on the collated evidence, what is the current assessment of the effectiveness of interventions for hunting and trapping (for pest control, food or sport)?	
Unknown effectiveness (limited evidence)	• Enforce legislation to protect birds against persecution • Provide 'sacrificial' grasslands to reduce the impact of wild geese on crops
No evidence found (no assessment)	• Avoid use of lead shot • Use alerts to reduce grey partridge by-catch during shoots
Evidence not assessed	• Use scaring devices (e.g. gas guns) and other deterrents to reduce persecution of native species

Unknown effectiveness (limited evidence)

Enforce legislation to protect birds against persecution

Two before-and-after studies from Denmark and the UK found increased numbers or survival of raptors under legislative protection. *Assessment: unknown effectiveness — limited evidence (effectiveness 90%; certainty 18%).*

http://www.conservationevidence.com/actions/101

Provide 'sacrificial' grasslands to reduce the impact of wild geese on crops

All six studies from the UK (including four replicated, controlled trials) found that managing grasslands for geese increased the number of geese using these areas. Four of these studies found geese were moving within the study sites. *Assessment: unknown effectiveness — limited evidence (effectiveness 20%; certainty 5%).*

http://www.conservationevidence.com/actions/641

No evidence found (no assessment)

We have captured no evidence for the following interventions:

- Avoid use of lead shot
- Use alerts to reduce grey partridge by-catch during shoots

Evidence not assessed

Use scaring devices (e.g. gas guns) and other deterrents to reduce persecution of native species

A replicated, controlled trial in Germany found phosphorescent tape was more effective than normal yellow tape at deterring one of three mammal species. *Assessment: this intervention has not been assessed.*

http://www.conservationevidence.com/actions/645

4.9 Threat: Natural system modification

Based on the collated evidence, what is the current assessment of the effectiveness of interventions for natural system modification?	
Likely to be beneficial	• Raise water levels in ditches or grassland
Unknown effectiveness (limited evidence)	• Create scrapes and pools • Manage heather by swiping to simulate burning • Mange heather, gorse or grass by burning • Remove flood defence banks to allow inundation
No evidence found (no assessment)	• Re-wet moorland

Likely to be beneficial

Raise water levels in ditches or grassland

Eight studies (including two replicated, controlled trials) from Denmark, the Netherlands and the UK found raising water levels increased numbers of birds, invertebrates or plants or allowed wet grassland plant species to establish more rapidly. Three studies (two replicated) from the Netherlands and the UK found raising water levels had negative, limited or no effects on plants or birds. A replicated study from the UK found unflooded pastures had a greater weight of soil invertebrates than flooded pastures. *Assessment: likely to be beneficial (effectiveness 100%; certainty 55%).*

http://www.conservationevidence.com/actions/121

Unknown effectiveness (limited evidence)

Create scrapes and pools

Five studies (including a replicated, controlled, paired trial) from Sweden and the UK found creating scrapes and pools provided habitat for birds, invertebrates or plants or increased invertebrate diversity. Two replicated studies (one controlled, paired) from Ireland and the UK found mixed or no differences in invertebrate numbers between created ponds and controls or natural ponds. A study in Sweden found fewer fish species in constructed than natural wetlands. *Assessment: unknown effectiveness — limited evidence (effectiveness 100%; certainty 28%).*

http://www.conservationevidence.com/actions/153

Manage heather by swiping to simulate burning

A replicated, controlled trial from the UK found heather moorland subject to flailing had fewer plant species than burned plots but more species than unflailed plots. *Assessment: unknown effectiveness — limited evidence (effectiveness 40%; certainty 9%).*

http://www.conservationevidence.com/actions/151

Manage heather, gorse or grass by burning

A long-term replicated, controlled trial in Switzerland found burning of chalk grassland did not increase the number of plant species. A replicated, controlled trial in the UK found more plant species on burned than unburned heather moorland. *Assessment: unknown effectiveness — limited evidence (effectiveness 10%; certainty 5%).*

http://www.conservationevidence.com/actions/152

Remove flood defence banks to allow inundation

A controlled before-and-after study from the UK found a stretch of river that was allowed to flood had more bird species and territories than a channelized section. A study from Belgium found flooding and mowing increased plant species richness in meadow plots. *Assessment: unknown effectiveness — limited evidence (effectiveness 80%; certainty 10%).*

http://www.conservationevidence.com/actions/122

No evidence found (no assessment)

We have captured no evidence for the following intervention:

- Re-wet moorland

4.10 Threat: Invasive and other problematic species

Based on the collated evidence, what is the current assessment of the effectiveness of interventions for invasive and other problematic species?	
Likely to be beneficial	• Control predatory mammals and birds (foxes, crows, stoats and weasels)
Unknown effectiveness (limited evidence)	• Control scrub • Control weeds without damaging other plants in conservation areas • Protect individual nests of ground-nesting birds
No evidence found (no assessment)	• Control grey squirrels • Erect predator-proof fencing around important breeding sites for waders • Manage wild deer numbers • Remove coarse fish
Evidence not assessed	• Control bracken • Control invasive non-native plants on farmland (such as Himalayan balsam, Japanese knotweed) • Control mink • Provide medicated grit for grouse

Likely to be beneficial

Control predatory mammals and birds (foxes, crows, stoats and weasels)

Eight studies (including a systematic review) from France and the UK found predator control (sometimes alongside other interventions) increased the abundance, population size or productivity of some birds. A randomized, replicated, controlled study from the UK did not. *Assessment: likely to be beneficial (effectiveness 90%; certainty 60%).*

http://www.conservationevidence.com/actions/699

Unknown effectiveness (limited evidence)

Control scrub

A replicated site comparison from the UK found the number of young grey partridge per adult was negatively associated with management that included scrub control. *Assessment: unknown effectiveness — limited evidence (effectiveness 0%; certainty 2%).*

http://www.conservationevidence.com/actions/127

Control weeds without damaging other plants in conservation areas

Two studies (one randomized, replicated, controlled) from the UK found that after specific plants were controlled, new plants established or diversity increased. A replicated, controlled laboratory and grassland study found a specific herbicide had negative impacts on one beetle species. Eleven studies investigated different methods of controlling plants. *Assessment: unknown effectiveness — limited evidence (effectiveness 90%; certainty 28%).*

http://www.conservationevidence.com/actions/123

Protect individual nests of ground-nesting birds

Two randomized, replicated, controlled studies from Sweden found nest exclosures increased measures of ground-nesting bird productivity, however both found bird numbers or adult predation rates were unaffected or negatively affected by exclosures. *Assessment: unknown effectiveness — limited evidence (effectiveness 30%; certainty 13%).*

http://www.conservationevidence.com/actions/108

No evidence found (no assessment)

We have captured no evidence for the following interventions:

- Control grey squirrels
- Erect predator-proof fencing around important breeding sites for waders
- Manage wild deer numbers
- Remove coarse fish

Evidence not assessed

Control bracken

A systematic review found repeated herbicide applications reduced bracken abundance but cutting may be equally effective. A laboratory trial found the same herbicide could inhibit the growth of mosses under certain conditions. *Assessment: this intervention has not been assessed.*

http://www.conservationevidence.com/actions/105

Control invasive non-native plants on farmland (such as Himalayan balsam, Japanese knotweed)

Two randomized, replicated, controlled trials in the Czech Republic found removing all giant hogweed flower heads at peak flowering time reduced seed production. *Assessment: this intervention has not been assessed.*

http://www.conservationevidence.com/actions/104

Control mink

A systematic review found trapping may be an effective method of reducing American mink populations. A study in the UK found mink were successfully eradicated from a large area by systematic trapping. *Assessment: this intervention has not been assessed.*

http://www.conservationevidence.com/actions/107

Provide medicated grit for grouse

A controlled study from the UK found higher red grouse productivity where medicated grit was provided. *Assessment: this intervention has not been assessed.*

http://www.conservationevidence.com/actions/112

4.11 Threat: Education and awareness

Based on the collated evidence, what is the current assessment of the effectiveness of interventions for education and awareness?	
No evidence found (no assessment)	• Provide specialist advice, assistance preparing conservation plans
Evidence not assessed	• Provide training for land managers, farmers and farm advisers

No evidence found (no assessment)

We have captured no evidence for the following intervention:

- Provide specialist advice, assistance preparing conservation plans

Evidence not assessed

Provide training for land managers, farmers and farm advisers

A study from the UK found farmers who were trained in how to implement agri-environment schemes created better quality wildlife habitat over five years. *Assessment: this intervention has not been assessed.*

http://www.conservationevidence.com/actions/113

5. FOREST CONSERVATION

Har'el Agra, Simon Schowanek, Yohay Carmel, Rebecca K. Smith & Gidi Ne'eman

Expert assessors

Rhett Harrison, Consultative Group on International Agricultural Research, Zambia
Keith Kirby, University of Oxford, UK
Gillian Petrokofsky, Biodiversity Institute Oxford, UK
Rebecca K. Smith, University of Cambridge, UK
William J. Sutherland, University of Cambridge, UK
Tom Swinfield, Royal Society for the Protection of Birds, UK

Scope of assessment: for the conservation of forest habitat (not specific species within forests), including tropical forests, temperate forests, woodland, scrubland, shrubland and dry forests.

Assessed: 2016.

Effectiveness measure is the median % score.

Certainty measure is the median % certainty of evidence, determined by the quantity and quality of the evidence in the synopsis.

Harm measure is the median % score for negative side-effects on the forest habitat of concern.

 https://doi.org/10.11647/OBP.0267.05

This book is meant as a guide to the evidence available for different conservation interventions and as a starting point in assessing their effectiveness. The assessments are based on the available evidence for the target habitat for each intervention. The assessment may therefore refer to different habitat to the one(s) you are considering. Before making any decisions about implementing interventions it is vital that you read the more detailed accounts of the evidence in order to assess their relevance for your study species or system.

Full details of the evidence are available at**www. conservationevidence.com**

There may also be significant negative side-effects on the target habitats or other species or communities that have not been identified in this assessment.

A lack of evidence means that we have been unable to assess whether or not an intervention is effective or has any harmful impacts.

5.1 Threat: Residential and commercial development

5.1.1 Housing and urban areas

Based on the collated evidence, what is the current assessment of the effectiveness of interventions for residential and commercial development in housing and urban areas?	
No evidence found (no assessment)	• Compensate for woodland removal with compensatory planting • Incorporate existing trees or woods into the landscape of new developments • Provide legal protection of forests from development

No evidence found (no assessment)

We have captured no evidence for the following interventions:

- Compensate for woodland removal with compensatory planting
- Incorporate existing trees or woods into the landscape of new developments
- Provide legal protection of forests from development

5.1.2 Tourism and recreation areas

Based on the collated evidence, what is the current assessment of the effectiveness of interventions for residential and commercial development in tourism and recreation areas?	
No evidence found (no assessment)	• Adopt ecotourism • Create managed paths/signs to contain disturbance • Re-route paths, control access or close paths • Use warning signs to prevent fire

No evidence found (no assessment)

We have captured no evidence for the following interventions:

- Adopt ecotourism
- Create managed paths/signs to contain disturbance
- Re-route paths, control access or close paths
- Use warning signs to prevent fire.

5.2 Threat: Agriculture

5.2.1 Livestock farming

Based on the collated evidence, what is the current assessment of the effectiveness of interventions for livestock farming?	
Likely to be beneficial	• Use wire fences within grazing areas to exclude livestock from specific forest sections
Trade-off between benefit and harms	• Prevent livestock grazing in forests
Unknown effectiveness (limited evidence)	• Reduce the intensity of livestock grazing in forests • Shorten livestock grazing period or control grazing season in forests
No evidence found (no assessment)	• Provide financial incentives not to graze

Likely to be beneficial

Use wire fences within grazing areas to exclude livestock from specific forest sections

Three of four studies, including one replicated, randomized, controlled study in Kenya, Israel, Mexico and Panama found that excluding livestock using wire fences increased the size, density or number of regenerating trees. One study found no effect on tree size and decreased tree density. Four of eight studies, including two replicated, randomized, controlled studies across the world found that excluding livestock using increased biomass, species richness, density or cover of understory plants. Four studies found mixed or no effects on understory plants. *Assessment: likely to be beneficial (effectiveness 58%; certainty 63%; harms 18%).*

http://www.conservationevidence.com/actions/1205

Trade-off between benefit and harms

Prevent livestock grazing in forests

One site comparison study in Israel found that preventing cattle grazing increased the density of seedlings and saplings. Two of three studies, including one replicated, controlled study, in Brazil, Costa Rica and the UK found that preventing livestock grazing increased survival, species richness or diversity of understory plants. One study found mixed effects. *Assessment: trade-offs between benefits and harms (effectiveness 69%; certainty 45%; harms 20%).*

http://www.conservationevidence.com/actions/1206

Unknown effectiveness (limited evidence)

Reduce the intensity of livestock grazing in forests

Two studies, including one replicated, randomized, controlled study, in the UK and Greece found that reducing grazing intensity increased the number of tree saplings or understory total weight. *Assessment: unknown effectiveness (effectiveness 78%; certainty 34%; harms 0%).*

http://www.conservationevidence.com/actions/1207

Shorten livestock grazing period or control grazing season in forests

One of two studies, including one replicated, controlled study, in Spain and Australia found that shortening the grazing period increased the abundance and size of regenerating trees. One found no effect native plant species richness. One replicated study in the UK found that numbers of tree seedlings were higher following summer compared to winter grazing. *Assessment: unknown effectiveness (effectiveness 58%; certainty 33%; harms 0%).*

http://www.conservationevidence.com/actions/1208

No evidence found (no assessment)

We have captured no evidence for the following interventions:

- Provide financial incentives not to graze.

5.3 Threat: Transport and service corridors

Based on the collated evidence, what is the current assessment of the effectiveness of interventions for transport and service corridors?	
No evidence found (no assessment)	• Maintain/create habitat corridors

No evidence found (no assessment)

We have captured no evidence for the following interventions:

- Maintain/create habitat corridors.

5.4 Threat: Biological resource use

5.4.1 Thinning and wood harvesting

Based on the collated evidence, what is the current assessment of the effectiveness of interventions for thinning and wood harvesting?	
Beneficial	• Log/remove trees within forests: effect on understory plants
Likely to be beneficial	• Thin trees within forests: effects on understory plants • Thin trees within forests: effects on young trees • Use shelterwood harvest instead of clearcutting
Trade-off between benefit and harms	• Thin trees within forests: effects on mature trees
Unknown effectiveness (limited evidence)	• Log/remove trees within forests: effects on young trees • Use partial retention harvesting instead of clearcutting • Use summer instead of winter harvesting
Unlikely to be beneficial	• Remove woody debris after timber harvest
Likely to be ineffective or harmful	• Log/remove trees within forests: effect on mature trees • Log/remove trees within forests: effect on non-vascular plants • Thin trees within forests: effect on non-vascular plants
No evidence found (no assessment)	• Adopt continuous cover forestry • Use brash mats during harvesting to avoid soil compaction

Beneficial

Log/remove trees within forests: effects on understory plants

Eight of 12 studies, including four replicated, randomized, controlled studies, in India, Australia, Bolivia, Canada and the USA found that logging increased the density and cover or species richness and diversity of understory plants. Two studies found mixed and three found no effect. *Assessment: beneficial (effectiveness 65%; certainty 65%; harms 10%).*

http://www.conservationevidence.com/actions/1273

Likely to be beneficial

Thin trees within forests: effects on understory plants

Twenty five of 38 studies, including 12 replicated, randomized, controlled studies, across the world found that thinning trees increased the density and cover or species richness and diversity of understory plants. Nine studies found mixed and two no effects, and one found a decrease the abundance of herbaceous species. *Assessment: likely to be beneficial (effectiveness 58%; certainty 73%; harms 13%).*

http://www.conservationevidence.com/actions/1211

Thin trees within forests: effects on young trees

Six of 12 studies, including two replicated, randomized, controlled studies, in Japan and the USA found that thinning increased the density of young trees and a study in Peru found it increased the growth rate of young trees. One study found thinning decreased the density and five found mixed or no effect on young trees. One replicated, controlled study in the USA found no effect on the density of oak acorns. *Assessment: likely to be beneficial (effectiveness 60%; certainty 65%; harms 15%).*

http://www.conservationevidence.com/actions/1210

Use shelterwood harvest instead of clearcutting

Three replicated, controlled studies in Sweden and the USA found that shelterwood harvesting increased density of trees or plant diversity, or decreased grass cover compared with clearcutting. *Assessment: likely to be beneficial (effectiveness 75%; certainty 55%; harms 15%).*

http://www.conservationevidence.com/actions/1214

Trade-off between benefit and harms

Thin trees within forests: effects on mature trees

Eleven of 12 studies, including two replicated, randomized, controlled studies, in Brazil, Canada, and the USA found that thinning trees decreased the density and cover of mature trees and in one case tree species diversity. Five of six studies, including one replicated, controlled, before-and-after study, in Australia, Sweden and the USA found that thinning increased mature tree size, the other found mixed effects. One of three studies, including two replicated controlled studies, in the USA found that thinning reduced the number of trees killed by beetles. *Assessment: trade-offs between benefits and harms (effectiveness 47%; certainty 55%; harms 35%).*

http://www.conservationevidence.com/actions/1209

Unknown effectiveness (limited evidence)

Log/remove trees within forests: effects on young trees

One of two replicated controlled studies in Canada and Costa Rica found that logging increased the density of young trees, the other found mixed effects. *Assessment: unknown effectiveness (effectiveness 50%; certainty 18%; harms 10%).*

http://www.conservationevidence.com/actions/1272

Use partial retention harvesting instead of clearcutting

Three studies, including one replicated, randomized, controlled study, in Canada found that using partial retention harvesting instead of clearcutting decreased the density of young trees. *Assessment: unknown effectiveness (effectiveness 5%; certainty 35%; harms 45%).*

http://www.conservationevidence.com/actions/1215

Use summer instead of winter harvesting

One replicated study in the USA found no effect of logging season on plant species richness and diversity. *Assessment: unknown effectiveness (effectiveness 0%; certainty 13%; harms 0%).*

http://www.conservationevidence.com/actions/1216

Unlikely to be beneficial

Remove woody debris after timber harvest

Two studies, including one replicated, randomized, controlled study, in France and the USA found no effect of woody debris removal on cover or species diversity of trees. One of six studies, including two replicated, randomized, controlled studies, in Ethiopia, Spain, Canada and the USA found that woody debris removal increased young tree density. One found that it decreased young tree density and three found mixed or no effect on density or survival. One of six studies, including two replicated, randomized, controlled studies, in the USA and France found that woody debris removal increased understory vegetation cover. Five studies found mixed or no effects on understory vegetation cover or species richness and diversity. *Assessment: unlikely to be beneficial (effectiveness 23%; certainty 50%; harms 10%).*

http://www.conservationevidence.com/actions/1213

Likely to be ineffective or harmful

Log/remove trees within forests: effect on mature trees

Three of seven studies, including two replicated, controlled studies, across the world found that logging trees decreased the density and cover of mature trees. Two found it increased tree density and two found no effect. Four of nine studies, including one replicated, randomized, controlled study, across the world found that logging increased mature tree size or diversity. Four found it decreased tree size or species richness and diversity, and two found no effect on mature tree size or diversity. One replicated, controlled study in Canada found that logging increased mature tree mortality rate. *Assessment: likely to be ineffective or harmful (effectiveness 35%; certainty 50%; harms 30%).*

http://www.conservationevidence.com/actions/1271

Log/remove trees within forests: effect on effects on non-vascular plants

Two of three studies, including one replicated, paired sites study, in Australia, Norway and Sweden found that logging decreased epiphytic plant abundance and fern fertility. One found mixed effects depending on species. *Assessment: likely to be ineffective or harmful (effectiveness 18%; certainty 40%; harms 50%).*

http://www.conservationevidence.com/actions/1270

Thin trees within forests: effects on non-vascular plants

Three of four studies, including one replicated, randomized, controlled study, in Canada, Finland and Sweden found that thinning decreased epiphytic plant abundance and species richness. Three found mixed effects depending on thinning method and species. *Assessment: likely to be ineffective or harmful (effectiveness 20%; certainty 48%; harms 50%).*

http://www.conservationevidence.com/actions/1212

No evidence found (no assessment)

We have captured no evidence for the following interventions:

- Adopt continuous cover forestry
- Use brash mats during harvesting to avoid soil compaction

5.4.2 Harvest forest products

Based on the collated evidence, what is the current assessment of the effectiveness of interventions for harvesting forest products?	
Unknown effectiveness (limited evidence)	• Adopt certification
No evidence found (no assessment)	• Sustainable management of non-timber products

Unknown effectiveness (limited evidence)

Adopt certification

One replicated, site comparison study in Ethiopia found that deforestation risk was lower in certified than uncertified forests. One controlled, before-and-after trial in Gabon found that, when corrected for logging intensity, although tree damage did not differ, changes in above-ground biomass were smaller in certified than in uncertified forests. *Assessment: unknown effectiveness (effectiveness 50%; certainty 20%; harms 3%).*

http://www.conservationevidence.com/actions/1150

No evidence found (no assessment)

We have captured no evidence for the following interventions:

- Sustainable management of non-timber products

5.4.3 Firewood

Based on the collated evidence, what is the current assessment of the effectiveness of interventions for firewood?	
No evidence found (no assessment)	• Provide fuel efficient stoves • Provide paraffin stoves

No evidence found (no assessment)

We have captured no evidence for the following interventions:

- Provide fuel efficient stoves
- Provide paraffin stoves.

5.5 Habitat protection

5.5.1 Changing fire frequency

Based on the collated evidence, what is the current assessment of the effectiveness of interventions for changing fire frequency?	
Trade-off between benefit and harms	• Use prescribed fire: effect on understory plants • Use prescribed fire: effect on young trees
Likely to be ineffective or harmful	• Use prescribed fire: effect on mature trees
No evidence found (no assessment)	• Mechanically remove understory vegetation to reduce wildfires • Use herbicides to remove understory vegetation to reduce wildfires

Trade-off between benefit and harms

Use prescribed fire: effect on understory plants

Eight of 22 studies, including seven replicated, randomized, controlled studies, in Australia, Canada and the USA found that prescribed fire increased the cover, density or biomass of understory plants. Six found it decreased plant cover and eight found mixed or no effect on cover or density. Fourteen of 24 studies, including 10 replicated, randomized, controlled studies, in Australia, France, West Africa and the USA found that fire increased species richness and diversity of understory plants. One found it decreased species richness and nine found mixed or no effect on understory plants. *Assessment: trade-offs between benefits and harms (effectiveness 55%; certainty 70%; harms 25%).*

http://www.conservationevidence.com/actions/1221

Use prescribed fire: effect on young trees

Five of 15 studies, including four replicated, randomized, controlled studies, in France, Canada and the USA found that prescribed fire increased the density and biomass of young trees. Two found that fire decreased young tree density. Eight found mixed or no effect on density and two found mixed effects on species diversity of young trees. Two replicated, controlled studies in the USA found mixed effects of prescribed fire on young tree survival. *Assessment: trade-offs between benefits and harms (effectiveness 45%; certainty 55%; harms 23%).*

http://www.conservationevidence.com/actions/1220

Likely to be ineffective or harmful

Use prescribed fire: effect on mature trees

Four of nine studies, including two replicated, randomized, controlled studies, in the USA found that prescribed fire decreased mature tree cover, density or diversity. Two studies found it increased tree cover or size, and four found mixed or no effect. Seven studies, including one replicated, randomized, controlled study, in the USA found that fire increased mature tree mortality. *Assessment: likely to be ineffective or harmful (effectiveness 25%; certainty 50%; harms 50%).*

http://www.conservationevidence.com/actions/1217

No evidence found (no assessment)

We have captured no evidence for the following interventions:

- Mechanically remove understory vegetation to reduce wildfires
- Use herbicides to remove understory vegetation to reduce wildfires

5.5.2 Water management

Based on the collated evidence, what is the current assessment of the effectiveness of interventions for water management?	
No evidence found (no assessment)	• Construct water detention areas to slow water flow and restore riparian forests • Introduce beavers to impede water flow in forest watercourses • Recharge groundwater to restore wetland forest

No evidence found (no assessment)

We have captured no evidence for the following interventions:

- Construct water detention areas to slow water flow and restore riparian forests
- Introduce beavers to impede water flow in forest watercourses
- Recharge groundwater to restore wetland forest

5.5.3 Changing disturbance regime

Based on the collated evidence, what is the current assessment of the effectiveness of interventions for changing the disturbance regime?	
Trade-off between benefit and harms	• Use clearcutting to increase understory diversity • Use group-selection harvesting • Use shelterwood harvesting
Unknown effectiveness (limited evidence)	• Thin trees by girdling (cutting rings around tree trunks) • Use herbicides to thin trees
Unlikely to be beneficial	• Use thinning followed by prescribed fire

No evidence found (no assessment)	• Adopt conservation grazing of woodland • Coppice trees • Halo ancient trees • Imitate natural disturbances by pushing over trees • Pollard trees (top cutting or top pruning) • Reintroduce large herbivores • Retain fallen trees

Trade-off between benefit and harms

Use clearcutting to increase understory diversity

Three of nine studies, including four replicated, randomized, controlled studies, in Australia, Japan, Brazil, Canada and the USA found that clearcutting decreased density, species richness or diversity of mature trees. One study found it increased trees species richness and six found mixed or no effect or mixed effect on density, size, species richness or diversity. One replicated, randomized, controlled study in Finland found that clearcutting decreased total forest biomass, particularly of evergreen shrubs. Three of six studies, including five replicated, randomized, controlled studies, in Brazil, Canada and Spain found that clearcutting increased the density and species richness of young trees. One found it decreased young tree density and two found mixed or no effect. Eight of 12 studies, including three replicated, randomized, controlled studies, across the world found that clearcutting increased the cover or species richness of understory plants. Two found it decreased density or species richness, and two found mixed or no effect. *Assessment: trade-offs between benefits and harms (effectiveness 63%; certainty 65%; harms 30%).*

http://www.conservationevidence.com/actions/1222

Use group-selection harvesting

Four of eight studies, including one replicated, controlled study, in Australia, Canada, Costa Rica and the USA found that group-selection harvesting increased cover or diversity of understory plants, or the density of young trees. Two studies found it decreased understory species richness or and biomass. Three studies found no effect on understory species richness or diversity or tree density or growth-rate. *Assessment: trade-offs between benefits and harms (effectiveness 50%; certainty 58%; harms 30%).*

http://www.conservationevidence.com/actions/1224

Use shelterwood harvesting

Six of seven studies, including five replicated, controlled studies, in Australia, Iran, Nepal and the USA found that shelterwood harvesting increased abundance, species richness or diversity or understory plants, as well as the growth and survival rate of young trees. One study found shelterwood harvesting decreased plant species richness and abundance and one found no effect on abundance. One replicated, controlled study in Canada found no effect on oak acorn production. *Assessment: trade-offs between benefits and harms (effectiveness 78%; certainty 70%; harms 28%).*

http://www.conservationevidence.com/actions/1223

Unknown effectiveness (limited evidence)

Thin trees by girdling (cutting rings around tree trunks)

One before-and-after study in Canada found that thinning trees by girdling increased understory plant species richness, diversity and cover. *Assessment: unknown effectiveness — limited evidence (effectiveness 58%; certainty 13%; harms 0%).*

http://www.conservationevidence.com/actions/1226

Use herbicides to thin trees

One replicated, controlled study in Canada found no effect of using herbicide to thin trees on total plant species richness. *Assessment: unknown effectiveness — limited evidence (effectiveness 5%; certainty 13%; harms 0%).*

http://www.conservationevidence.com/actions/1225

Unlikely to be beneficial

Use thinning followed by prescribed fire

Three of six studies, including one replicated, randomized, controlled study, in the USA found that thinning followed by prescribed fire increased cover or abundance of understory plants, and density of deciduous trees. One study found it decreased tree density and species richness. Three studies found mixed or no effect or mixed effect on tree growth rate or density of young trees. One replicated, controlled study Australia found no effect of thinning then burning on the genetic diversity of black ash. *Assessment: unlikely to be beneficial (effectiveness 35%; certainty 40%; harms 15%).*

http://www.conservationevidence.com/actions/1227

No evidence found (no assessment)

We have captured no evidence for the following interventions:

- Adopt conservation grazing of woodland
- Coppice trees
- Halo ancient trees
- Imitate natural disturbances by pushing over trees
- Pollard trees (top cutting or top pruning)
- Reintroduce large herbivores
- Retain fallen trees.

5.6 Threat: Invasive and other problematic species

5.6.1 Invasive plants

Based on the collated evidence, what is the current assessment of the effectiveness of interventions for invasive plants?	
Unknown effectiveness (limited evidence)	• Manually/mechanically remove invasive plants • Use herbicides to remove invasive plant species
No evidence found (no assessment)	• Use grazing to remove invasive plant species • Use prescribed fire to remove invasive plant species

Unknown effectiveness (limited evidence)

Manually/mechanically remove invasive plants

Two replicated, controlled studies in Hawaii and Ghana found that removing invasive grass or weed species increased understory plant biomass or tree seedling height. Two replicated, controlled studies in the USA and Hawaii found no effect of removing invasive shrubs or plants on understory plant diversity or growth rate of native species. *Assessment: unknown effectiveness — limited evidence (effectiveness 40%; certainty 33%; harms 15%).*

http://www.conservationevidence.com/actions/1228

Use herbicides to remove invasive plant species

One replicated, randomized, controlled study in the USA found no effect of controlling invasive plants using herbicide on native plant species richness.

Assessment: unknown effectiveness — limited evidence (effectiveness 5%; certainty 10%; harms 0%).

http://www.conservationevidence.com/actions/1229

No evidence found (no assessment)

We have captured no evidence for the following interventions:

- Use grazing to remove invasive plant species
- Use prescribed fire to remove invasive plant species

5.6.2 Native plants

Based on the collated evidence, what is the current assessment of the effectiveness of interventions for native plants?	
No evidence found (no assessment)	• Manually/mechanically remove native plants

No evidence found (no assessment)

We have captured no evidence for the following interventions:

- Manually/mechanically remove native plants

5.6.3 Herbivores

Based on the collated evidence, what is the current assessment of the effectiveness of interventions for herbivores?	
Likely to be beneficial	• Use wire fences to exclude large native herbivores
Unknown effectiveness (limited evidence)	• Use electric fencing to exclude large native herbivores

No evidence found (no assessment)	• Control large herbivore populations • Control medium-sized herbivores • Use fencing to enclose large herbivores (e.g. deer)

Likely to be beneficial

Use wire fences to exclude large native herbivores

Two replicated, controlled studies in the USA found that excluding large herbivores increased tree density. One of three studies, including two replicated, paired-sites, before-and-after studies, in Canada, Bhutan and Ireland found that excluding large herbivores increased the biomass of young trees. One found it decreased the density of young trees and one found mixed effects on species. Five of 10 studies, including two replicated, randomized, controlled studies, across the world found that excluding large herbivores increased the cover or and size of understory plants. Six found no effect on the cover, seed density, species richness or diversity of understory plants. *Assessment: likely to be beneficial (effectiveness 50%; certainty 65%; harms 10%).*

http://www.conservationevidence.com/actions/1230

Unknown effectiveness (limited evidence)

Use electric fencing to exclude large native herbivores

One controlled study in South Africa found that using electric fencing to exclude elephants and nyalas increased tree density. *Assessment: Unknown effectiveness (effectiveness 65%; certainty 10%; harms 0%).*

http://www.conservationevidence.com/actions/1231

No evidence found (no assessment)

We have captured no evidence for the following interventions:

- Control large herbivore populations
- Control medium-sized herbivores
- Use fencing to enclose large herbivores (e.g. deer)

5.6.4 Rodents

Based on the collated evidence, what is the current assessment of the effectiveness of interventions for rodents?	
Unknown effectiveness (limited evidence)	• Control rodents

Unknown effectiveness (limited evidence)

Control rodents

One controlled study in New Zealand found that rodent control decreased native plant species richness and had no effect on total plant species richness. *Assessment: unknown effectiveness — limited evidence (effectiveness 10%; certainty 10%; harms 50%).*

http://www.conservationevidence.com/actions/1232

5.6.5 Birds

Based on the collated evidence, what is the current assessment of the effectiveness of interventions for birds?	
Unknown effectiveness (limited evidence)	• Control birds

Unknown effectiveness (limited evidence)

Control birds

One controlled study in Australia found that removing birds did not improve the health of the trees in a narrow-leaved peppermint forest. *Assessment: unknown effectiveness — limited evidence (effectiveness 0%; certainty 15%; harms 0%).*

http://www.conservationevidence.com/actions/1151

5.7 Threat: Pollution

Based on the collated evidence, what is the current assessment of the effectiveness of interventions for pollution?	
Unknown effectiveness (limited evidence)	• Maintain/create buffer zones
No evidence found (no assessment)	• Remove nitrogen and phosphorus using harvested products

Unknown effectiveness (limited evidence)

Maintain/create buffer zones

One site comparison study in Australia found that a forest edge protected by a planted buffer strip had higher canopy cover and lower stem density, but similar understory species richness to an unbuffered forest edge. *Assessment: unknown effectiveness — limited evidence (effectiveness 50%; certainty 10%; harms 0%).*

http://www.conservationevidence.com/actions/1168

No evidence found (no assessment)

We have captured no evidence for the following interventions:

- Remove nitrogen and phosphorus using harvested products.

5.8 Threat: Climate change and severe weather

Based on the collated evidence, what is the current assessment of the effectiveness of interventions for climate change and severe weather?	
No evidence found (no assessment)	• Prevent damage from strong winds

No evidence found (no assessment)

We have captured no evidence for the following interventions:

- Prevent damage from strong winds.

5.9 Habitat protection

Based on the collated evidence, what is the current assessment of the effectiveness of interventions for habitat protection?	
Unknown effectiveness (limited evidence)	• Adopt community-based management to protect forests • Legal protection of forests
No evidence found (no assessment)	• Adopt Protected Species legislation (impact on forest management)

Unknown effectiveness (limited evidence)

Adopt community-based management to protect forests

Two studies, including one replicated, before-and-after, site comparison, in Ethiopia and Nepal found that forest cover increased more in community-managed forests than in forests not managed by local communities. However, one replicated, site comparison study in Colombia found that deforestation rates in community-managed forests did not differ from deforestation rates in unmanaged forests. *Assessment: unknown effectiveness — limited evidence (effectiveness 60%; certainty 35%; harms 0%).*

http://www.conservationevidence.com/actions/1152

Legal protection of forests

Two site comparison studies in Nigeria and Iran found that legal protection of forest increased tree species richness and diversity or the density of young trees. One replicated, paired site study in Mexico found no effect of forest protection on seed density and diversity of trees and shrubs. *Assessment: unknown effectiveness — limited evidence (effectiveness 50%; certainty 20%; harms 0%).*

http://www.conservationevidence.com/actions/1233

No evidence found (no assessment)

We have captured no evidence for the following interventions:

- Adopt Protected Species legislation (impact on forest management).

5.10 Habitat restoration and creation

5.10.1 Restoration after wildfire

Based on the collated evidence, what is the current assessment of the effectiveness of interventions for restoration after wildfire?	
Trade-off between benefit and harms	• Thin trees after wildfire
Unknown effectiveness (limited evidence)	• Remove burned trees
Likely to be ineffective or harmful	• Sow tree seeds after wildfire
No evidence found (no assessment)	• Plant trees after wildfire

Trade-off between benefit and harms

Thin trees after wildfire

Four of five replicated, controlled studies in Spain, Israel, Cananda and the USA found that thinning trees in burnt forest areas increased plant species richness, cover or survival of saplings. One study found thinning decreased plant biomass. One paired-site study in Canada found that logging after wildfire decreased species richness and diversity of mosses. *Assessment: trade-offs between benefits and harms (effectiveness 50%; certainty 50%; harms 38%).*

http://www.conservationevidence.com/actions/1234

Unknown effectiveness (limited evidence)

Remove burned trees

Two replicated, controlled studies in Israel and Spain found that removing burned trees increased total plant species richness or the cover and species richness of some plant species. *Assessment: unknown effectiveness (effectiveness 60%; certainty 20%; harms 25%).*

http://www.conservationevidence.com/actions/1237

Likely to be ineffective or harmful

Sow tree seeds after wildfire

Three studies, including one replicated, randomized, controlled study, in the USA found that sowing herbaceous plant seeds in burnt forest areas decreased the density of tree seedlings or the number and cover of native species. All three found no effect of seeding on total plant cover or species richness. *Assessment: likely to be ineffective or harmful (effectiveness 0%; certainty 43%; harms 40%).*

http://www.conservationevidence.com/actions/1236

No evidence found (no assessment)

We have captured no evidence for the following interventions:

- Plant trees after wildfire

5.10.2 Restoration after agriculture

Based on the collated evidence, what is the current assessment of the effectiveness of interventions for restoration after agriculture?	
Unknown effectiveness (limited evidence)	• Restore wood pasture (e.g. introduce grazing)

Unknown effectiveness (limited evidence)

Restore wood pasture (e.g. introduce grazing)

One replicated paired study in Sweden found that partial harvesting in abandoned wood pastures increased tree seedling density, survival and growth. *Assessment: unknown effectiveness (effectiveness 65%; certainty 25%; harms 0%).*

http://www.conservationevidence.com/actions/1164

5.10.3 Manipulate habitat to increase planted tree survival during restoration

Based on the collated evidence, what is the current assessment of the effectiveness of interventions for manipulating habitat to increase planted tree survival during restoration?	
Unknown effectiveness (limited evidence)	• Apply herbicides after restoration planting • Cover the ground using techniques other than plastic mats after restoration planting • Cover the ground with plastic mats after restoration planting • Use selective thinning after restoration planting

Unknown effectiveness (limited evidence)

Apply herbicides after restoration planting

One replicated, randomized, controlled study in the USA found that controlling vegetation using herbicides after restoration planting decreased plant species richness and diversity. *Assessment: unknown effectiveness (effectiveness 45%; certainty 25%; harms 40%).*

http://www.conservationevidence.com/actions/1241

Cover the ground using techniques other than plastic mats after restoration planting

One replicated, randomized, controlled study in the USA found that covering the ground with mulch after planting increased total plant cover. *Assessment: unknown effectiveness (effectiveness 30%; certainty 15%; harms 10%).*

http://www.conservationevidence.com/actions/1240

Cover the ground with plastic mats after restoration planting

One replicated study in Canada found that covering the ground with plastic mats after restoration planting decreased the cover of herbecous plants and grasses. *Assessment: unknown effectiveness (effectiveness 40%; certainty 20%; harms 0%).*

http://www.conservationevidence.com/actions/1239

Use selective thinning after restoration planting

One replicated, paired sites study in Canada found that selective thinning after restoration planting conifers increased the abundance of herbaceous species. *Assessment: unknown effectiveness (effectiveness 43%; certainty 18%; harms 0%).*

http://www.conservationevidence.com/actions/1238

5.10.4 Restore forest community

Based on the collated evidence, what is the current assessment of the effectiveness of interventions for restoring a forest community?	
Unknown effectiveness (limited evidence)	• Build bird-perches to enhance natural seed dispersal • Plant a mixture of tree species to enhance diversity • Sow tree seeds • Water plants to preserve dry tropical forest species
No evidence found (no assessment)	• Restore woodland herbaceous plants using transplants and nursery plugs • Use rotational grazing to restore oak savannas

Unknown effectiveness (limited evidence)

Build bird-perches to enhance natural seed dispersal

One replicated, randomized, controlled, before-and-after study in Brazil found that sowing tree seeds increased the density and species richness of new trees. *Assessment: unknown effectiveness (effectiveness 50%; certainty 13%; harms 0%).*

http://www.conservationevidence.com/actions/1245

Plant a mixture of tree species to enhance diversity

One replicated, randomized, controlled study in Brazil found that planting various tree species increased species richness, but had no effect on the density of new trees. One replicated, controlled study in Greece found that planting native tree species increased total plant species richness, diversity and cover. *Assessment: unknown effectiveness (effectiveness 50%; certainty 28%; harms 0%).*

http://www.conservationevidence.com/actions/1243

Sow tree seeds

One replicated, randomized, controlled, before-and-after study in Brazil found that sowing tree seeds increased the density and species richness of new trees. *Assessment: unknown effectiveness (effectiveness 60%; certainty 13%; harms 0%).*

http://www.conservationevidence.com/actions/1244

Water plants to preserve dry tropical forest species

One replicated, controlled study in Hawaii found that watering plants increased the abundance and biomass of forest plants. *Assessment: unknown effectiveness (effectiveness 65%; certainty 18%; harms 0%).*

http://www.conservationevidence.com/actions/1242

No evidence found (no assessment)

We have captured no evidence for the following interventions:

- Restore woodland herbaceous plants using transplants and nursery plugs
- Use rotational grazing to restore oak savannas

5.10.5 Prevent/encourage leaf litter accumulation

Based on the collated evidence, what is the current assessment of the effectiveness of interventions for preventing/encouraging leaf litter accumulation?	
Unknown effectiveness (limited evidence)	• Remove or disturb leaf litter to enhance germination
No evidence found (no assessment)	• Encourage leaf litter development in new planting

Unknown effectiveness (limited evidence)

Remove or disturb leaf litter to enhance germination

One of two replicated, controlled studies in Poland and Costa Rica found that removing leaf litter increased understory plant species richness. The two studies found that removal decreased understory plant cover or the density of new tree seedlings. *Assessment: unknown effectiveness (effectiveness 40%; certainty 25%; harms 23%).*

http://www.conservationevidence.com/actions/1246

No evidence found (no assessment)

We have captured no evidence for the following interventions:

- Encourage leaf litter development in new planting

5.10.6 Increase soil fertility

Based on the collated evidence, what is the current assessment of the effectiveness of interventions for increasing soil fertility?	
Likely to be beneficial	• Use vegetation removal together with mechanical disturbance to the soil

Trade-off between benefit and harms	• Add organic matter • Use fertilizer • Use soil scarification or ploughing to enhance germination
Unknown effectiveness (limited evidence)	• Add lime to the soil to increase fertility • Use soil disturbance to enhance germination (excluding scarification or ploughing)
Likely to be ineffective or harmful	• Enhance soil compaction

Likely to be beneficial

Use vegetation removal together with mechanical disturbance to the soil

Three studies, including one replicated, randomized, controlled study, in Portugal and France found that vegetation removal together with mechanical disturbance of the soil increased the cover or diversity of understory plants, or density of young trees. One of the studies found it decreased understory shrub cover. *Assessment: likely to be beneficial (effectiveness 61%; certainty 40%; harms 15%).*

http://www.conservationevidence.com/actions/1274

Trade-off between benefit and harms

Add organic matter

One of two studies, including one replicated, randomized, controlled study, in Brazil and Costa Rica found that adding leaf litter increased species richness of young trees. One found it decreased young tree density in artificial forest gaps and both found no effect on the density of tree regenerations under intact forest canopy. One of two replicated, controlled study in Portugal and the USA found that adding plant material increased total plant cover. One found mixed effects on cover depending on plant group. *Assessment: trade-offs between benefits and harms (effectiveness 45%; certainty 43%; harms 28%).*

http://www.conservationevidence.com/actions/1250

Use fertilizer

Six of eight studies, including five replicated, randomized, controlled, in Europe, Brazil, Australia and the USA found that applying fertilizer increased total plant cover, understory plant biomass, size of young trees, biomass of grasses or cover of artificially seeded plant species. Five of the studies found no effect on plant biomass, cover, seedling abundance, tree growth or tree seedling diversity. *Assessment: trade-offs between benefits and harms (effectiveness 55%; certainty 65%; harms 25%).*

http://www.conservationevidence.com/actions/1248

Use soil scarification or ploughing to enhance germination

Two studies, including one replicated, randomized, controlled study, in Portugal and the USA found that ploughing increased the cover or diversity of understory plants. Two of five studies, including two replicated, randomized, controlled, in Canada, Brazil, Ethiopia and Sweden found that ploughing increased the density of young trees. One found a decrease in density and two found mixed effects depending on tree species. One replicated, before-and-after trial in Finland found that ploughing decreased the cover of plants living on wood surface. One replicated, controlled study in the USA found that ploughing did not decrease the spreading distance and density of invasive grass seedlings. *Assessment: unknown effectiveness (effectiveness 60%; certainty 50%; harms 25%).*

http://www.conservationevidence.com/actions/1251

Unknown effectiveness (limited evidence)

Add lime to the soil to increase fertility

One replicated, randomized controlled study in the USA found that adding lime increased vegetation cover. *Assessment: unknown effectiveness (effectiveness 80%; certainty 18%; harms 0%).*

http://www.conservationevidence.com/actions/1249

Use soil disturbance to enhance germination (excluding scarification or ploughing)

Two replicated, controlled studies in Canada and Finland found that disturbance of the forest floor decreased understory vegetation cover. *Assessment: unknown effectiveness (effectiveness 30%; certainty 35%; harms 40%).*

http://www.conservationevidence.com/actions/1252

Likely to be ineffective or harmful

Enhance soil compaction

Two of three studies, including two replicated, randomized, controlled studies in Canada and the USA found that soil compaction increased understory plant cover and density. Two found it decreased tree regeneration height or density and understory plant species richness. *Assessment: likely to be ineffective or harmful (effectiveness 28%; certainty 40%; harms 45%).*

http://www.conservationevidence.com/actions/1253

5.11 Actions to improve survival and growth rate of planted trees

Based on the collated evidence, what is the current assessment of the effectiveness of interventions to improve the survival and growth rate of planted trees?	
Beneficial	• Prepare the ground before tree planting • Use mechanical thinning before or after planting
Likely to be beneficial	• Fence to prevent grazing after tree planting • Use herbicide after tree planting
Trade-off between benefit and harms	• Use prescribed fire after tree planting
Unknown effectiveness (limited evidence)	• Apply insecticide to protect seedlings from invertebrates • Add lime to the soil after tree planting • Add organic matter after tree planting • Cover the ground with straw after tree planting • Improve soil quality after tree planting (excluding applying fertilizer) • Manage woody debris before tree planting • Use shading for planted trees • Use tree guards or shelters to protect planted trees • Use weed mats to protect planted trees • Water seedlings

Unlikely to be beneficial	• Mechanically remove understory vegetation after tree planting • Use different planting or seeding methods • Use fertilizer after tree planting
No evidence found (no assessment)	• Apply fungicide to protect seedlings from fungal diseases • Infect tree seedlings with mycorrhizae • Introduce leaf litter to forest stands • Plant a mixture of tree species to enhance the survival and growth of planted trees • Reduce erosion to increase seedling survival • Transplant trees • Use pioneer plants or crops as nurse-plants

Beneficial

Prepare the ground before tree planting

Six of seven studies, including five replicated, randomized, controlled studies, in Canada and Sweden found that ground preparation increased the survival or growth rate of planted trees. One study found no effect of creating mounds on frost damage to seedlings. *Assessment: beneficial (effectiveness 78%; certainty 73%; harms 0%).*

http://www.conservationevidence.com/actions/1263

Use mechanical thinning before or after planting

Five of six studies, including two replicated, randomized, controlled studies, in Brazil, Canada, Finland, France and the USA found that thinning trees after planting increased survival or size of planted trees. One study found mixed effects on survival and size and one found it decreased their density. One replicated study in the USA found that seedling survival rate increased with the size of the thinned area. *Assessment: beneficial (effectiveness 75%; certainty 63%; harms 10%).*

http://www.conservationevidence.com/actions/1261

Likely to be beneficial

Fence to prevent grazing after tree planting

Four of five studies, including two replicated, randomized, controlled studies, in Finland, Australia, Canada and the USA found that using fences to exclude grazing increased the survival, size or cover of planted trees. Two studies found no effect on survival rate and one found mixed effects on planted tree size. *Assessment: likely to be beneficial (effectiveness 70%; certainty 50%; harms 0%).*

http://www.conservationevidence.com/actions/1254

Use herbicide after tree planting

Two of three studies, including two replicated, randomized, controlled studies, in Sweden and the USA found that using herbicide increased the size of planted trees. One study found no effect. One replicated, randomized, controlled study in Sweden found no effect of using herbicide on frost damage to seedlings. *Assessment: unlikely to be beneficial (effectiveness 58%; certainty 45%; harms 0%).*

http://www.conservationevidence.com/actions/1262

Trade-off between benefit and harms

Use prescribed fire after tree planting

Two of four studies, including one replicated, randomized, controlled study, in Finland, France and the USA found that using prescribed fire after planting increased the survival and sprouting rate of planted trees. One study found fire decreased planted tree size and one found no effect on the size and survival rate. *Assessment: trade-offs between benefits and harms (effectiveness 50%; certainty 43%; harms 20%).*

http://www.conservationevidence.com/actions/1255

Unknown effectiveness (limited evidence)

Apply insecticide to protect seedlings from invertebrates

One randomized, replicated, controlled study in the USA found that applying insecticide increased tree seedling emergence and survival. *Assessment: unknown effectiveness (effectiveness 70%; certainty 13%; harms 0%).*

http://www.conservationevidence.com/actions/1149

Add lime to the soil after tree planting

One of two replicated, randomized, controlled studies in the USA found that adding lime before restoration planting decreased the survival of pine seedlings. One found no effect on seedling growth. *Assessment: unknown effectiveness (effectiveness 0%; certainty 30%; harms 50%).*

http://www.conservationevidence.com/actions/1259

Add organic matter after tree planting

Two replicated, randomized, controlled studies in the USA found that adding organic matter before restoration planting increased seedling biomass, but decreased seedling emergence or survival. *Assessment: unknown effectiveness (effectiveness 20%; certainty 25%; harms 50%).*

http://www.conservationevidence.com/actions/1258

Cover the ground with straw after tree planting

One replicated, randomized, controlled study in the Czech Republic found that covering the ground with straw, but not bark or fleece, increased the growth rate of planted trees and shrubs. *Assessment: unknown effectiveness (effectiveness 75%; certainty 20%; harms 0%).*

http://www.conservationevidence.com/actions/1266

Improve soil quality after tree planting (excluding applying fertilizer)

Two randomized, replicated, controlled studies in Australia found that different soil enhancers had mixed or no effects on tree seedling survival and height, and no effect on diameter or health. *Assessment: unknown effectiveness (effectiveness 25%; certainty 23%; harms 13%).*

http://www.conservationevidence.com/actions/1153

Manage woody debris before tree planting

One replicated, randomized, controlled study in Canada found that removing woody debris increased the survival rate of planted trees. One replicated, controlled study in the USA found mixed effects on the size of planted trees. *Assessment: unknown effectiveness (effectiveness 40%; certainty 25%; harms 13%).*

http://www.conservationevidence.com/actions/1257

Use shading for planted trees

One replicated, controlled study in Panama found that shading increased the survival rate of planted native tree seedlings. *Assessment: unknown effectiveness (effectiveness 85%; certainty 23%; harms 0%).*

http://www.conservationevidence.com/actions/1269

Use tree guards or shelters to protect planted trees

One replicated, randomized, controlled study in the USA found that using light but not dark coloured plastic tree shelters increased the survival rate of planted tree seedlings. One replicated, controlled study in Hong Kong found that tree guards increased tree height after 37 but not 44 months. *Assessment: unknown effectiveness (effectiveness 60%; certainty 28%; harms 20%).*

http://www.conservationevidence.com/actions/1268

Use weed mats to protect planted trees

One replicated, controlled study in Hong Kong found no effect of using weed mats on seedling height. *Assessment: unknown effectiveness (effectiveness 0%; certainty 18%; harms 0%).*

http://www.conservationevidence.com/actions/1267

Water seedlings

One replicated, randomized, controlled study in Spain found that watering seedlings increased or had no effect on seedling emergence and survival, depending on habitat and water availability. *Assessment: unknown effectiveness (effectiveness 45%; certainty 20%; harms 0%).*

http://www.conservationevidence.com/actions/1154

Unlikely to be beneficial

Mechanically remove understory vegetation after tree planting

Four of five studies, including three replicated, randomized, controlled studies in France, Sweden, Panama, Canada and the USA found no effect of controlling understory vegetation on the emergence, survival, growth rate or frost damage of planted seedlings. One found that removing shrubs increased the growth rate and height of planted seedlings, and another that removing competing herbs increased seedling biomass. *Assessment: unlikely to be beneficial (effectiveness 20%; certainty 50%; harms 0%).*

http://www.conservationevidence.com/actions/1256

Use different planting or seeding methods

Four studies, including one replicated, randomized study, in Australia, Brazil, Costa Rica and Mexico found no effect of planting or seeding methods on the size and survival rate of seedlings. One replicated, controlled study in Brazil found that planting early succession pioneer tree species decreased the height of other planted species. *Assessment: unlikely to be beneficial (effectiveness 0%; certainty 43%; harms 13%).*

http://www.conservationevidence.com/actions/1264

Use fertilizer after tree planting

Two of five studies, including two randomized, replicated, controlled studies, in Canada, Australia, France and Portugal found that applying fertilizer after planting increased the size of the planted trees. Three studies found no effect on the size, survival rate or health of planted trees. One randomized, replicated, controlled study in Australia found that soil enhancers including fertilizer had mixed effects on seedling survival and height. *Assessment: unlikely to be beneficial (effectiveness 38%; certainty 45%; harms 3%).*

http://www.conservationevidence.com/actions/1260

No evidence found (no assessment)

We have captured no evidence for the following interventions:

- Apply fungicide to protect seedlings from fungal diseases
- Infect tree seedlings with mycorrhizae
- Introduce leaf litter to forest stands
- Plant a mixture of tree species to enhance the survival and growth of planted trees
- Reduce erosion to increase seedling survival
- Transplant trees
- Use pioneer plants or crops as nurse-plants.

5.12 Education and awareness raising

Based on the collated evidence, what is the current assessment of the effectiveness of interventions to improve education and awareness raising?	
No evidence found (no assessment)	• Provide education programmes about forests • Raise awareness amongst the general public through campaigns and public information

No evidence found (no assessment)

We have captured no evidence for the following interventions:

- Provide education programmes about forests
- Raise awareness amongst the general public through campaigns and public information.

6. PEATLAND CONSERVATION

Nigel G. Taylor, Patrick Grillas & William J. Sutherland

Global evidence for the effects of interventions to conserve peatland vegetation

Expert assessors
Stephanie Boudreau, Canadian Sphagnum Peat Moss Association, Canada
Emma Goodyer, IUCN UK Peatlands Programme, UK
Laura Graham, Borneo Orangutan Survival Foundation, Indonesia
Richard Lindsay, University of East London, UK
Edgar Karofeld, University of Tartu, Estonia
David Locky, MacEwan University, Canada
Nancy Ockendon, University of Cambridge, UK
Anabel Rial, Independent Consultant & IUCN Species Survival Commission, Colombia
Sarah Ross, Penny Anderson Associates, UK
Nigel Taylor, Tour du Valat, France
Tim Thom, Yorkshire Peat Partnership, UK
Jennie Whinam, University of Tasmania, Australia

Scope of assessment: for the conservation of vegetation in wet peatlands, including bogs, fens, fen meadows and tropical peat swamps. The focus is on overall communities and habitat-defining species, rather than rare species.
Assessed: 2018.
Effectiveness measure is the median % score. How effective is the intervention at conserving peatland vegetation in the collated evidence?
Certainty measure is the median % certainty for the effectiveness score across all peatlands that are appropriate targets of the intervention, determined by the quantity and quality of the evidence in the synopsis.
Harm measure is the median % score. Are there any negative side effects of the intervention, on peatland vegetation, in the collated evidence?
Each **effectiveness category** assumes that the aims of the intervention match your management goals. For example, planting trees/shrubs is likely to be beneficial assuming that you want to create forested/shrubby peatland. This might not be a desirable outcome on all peatland types or in all locations.

 https://doi.org/10.11647/OBP.0267.06

This book is meant as a guide to the evidence available for different conservation interventions and as a starting point in assessing their effectiveness. The assessments are based on the available evidence for the target habitat for each intervention. The assessment may therefore refer to different habitat to the one(s) you are considering. Before making any decisions about implementing interventions it is vital that you read the more detailed accounts of the evidence in order to assess their relevance for your study species or system.

Full details of the evidence are available at**www. conservationevidence.com**

There may also be significant negative side-effects on the target habitats or other species or communities that have not been identified in this assessment.

A lack of evidence means that we have been unable to assess whether or not an intervention is effective or has any harmful impacts.

6.1 Threat: Residential and commercial development

Based on the collated evidence, what is the current assessment of the effectiveness of interventions for residential/commercial development?	
No evidence found (no assessment)	• Remove residential or commercial development from peatlands • Retain/create habitat corridors in developed areas

No evidence found (no assessment)

We have captured no evidence for the following interventions:

- Remove residential or commercial development from peatlands
- Retain/create habitat corridors in developed areas.

6.2 Threat: Agriculture and aquaculture

6.2.1 Multiple farming systems

Based on the collated evidence, what is the current assessment of the effectiveness of interventions for multiple farming systems?	
Unknown effectiveness (limited evidence)	• Retain/create habitat corridors in farmed areas
No evidence found (no assessment)	• Implement 'mosaic management' of agriculture

Unknown effectiveness (limited evidence)

Retain/create habitat corridors in farmed areas

- *Vegetation structure:* One study in Indonesia found that a peat swamp forest corridor contained 5,819 trees/ha: 331 large trees, 1,360 saplings and 4,128 seedlings.
- *Overall plant richness/diversity:* The same study recorded 18–29 tree species (depending on size class) in the peat swamp forest corridor.
- *Assessment: unknown effectiveness – limited evidence (effectiveness 45%; certainty 15%; harms 4%). Based on evidence from: tropical peat swamps (one study).*

https://www.conservationevidence.com/actions/1730

No evidence found (no assessment)

We have captured no evidence for the following intervention:

- Implement 'mosaic management' of agriculture.

6.2.2 Wood and pulp plantations

Based on the collated evidence, what is the current assessment of the effectiveness of interventions for wood and pulp plantations?	
Likely to be beneficial	• Cut/remove/thin forest plantations • Cut/remove/thin forest plantations and rewet peat

Likely to be beneficial

Cut/remove/thin forest plantations

- *Herb cover:* Three replicated studies (two also paired and controlled) in bogs in the UK and fens in Sweden reported that tree removal increased cover of some herbs, including cottongrasses *Eriophorum* spp. and sedges overall. One of the studies reported no effect on other herb species, including purple moor grass *Molinia caerulea.*
- *Moss cover:* Two replicated studies, in bogs in the UK and a drained rich fen in Sweden, reported that tree removal reduced moss cover after 3–5 years (specifically fen-characteristic mosses or *Sphagnum* moss). However, one replicated, paired, controlled study in partly rewetted rich fens in Sweden reported that tree removal increased *Sphagnum* moss cover after eight years.
- *Overall plant richness/diversity:* Two replicated, paired, controlled studies in rich fens in Sweden reported that tree removal increased total plant species richness, especially in rewetted plots.
- *Assessment: likely to be beneficial (effectiveness 60%; certainty 50%; harms 10%). Based on evidence from: fens (three studies); bogs (one study).*

https://www.conservationevidence.com/actions/1731

Cut/remove/thin forest plantations and rewet peat

- *Plant community composition:* Of three replicated studies in fens in Finland and Sweden, two found that removing trees/rewetting did not affect the overall plant community composition. One reported only a small effect. Two site comparison studies, in bogs and fens in Finland, found that removing trees/rewetting changed the community composition: it became less like forested/drained sites.
- *Characteristic plants:* Two before-and-after studies (one site comparison, one controlled) in bogs and fens in Finland and Sweden reported that removing trees/rewetting increased the abundance of wetland-characteristic plants.
- *Moss cover:* Five studies (four replicated, three site comparisons) in Sweden and Finland examined the effect of removing trees/rewetting on *Sphagnum* moss cover. Of these, two studies in bogs and fens found that removing trees/rewetting increased *Sphagnum* cover. One study in forested fens found no effect. Two studies in a bog and a fen found mixed effects amongst sites or species. Four studies (three replicated, two paired) in the UK and Finland examined the effect of removing trees/rewetting on other moss cover. Of these, three found that removing trees/rewetting reduced moss cover, but one study in forested fens found no effect.
- *Herb cover:* Seven studies (two replicated, paired, controlled) in bogs and fens in the UK, Finland and Sweden reported that removing trees/rewetting increased cover of at least one group of herbs. This included cottongrasses *Eriophorum* spp. in four of five studies and other/total sedges in three of three studies. One study reported that tree removal/rewetting reduced cover of cottongrass (where it was rare before intervention) and purple moor grass *Molinia caerulea*.
- *Vegetation structure:* One replicated study in a bog in the UK found that removing trees/rewetting increased ground vegetation height, but another in a fen in Sweden reported no effect on canopy height after eight years. Two replicated, paired, site comparison studies in bogs and fens in Finland reported that thinning trees/rewetting reduced the number of tall trees present for 1–3 years after intervention (but not to the level of natural peatlands).
- *Overall plant richness/diversity:* Of four replicated studies in fens in Sweden and Finland, two (also paired and controlled) reported that

removing trees/rewetting increased plant species richness. The other two studies found that removing trees/rewetting had no effect on plant species richness or diversity.

- *Assessment: likely to be beneficial (effectiveness 60%; certainty 60%; harms 10%). Based on evidence from: fens (six studies); bogs (two studies); mixed peatlands (three studies).*

https://www.conservationevidence.com/actions/1732

6.2.3 Livestock farming and ranching

Based on the collated evidence, what is the current assessment of the effectiveness of interventions for livestock farming and ranching?	
Likely to be beneficial	• Exclude or remove livestock from degraded peatlands
Unknown effectiveness (limited evidence)	• Reduce intensity of livestock grazing
No evidence found (no assessment)	• Use barriers to keep livestock off ungrazed peatlands • Change type of livestock • Change season/timing of livestock grazing

Likely to be beneficial

Exclude or remove livestock from degraded peatlands

- *Plant community composition:* Of two replicated, paired, controlled studies in bogs in the UK, one found that excluding sheep had no effect on the plant community. The other found that excluding sheep only affected the community in drier areas of the bog, favouring plants typically found on dry moorlands.
- *Herb cover:* Seven studies (six replicated, paired, controlled) in bogs and fens in the UK, Australia and the USA found that excluding/removing livestock did not affect cover of key herb groups: cottongrasses *Eriophorum* spp. in five of five studies and true sedges

Carex spp. in two of two studies. However, one before-and-after study in a poor fen in Spain reported that rush cover increased after cattle were excluded (along with rewetting). One site comparison study in Chile found that excluding livestock, along with other interventions, increased overall herb cover but one replicated, paired, controlled study in bogs in Australia found that excluding livestock had no effect on herb cover.

- *Moss cover:* Five replicated, paired, controlled studies in bogs in the UK and Australia found that excluding livestock typically had no effect on *Sphagnum* moss cover. Three of the studies in the UK also found no effect on cover of other mosses. One before-and-after study in a poor fen in Spain reported that *Sphagnum* moss appeared after excluding cattle (along with rewetting).
- *Tree/shrub cover:* Five replicated, paired, controlled studies in bogs in the UK and Australia found that excluding livestock typically had no effect on shrub cover (specifically heather *Calluna vulgaris* or heathland plants). However, one of these studies found that heather cover increased in drier areas. Three studies (two site comparisons) in bogs in the UK, fens in the USA and a peatland in Chile found that excluding/removing livestock increased shrub cover.
- *Vegetation structure:* One replicated, paired, controlled study in a bog in the UK found that excluding sheep increased total vegetation, shrub and bryophyte biomass, but had no effect on grass-like plants.
- *Assessment: likely to be beneficial (effectiveness 40%; certainty 50%; harms 12%). Based on evidence from: bogs (seven studies); fens (two studies); unspecified peatlands (one study).*

https://www.conservationevidence.com/actions/1734

Unknown effectiveness (limited evidence)

Reduce intensity of livestock grazing

- *Vegetation cover:* One replicated, paired, controlled study in bogs in the UK found greater cover of total vegetation, shrubs and sheathed cottongrass *Eriophorum vaginatum* under lower grazing intensities.
- *Vegetation structure:* The same study found that vascular plant biomass was higher under lower grazing intensities.

- *Assessment: unknown effectiveness – limited evidence (effectiveness 60%; certainty 25%; harms 1%). Based on evidence from: bogs (one study).*

https://www.conservationevidence.com/actions/1735

No evidence found (no assessment)

We have captured no evidence for the following interventions:

- Use barriers to keep livestock off ungrazed peatlands
- Change type of livestock
- Change season/timing of livestock grazing.

6.3 Threat: Energy production and mining

Based on the collated evidence, what is the current assessment of the effectiveness of interventions for energy production and mining?	
Unknown effectiveness (limited evidence)	• Replace blocks of vegetation after mining or peat extraction
No evidence found (no assessment)	• Retain/create habitat corridors in areas of energy production or mining

Unknown effectiveness (limited evidence)

Replace blocks of vegetation after mining or peat extraction

- *Plant community composition:* Two studies, in a bog in the UK and a fen in Canada, reported that transplanted blocks of peatland vegetation retained their overall community composition: over time in the UK, or relative to an undisturbed fen in Canada.
- *Vegetation cover:* One before-and-after study in the UK reported that bare peat next to translocated bog vegetation developed vegetation cover (mainly grasses/rushes). *Sphagnum* moss cover declined in the translocated blocks. One site comparison study in a fen in Canada reported that replaced vegetation blocks retained similar *Sphagnum* and shrub cover to an undisturbed fen.

- *Assessment: unknown effectiveness – limited evidence (effectiveness 60%; certainty 35%; harms 10%). Based on evidence from: bogs (one study); fens (one study).*

https://www.conservationevidence.com/actions/1738

No evidence found (no assessment)

We have captured no evidence for the following intervention:

- Retain/create habitat corridors in areas of energy production or mining.

6.4 Threat: Transportation and service corridors

Based on the collated evidence, what is the current assessment of the effectiveness of interventions for transportation and service corridors?	
Unknown effectiveness (limited evidence)	• Maintain/restore water flow across service corridors
No evidence found (no assessment)	• Backfill trenches dug for pipelines • Retain/create habitat corridors across service corridors

Unknown effectiveness (limited evidence)

Maintain/restore water flow across service corridors

- *Characteristic plants:* One before-and-after study in a fen in the USA found that after restoring water inflow across a road, along with general rewetting, cover of wet peatland sedges increased whilst cover of grasses preferring drier conditions decreased.
- *Assessment: unknown effectiveness – limited evidence (effectiveness 60%; certainty 20%; harms 1%). Based on evidence from: fens (one study).*

https://www.conservationevidence.com/actions/1741

No evidence found (no assessment)

We have captured no evidence for the following interventions:

- Backfill trenches dug for pipelines
- Retain/create habitat corridors across service corridors.

6.5 Threat: Biological resource use

Based on the collated evidence, what is the current assessment of the effectiveness of interventions for biological resource use?	
Unknown effectiveness (limited evidence)	• Reduce intensity of harvest
No evidence found (no assessment)	• Reduce frequency of harvest • Use low impact harvesting techniques • Use low impact vehicles for harvesting • Implement 'mosaic management' when harvesting wild biological resources • Provide new technologies to reduce pressure on wild biological resources

Unknown effectiveness (limited evidence)

Reduce intensity of harvest

- *Moss cover:* One replicated, controlled study in a bog in New Zealand reported that *Sphagnum* moss cover was higher, three years after harvesting, when some *Sphagnum* was left in plots than when it was completely harvested.
- *Assessment: unknown effectiveness – limited evidence (effectiveness 70%; certainty 25%; harms 0%). Based on evidence from: bogs (one study).*

https://www.conservationevidence.com/actions/1744

No evidence found (no assessment)

We have captured no evidence for the following interventions:

- Reduce frequency of harvest
- Use low impact harvesting techniques
- Use low impact vehicles for harvesting
- Implement 'mosaic management' when harvesting wild biological resources
- Provide new technologies to reduce pressure on wild biological resources.

6.6 Threat: Human intrusions and disturbance

Based on the collated evidence, what is the current assessment of the effectiveness of interventions for human intrusions and disturbance?	
Unknown effectiveness (limited evidence)	• Physically exclude vehicles from peatlands
No evidence found (no assessment)	• Restrict vehicle use on peatlands • Restrict pedestrian access to peatlands • Physically exclude pedestrians from peatlands • Install boardwalks/paths to prevent trampling • Wear snowshoes to prevent trampling • Adopt ecotourism principles/create an ecotourism site

Unknown effectiveness (limited evidence)

Physically exclude vehicles from peatlands

- *Vegetation structure:* One replicated, paired, controlled, site comparison study in a floating fen in the USA reported that fencing off airboat trails allowed total and non-woody vegetation biomass to increase, up to levels recorded in undisturbed fen. Woody plant biomass did not recover.

- *Overall plant richness/diversity:* The same study reported that fencing off airboat trails allowed overall plant diversity to increase, recovering to levels recorded in undisturbed fen.
- *Assessment: unknown effectiveness – limited evidence (effectiveness 70%; certainty 35%; harms 0%). Based on evidence from: fens (one study).*

https://www.conservationevidence.com/actions/1750

No evidence found (no assessment)

We have captured no evidence for the following interventions:

- Restrict vehicle use on peatlands
- Restrict pedestrian access to peatlands
- Physically exclude pedestrians from peatlands
- Install boardwalks/paths to prevent trampling
- Wear snowshoes to prevent trampling
- Adopt ecotourism principles/create an ecotourism site.

6.7 Threat: Natural system modifications

6.7.1 Modified water management

Based on the collated evidence, what is the current assessment of the effectiveness of interventions for modified water management?	
Beneficial	• Rewet peatland (raise water table)
Unknown effectiveness (limited evidence)	• Irrigate peatland
No evidence found (no assessment)	• Reduce water level of flooded peatlands • Restore natural water level fluctuations

Beneficial

Rewet peatland (raise water table)

- *Plant community composition:* Ten of thirteen studies reported that rewetting affected the overall plant community composition. Six before-and-after studies (four also replicated) in peatlands in Finland, Hungary, Sweden, Poland and Germany reported development of wetland- or peatland-characteristic communities following rewetting. One replicated, paired, controlled study in the Czech Republic found differences between rewetted and drained parts of a bog. Three site comparison studies in Finland and Canada reported differences between rewetted and natural peatlands. In contrast, three replicated studies in peatlands in the UK and fens in Germany reported that rewetting typically had no effect, or insignificant effects, on the plant community.

- *Characteristic plants:* Five studies (including one replicated site comparison) in peatlands in Canada, the UK, China and Poland reported that rewetting, sometimes along with other interventions, increased the abundance of wetland- or peatland-characteristic plants. Two replicated site comparison studies, in fens and fen meadows in Europe, found that rewetting reduced the number of fen-characteristic plant species. Two studies (one replicated, paired, controlled, before-and-after) in fens in Sweden reported that rewetting had no effect on cover of fen-characteristic plants.
- *Moss cover:* Twelve studies (two replicated, paired, controlled) in peatlands in Europe and Canada reported that rewetting, sometimes along with other interventions, increased *Sphagnum* moss cover or abundance. However two replicated studies, in bogs in Latvia and forested fens in Finland, reported that rewetting did not affect *Sphagnum* cover. Five studies (one paired, controlled, before-and-after) in bogs and fens in Finland, Sweden and Canada reported that rewetting did not affect cover of non-*Sphagnum* mosses/lichens. However two controlled studies, in bogs in Ireland and the UK, reported that rewetting reduced cover of non-*Sphagnum* bryophytes. One study in Finland reported similar moss cover in rewetted and natural peatlands, but one study in Canada reported that a rewetted bog had lower moss cover than target peatlands.
- *Herb cover:* Twenty-one studies (four replicated, paired, controlled) reported that rewetting, sometimes along with other interventions, increased cover of at least one group of herbs: reeds/rushes in five of seven studies, cottongrasses *Eriophorum* spp. in eight of nine studies, and other/total sedges in 13 of 15 studies. The studies were in bogs, fens or other peatlands in Europe, North America and China. Of four before-and-after studies in peatlands in the UK and Sweden, three reported that rewetting reduced cover of purple moor grass *Molinia caerulea* but one reported no effect. One replicated site comparison study, in forested fens in Finland, reported that rewetting had no

effect on total herb cover. Two site comparison studies in Europe reported that rewetted peatlands had greater herb cover (total or sedges/rushes) than natural peatlands.

- *Tree/shrub cover:* Ten studies (two paired and controlled) in peatlands in Finland, the UK, Germany, Latvia and Canada reported that rewetting typically reduced or had no effect on tree and/or shrub cover. Two before-and-after studies in fens in Sweden and Germany reported that tree/shrub cover increased following rewetting. One before-and-after study in a bog in the UK reported mixed effects of rewetting on different tree/shrub species.
- *Overall vegetation cover:* Of four before-and-after studies (including three controlled), two in bogs in Ireland and Sweden reported that rewetting increased overall vegetation cover. One study in a fen in New Zealand reported that rewetting reduced vegetation cover. One study in a peatland in Finland reported no effect.
- *Overall plant richness/diversity:* Six studies (including one replicated, paired, controlled, before-and-after) in Sweden, Germany and the UK reported that rewetting increased total plant species richness or diversity in peatlands. However, five studies found no effect: in bogs in the Czech Republic and Latvia, fens in Sweden and Germany, and forested fens in Finland. One study in fen meadows in the Netherlands found scale-dependent effects. One paired, controlled, before-and-after study in a peatland in Finland reported that rewetting reduced plant diversity. Of four studies that compared rewetted and natural peatlands, two in Finland and Germany reported lower species richness in rewetted peatlands, one in Sweden found higher species richness in rewetted fens, and one in Europe found similar richness in rewetted and natural fens.
- *Growth:* One replicated site comparison study, in forested fens in Finland, found that rewetting increased *Sphagnum* moss growth to natural levels.
- *Assessment: beneficial (effectiveness 80%; certainty 80%; harms 10%). Based on evidence from: bogs (fifteen studies); fens (fourteen studies); fen meadows (one study); mixed or unspecified peatlands (six studies).*

https://www.conservationevidence.com/actions/1756

Unknown effectiveness (limited evidence)

Irrigate peatland

- *Vegetation cover:* One replicated, paired, controlled, before-and-after study in a bog in Canada found that irrigation increased the number of *Sphagnum* moss shoots present after one growing season, but had no effect after two. One before-and-after study in Germany reported that an irrigated fen was colonized by wetland- and fen-characteristic herbs, whilst cover of dryland grasses decreased.
- *Assessment: unknown effectiveness – limited evidence (effectiveness 55%; certainty 30%; harms 1%). Based on evidence from: bogs (one study); fens (one study).*

https://www.conservationevidence.com/actions/1859

No evidence found (no assessment)

We have captured no evidence for the following interventions:

- Reduce water level of flooded peatlands
- Restore natural water level fluctuations.

6.7.2 Modified vegetation management

Based on the collated evidence, what is the current assessment of the effectiveness of interventions for modified vegetation management?	
Likely to be beneficial	• Cut/mow herbaceous plants to maintain or restore disturbance • Cut large trees/shrubs to maintain or restore disturbance
Trade-off between benefit and harms	• Use grazing to maintain or restore disturbance
Unknown effectiveness (limited evidence)	• Remove plant litter to maintain or restore disturbance • Use prescribed fire to maintain or restore disturbance

Likely to be beneficial

Cut/mow herbaceous plants to maintain or restore disturbance

- *Plant community composition:* Six replicated studies in fens and fen meadows in the UK, Belgium, Germany and the Czech Republic reported that mowing altered the overall plant community composition (vs no mowing, before mowing or grazing). One site comparison study in Poland reported that mowing a degraded fen, along with other interventions, made the plant community more similar to target fen meadow vegetation.
- *Characteristic plants:* Four studies (including one replicated, paired, controlled, before-and-after) in fens and fen meadows in Switzerland, Germany, the Czech Republic and Poland found that cutting/mowing increased cover of fen meadow- or wet meadow-characteristic plants. One replicated before-and-after study, in fens in the UK, found that a single mow typically did not affect cover of fen-characteristic plants. In Poland and the UK, the effect of mowing was not separated from the effects of other interventions.
- *Moss cover:* Four replicated, paired studies (three also controlled) in fens and fen meadows in Belgium, Switzerland and the Czech Republic found that mowing increased total moss or bryophyte cover. Two replicated studies (one also controlled) in fens in Poland and the UK found that a single mow typically had no effect on bryophyte cover (total or hollow-adapted mosses).
- *Herb cover:* Six replicated studies (three also randomized and controlled) in fens and fen meadows in Belgium, Germany, Poland and the UK found that mowing reduced cover or abundance of at least one group of herbs (including bindweed *Calystegia sepium,* purple moor grass *Molinia caerulea,* reeds, sedges, and grass-like plants overall). One before-and-after study in a fen in Poland found that mowing, along with other interventions, increased sedge cover. One replicated, randomized, paired, controlled study in fen meadows in Switzerland found that mowing had no effect on overall herb cover.
- *Tree/shrub cover:* Of three replicated studies in fens, two in the UK found that a single mow, sometimes along with other interventions,

reduced overall shrub cover. The other study, in Poland, found that a single mow had no effect on overall shrub cover.

- *Vegetation structure:* In the following studies, vegetation structure was measured 6–12 months after the most recent cut/mow. Three replicated studies in fens in Poland and the UK reported that a single mow, sometimes along with other interventions, had no (or no consistent) effect on vegetation height. One replicated, paired, site comparison study in fen meadows in Switzerland found that mowing reduced vegetation height. Three studies in fen meadows in Switzerland, Poland and Italy found mixed effects of mowing on vegetation biomass (total, moss, sedge/rush, or common reed *Phragmites australis*). One replicated, paired, site comparison study in Germany reported that vegetation structure was similar in mown and grazed fen meadows.
- *Overall plant richness/diversity:* Eight studies in fens and fen meadows in the UK, Belgium, Switzerland, Germany, the Czech Republic and Poland found that mowing/cutting increased plant species richness (vs no mowing, before mowing or grazing). Three studies (two replicated, randomized, paired, controlled) in fens in Poland and the UK found that a single mow, sometimes along with other interventions, typically did not affect plant richness/diversity.
- *Assessment: likely to be beneficial (effectiveness 70%; certainty 60%; harms 10%). Based on evidence from: fens (seven studies); fen meadows (seven studies).*

https://www.conservationevidence.com/actions/1759

Cut large trees/shrubs to maintain or restore disturbance

- *Plant community composition:* One study in a fen in Poland found that where shrubs were removed, along with other interventions, the plant community became more like a target fen meadow over time.
- *Characteristic plants:* One study in a fen in Poland found that where shrubs were removed, along with other interventions, the abundance of fen meadow plant species increased over time.
- *Vegetation cover:* One replicated, paired, controlled study in a forested fen in the USA found that cutting and removing trees increased herb cover, but did not affect shrub cover.

- *Vegetation structure:* One replicated, paired, controlled study in a forested fen in the USA found that cutting and removing trees increased herb biomass and height.
- *Assessment: likely to be beneficial (effectiveness 60%; certainty 45%; harms 5%). Based on evidence from: fens (two studies).*

https://www.conservationevidence.com/actions/1761

Trade-off between benefit and harms

Use grazing to maintain or restore disturbance

- *Plant community composition:* One replicated, paired, site comparison study in Germany found that the overall plant community composition differed between grazed and mown fen meadows.
- *Characteristic plants:* One replicated, paired, controlled study in Germany reported that the abundance of bog/fen-characteristic plants was similar in grazed and ungrazed fen meadows. One replicated before-and-after study, in a fen in the UK, reported that cover of fen-characteristic mosses did not change after grazers were introduced. One replicated, paired, site comparison study in Germany found that grazed fen meadows contained fewer fen-characteristic plant species than mown meadows.
- *Herb cover:* Two before-and-after studies in fens in the UK reported that grazing increased cover of some herb species/groups (common cottongrass *Eriophorum angustifolium*, carnation sedge *Carex panicea* or grass-like plants overall). One of the studies found that grazing reduced cover of purple moor grass *Molinia caerulea*, but the other found that grazing typically had no effect on this species.
- *Moss cover:* One replicated before-and-after study, in a fen in the UK, reported that cover of fen-characteristic mosses did not change after grazers were introduced. One controlled, before-and-after study in a fen in the UK found that grazing reduced *Sphagnum* moss cover.
- *Tree/shrub cover:* Of two before-and-after studies in fens in the UK, one found that grazing reduced overall shrub cover but the other found that grazing typically had no effect on overall shrub cover.

- *Overall plant richness/diversity:* Of two before-and-after studies in fens in the UK, one (also controlled) reported that grazing increased plant species richness but the other (also replicated) found that grazing had no effect. One replicated, paired, site comparison study in Germany found that grazed fen meadows contained fewer plant species than mown meadows.
- *Assessment: trade-off between benefit and harms (effectiveness 40%; certainty 40%; harms 25%). Based on evidence from: fens (two studies); fen meadows (two studies).*

https://www.conservationevidence.com/actions/1762

Unknown effectiveness (limited evidence)

Remove plant litter to maintain or restore disturbance

- *Plant community composition:* Two studies (including one replicated, paired, controlled, before-and-after) in a fen meadow in Germany and a fen in Czech Republic found that removing plant litter did not affect plant community composition.
- *Vegetation cover:* One replicated, paired, controlled, before-and-after study in a fen in the Czech Republic found that removing plant litter did not affect cover of bryophytes or tall moor grass *Molinia arundinacea*.
- *Overall plant richness/diversity:* Of two replicated, controlled studies, one (also randomized) in a fen meadow in Germany reported that removing plant litter increased plant species richness and diversity. The other study (also paired and before-and-after) in a fen in the Czech Republic found that removing litter did not affect vascular plant diversity.
- *Assessment: unknown effectiveness – limited evidence (effectiveness 35%; certainty 38%; harms 7%). Based on evidence from: fens (one study); fen meadows (one study).*

https://www.conservationevidence.com/actions/1760

Use prescribed fire to maintain or restore disturbance

- *Characteristic plants:* One replicated before-and-after study in a fen in the UK reported that burning, along with other interventions, did not affect cover of fen-characteristic mosses or herbs.
- *Herb cover:* One replicated, controlled study in a fen in the USA reported that burning reduced forb cover and increased sedge/rush cover, but had no effect on grass cover. One replicated before-and-after study in a fen in the UK reported that burning, along with other interventions, reduced grass/sedge/rush cover.
- *Tree/shrub cover:* Two replicated studies in fens in the USA and the UK reported that burning, sometimes along with other interventions, reduced overall tree/shrub cover.
- *Overall plant richness/diversity:* Two replicated, controlled studies in a fen in the USA and a bog in New Zealand found that burning increased plant species richness or diversity. However, one replicated before-and-after study in a fen in the UK reported that burning, along with other interventions, typically had no effect on plant species richness and diversity.
- *Assessment: unknown effectiveness – limited evidence (effectiveness 40%; certainty 35%; harms 20%). Based on evidence from: fens (two studies); bogs (one study).*

https://www.conservationevidence.com/actions/1763

6.7.3 Modified wild fire regime

Based on the collated evidence, what is the current assessment of the effectiveness of interventions for modified wild fire regime?	
No evidence found (no assessment)	• Thin vegetation to prevent wild fires • Rewet peat to prevent wild fires • Build fire breaks • Adopt zero burning policies near peatlands

No evidence found (no assessment)

We have captured no evidence for the following interventions:

- Thin vegetation to prevent wild fires
- Rewet peat to prevent wild fires
- Build fire breaks
- Adopt zero burning policies near peatlands.

6.8 Threat: Invasive and other problematic species

This section includes evidence for the effects of interventions on peatland vegetation overall. Studies that only report effects on the target problematic species are, or will be, summarized in separate chapters (like Chapter 10).

6.8.1 All problematic species

Based on the collated evidence, what is the current assessment of the effectiveness of interventions for all problematic species?	
No evidence found (no assessment)	• Implement biosecurity measures to prevent introductions of problematic species

No evidence found (no assessment)

We have captured no evidence for the following intervention:

- Implement biosecurity measures to prevent introductions of problematic species.

6.8.2 Problematic plants

Based on the collated evidence, what is the current assessment of the effectiveness of interventions for problematic plants?	
Trade-off between benefit and harms	• Use prescribed fire to control problematic plants

Unknown effectiveness (limited evidence)	• Physically remove problematic plants • Use cutting/mowing to control problematic herbaceous plants • Change season/timing of cutting/mowing • Use cutting to control problematic large trees/ shrubs • Use herbicide to control problematic plants • Introduce an organism to control problematic plants
No evidence found (no assessment)	• Physically damage problematic plants • Use grazing to control problematic plants • Use covers/barriers to control problematic plants

Trade-off between benefit and harms

Use prescribed fire to control problematic plants

- *Plant community composition:* One replicated, paired, site comparison study in Germany found that the overall plant community composition differed between grazed and mown fen meadows.
- *Moss cover:* One replicated, paired, controlled study in bogs in Germany found that burning increased moss/lichen/bare ground cover in the short term (2–7 months after burning). Three replicated, paired studies in one bog in the UK found that moss cover (including *Sphagnum*) was higher in plots burned more often.
- *Herb cover:* Four replicated, paired studies (two also controlled) in bogs in Germany and the UK examined the effect of prescribed fire on cottongrass *Eriophorum* spp. cover. One found that burning had no effect on cottongrass cover after 2–7 months. One found that burning increased cottongrass cover after 8–18 years. Two reported that cottongrass cover was similar in plots burned every 10 or 20 years. The study in Germany also found that burning reduced cover of purple moor grass *Molinia caerulea* after 2–7 months but had mixed effects, amongst sites, on cover of other grass-like plants and forbs.
- *Tree/shrub cover:* Four replicated, paired studies (two also controlled) in bogs in Germany and the UK found that burning, or burning

more often, reduced heather *Calluna vulgaris* cover. Two replicated, controlled studies in the bogs in Germany and fens in the USA found that burning, sometimes along with other interventions, had no effect on cover of other woody plants.

- *Vegetation structure:* One replicated, paired, controlled study in a bog in the UK found that plots burned more frequently contained more biomass of grass-like plants than plots burned less often, but contained less total vegetation, shrub and bryophyte biomass.
- *Overall plant richness/diversity:* Two replicated, controlled studies in fens in the USA and a bog in the UK found that burning reduced or limited plant species richness. In the USA, burning was carried out along with other interventions.
- *Assessment: trade-off between benefit and harms (effectiveness 45%; certainty 40%; harms 20%). Based on evidence from: bogs (five studies); fens (one study).*

https://www.conservationevidence.com/actions/1774

Unknown effectiveness (limited evidence)

Physically remove problematic plants

- *Characteristic plants:* One replicated, randomized, controlled study in a fen in Ireland reported that cover of fen-characteristic plants increased after mossy vegetation was removed.
- *Herb cover:* Three replicated, controlled studies in fens in the Netherlands and Ireland reported mixed effects of moss removal on herb cover after 2–5 years. Results varied between species or between sites, and sometimes depended on other treatments applied to plots.
- *Moss cover:* One replicated, randomized, controlled study in a fen in Ireland reported that removing the moss carpet reduced total bryophyte and *Sphagnum* moss cover for three years. Two replicated, controlled, before-and-after studies in fens in the Netherlands reported that removing the moss carpet had no effect on moss cover 2–5 years later in wet plots, but reduced total moss and *Sphagnum* cover in drained plots.
- *Overall plant richness/diversity:* One replicated, controlled, before-and-after study in a fen in the Netherlands reported that removing

moss from a drained area increased plant species richness, but that there was no effect in a wetter area.

- *Assessment: unknown effectiveness – limited evidence (effectiveness 48%; certainty 35%; harms 12%). Based on evidence from: fens (three studies).*

https://www.conservationevidence.com/actions/1768

Use cutting/mowing to control problematic herbaceous plants

- *Plant community composition:* Two replicated, randomized, paired, controlled, before-and-after studies in rich fens in Sweden found that mowing typically did not affect plant community composition. One controlled study in a fen meadow in the UK reported that mown plots developed different communities to unmown plots.
- *Characteristic plants:* One replicated, randomized, paired, controlled, before-and-after study in a fen in Sweden found that mown plots contained more fen-characteristic plant species than unmown plots, although their overall cover did not differ significantly between treatments.
- *Vegetation cover:* Of two replicated, randomized, paired, controlled, before-and-after studies in rich fens in Sweden, one found that mowing had no effect on vascular plant or bryophyte cover over five years. The other study reported that mowing typically increased cover of *Sphagnum* moss and reduced cover of purple moor grass *Molinia caerulea*, but had mixed effects on cover of other plant species.
- *Growth:* One replicated, controlled, before-and-after study in a bog in Estonia found that clipping competing vegetation did not affect *Sphagnum* moss growth.
- *Assessment: unknown effectiveness – limited evidence (effectiveness 40%; certainty 35%; harms 10%). Based on evidence from: fens (two studies); fen meadows (one study); bogs (one study).*

https://www.conservationevidence.com/actions/1770

Change season/timing of cutting/mowing

- *Plant community composition:* One replicated, randomized, paired, before-and after study in a fen meadow in the UK reported that changes in plant community composition over time were similar in spring-, summer- and autumn-mown plots. One study in a peatland in the Netherlands reported that summer- and winter-mown areas developed different plant community types.

- *Overall plant richness/diversity:* One replicated, randomized, paired, before-and after study in a fen meadow in the UK found that plant species richness increased more, over two years, in summer-mown plots than spring- or autumn-mown plots.
- *Assessment: unknown effectiveness – limited evidence (effectiveness 50%; certainty 25%; harms 10%). Based on evidence from: fen meadows (one study); mixed peatlands (one study).*

https://www.conservationevidence.com/actions/1771

Use cutting to control problematic large trees/shrubs

- *Plant community composition:* Two studies (one replicated, controlled, before-and-after) in fens in the USA and Sweden reported that the plant community composition changed after removing trees/shrubs to less like unmanaged fens or more like undegraded, open fen.
- *Characteristic plants:* One study in a fen in Sweden found that species richness and cover of fen-characteristic plants increased after trees/shrubs were removed.
- *Vegetation cover:* One study in a fen in Sweden found that bryophyte and vascular plant cover increased after trees/shrubs were removed. One replicated, controlled, before-and-after study in fens in the USA found that removing shrubs, along with other interventions, could not prevent increases in total woody plant cover over time.
- *Overall plant richness/diversity:* One study in a fen in Sweden found that moss and vascular plant species richness increased after trees/shrubs were removed. However, one replicated, controlled, before-and-after study in fens in the USA found that removing shrubs, along with other interventions, prevented increases in total plant species richness.
- *Assessment: unknown effectiveness – limited evidence (effectiveness 60%; certainty 30%; harms 15%). Based on evidence from: fens (two studies).*

https://www.conservationevidence.com/actions/1772

Use herbicide to control problematic plants

- *Plant community composition:* One replicated, controlled, before-and-after study in fens in the USA found that applying herbicide to shrubs, along with other interventions, changed the overall plant community composition.
- *Tree/shrub cover:* The same study found that applying herbicide to shrubs, along with other interventions, could not prevent increases in total woody plant cover over time.
- *Overall plant richness/diversity:* The same study found that applying herbicide to shrubs, along with other interventions, prevented increases in plant species richness.
- *Assessment: unknown effectiveness – limited evidence (effectiveness 20%; certainty 20%; harms 30%). Based on evidence from: fens (one study).*

https://www.conservationevidence.com/actions/1776

Introduce an organism to control problematic plants

- *Plant community composition:* One controlled, before-and-after study in a fen meadow in Belgium found that introducing a parasitic plant altered the plant community composition.
- *Vegetation cover:* The same study found that introducing a parasitic plant reduced cover of the dominant sedge *Carex acuta* but increased moss cover.
- *Overall plant richness/diversity:* The same study found that introducing a parasitic plant increased overall plant species richness.
- *Assessment: unknown effectiveness – limited evidence (effectiveness 40%; certainty 20%; harms 15%). Based on evidence from: fen meadows (one study).*

https://www.conservationevidence.com/actions/1777

No evidence found (no assessment)

We have captured no evidence for the following interventions:

- Physically damage problematic plants
- Use grazing to control problematic plants
- Use covers/barriers to control problematic plants.

6.8.3 Problematic animals

Based on the collated evidence, what is the current assessment of the effectiveness of interventions for problematic animals?	
Unknown effectiveness (limited evidence)	• Exclude wild herbivores using physical barriers
No evidence found (no assessment)	• Control populations of wild herbivores

Unknown effectiveness (limited evidence)

Exclude wild herbivores using physical barriers

- *Vegetation cover:* One replicated, paired, controlled study in a fen meadow in Poland reported that the effect of boar- and deer exclusion on vascular plant and moss cover depended on other treatments applied to plots.
- *Vegetation structure:* The same study reported that the effect of boar- and deer exclusion on total vegetation biomass depended on other treatments applied to plots.
- *Overall plant richness/diversity:* The same study reported that the effect of boar- and deer exclusion on plant species richness depended on other treatments applied to plots.
- *Assessment: unknown effectiveness – limited evidence (effectiveness 30%; certainty 25%; harms 10%). Based on evidence from: fen meadows (one study).*

https://www.conservationevidence.com/actions/1860

No evidence found (no assessment)

We have captured no evidence for the following intervention:

- Control populations of wild herbivores.

6.9 Threat: Pollution

6.9.1 Multiple sources of pollution

Based on the collated evidence, what is the current assessment of the effectiveness of interventions for multiple sources of pollution?	
Likely to be beneficial	• Divert/replace polluted water source(s)
Unknown effectiveness (limited evidence)	• Clean waste water before it enters the environment • Slow down input water to allow more time for pollutants to be removed
No evidence found (no assessment)	• Retain or create buffer zones between pollution sources and peatlands • Use artificial barriers to prevent pollution entering peatlands • Reduce fertilizer or herbicide use near peatlands • Manage fertilizer or herbicide application near peatlands

Likely to be beneficial

Divert/replace polluted water source(s)

- *Characteristic plants:* One study in a fen in the Netherlands found that after a nutrient-enriched water source was replaced, along with other interventions to reduce pollution, cover of mosses characteristic of low nutrient levels increased.

- *Vegetation cover:* Two studies in bogs in the UK and Japan reported that after polluting water sources were diverted, sometimes along with other interventions, *Sphagnum* moss cover increased. Both studies reported mixed effects on different species of herbs.
- *Assessment: likely to be beneficial (effectiveness 70%; certainty 50%; harms 10%). Based on evidence from: bogs (two studies); fens (one study).*

https://www.conservationevidence.com/actions/1779

Unknown effectiveness (limited evidence)

Clean waste water before it enters the environment

- *Characteristic plants:* One study in the Netherlands found that cleaning water entering a floating fen, along with other interventions to reduce pollution, allowed cover of mosses characteristic of low nutrient levels to increase.
- *Vegetation structure:* The same study found that after the input water began to be cleaned, along with other interventions to reduce pollution, vascular plant biomass decreased.
- *Assessment: unknown effectiveness – limited evidence (effectiveness 60%; certainty 25%; harms 0%). Based on evidence from: fens (one study).*

https://www.conservationevidence.com/actions/1778

Slow down input water to allow more time for pollutants to be removed

- *Characteristic plants:* One before-and-after study in a floating fen in the Netherlands found that after input water was rerouted on a longer path, along with other interventions to reduce pollution, cover of mosses characteristic of low nutrient levels increased.
- *Vegetation structure:* The same study found that after the input water was rerouted on a longer path, along with other interventions to reduce pollution, vascular plant biomass decreased.
- *Assessment: unknown effectiveness – limited evidence (effectiveness 50%; certainty 20%; harms 5%). Based on evidence from: fens (one study).*

https://www.conservationevidence.com/actions/1780

No evidence found (no assessment)

We have captured no evidence for the following interventions:

- Retain or create buffer zones between pollution sources and peatlands
- Use artificial barriers to prevent pollution entering peatlands
- Reduce fertilizer or herbicide use near peatlands
- Manage fertilizer or herbicide application near peatlands.

6.9.2 Agricultural and aquacultural effluents

Based on the collated evidence, what is the current assessment of the effectiveness of interventions for agricultural/aquacultural effluents?	
No evidence found (no assessment)	• Convert to organic agriculture or aquaculture near peatlands • Limit the density of livestock on farmland near peatlands • Use biodegradable oil in farming machinery

No evidence found (no assessment)

We have captured no evidence for the following interventions:

- Convert to organic agriculture or aquaculture near peatlands
- Limit the density of livestock on farmland near peatlands
- Use biodegradable oil in farming machinery.

6.9.3 Industrial and military effluents

Based on the collated evidence, what is the current assessment of the effectiveness of interventions for industrial and military effluents?	
No evidence found (no assessment)	• Remove oil from contaminated peatlands

No evidence found (no assessment)

We have captured no evidence for the following intervention:

- Remove oil from contaminated peatlands.

6.9.4 Airborne pollutants

Based on the collated evidence, what is the current assessment of the effectiveness of interventions for airborne pollutants?	
Unknown effectiveness (limited evidence)	• Remove pollutants from waste gases before they enter the environment • Add lime to reduce acidity and/or increase fertility • Drain/replace acidic water

Unknown effectiveness (limited evidence)

Remove pollutants from waste gases before they enter the environment

- *Plant richness/diversity:* One study in bogs in Estonia reported that after dust filters were installed in industrial plants, along with a general reduction in emissions, the number of *Sphagnum* moss species increased but the total number of plant species decreased.
- *Assessment: unknown effectiveness – limited evidence (effectiveness 50%; certainty 20%; harms 0%). Based on evidence from: bogs (one study).*

https://www.conservationevidence.com/actions/1789

Add lime to reduce acidity and/or increase fertility

- *Vegetation structure:* One replicated, controlled study in a fen meadow in the Netherlands found that liming increased overall vegetation biomass (mostly velvety bentgrass *Agrostis canina*).
- *Assessment: unknown effectiveness – limited evidence (effectiveness 50%; certainty 15%; harms 20%). Based on evidence from: fen meadows (one study).*

https://www.conservationevidence.com/actions/1790

Drain/replace acidic water

- *Vegetation cover:* Two controlled studies in fens in the Netherlands reported that draining acidic water had mixed effects on cover of *Sphagnum* moss and herbs after 4–5 years, depending on the species and whether moss was also removed.
- *Overall plant richness/diversity:* One controlled, before-and-after study in a fen in the Netherlands reported that draining and replacing acidic water increased plant species richness.
- *Assessment: unknown effectiveness – limited evidence (effectiveness 40%; certainty 35%; harms 10%). Based on evidence from: fens (two studies).*

https://www.conservationevidence.com/actions/1791

6.10 Threat: Climate change and severe weather

Based on the collated evidence, what is the current assessment of the effectiveness of interventions for climate change and severe weather?	
No evidence found (no assessment)	• Add water to peatlands to compensate for drought • Plant shelter belts to protect peatlands from wind • Build barriers to protect peatlands from the sea • Restore/create peatlands in areas that will be climatically suitable in the future

No evidence found (no assessment)

We have captured no evidence for the following interventions:

- Add water to peatlands to compensate for drought
- Plant shelter belts to protect peatlands from wind
- Build barriers to protect peatlands from the sea
- Restore/create peatlands in areas that will be climatically suitable in the future.

6.11 Habitat creation and restoration

Remember, the effectiveness category for each intervention assumes that the aims of the intervention match your management goals. You should consider whether each intervention is necessary and appropriate in your focal peatland.

6.11.1 General habitat creation and restoration

Based on the collated evidence, what is the current assessment of the effectiveness of general habitat creation and restoration interventions?	
Likely to be beneficial	• Restore/create peatland vegetation (multiple interventions) • Restore/create peatland vegetation using the moss layer transfer technique

Likely to be beneficial

Restore/create peatland vegetation (multiple interventions)

- *Plant community composition:* One replicated, controlled, before-and-after study in the UK reported that the overall plant community composition differed between restored and unrestored bogs. One replicated, controlled, site comparison study in Estonia found that restored and natural bogs contained more similar plant communities than unrestored and natural bogs. However, one site comparison study in Canada reported that after five years, bogs being restored as fens contained a different plant community to natural fens.

- *Characteristic plants:* One controlled study, in a fen in France, reported that restoration interventions increased cover of fen-characteristic plants.
- *Moss cover:* Five studies (one replicated, paired, controlled, before-and-after) in bogs or other peatlands in the UK, Estonia and Canada found that restoration interventions increased total moss or bryophyte cover. Two studies (one replicated and controlled) in bogs in the Czech Republic and Estonia reported that restoration interventions increased *Sphagnum* moss cover, but one replicated before-and-after study in bogs in the UK reported no change in *Sphagnum* cover following intervention. Two site comparison studies in Canada reported that after 1–15 years, restored areas had lower moss cover than natural fens.
- *Herb cover:* Five studies (one replicated, paired, controlled, before-and-after) in peatlands in the Czech Republic, the UK, Estonia and Canada reported that restoration interventions increased cover of herbs, including cottongrasses *Eriophorum* spp. and other grass-like plants.
- *Overall vegetation cover:* Three studies (one replicated, controlled, before-and-after) in bogs in the UK and France reported that restoration interventions increased overall vegetation cover.
- *Assessment: likely to be beneficial (effectiveness 75%; certainty 60%; harms 5%). Based on evidence from: bogs (six studies); fens (one study); mixed or unspecified peatlands (two studies).*

https://www.conservationevidence.com/actions/1803

Restore/create peatland vegetation using the moss layer transfer technique

- *Plant community composition:* One replicated study in bogs in Canada reported that the majority of restored areas developed a community of bog-characteristic plant species within eleven years. One controlled, before-and-after study in a bog in Canada reported that a restored area (included in the previous study) developed a more peatland-characteristic plant community over time, and relative to an unrestored area.

- *Vegetation cover:* Two controlled studies in one bog in Canada reported that after 4–8 years, a restored area had greater cover than an unrestored area of mosses and bryophytes (including *Sphagnum* spp.) and herbs (including cottongrasses *Eriophorum* spp.), but less cover of shrubs. One of the studies reported that vegetation in the restored area became more similar to local natural bogs.
- *Overall plant richness/diversity:* One controlled, before-and-after study in a bog in Canada reported that after eight years, a restored area contained more plant species than an unrestored area.
- *Assessment: likely to be beneficial (effectiveness 70%; certainty 60%; harms 1%). Based on evidence from: bogs (four studies).*

https://www.conservationevidence.com/actions/1804

6.11.2 Modify physical habitat only

Based on the collated evidence, what is the current assessment of the effectiveness of interventions that modify the physical habitat only?	
Likely to be beneficial	• Fill/block ditches to create conditions suitable for peatland plants • Remove upper layer of peat/soil
Unknown effectiveness (limited evidence)	• Excavate pools • Reprofile/relandscape peatland • Disturb peatland surface to encourage growth of desirable plants • Add inorganic fertilizer • Cover peatland with organic mulch • Cover peatland with something other than mulch • Stabilize peatland surface to help plants colonize • Build artificial bird perches to encourage seed dispersal
No evidence found (no assessment)	• Roughen peat surface to create microclimates • Bury upper layer of peat/soil • Introduce nurse plants

Likely to be beneficial

Fill/block ditches to create conditions suitable for peatland plants

- *Vegetation cover:* Two studies, in a bog in the UK and a fen in the USA, reported that blocked or filled ditches were colonized by peatland vegetation within 2–3 years. In the USA, vegetation cover was restored to natural, undisturbed levels. One replicated study in bogs in the UK reported that plants had not colonized blocked gullies after six months.
- *Overall plant richness/diversity:* One site comparison study in a fen in the USA found that after two years, a filled ditch contained more plant species than adjacent undisturbed fen.
- *Assessment: likely to be beneficial (effectiveness 60%; certainty 50%; harms 0%). Based on evidence from: bogs (two studies); fens (one study).*

https://www.conservationevidence.com/actions/1805

Remove upper layer of peat/soil

- *Plant community composition:* Five studies (one replicated, randomized, paired, controlled) in a peatland in the USA and fens or fen meadows in the Netherlands and Poland reported that plots stripped of topsoil developed different plant communities to unstripped peatlands. In one study, the effect of stripping was not separated from the effect of rewetting. Two studies in fen meadows in Germany and Poland reported that the depth of soil stripping affected plant community development.
- *Characteristic plants:* Four studies (one replicated, randomized, paired, controlled) in fen meadows in Germany and the Netherlands, and a peatland in the USA, reported that stripping soil increased cover of wetland- or peatland-characteristic plants after 4–13 years. In the Netherlands, the effect of stripping was not separated from the effect of rewetting. One replicated site comparison study in fens in Belgium and the Netherlands found that stripping soil increased fen-characteristic plant richness.

- *Herb cover:* Three studies (one replicated, paired, controlled) in fens or fen meadows in Germany, the UK and Poland found that stripping soil increased rush, reed or sedge cover after 2–6 years. One controlled study in a fen meadow in the Netherlands reported that stripping soil had no effect on cover of true sedges *Carex* spp. or velvety bentgrass *Agrostis canina* after five years. Two controlled studies, in fens or fen meadows in the Netherlands and the UK, found that stripping soil reduced cover of purple moor grass *Molinia caerulea* for 2–5 years.
- *Vegetation structure:* Two studies, in fens or fen meadows in the Netherlands and Belgium, found that stripping soil reduced vegetation biomass (total or herbs) for up to 18 years. One replicated, randomized, paired, controlled study in a peatland in the USA found that stripping soil did not affect vegetation biomass after four years.
- *Overall plant richness/diversity:* Three studies (one replicated, paired, controlled) in fens or fen meadows in the UK, Belgium and the Netherlands reported that stripping soil increased total plant species richness over 2–18 years. In one study, the effect of stripping was not separated from the effect of rewetting. One replicated, controlled study in a fen in Poland found that stripping soil had no effect on plant species richness after three years. One replicated, randomized, paired, controlled study in a peatland in the USA found that stripping soil increased plant species richness and diversity, after four years, in one field but decreased it in another. One replicated study in a fen meadow in Poland reported that plant species richness increased after soil was stripped.
- *Assessment: likely to be beneficial (effectiveness 55%; certainty 50%; harms 10%). Based on evidence from: fen meadows (six studies); fens (three studies); unspecified peatlands (one study).*

https://www.conservationevidence.com/actions/1809

Unknown effectiveness (limited evidence)

Excavate pools

- *Plant community composition:* One replicated, before-and-after, site comparison study in bogs in Canada reported that excavated pools were colonized by some peatland vegetation over 4–6 years, but contained different plant communities to natural pools. In particular, cattail *Typha latifolia* was more common in created pools.

- *Vegetation cover:* One replicated, before-and-after, site comparison study in bogs in Canada reported that after four years, created pools had less cover than natural pools of *Sphagnum* moss, herbs and shrubs.
- *Overall plant richness/diversity:* One replicated, before-and-after, site comparison study in bogs in Canada reported that after six years, created pools contained a similar number of plant species to natural pools.
- *Assessment: unknown effectiveness – limited evidence (effectiveness 45%; certainty 38%; harms 5%). Based on evidence from: bogs (two studies).*

https://www.conservationevidence.com/actions/1806

Reprofile/relandscape peatland

- *Plant community composition:* One site comparison study in Canada reported that after five years, reprofiled and rewetted bogs (being restored as fens) contained a different plant community to nearby natural fens.
- *Vegetation cover:* The same study reported that after five years, reprofiled and rewetted bogs (being restored as fens) had lower vegetation cover than nearby natural fens (specifically *Sphagnum* moss, other moss and vascular plants).
- *Assessment: unknown effectiveness – limited evidence (effectiveness 40%; certainty 20%; harms 10%). Based on evidence from: bogs (one study).*

https://www.conservationevidence.com/actions/1807

Disturb peatland surface to encourage growth of desirable plants

- *Plant community composition:* Two replicated, paired, controlled, before-and-after studies (one also randomized) in fens in Germany and Sweden reported that soil disturbance affected development of the plant community over 2–3 years. In Germany, disturbed plots developed greater cover of weedy species from the seed bank than undisturbed plots. In Sweden, the community in disturbed and undisturbed plots became less similar over time.

- *Characteristic plants:* The same two studies reported that wetland- or fen-characteristic plants colonized plots that had been disturbed (along with other interventions). The study in Germany noted that no peat-forming species colonized the fen.
- *Assessment: unknown effectiveness – limited evidence (effectiveness 45%; certainty 30%; harms 20%). Based on evidence from: fens (two studies).*

https://www.conservationevidence.com/actions/1811

Add inorganic fertilizer

- *Vegetation cover:* One replicated, randomized, paired, controlled, before-and-after study in a bog in New Zealand reported that fertilizing typically increased total vegetation cover.
- *Vegetation structure:* One replicated, paired, controlled study in a fen meadow in the Netherlands found that fertilizing with phosphorous typically increased total above-ground vegetation biomass, but other chemicals typically had no effect.
- *Overall plant richness/diversity:* One replicated, randomized, paired, controlled, before-and-after study in a bog in New Zealand reported that fertilizing typically increased plant species richness.
- *Growth:* One replicated, controlled, before-and-after study in a bog in Germany found that fertilizing with phosphorous typically increased herb and shrub growth rate, but other chemicals had no effect.
- *Other:* Three replicated, controlled studies in a fen meadow in Germany and bogs in Germany and New Zealand reported that effects of fertilizer on peatland vegetation were more common when phosphorous was added, than when nitrogen or potassium were added.
- *Assessment: unknown effectiveness – limited evidence (effectiveness 50%; certainty 30%; harms 15%). Based on evidence from: bogs (two studies); fen meadows (one study).*

https://www.conservationevidence.com/actions/1812

Cover peatland with organic mulch

- *Vegetation cover:* One replicated, randomized, paired, controlled, before-and-after study in a bog (being restored as a fen) in Canada found that mulching bare peat did not affect cover of fen-characteristic plants. One replicated, controlled, before-and-after study in a bog in Australia reported that plots mulched with straw had similar *Sphagnum* moss cover to unmulched plots.

- *Characteristic plants:* One replicated, randomized, paired, controlled, before-and-after study in a bog (being restored as a fen) in Canada found that covering bare peat with straw mulch increased the number of fen characteristic plants, but not their overall cover.
- *Assessment: unknown effectiveness – limited evidence (effectiveness 40%; certainty 30%; harms 5%). Based on evidence from: bogs (two studies).*

https://www.conservationevidence.com/actions/1813

Cover peatland with something other than mulch

- *Vegetation cover:* One replicated, controlled, before-and-after study in a bog in Germany reported that covering bare peat with fleece or fibre mats did not affect the number of seedlings of five herb/shrub species. One replicated, controlled, before-and-after study in bogs in Australia reported that recently-burned plots shaded with plastic mesh developed greater cover of native plants, forbs and *Sphagnum* moss than unshaded plots.
- *Assessment: unknown effectiveness – limited evidence (effectiveness 40%; certainty 30%; harms 5%). Based on evidence from: bogs (two studies).*

https://www.conservationevidence.com/actions/1814

Stabilize peatland surface to help plants colonize

- *Vegetation cover:* One controlled, before-and-after study in a bog in the UK found that pegging coconut fibre rolls onto almost-bare peat did not affect the development of vegetation cover (total, mosses, shrubs or common cottongrass *Eriophorum angustifolium*).
- *Assessment: unknown effectiveness – limited evidence (effectiveness 20%; certainty 20%; harms 5%). Based on evidence from: bogs (one study).*

https://www.conservationevidence.com/actions/1815

Build artificial bird perches to encourage seed dispersal

- *Vegetation cover:* One replicated, paired, controlled study in a peat swamp forest in Indonesia found that artificial bird perches had no significant effect on tree seedling abundance.

- *Assessment: unknown effectiveness – limited evidence (effectiveness 20%; certainty 20%; harms 1%). Based on evidence from: tropical peat swamps (one study).*

https://www.conservationevidence.com/actions/1817

No evidence found (no assessment)

We have captured no evidence for the following interventions:

- Roughen peat surface to create microclimates
- Bury upper layer of peat/soil
- Introduce nurse plants.

6.11.3 Introduce peatland vegetation

Based on the collated evidence, what is the current assessment of the effectiveness of interventions that introduce peatland vegetation?	
Beneficial	• Add mosses to peatland surface • Add mixed vegetation to peatland surface
Likely to be beneficial	• Directly plant peatland mosses • Directly plant peatland herbs • Directly plant peatland trees/shrubs • Introduce seeds of peatland herbs • Introduce seeds of peatland trees/shrubs

Beneficial

Add mosses to peatland surface

- *Sphagnum moss cover:* Eleven studies in bogs in the UK, Canada, Finland and Germany and fens in the USA reported that *Sphagnum* moss was present, after 1–4 growing seasons, in at least some plots sown with *Sphagnum*. Cover ranged from negligible to >90%. Six of these studies were controlled and found that there was more *Sphagnum* in sown than unsown plots. One additional study in

Canada found that adding *Sphagnum* to bog pools did not affect *Sphagnum* cover.

- *Other moss cover:* Four studies (including one replicated, randomized, paired, controlled, before-and-after) in bogs in Canada and fens in Sweden and the USA reported that mosses other than *Sphagnum* were present, after 2–3 growing seasons, in at least some plots sown with moss fragments. Cover ranged from negligible to 76%. In the fens in Sweden and the USA, moss cover was low (<1%) unless the plots were mulched, shaded or limed.
- *Assessment: beneficial (effectiveness 78%; certainty 70%; harms 1%). Based on evidence from: bogs (eleven studies); fens (two studies).*

https://www.conservationevidence.com/actions/1821

Add mixed vegetation to peatland surface

- *Characteristic plants:* One replicated, randomized, paired, controlled, before-and-after study in a degraded bog (being restored as a fen) in Canada found that adding fen vegetation increased the number and cover of fen-characteristic plant species.
- *Sphagnum moss cover:* Seventeen replicated studies (five also randomized, paired, controlled, before-and-after) in bogs in Canada, the USA and Estonia reported that *Sphagnum* moss was present, after 1–6 growing seasons, in at least some plots sown with vegetation containing *Sphagnum*. Cover ranged from <1 to 73%. Six of the studies were controlled and found that *Sphagnum* cover was higher in sown than unsown plots. Five of the studies reported that *Sphagnum* cover was very low (<1%) unless plots were mulched after spreading fragments.
- *Other moss cover:* Eight replicated studies (seven before-and-after, one controlled) in bogs in Canada, the USA and Estonia reported that mosses or bryophytes other than *Sphagnum* were present, after 1–6 growing seasons, in at least some plots sown with mixed peatland vegetation. Cover ranged from <1 to 65%.
- *Vascular plant cover:* Ten replicated studies in Canada, the USA and Estonia reported that vascular plants appeared following addition of mixed vegetation fragments to bogs. Two of the studies were controlled: one found that vascular plant cover was significantly

higher in sown than unsown plots, but one found that sowing peatland vegetation did not affect herb cover.

- *Assessment: beneficial (effectiveness 78%; certainty 68%; harms 1%). Based on evidence from: bogs (eighteen studies).*

https://www.conservationevidence.com/actions/1822

Likely to be beneficial

Directly plant peatland mosses

- *Survival:* One study in Lithuania reported that 47 of 50 *Sphagnum*-dominated sods planted into a rewetted bog survived for one year.
- *Growth:* Two before-and-after studies, in a fen in the Netherlands and bog pools in the UK, reported that mosses grew after planting.
- *Moss cover:* Five before-and-after studies in a fen in the Netherlands and bogs in Germany, Ireland, Estonia and Australia reported that after planting mosses, the area covered by moss increased in at least some cases. The study in the Netherlands reported spread of planted moss beyond the introduction site. The study in Australia was controlled and reported that planted plots developed greater *Sphagnum* moss cover than unplanted plots.
- *Assessment: likely to be beneficial (effectiveness 75%; certainty 60%; harms 0%). Based on evidence from: bogs (six studies); fens (one study).*

https://www.conservationevidence.com/actions/1818

Directly plant peatland herbs

- *Survival:* Three replicated studies, in a fen meadow in the Netherlands and fens in the USA, reported that planted herbs survived over 2–3 years. However, for six of nine species only a minority of individuals survived.
- *Growth:* Two replicated before-and-after studies, in a bog in Germany and fens in the USA, reported that planted herbs grew.
- *Vegetation cover:* One replicated, controlled, before-and-after study in Canada found that planting herbs had no effect on moss, herb or shrub cover in created bog pools relative to natural colonization.

- *Assessment: likely to be beneficial (effectiveness 50%; certainty 40%; harms 0%). Based on evidence from: bogs (two studies); fens (two studies); fen meadows (one study).*

https://www.conservationevidence.com/actions/1819

Directly plant peatland trees/shrubs

- *Survival:* Eight studies (seven replicated) in peat swamp forests in Thailand, Malaysia and Indonesia and bogs in Canada reported that the majority of planted trees/shrubs survived over periods between 10 weeks and 13 years. One study in a peat swamp forest in Indonesia reported <5% survival of planted trees after five months, following unusually deep flooding. One replicated study in a fen in the USA reported that most planted willow *Salix* spp. cuttings died within two years.
- *Growth:* Four studies (including two replicated, before-and-after) in peat swamp forests in Thailand, Indonesia and Malaysia reported that planted trees grew. One replicated before-and-after study in bogs in Canada reported that planted shrubs grew.
- *Assessment: likely to be beneficial (effectiveness 70%; certainty 50%; harms 0%). Based on evidence from: tropical peat swamps (seven studies); bogs (three studies); fens (one study).*

https://www.conservationevidence.com/actions/1820

Introduce seeds of peatland herbs

- *Germination:* Two replicated studies (one also controlled, before-and-after) reported that some planted herb seeds germinated. In a bog in Germany three of four species germinated, but in a fen in the USA only one of seven species germinated.
- *Characteristic plants:* Three studies (two controlled) in fen meadows in Germany and a peatland in China reported that wetland-characteristic or peatland-characteristic plants colonized plots where herb seeds were sown (sometimes along with other interventions).
- *Herb cover:* Three before-and-after studies (one also replicated, randomized, paired, controlled) in a bog in New Zealand, fen meadows in Switzerland and a peatland in China reported that plots sown with herb seeds developed cover of the sown herbs (and, in

New Zealand, greater cover than unsown plots). In China, the effect of sowing was not separated from the effects of other interventions. One replicated, randomized, paired, controlled study in a fen in the USA found that plots sown with herb (and shrub) seeds developed similar herb cover to plots that were not sown.

- *Overall vegetation cover:* Of three replicated, controlled studies, one in a fen in the USA found that sowing herb (and shrub) seeds increased total vegetation cover. One study in a bog in New Zealand found that sowing herb seeds had no effect on total vegetation cover. One study in a fen meadow in Poland found that the effect of adding seed-rich hay depended on other treatments applied to plots.
- *Overall plant richness/diversity:* Two replicated, controlled studies in fens in the USA and Poland found that sowing herb seeds had no effect on plant species richness (total or vascular). Two replicated, controlled, before-and-after studies in a bog in New Zealand and a fen meadow in Poland each reported inconsistent effects of herb sowing on total plant species richness.
- *Assessment: likely to be beneficial (effectiveness 50%; certainty 50%; harms 0%). Based on evidence from: fen meadows (four studies); fens (three studies); bogs (two studies); unspecified peatlands (one study).*

https://www.conservationevidence.com/actions/1823

Introduce seeds of peatland trees/shrubs

- *Germination:* Two replicated studies in a bog in Germany and a fen in the USA reported germination of heather *Calluna vulgaris* and hoary willow *Salix candida* seeds, respectively, in at least some sown plots.
- *Survival:* The study in the bog Germany reported survival of some heather seedlings over two years. The study in the fen in the USA reported that all germinated willow seedlings died within one month.
- *Shrub cover:* Two studies (one replicated, randomized, paired, controlled) in bogs in New Zealand and Estonia reported that plots sown with shrub seeds, sometimes along with other interventions, developed greater cover of some shrubs than plots that were not sown: sown manuka *Leptospermum scoparium* or naturally colonizing heather *Calluna vulgaris* (but not sown cranberry *Oxycoccus palustris*).

One replicated, randomized, paired, controlled study in a fen in the USA found that plots sown with shrub (and herb) seeds developed similar overall shrub cover to unsown plots within two years.

- *Overall vegetation cover:* Two replicated, randomized, paired, controlled studies in a bog in New Zealand and a fen in the USA reported that plots sown with shrub (and herb) seeds developed greater total vegetation cover than unsown plots after two years. One site comparison study in bogs in Estonia reported that sowing shrub seeds, along with fertilization, had no effect on total vegetation cover after 25 years.
- *Overall plant richness/diversity:* One site comparison study in bogs in Estonia reported that sowing shrub seeds, along with fertilization, increased plant species richness. However, one replicated, randomized, paired, controlled study in a bog in New Zealand reported that plots sown with shrub seeds typically contained fewer plant species than plots that were not sown. One replicated, randomized, paired, controlled study in a fen in the USA found that sowing shrub (and herb) seeds had no effect on plant species richness.
- *Assessment: likely to be beneficial (effectiveness 45%; certainty 40%; harms 5%). Based on evidence from: bogs (three studies); fens (two studies).*

https://www.conservationevidence.com/actions/1824

6.12 Actions to complement planting

Based on the collated evidence, what is the current assessment of the effectiveness of actions to complement planting peatland vegetation?	
Likely to be beneficial	• Cover peatland with organic mulch (after planting) • Cover peatland with something other than mulch (after planting) • Reprofile/relandscape peatland (before planting)
Trade-off between benefit and harms	• Add inorganic fertilizer (before/after planting)
Unknown effectiveness (limited evidence)	• Introduce nurse plants (to aid focal peatland plants) • Irrigate peatland (before/after planting) • Create mounds or hollows (before planting) • Add fresh peat to peatland (before planting) • Remove vegetation that could compete with planted peatland vegetation • Add root-associated fungi to plants (before planting)
Likely to be ineffective or harmful	• Add lime (before/after planting)
No evidence found (no assessment)	• Add organic fertilizer (before/after planting) • Rewet peatland (before/after planting) • Remove upper layer of peat/soil (before planting) • Bury upper layer of peat/soil (before planting) • Encapsulate planted moss fragments in beads/gel • Use fences or barriers to protect planted vegetation • Protect or prepare vegetation before planting (other interventions)

Likely to be beneficial

Cover peatland with organic mulch (after planting)

- *Germination:* One replicated, controlled, before-and-after study in a bog in Germany found that mulching after sowing seeds increased germination of two species (a grass and a shrub), but had no effect on three other herb species.
- *Survival:* Two replicated, paired, controlled studies in a fen in Sweden and a bog in the USA reported that mulching increased survival of planted vegetation (mosses or sedges). One replicated, paired, controlled study in Indonesia reported that mulching with oil palm fruits reduced survival of planted peat swamp tree seedlings.
- *Growth:* One replicated, randomized, paired, controlled, before-and-after study in a fen in the USA reported that mulching increased growth of transplanted water sedge *Carex aquatilis*.
- *Cover:* Six studies (including four replicated, randomized, paired, controlled, before-and-after) in bogs in Canada and the USA, and a fen in Sweden, found that mulching after planting increased vegetation cover (specifically total vegetation, total mosses/bryophytes, *Sphagnum* mosses or vascular plants after 1–3 growing seasons). Three replicated, randomized, paired, controlled, before-and-after studies in bogs in Canada found that mulching after planting had no effect on vegetation cover (*Sphagnum* mosses or fen-characteristic plants).
- *Assessment: likely to be beneficial (effectiveness 60%; certainty 60%; harms 10%). Based on evidence from: bogs (nine studies); fens (two studies); tropical peat swamps (one study).*

https://www.conservationevidence.com/actions/1828

Cover peatland with something other than mulch (after planting)

- *Germination:* One replicated, controlled, before-and-after study in a bog in Germany reported mixed effects of fleece and fibre mats on germination of sown herb and shrub seeds (positive or no effect, depending on species).

- *Survival:* Two replicated, randomized, controlled studies examined the effect, on plant survival, of covering planted areas. One study in a fen in Sweden reported that shading increased survival of planted mosses. One study in a nursery in Indonesia reported that shading did not affect survival of most studied peat swamp tree species, but increased survival of some.
- *Growth:* Three replicated, randomized, controlled, before-and-after studies examined the effect, on plant growth, of covering planted areas. One study in a greenhouse in Switzerland found that covers, either transparent plastic or shading mesh, increased growth of planted *Sphagnum* moss. One study in a fen in Sweden found that shading with plastic mesh reduced growth of planted fen mosses. One study in a nursery in Indonesia reported that seedlings shaded with plastic mesh grew taller and thinner than unshaded seedlings.
- *Cover:* Two replicated and paired studies, in a fen in Sweden and a bog in Australia, reported that shading plots with plastic mesh increased planted moss cover. One study in a bog in Canada found that covering sown plots with plastic mesh, but not transparent sheets, increased *Sphagnum* moss abundance. Another study in a bog in Canada reported that shading sown plots with plastic mesh did not affect cover of vegetation overall, vascular plants or mosses.
- *Assessment: likely to be beneficial (effectiveness 50%; certainty 50%; harms 10%). Based on evidence from: bogs (five studies); fens (two studies); tropical peat swamps (one study).*

https://www.conservationevidence.com/actions/1829

Reprofile/relandscape peatland (before planting)

- *Survival:* One replicated, paired, controlled study in a bog in Canada found that over one growing season, survival of sown *Sphagnum* mosses was higher in reprofiled basins than on raised plots.
- *Cover:* Two replicated, controlled, before-and-after studies in bogs in Canada found that reprofiled basins had higher *Sphagnum* cover than raised plots, 3–4 growing seasons after sowing *Sphagnum*-dominated vegetation fragments. One controlled study in a bog in Estonia reported that reprofiled and raised plots had similar *Sphagnum* cover, 1–2 years after sowing. All three studies found that

reprofiled and raised plots developed similar cover of other mosses/bryophytes and vascular plants.

- *Assessment: likely to be beneficial (effectiveness 60%; certainty 40%; harms 5%). Based on evidence from: bogs (four studies).*

https://www.conservationevidence.com/actions/1833

Trade-off between benefit and harms

Add inorganic fertilizer (before/after planting)

- *Survival:* Two replicated, randomized, paired, controlled studies in bogs in Canada examined the effect, on plant survival, of adding inorganic fertilizer to areas planted with peatland plants. One study reported that fertilizer increased survival of two planted tree species. The other study found that fertilizer had no effect on three planted tree species and reduced survival of one.
- *Growth:* Five studies (three replicated, randomized, paired, controlled) in bogs in the UK, Germany and Canada found that fertilizer typically increased growth of planted mosses, herbs or trees. However, for some species or in some conditions, fertilizer had no effect on growth. One replicated, randomized, controlled, before-and-after study in a nursery in Indonesia found that fertilizer typically had no effect on growth of peat swamp tree seedlings.
- *Cover:* Three replicated, randomized, paired, controlled studies in bogs examined the effect, on vegetation cover, of adding inorganic fertilizer to areas planted with peatland plants. One study in Canada found that fertilizer increased total vegetation, vascular plant and bryophyte cover. Another study in Canada found that fertilizer increased cover of true sedges *Carex* spp. but had no effect on other vegetation. One study in New Zealand reported that fertilizer typically increased cover of a sown shrub and rush, but this depended on the chemical used and preparation of the peat.
- *Assessment: trade-off between benefit and harms (effectiveness 45%; certainty 40%; harms 20%). Based on evidence from: bogs (eight studies); tropical peat swamps (one study).*

https://www.conservationevidence.com/actions/1826

Unknown effectiveness (limited evidence)

Introduce nurse plants (to aid focal peatland plants)

- *Survival:* One replicated, paired, controlled study in Malaysia reported that planting nurse trees did not affect survival of planted peat swamp tree seedlings (averaged across six species).
- *Cover:* Two replicated, randomized, paired, controlled, before-and-after studies in bogs in the USA and Canada found that planting nurse herbs had no effect on cover, after 2–3 years, of other planted vegetation (mosses/bryophytes, vascular plants or total cover).
- *Assessment: unknown effectiveness – limited evidence (effectiveness 30%; certainty 38%; harms 1%). Based on evidence from: bogs (two studies); tropical peat swamps (one study).*

https://www.conservationevidence.com/actions/1830

Irrigate peatland (before/after planting)

- *Cover:* One replicated, paired, controlled, before-and-after study in a bog in Canada found that irrigation increased the number of *Sphagnum* moss shoots present 1–2 growing seasons after sowing *Sphagnum* fragments.
- *Assessment: unknown effectiveness – limited evidence (effectiveness 60%; certainty 20%; harms 5%). Based on evidence from: bogs (one study).*

https://www.conservationevidence.com/actions/1832

Create mounds or hollows (before planting)

- *Growth:* One controlled study, in a peat swamp in Thailand, reported that trees planted into mounds of peat grew thicker stems than trees planted at ground level.
- *Cover:* Two replicated, randomized, paired, controlled, before-and-after studies in bogs in Canada found that roughening the peat surface (e.g. by harrowing or adding peat blocks) did not significantly affect cover of planted *Sphagnum* moss, after 1–3 growing seasons.
- *Assessment: unknown effectiveness – limited evidence (effectiveness 30%; certainty 38%; harms 5%). Based on evidence from: bogs (two studies); tropical peat swamps (one study).*

https://www.conservationevidence.com/actions/1834

Add fresh peat to peatland (before planting)

- *Cover:* One replicated, controlled, before-and-after study in New Zealand reported that plots amended with fine peat supported higher cover of two sown plant species than the original (tilled) bog surface.
- *Assessment: unknown effectiveness – limited evidence (effectiveness 45%; certainty 25%; harms 5%). Based on evidence from: bogs (one study).*

https://www.conservationevidence.com/actions/1837

Remove vegetation that could compete with planted peatland vegetation

- *Survival:* One controlled study in a bog the UK reported that some *Sphagnum* moss survived when sown, in gel beads, into a plot where purple moor grass *Molinia caerulea* had previously been cut. No moss survived in a plot where grass had not been cut.
- *Assessment: unknown effectiveness – limited evidence (effectiveness 60%; certainty 20%; harms 2%). Based on evidence from: bogs (one study).*

https://www.conservationevidence.com/actions/1840

Add root-associated fungi to plants (before planting)

- *Survival:* Two controlled studies (one also replicated, paired, before-and-after) in peat swamps in Indonesia found that adding root fungi did not affect survival of planted red balau *Shorea balangeran* or jelutong *Dyera polyphylla* in all or most cases. However, one fungal treatment increased red balau survival.
- *Growth:* Two replicated, controlled, before-and-after studies of peat swamp trees in Indonesia found that adding root fungi to seedlings, before planting, typically had no effect on their growth. However, one controlled study in Indonesia found that adding root fungi increased growth of red balau seedlings.
- *Assessment: unknown effectiveness – limited evidence (effectiveness 30%; certainty 35%; harms 0%). Based on evidence from: tropical peat swamps (three studies).*

https://www.conservationevidence.com/actions/1841

Likely to be ineffective or harmful

Add lime (before/after planting)

- *Survival:* One replicated, controlled study in the Netherlands reported that liming reduced survival of planted fen herbs after two growing seasons. One replicated, randomized, paired, controlled study in Sweden found that liming increased survival of planted fen mosses over one season.
- *Growth:* Two controlled, before-and-after studies found that liming did not increase growth of planted peatland vegetation: for two *Sphagnum* moss species in bog pools in the UK, and for most species of peat swamp tree in a nursery in Indonesia. One replicated, controlled, before-and-after study in Sweden found that liming increased growth of planted fen mosses.
- *Cover:* Of two replicated, randomized, paired, controlled studies, one in a fen in Sweden found that liming increased cover of sown mosses. The other, in a bog in Canada, found that liming plots sown with mixed fen vegetation did not affect vegetation cover (total, vascular plants or bryophytes).
- *Assessment: likely to be ineffective or harmful (effectiveness 35%; certainty 40%; harms 20%). Based on evidence from: bogs (two studies); fens (two studies); fen meadows (one study); tropical peat swamps (one study).*

https://www.conservationevidence.com/actions/1825

No evidence found (no assessment)

We have captured no evidence for the following interventions:

- Add organic fertilizer (before/after planting)
- Rewet peatland (before/after planting)
- Remove upper layer of peat/soil (before planting)
- Bury upper layer of peat/soil (before planting)
- Encapsulate planted moss fragments in beads/gel
- Use fences or barriers to protect planted vegetation
- Protect or prepare vegetation before planting (other interventions).

6.13 Habitat protection

Based on the collated evidence, what is the current assessment of the effectiveness of actions to protect peatland habitats?	
Likely to be beneficial	• Legally protect peatlands
Unknown effectiveness (limited evidence)	• Pay landowners to protect peatlands • Increase 'on-the-ground' protection (e.g. rangers)
No evidence found (no assessment)	• Create legislation for 'no net loss' of wetlands • Adopt voluntary agreements to protect peatlands • Allow sustainable use of peatlands

Likely to be beneficial

Legally protect peatlands

- *Peatland habitat:* Two studies in Indonesia reported that peat swamp forest was lost from within the boundaries of national parks. However, one of these studies reported that forest loss was greater outside the national park. One before-and-after study in China reported that peatland area initially decreased following legal protection, but increased in the longer term.
- *Plant community composition:* One before-and-after study in a bog in Denmark reported that the plant community composition changed over 161 years of protection. Woody plants became more abundant.
- *Vegetation cover:* One site comparison study in Chile found that protected peatland had greater vegetation cover (total, herbs and shrubs) than adjacent grazed and moss-harvested peatland.

- *Overall plant richness/diversity:* One before-and-after study in Denmark reported that the number of plant species in a protected bog fluctuated over time, with no clear trend. One site comparison study in Chile found that protected peatland had lower plant richness and diversity, but also fewer non-native species, than adjacent grazed and harvested peatland.
- *Assessment: likely to be beneficial (effectiveness 60%; certainty 40%; harms 1%). Based on evidence from: tropical peat swamps (two studies); bogs (one study); unspecified peatlands (two studies).*

https://www.conservationevidence.com/actions/1796

Unknown effectiveness (limited evidence)

Pay landowners to protect peatlands

- *Peatland habitat:* One review reported that agri-environment schemes in the UK had mixed effects on bogs, protecting the area of bog habitat in three of six cases.
- *Assessment: unknown effectiveness – limited evidence (effectiveness 50%; certainty 20%; harms 10%). Based on evidence from: bogs (one study).*

https://www.conservationevidence.com/actions/1799

Increase 'on the ground' protection (e.g. rangers)

- *Behaviour change:* One before-and-after study in a peat swamp forest in Indonesia reported that the number of illegal sawmills decreased over two years of anti-logging patrols.
- *Assessment: unknown effectiveness – limited evidence (effectiveness 60%; certainty 20%; harms 0%). Based on evidence from: tropical peat swamps (one study).*

https://www.conservationevidence.com/actions/1800

No evidence found (no assessment)

We have captured no evidence for the following interventions:

- Create legislation for 'no net loss' of wetlands
- Adopt voluntary agreements to protect peatlands
- Allow sustainable use of peatlands.

6.14 Education and awareness

Based on the collated evidence, what is the current assessment of the effectiveness of actions to educate/raise awareness about peatlands?	
Unknown effectiveness (limited evidence)	• Raise awareness amongst the public (general) • Provide education or training programmes about peatlands or peatland management • Lobby, campaign or demonstrate to protect peatlands
No evidence found (no assessment)	• Raise awareness amongst the public (wild fire) • Raise awareness amongst the public (problematic species) • Raise awareness through engaging volunteers in peatland management or monitoring

Unknown effectiveness (limited evidence)

Raise awareness amongst the public (general)

- *Behaviour change:* One before-and-after study in the UK reported that following awareness-raising activities (e.g. publishing reports, organizing seminars and using education volunteers in garden centres), the percentage of the public buying peat-free compost increased.
- *Assessment: unknown effectiveness – limited evidence (effectiveness 60%; certainty 25%; harms 0%). Based on evidence from: unspecified peatlands (one study).*

https://www.conservationevidence.com/actions/1844

Provide education or training programmes about peatlands or peatland management

- *Behaviour change:* One study in peat swamps in Indonesia reported that over 3,500 households adopted sustainable farming practices following workshops about sustainable farming. One before-and-after study in peat swamps in Indonesia reported that a training course increased the quality of rubber produced by local farmers.
- *Assessment: unknown effectiveness – limited evidence (effectiveness 60%; certainty 30%; harms 0%). Based on evidence from: tropical peat swamps (two studies).*

https://www.conservationevidence.com/actions/1848

Lobby, campaign or demonstrate to protect peatlands

- *Peatland protection:* Two studies in the UK reported that the area of protected peatland increased following pressure from a campaign group (including business meetings, parliamentary debates, publishing reports and public engagement).
- *Behaviour change:* One study in the UK reported that following pressure from the same campaign group, major retailers stopped buying compost containing peat from important peatland areas and horticultural companies began marketing peat-free compost.
- *Attitudes/awareness:* One study in the UK reported that following pressure from the same campaign group, garden centres and local governments signed voluntary peatland conservation agreements.
- *Assessment: unknown effectiveness – limited evidence (effectiveness 60%; certainty 35%; harms 0%). Based on evidence from: unspecified peatlands (two studies).*

https://www.conservationevidence.com/actions/1849

No evidence found (no assessment)

We have captured no evidence for the following interventions:

- Raise awareness amongst the public (wild fire)
- Raise awareness amongst the public (problematic species)
- Raise awareness through engaging volunteers in peatland management or monitoring.

7. PRIMATE CONSERVATION

Jessica Junker, Hjalmar S. Kühl, Lisa Orth, Rebecca K. Smith, Silviu O. Petrovan & William J. Sutherland

Expert assessors
Graham L. Banes, University of Wisconsin-Madison, USA
Sergio Marrocoli, Max Planck Institute for Evolutionary Anthropology, Germany
Sarah Papworth, Royal Holloway University of London, UK
Silviu O. Petrovan, University of Cambridge, UK
Andrew J. Plumptre, Wildlife Conservation Society, Uganda
Ricardo Rocha, University of Cambridge, UK
Joanna M. Setchell, Durham University, UK
Kathy Slater, Operation Wallacea, UK
Erin Wessling, Max Planck Institute for Evolutionary Anthropology, Germany
Liz Williamson, University of Stirling, UK

Scope of assessment: for wild primate species across the world.

Assessed: 2017.

Effectiveness measure is the median % score for effectiveness.

Certainty measure is the median % certainty of evidence for effectiveness, determined by the quantity and quality of the evidence in the synopsis.

Harm measure is the median % score for negative side-effects to the group of species of concern.

 https://doi.org/10.11647/OBP.0267.07

This book is meant as a guide to the evidence available for different conservation interventions and as a starting point in assessing their effectiveness. The assessments are based on the available evidence for the target group of species for each intervention. The assessment may therefore refer to different species or habitat to the one(s) you are considering. Before making any decisions about implementing interventions it is vital that you read the more detailed accounts of the evidence in order to assess their relevance for your study species or system.

Full details of the evidence are available at**www. conservationevidence.com**

There may also be significant negative side-effects on the target groups or other species or communities that have not been identified in this assessment.

A lack of evidence means that we have been unable to assess whether or not an intervention is effective or has any harmful impacts.

7.1 Threat: Residential and commercial development

Based on the collated evidence, what is the current assessment of the effectiveness of interventions for residential and commercial development?	
Likely to be beneficial	• Remove and relocate 'problem' animals
No evidence found (no assessment)	• Relocate primates to non-residential areas • Discourage the planting of fruit trees and vegetable gardens on the urban edge

Likely to be beneficial

Remove and relocate 'problem' animals

Three studies, including one replicated, before-and-after trial, in India, Kenya, the Republic of Congo and Gabon found that most primates survived the translocation. One study found that all translocated rhesus monkeys remained at the release site for at least four years. Another study showed that after 16 years, 66% of olive baboons survived and survival rate was similar to wild study groups. The third study showed that 84% of gorillas released in the Republic of Congo and Gabon survived for at least four years. *Assessment: likely to be beneficial (effectiveness 60%; certainty 50%; harms 10%).*

https://www.conservationevidence.com/actions/1422

No evidence found (no assessment)

We have captured no evidence for the following interventions:

- Relocate primates to non-residential areas
- Discourage the planting of fruit trees and vegetable gardens on the urban edge biodiversity-friendly farming.

7.2 Threat: Agriculture

Based on the collated evidence, what is the current assessment of the effectiveness of interventions for agriculture?	
Likely to be beneficial	• Humans chase primates using random loud noises
Unknown effectiveness (limited evidence)	• Prohibit (livestock) farmers from entering protected areas • Use nets to keep primates out of fruit trees
No evidence found (no assessment)	• Create natural habitat islands within agricultural land • Use fences as biological corridors for primates • Provide sacrificial rows of crops on outer side of fields • Compensate farmers for produce loss caused by primates • Pay farmers to cover the costs of non-harmful strategies to deter primates • Retain nesting trees/shelter for primates within agricultural fields • Plant nesting trees/shelter for primates within agricultural fields • Regularly remove traps and snares around agricultural fields • Certify farms and market their products as 'primate friendly' • Farm more intensively and effectively in selected areas and spare more natural land • Install mechanical barriers to deter primates (e.g. fences, ditches)

	• Use of natural hedges to deter primates • Use of unpalatable buffer crops • Change of crop (i.e. to a crop less palatable to primates) • Plant crops favoured by primates away from primate areas • Destroy habitat within buffer zones to make them unusable for primates • Use GPS and/or VHF tracking devices on individuals of problem troops to provide farmers with early warning of crop raiding • Chase crop-raiding primates using dogs • Train langur monkeys to deter rhesus macaques • Use loud-speakers to broadcast sounds of potential threats (e.g. barking dogs, explosions, gunshots)
	• Use loud-speakers to broadcast primate alarm calls • Strategically lay out the scent of a primate predator (e.g. leopard, lion) • Humans chase primates using bright light

Likely to be beneficial

Humans chase primates using random loud noise

One controlled, replicated, before-and-after study in Indonesia found that in areas where noise deterrents were used, along with tree nets, crop raiding by orangutans was reduced. One study in the Democratic Republic Congo found that chasing gorillas and using random noise resulted in the return of gorillas from plantation to areas close to protected forest. *Assessment: likely to be beneficial (effectiveness 50%; certainty 40%; harms 0%).*

https://www.conservationevidence.com/actions/1449

Unknown effectiveness (limited evidence)

Prohibit (livestock) farmers from entering protected areas

One before-and-after site comparison study in Rwanda found that numbers of young gorillas increased after removal of cattle from a protected area, alongside other interventions. One before-and-after study in Rwanda, Uganda, and the Democratic Republic of Congo found that gorilla numbers declined following the removal of livestock, alongside other interventions. *Assessment: unknown effectiveness — limited evidence (effectiveness 50%; certainty 30%; harms 0%).*

https://www.conservationevidence.com/actions/1432

Use nets to keep primates out of fruit trees

A controlled, replicated, before-and-after study in Indonesia found that areas where nets were used to protect crop trees, crop-raiding by orangutans was reduced. *Assessment: unknown effectiveness — limited evidence (effectiveness 40%; certainty 30%; harms 20%).*

https://www.conservationevidence.com/actions/1442

No evidence found (no assessment)

We have captured no evidence for the following interventions:

- Create natural habitat islands within agricultural land
- Use fences as biological corridors for primates
- Provide sacrificial rows of crops on outer side of fields
- Compensate farmers for produce loss caused by primates
- Pay farmers to cover the costs of non-harmful strategies to deter primates
- Retain nesting trees/shelter for primates within agricultural fields
- Plant nesting trees/shelter for primates within agricultural fields
- Regularly remove traps and snares around agricultural fields
- Certify farms and market their products as 'primate friendly'
- Farm more intensively and effectively in selected areas and spare more natural land

- Install mechanical barriers to deter primates (e.g. fences, ditches)
- Use of natural hedges to deter primates
- Use of unpalatable buffer crops
- Change of crop (i.e. to a crop less palatable to primates)
- Plant crops favoured by primates away from primate areas
- Destroy habitat within buffer zones to make them unusable for primates
- Use GPS and/or VHF tracking devices on individuals of problem troops to provide farmers with early warning of crop raiding
- Chase crop-raiding primates using dogs
- Train langur monkeys to deter rhesus macaques
- Use loud-speakers to broadcast sounds of potential threats (e.g. barking dogs, explosions, gunshots)
- Use loud-speakers to broadcast primate alarm calls
- Strategically lay out the scent of a primate predator (e.g. leopard, lion)
- Humans chase primates using bright light.

7.3 Threat: Energy production and mining

Based on the collated evidence, what is the current assessment of the effectiveness of interventions for energy and production mining?	
No evidence found (no assessment)	• Minimize ground vibrations caused by open cast mining activities • Establish no-mining zones in/near watersheds so as to preserve water levels and water quality • Use 'set-aside' areas of natural habitat for primate protection within mining area • Certify mines and market their products as 'primate friendly' (e.g. ape-friendly cellular phones) • Create/preserve primate habitat on islands before dam construction

No evidence found (no assessment)

We have captured no evidence for the following interventions:

- Minimize ground vibrations caused by open cast mining activities
- Establish no-mining zones in/near watersheds so as to preserve water levels and water quality
- Use 'set-aside' areas of natural habitat for primate protection within mining area
- Certify mines and market their products as 'primate friendly' (e.g. ape-friendly cellular phones)
- Create/preserve primate habitat on islands before dam construction.

7.4 Threat: Transportation and service corridors

Based on the collated evidence, what is the current assessment of the effectiveness of interventions for transportation and service corridors?	
Likely to be beneficial	• Install rope or pole (canopy) bridges
No evidence found (no assessment)	• Install green bridges (overpasses) • Implement speed limits in particular areas (e.g. with high primate densities) to reduce vehicle collisions with primates • Reduce road widths • Impose fines for breaking the speed limit or colliding with primates • Avoid building roads in key habitat or migration routes • Implement a minimum number of roads (and minimize secondary roads) needed to reach mining extraction sites • Re-use old roads rather than building new roads • Re-route vehicles around protected areas • Install speed bumps to reduce vehicle collisions with primates • Provide adequate signage of presence of primates on or near roads

Likely to be beneficial

Install rope or pole (canopy) bridges

One before-and-after study in Belize study found that howler monkey numbers increased after pole bridges were constructed over man-made gaps. Two studies in Brazil and Madagascar found that primates used pole bridges to cross roads and pipelines. *Assessment: likely to be beneficial (effectiveness 50%; certainty 50%; harms 0%).*

https://www.conservationevidence.com/actions/1457

No evidence found (no assessment)

We have captured no evidence for the following interventions:

- Install green bridges (overpasses)
- Implement speed limits in particular areas (e.g. with high primate densities) to reduce vehicle collisions with primates
- Reduce road widths
- Impose fines for breaking the speed limit or colliding with primates
- Avoid building roads in key habitat or migration routes
- Implement a minimum number of roads (and minimize secondary roads) needed to reach mining extraction sites
- Re-use old roads rather than building new roads
- Re-route vehicles around protected areas
- Install speed bumps to reduce vehicle collisions with primates
- Provide adequate signage of presence of primates on or near roads.

7.5 Threat: Biological resource use

7.5.1 Hunting

Based on the collated evidence, what is the current assessment of the effectiveness of interventions for hunting?	
Likely to be beneficial	• Conduct regular anti-poaching patrols • Regularly de-activate/remove ground snares • Provide better equipment (e.g. guns) to anti-poaching ranger patrols • Implement local no-hunting community policies/traditional hunting ban • Implement community control of patrolling, banning hunting and removing snares
Unknown effectiveness (limited evidence)	• Strengthen/support/re-install traditions/taboos that forbid the killing of primates • Implement monitoring surveillance strategies (e.g. SMART) or use monitoring data to improve effectiveness of wildlife law enforcement patrols • Provide training to anti-poaching ranger patrols
No evidence found (no assessment)	• Implement no-hunting seasons for primates • Implement sustainable harvesting of primates (e.g. with permits, resource access agreements) • Encourage use of traditional hunting methods rather than using guns • Implement road blocks to inspect cars for illegal primate bushmeat

	• Provide medicine to local communities to control killing of primates for medicinal purposes • Introduce ammunition tax • Inspect bushmeat markets for illegal primate species • Inform hunters of the dangers (e.g., disease transmission) of wild primate meat

Likely to be beneficial

Conduct regular anti-poaching patrols

Two of three studies found that gorilla populations increased after regular anti-poaching patrols were conducted, alongside other interventions. One study in Ghana found a decline in gorilla populations. One review on gorillas in Uganda found that no gorillas were killed after an increase in anti-poaching patrols. *Assessment: likely to be beneficial (effectiveness 70%; certainty 50%; harms 0%).*

https://www.conservationevidence.com/actions/1471

Regularly de-activate/remove ground snares

One of two studies found that the number of gorillas increased in an area patrolled for removing snares, alongside other interventions. One study in the Democratic Republic of Congo, Rwanda, and Uganda found that gorilla populations declined despite snare removal. *Assessment: likely to be beneficial (effectiveness 60%; certainty 40%; harms 0%).*

https://www.conservationevidence.com/actions/1475

Provide better equipment (e.g. guns) to anti-poaching ranger patrols

Two studies in the Democratic Republic of Congo and Rwanda found that gorilla populations increased after providing anti-poaching guards with better equipment, alongside other interventions. One study in Uganda found that no gorillas were killed after providing game guards with better equipment. *Assessment: likely to be beneficial (effectiveness 50%; certainty 40%; harms 0%).*

https://www.conservationevidence.com/actions/1476

Implement local no-hunting community policies/traditional hunting ban

Four studies, one of which had multiple interventions, in the Democratic Republic of Congo, Belize, Cameroon and Nigeria found that primate populations increased in areas where there were bans on hunting or where hunting was reduced due to local taboos. One study found that very few primates were killed in a sacred site in China where it is forbidden to kill wildlife. *Assessment: likely to be beneficial (effectiveness 60%; certainty 40%; harms 0%).*

https://www.conservationevidence.com/actions/1478

Implement community control of patrolling, banning hunting and removing snares

Two site comparison studies found that there were more gorillas and chimpanzees in an area managed by a community conservation organisation than in areas not managed by local communities and community control was more effective at reducing illegal primate hunting compared to the nearby national park. A before-and-after study in Cameroon found that no incidents of gorilla poaching occurred over three years after implementation of community control and monitoring of illegal activities. *Assessment: likely to be beneficial (effectiveness 70%; certainty 50%; harms 0%).*

https://www.conservationevidence.com/actions/1482

Unknown effectiveness (limited evidence)

Strengthen/support/re-install traditions/taboos that forbid the killing of primates

One site comparison study in Laos found that Laotian black crested gibbons occurred at higher densities in areas where they were protected by a local hunting taboo compared to sites were there was no taboo. *Assessment: unknown effectiveness — limited evidence (effectiveness 60%; certainty 10%; harms 0%).*

https://www.conservationevidence.com/actions/1479

Implement monitoring surveillance strategies (e.g. SMART) or use monitoring data to improve effectiveness of wildlife law enforcement patrols

One before-and-after study in Nigeria found that more gorillas and chimpanzees were observed after the implementation of law enforcement and a monitoring system. *Assessment: unknown effectiveness — limited evidence (effectiveness 60%; certainty 30%; harms 0%).*

https://www.conservationevidence.com/actions/1481

Provide training to anti-poaching ranger patrols

Two before-and-after studies in Rwanda and India found that primate populations increased in areas where anti-poaching staff received training, alongside other interventions. Two studies in Uganda and Cameroon found that no poaching occurred following training of anti-poaching rangers, alongside other interventions. *Assessment: unknown effectiveness — limited evidence (effectiveness 70%; certainty 30%; harms 0%).*

https://www.conservationevidence.com/actions/1477

No evidence found (no assessment)

We have captured no evidence for the following interventions:

- Implement no-hunting seasons for primates
- Implement sustainable harvesting of primates (e.g. with permits, resource access agreements)
- Encourage use of traditional hunting methods rather than using guns
- Implement road blocks to inspect cars for illegal primate bushmeat
- Provide medicine to local communities to control killing of primates for medicinal purposes
- Introduce ammunition tax
- Inspect bushmeat markets for illegal primate species
- Inform hunters of the dangers (e.g., disease transmission) of wild primate meat.

7.5.2 Substitution

Based on the collated evidence, what is the current assessment of the effectiveness of interventions for substitution?	
Unknown effectiveness (limited evidence)	• Use selective logging instead of clear-cutting • Avoid/minimize logging of important food tree species for primates
No evidence found (no assessment)	• Use patch retention harvesting instead of clear-cutting • Implement small and dispersed logging compartments • Use shelter wood cutting instead of clear-cutting • Leave hollow trees in areas of selective logging for sleeping sites • Clear open patches in the forest • Thin trees within forests • Coppice trees • Manually control or remove secondary mid-storey and ground-level vegetation • Avoid slashing climbers/lianas, trees housing them, hemi-epiphytic figs, and ground vegetation • Incorporate forested corridors or buffers into logged areas • Close non-essential roads as soon as logging operations are complete • Use 'set-asides' for primate protection within logging area • Work inward from barriers or boundaries (e.g. river) to avoid pushing primates toward an impassable barrier or inhospitable habitat • Reduce the size of forestry teams to include employees only (not family members) • Certify forest concessions and market their products as 'primate friendly' • Provide domestic meat to workers of the logging company to reduce hunting

Unknown effectiveness (limited evidence)

Use selective logging instead of clear-cutting

One of two site comparison studies in Africa found that primate abundance was higher in forests that had been logged at low intensity compared to forest logged at high intensity. One study in Uganda found that primate abundances were similar in lightly and heavily logged forests. One study in Madagascar found that the number of lemurs increased following selective logging. *Assessment: unknown effectiveness — limited evidence (effectiveness 60%; certainty 35%; harms 30%).*

https://www.conservationevidence.com/actions/1485

Avoid/minimize logging of important food tree species for primates

One before-and-after study in Belize found that black howler monkey numbers increased over a 13 year period after trees important for food for the species were preserved, alongside other interventions. *Assessment: unknown effectiveness — limited evidence (effectiveness 60%; certainty 20%; harms 0%).*

https://www.conservationevidence.com/actions/1494

No evidence found (no assessment)

We have captured no evidence for the following interventions:

- Use patch retention harvesting instead of clear-cutting
- Implement small and dispersed logging compartments
- Use shelter wood cutting instead of clear-cutting
- Leave hollow trees in areas of selective logging for sleeping sites
- Clear open patches in the forest
- Thin trees within forests
- Coppice trees
- Manually control or remove secondary mid-storey and ground-level vegetation.
- Avoid slashing climbers/lianas, trees housing them, hemi-epiphytic figs, and ground vegetation
- Incorporate forested corridors or buffers into logged areas
- Close non-essential roads as soon as logging operations are complete
- Use 'set-asides' for primate protection within logging area

- Work inward from barriers or boundaries (e.g. river) to avoid pushing primates toward an impassable barrier or inhospitable habitat
- Reduce the size of forestry teams to include employees only (not family members)
- Certify forest concessions and market their products as 'primate friendly'
- Provide domestic meat to workers of the logging company to reduce hunting.

7.6 Threat: Human intrusions and disturbance

Based on the collated evidence, what is the current assessment of the effectiveness of interventions for human intrusions and disturbance?	
Unknown effectiveness (limited evidence)	• Implement a 'no-feeding of wild primates' policy • Put up signs to warn people about not feeding primates • Resettle illegal human communities (i.e. in a protected area) to another location
No evidence found (no assessment)	• Build fences to keep humans out • Restrict number of people that are allowed access to the site • Install 'primate-proof' garbage bins • Do not allow people to consume food within natural areas where primates can view them

Unknown effectiveness (limited evidence)

Implement a 'no-feeding of wild primates' policy

A controlled before-and-after study in Japan found that reducing food provisioning of macaques progressively reduced productivity and reversed population increases and crop and forest damage. *Assessment: unknown effectiveness — limited evidence (effectiveness 40%; certainty 20%; harms 0%).*

https://www.conservationevidence.com/actions/1502

Put up signs to warn people about not feeding primates

One review study in Japan found that after macaque feeding by tourists was banned and advertised, the number of aggressive incidents between people and macaques decreased as well as the number of road collisions with macaques that used to be fed from cars. *Assessment: unknown effectiveness — limited evidence (effectiveness 30%; certainty 10%; harms 0%).*

https://www.conservationevidence.com/actions/1507

Resettle illegal human communities (i.e. in a protected area) to another location

One review on gorillas in Uganda found that no more gorillas were killed after human settlers were relocated outside the protected area, alongside other interventions. One before-and-after study in the Republic of Congo found that most reintroduced chimpanzees survived over five years after human communities were resettled, alongside other interventions. *Assessment: unknown effectiveness — limited evidence (effectiveness 65%; certainty 15%; harms 0%).*

https://www.conservationevidence.com/actions/1515

No evidence found (no assessment)

We have captured no evidence for the following interventions:

- Build fences to keep humans out
- Restrict number of people that are allowed access to the site
- Install 'primate-proof' garbage bins
- Do not allow people to consume food within natural areas where primates can view them.

7.7 Threat: Natural system modifications

Based on the collated evidence, what is the current assessment of the effectiveness of interventions for natural system modifications?	
No evidence found (no assessment)	• Use prescribed burning within the context of home range size and use • Protect important food/nest trees before burning

No evidence found (no assessment)

We have captured no evidence for the following interventions:

- Use prescribed burning within the context of home range size and use
- Protect important food/nest trees before burning.

7.8 Threat: Invasive and other problematic species and genes

7.8.1 Problematic animal/plant species and genes

Based on the collated evidence, what is the current assessment of the effectiveness of interventions for problematic animal/plant species and genes?	
No evidence found (no assessment)	• Reduce primate predation by non-primate species through exclusion (e.g. fences) or translocation • Reduce primate predation by other primate species through exclusion (e.g. fences) or translocation • Control habitat-altering mammals (e.g. elephants) through exclusion (e.g. fences) or translocation • Control inter-specific competition for food through exclusion (e.g. fences) or translocation • Remove alien invasive vegetation where the latter has a clear negative effect on the primate species in question • Prevent gene contamination by alien primate species introduced by humans, through exclusion (e.g. fences) or translocation

No evidence found (no assessment)

We have captured no evidence for the following interventions:

- Reduce primate predation by non-primate species through exclusion (e.g. fences) or translocation
- Reduce primate predation by other primate species through exclusion (e.g. fences) or translocation
- Control habitat-altering mammals (e.g. elephants) through exclusion (e.g. fences) or translocation
- Control inter-specific competition for food through exclusion (e.g. fences) or translocation
- Remove alien invasive vegetation where the latter has a clear negative effect on the primate species in question
- Prevent gene contamination by alien primate species introduced by humans, through exclusion (e.g. fences) or translocation.

7.8.2 Disease transmission

Based on the collated evidence, what is the current assessment of the effectiveness of interventions for disease transmission?	
Trade-off between benefit and harms	• Preventative vaccination of habituated or wild primates
Unknown effectiveness (limited evidence)	• Wear face-masks to avoid transmission of viral and bacterial diseases to primates • Keep safety distance to habituated animals • Limit time that researchers/tourists are allowed to spend with habituated animals • Implement quarantine for primates before reintroduction/translocation • Ensure that researchers/tourists are up-to-date with vaccinations and healthy • Regularly disinfect clothes, boots etc.

	• Treat sick/injured animals • Remove/treat external/internal parasites to increase reproductive success/survival • Conduct veterinary screens of animals before reintroducing/translocating them • Implement continuous health monitoring with permanent vet on site • Detect and report dead primates and clinically determine their cause of death to avoid disease transmission
No evidence found (no assessment)	• Implement quarantine for people arriving at, and leaving the site • Wear gloves when handling primate food, tool items, etc. • Control 'reservoir' species to reduce parasite burdens/pathogen sources • Avoid contact between wild primates and human-raised primates • Implement a health programme for local communities

Trade-off between benefit and harms

Preventative vaccination of habituated or wild primates

Three before-and-after studies in the Republic of Congo and Gabon, two focusing on chimpanzees and one on gorillas, found that most reintroduced individuals survived over 3.5-10 years after being vaccinated, alongside other interventions. One before-and-after study in Puerto Rico found that annual mortality of introduced rhesus macaques decreased after a preventive tetanus vaccine campaign, alongside other interventions. *Assessment: trade-offs between benefits and harms (effectiveness 70%; certainty 40%; harms 30%).*

https://www.conservationevidence.com/actions/1549

Unknown effectiveness (limited evidence)

Wear face-masks to avoid transmission of viral and bacterial diseases to primates

One before-and-after study in Rwanda, Uganda and the Democratic Republic of Congo found that gorilla numbers increased while being visited by researchers and visitors wearing face-masks, alongside other interventions. One study in Uganda found that a confiscated chimpanzee was successfully reunited with his mother after being handled by caretakers wearing face-masks, alongside other interventions. *Assessment: unknown effectiveness — limited evidence (effectiveness 50%; certainty 5%; harms 0%).*

https://www.conservationevidence.com/actions/1537

Keep safety distance to habituated animals

One before-and-after study in the Republic of Congo found that most reintroduced chimpanzees survived over five years while being routinely followed from a safety distance, alongside other interventions. One before-and-after study in Rwanda, Uganda and the Democratic Republic of Congo found that gorilla numbers increased while being routinely visited from a safety distance, alongside other interventions. However, one study in Malaysia found that orangutan numbers declined while being routinely visited from a safety distance. *Assessment: unknown effectiveness — limited evidence (effectiveness 40%; certainty 10%; harms 0%).*

https://www.conservationevidence.com/actions/1538

Limit time that researchers/tourists are allowed to spend with habituated animals

One before-and-after study in Rwanda, Uganda and the Democratic Republic of Congo found that gorilla numbers increased while being routinely visited during limited time, alongside other interventions. One controlled study in Indonesia found that the behaviour of orangutans that spent limited time with caretakers was more similar to the behaviour of wild orangutans than that of individuals that spent more time with caretakers. *Assessment: unknown effectiveness — limited evidence (effectiveness 40%; certainty 10%; harms 0%).*

https://www.conservationevidence.com/actions/1539

Implement quarantine for primates before reintroduction/ translocation

Six studies, including four before-and-after studies, in Brazil, Madagascar, Malaysia and Indonesia have found that most reintroduced primates did not survive or their population size decreased over periods ranging from months up to seven years post-release, despite being quarantined before release, alongside other interventions. However, two before-and-after studies in Indonesia, the Republic of Congo and Gabon found that most orangutans and gorillas that underwent quarantine survived over a period ranging from three months to 10 years. One before-and-after study in Uganda found that one reintroduced chimpanzee repeatedly returned to human settlements after being quarantined before release alongside other interventions. *Assessment: unknown effectiveness — limited evidence (effectiveness 50%; certainty 10%; harms 0%).*

https://www.conservationevidence.com/actions/1541

Ensure that researchers/tourists are up-to-date with vaccinations and healthy

One before-and-after study in Rwanda, Uganda and the Republic of Congo found that gorilla numbers increased while being visited by healthy researchers and visitors, alongside other interventions. However, one controlled study in Malaysia found that orangutan numbers decreased despite being visited by healthy researchers and visitors, alongside other interventions. *Assessment: unknown effectiveness — limited evidence (effectiveness 30%; certainty 10%; harms 0%).*

https://www.conservationevidence.com/actions/1546

Regularly disinfect clothes, boots etc.

One controlled, before-and-after study in Rwanda, Uganda and the Democratic Republic of Congo found that gorilla numbers increased while being regularly visited by researchers and visitors whose clothes were disinfected, alongside other interventions. *Assessment: unknown effectiveness — limited evidence (effectiveness 50%; certainty 10%; harms 0%).*

https://www.conservationevidence.com/actions/1547

Treat sick/injured animals

Eight studies, including four before-and-after studies, in Brazil, Malaysia, Liberia, the Democratic Republic of Congo, The Gambia and South Africa found that most reintroduced or translocated primates that were treated when sick or injured, alongside other interventions, survived being released and up to at least five years. However, five studies, including one review and four before-and-after studies, in Brazil, Thailand, Malaysia and Madagascar found that most reintroduced or translocated primates did not survive or their numbers declined despite being treated when sick or injured, alongside other interventions. One study in Uganda found that several infected gorillas were medically treated after receiving treatment, alongside other interventions. One study in Senegal found that one chimpanzee was reunited with his mother after being treated for injuries, alongside other interventions. *Assessment: unknown effectiveness — limited evidence (effectiveness 50%; certainty 20%; harms 0%).*

https://www.conservationevidence.com/actions/1550

Remove/treat external/internal parasites to increase reproductive success/survival

Five studies, including four before-and-after studies, in the Republic of Congo, The Gambia and Gabon found that most reintroduced or translocated primates that were treated for parasites, alongside other interventions, survived periods of at least five years. However, four studies, including one before-and-after study, in Brazil, Gabon and Vietnam found that most reintroduced primates did not survive or their numbers declined after being treated for parasites, alongside other interventions. *Assessment: unknown effectiveness — limited evidence (effectiveness 40%; certainty 5%; harms 0%).*

https://www.conservationevidence.com/actions/1551

Conduct veterinary screens of animals before reintroducing/ translocating them

Twelve studies, including seven before-and-after studies, in Brazil, Malaysia, Indonesia, Liberia, the Republic of Congo, Guinea, Belize, French Guiana and Madagascar found that most reintroduced or translocated primates that underwent pre-release veterinary screens, alongside other interventions, survived, in some situations, up to at least five years or increased in population size. However, 10 studies, including six before-and-after studies, in Brazil, Malaysia, French Guiana, Madagascar, Kenya, South Africa and Vietnam found that most reintroduced or translocated primates did not survive or their numbers declined after undergoing pre-release veterinary screens, alongside other interventions. One before-and-after study in Uganda, found that one reintroduced chimpanzee repeatedly returned to human settlements after undergoing pre-release veterinary screens, alongside other interventions. One controlled study in Indonesia found that gibbons that underwent pre-release veterinary screens, alongside other interventions, behaved similarly to wild gibbons. *Assessment: unknown effectiveness — limited evidence (effectiveness 50%; certainty 10%; harms 0%).*

https://www.conservationevidence.com/actions/1553

Implement continuous health monitoring with permanent vet on site

One controlled, before-and-after study in Rwanda, Uganda and the Republic of Congo found that numbers of gorillas that were continuously monitored by vets, alongside other interventions, increased over 41 years. *Assessment: unknown effectiveness — limited evidence (effectiveness 60%; certainty 20%; harms 0%).*

https://www.conservationevidence.com/actions/1554

Detect and report dead primates and clinically determine their cause of death to avoid disease transmission

One controlled, before-and-after study in Rwanda, Uganda and the Republic of Congo found that numbers of gorillas that were continuously monitored by vets, alongside other interventions, increased over 41 years. *Assessment: unknown effectiveness — limited evidence (effectiveness 40%; certainty 10%; harms 0%).*

https://www.conservationevidence.com/actions/1556

No evidence found (no assessment)

We have captured no evidence for the following interventions:

- Implement quarantine for people arriving at, and leaving the site
- Wear gloves when handling primate food, tool items, etc.
- Control 'reservoir' species to reduce parasite burdens/pathogen sources
- Avoid contact between wild primates and human-raised primates
- Implement a health programme for local communities.

7.9 Threat: Pollution

7.9.1 Garbage/solid waste

Based on the collated evidence, what is the current assessment of the effectiveness of interventions for garbage and solid waste?	
No evidence found (no assessment)	• Reduce garbage/solid waste to avoid primate injuries • Remove human food waste that may potentially serve as food sources for primates to avoid disease transmission and conflict with humans

No evidence found (no assessment)

We have captured no evidence for the following interventions:

- Reduce garbage/solid waste to avoid primate injuries
- Remove human food waste that may potentially serve as food sources for primates to avoid disease transmission and conflict with humans.

7.9.2 Excess energy

Based on the collated evidence, what is the current assessment of the effectiveness of interventions for excess energy?	
No evidence found (no assessment)	• Reduce noise pollution by restricting development activities to certain times of the day/night

No evidence found (no assessment)

We have captured no evidence for the following interventions:

- Reduce noise pollution by restricting development activities to certain times of the day/night.

7.10 Education and Awareness

Based on the collated evidence, what is the current assessment of the effectiveness of interventions for education and awareness?	
Unknown effectiveness (limited evidence)	• Educate local communities about primates and sustainable use • Involve local community in primate research and conservation management • Regularly play TV and radio announcements to raise primate conservation awareness • Implement multimedia campaigns using theatre, film, print media, discussions
No evidence found (no assessment)	• Install billboards to raise primate conservation awareness • Integrate local religion/taboos into conservation education

Unknown effectiveness (limited evidence)

Educate local communities about primates and sustainable use

One before-and-after study in Cameroon found that numbers of drills increased after the implementation of an education programme, alongside one other intervention. *Assessment: unknown effectiveness — limited evidence (effectiveness 50%; certainty 0%; harms 0%).*

https://www.conservationevidence.com/actions/1563

Involve local community in primate research and conservation management

One before-and-after study in Rwanda, Uganda and the Democratic Republic of Congo found that gorilla numbers decreased despite the implementation of an environmental education programme, alongside other interventions. However, one before-and-after study in Cameroon found that gorilla poaching stopped after the implementation of a community-based monitoring scheme, alongside other interventions. One before-and-after study in Belize found that numbers of howler monkeys increased while local communities were involved in the management of the sanctuary, alongside other interventions. One before-and-after study in Uganda found that a reintroduced chimpanzee repeatedly returned to human settlements despite the involvement of local communities in the reintroduction project, alongside other interventions. *Assessment: unknown effectiveness — limited evidence (effectiveness 50%; certainty 20%; harms 0%).*

https://www.conservationevidence.com/actions/1565

Regularly play TV and radio announcements to raise primate conservation awareness

One before-and-after study in Congo found that most reintroduced chimpanzees whose release was covered by media, alongside other interventions, survived over five years. *Assessment: unknown effectiveness — limited evidence (effectiveness 50%; certainty 5%; harms 0%).*

https://www.conservationevidence.com/actions/1569

Implement multimedia campaigns using theatre, film, print media, and discussions

Three before-and-after studies in Belize and India found that primate numbers increased after the implementation of education programs, alongside other interventions. Three before-and-after studies found that the knowledge about primates increased after the implementation of education programmes. One before-and-after study in Madagascar found that lemur poaching appeared to have ceased after the distribution of conservation books in schools. One study in four African countries found that large numbers of people were informed about gorillas through multimedia campaigns using theatre and film. *Assessment: unknown effectiveness — limited evidence (effectiveness 40%; certainty 10%; harms 0%).*

https://www.conservationevidence.com/actions/1571

No evidence found (no assessment)

We have captured no evidence for the following interventions:

- Install billboards to raise primate conservation awareness
- Integrate local religion/taboos into conservation education.

7.11 Habitat protection

7.11.1 Habitat protection

Based on the collated evidence, what is the current assessment of the effectiveness of interventions for habitat protection?	
Likely to be beneficial	• Create/protect habitat corridors
Unknown effectiveness (limited evidence)	• Legally protect primate habitat • Establish areas for conservation which are not protected by national or international legislation (e.g. private sector standards and codes) • Create/protect forest patches in highly fragmented landscapes
No evidence found (no assessment)	• Create buffer zones around protected primate habitat • Demarcate and enforce boundaries of protected areas

Likely to be beneficial

Create/protect habitat corridors

One before-and-after study in Belize found that howler monkey numbers increased after the protection of a forest corridor, alongside other interventions. *Assessment: likely to be beneficial (effectiveness 65%; certainty 41%; harms 0%).*

https://www.conservationevidence.com/actions/1580

Unknown effectiveness (limited evidence)

Legally protect primate habitat

Two reviews and a before-and-after study in China found that primate numbers increased or their killing was halted after their habitat became legally protected, alongside other interventions. However, one before-and-after study in Kenya found that colobus and mangabey numbers decreased despite the area being declared legally protected, alongside other interventions. Two before-and-after studies found that most chimpanzees and gorillas reintroduced to areas that received legal protection, alongside other interventions, survived over 4–5 years. However, one before-and-after study in Brazil found that most golden lion tamarins did not survive over seven years despite being reintroduced to a legally protected area, alongside other interventions, yet produced offspring that partly compensated the mortality. One controlled, site comparison study in Mexico found that howler monkeys in protected areas had lower stress levels than individuals living in unprotected forest fragments. *Assessment: unknown effectiveness — limited evidence (effectiveness 60%; certainty 30%; harms 0%).*

https://www.conservationevidence.com/actions/1578

Establish areas for conservation which are not protected by national or international legislation (e.g. private sector standards and codes)

Two before-and-after studies in Rwanda, Republic of Congo and Belize found that gorilla and howler monkey numbers increased after the implementation of a conservation project funded by a consortium of organizations or after being protected by local communities, alongside other interventions. *Assessment: unknown effectiveness — limited evidence (effectiveness 60%; certainty 10%; harms 0%).*

https://www.conservationevidence.com/actions/1579

Create/protect forest patches in highly fragmented landscapes

One before-and-after study in Belize found that howler monkey numbers increased after the protection of forest along property boundaries and across cleared areas, alongside other interventions. *Assessment: unknown effectiveness — limited evidence (effectiveness 40%; certainty 10%; harms 0%).*

https://www.conservationevidence.com/actions/1581

No evidence found (no assessment)

We have captured no evidence for the following interventions:

- Create buffer zones around protected primate habitat
- Demarcate and enforce boundaries of protected areas.

7.11.2 Habitat creation or restoration

Based on the collated evidence, what is the current assessment of the effectiveness of interventions for habitat creation or restoration?	
Unknown effectiveness (limited evidence)	• Plant indigenous trees to re-establish natural tree communities in clear-cut areas
No evidence found (no assessment)	• Restore habitat corridors • Plant indigenous fast-growing trees (will not necessarily resemble original community) in clear-cut areas • Use weeding to promote regeneration of indigenous tree communities

Unknown effectiveness (limited evidence)

Plant indigenous trees to re-establish natural tree communities in clear-cut areas

One site comparison study in Kenya found that group densities of two out of three primate species were lower in planted forests than in natural forests. *Assessment: unknown effectiveness — limited evidence (effectiveness 30%; certainty 5%; harms 0%).*

https://www.conservationevidence.com/actions/1584

No evidence found (no assessment)

We have captured no evidence for the following interventions:

- Restore habitat corridors
- Plant indigenous fast-growing trees (will not necessarily resemble original community) in clear-cut areas
- Use weeding to promote regeneration of indigenous tree communities.

7.12 Species management

7.12.1 Species management

Based on the collated evidence, what is the current assessment of the effectiveness of interventions for species management?	
Likely to be beneficial	• Guard habituated primate groups to ensure their safety/well-being
Unknown effectiveness (limited evidence)	• Habituate primates to human presence to reduce stress from tourists/researchers etc. • Implement legal protection for primate species under threat
No evidence found (no assessment)	• Implement birth control to stabilize primate community/population size

Likely to be beneficial

Guard habituated primate groups to ensure their safety/well-being

One study in Rwanda, Uganda and the Congo found that a population of mountain gorillas increased after being guarded against poachers, alongside other interventions. *Assessment: likely to be beneficial (effectiveness 60%; certainty 40%; harms 0%).*

https://www.conservationevidence.com/actions/1523

Unknown effectiveness (limited evidence)

Habituate primates to human presence to reduce stress from tourists/researchers etc.

Two studies in Central Africa and Madagascar found that primate populations increased or were stable following habituation to human presence, alongside other interventions. One study in Brazil found that golden lion tamarin populations declined following habituation to human presence, alongside other interventions. *Assessment: unknown effectiveness — limited evidence (effectiveness 40%; certainty 20%; harms 10%).*

https://www.conservationevidence.com/actions/1519

Implement legal protection for primate species under threat

Three of four studies in India, South East Asia, and West Africa found that primate populations declined after the respective species were legally protected, alongside other interventions. One of four studies in India found that following a ban on export of rhesus macaques, their population increased. One study in Malaysia found that a minority of introduced gibbons survived after implementing legal protection, along with other interventions. *Assessment: unknown effectiveness — limited evidence (effectiveness 40%; certainty 30%; harms 0%).*

https://www.conservationevidence.com/actions/1524

No evidence found (no assessment)

We have captured no evidence for the following interventions:

- Implement birth control to stabilize primate community/population size.

7.12.2 Species recovery

Based on the collated evidence, what is the current assessment of the effectiveness of interventions for species recovery?	
Unknown effectiveness (limited evidence)	• Regularly and continuously provide supplementary food to primates • Regularly provide supplementary food to primates during resource scarce periods only • Provide supplementary food for a certain period of time only • Provide additional sleeping platforms/nesting sites for primates • Provide artificial water sources
No evidence found (no assessment)	• Provide salt licks for primates • Provide supplementary food to primates through the establishment of prey populations

Unknown effectiveness (limited evidence)

Regularly and continuously provide supplementary food to primates

Two of four studies found that primate populations increased after regularly providing supplementary food, alongside other interventions, while two of four studies found that populations declined. Four of four studies found that the majority of primates survived after regularly providing supplementary food, alongside other interventions. One study found that introduced lemurs had different diets to wild primates after regularly being providing supplementary food, along with other interventions. *Assessment: unknown effectiveness — limited evidence (effectiveness 50%; certainty 30%; harms 60%).*

https://www.conservationevidence.com/actions/1526

Regularly provide supplementary food to primates during resource scarce periods only

Two studies found that the majority of primates survived after supplementary feeding in resource scarce periods, alongside other interventions. One study in Madagascar found that the diet of introduced lemurs was similar to that of wild lemurs after supplementary feeding in resource scarce periods, alongside other interventions. *Assessment: unknown effectiveness — limited evidence (effectiveness 40%; certainty 10%; harms 10%).*

https://www.conservationevidence.com/actions/1527

Provide supplementary food for a certain period of time only

Six of eleven studies found that a majority of primates survived after supplementary feeding, alongside other interventions. Five of eleven studies found that a minority of primates survived. One of two studies found that a reintroduced population of primates increased after supplementary feeding for two months immediately after reintroduction, alongside other interventions. One study found that a reintroduced population declined. Two studies found that abandoned primates rejoined wild groups after supplementary feeding, alongside other interventions. *Assessment: unknown effectiveness — limited evidence (effectiveness 40%; certainty 0%; harms 0%).*

https://www.conservationevidence.com/actions/1528

Provide additional sleeping platforms/nesting sites for primates

One study found that a translocated golden lion tamarin population declined despite providing artificial nest boxes, alongside other interventions. One of two studies found that the majority of gorillas survived for at least seven years after nesting platforms were provided, alongside other interventions. One of two studies found that a minority of tamarins survived for at least seven years after artificial nest boxes were provided, alongside other interventions. *Assessment: unknown effectiveness — limited evidence (effectiveness 20%; certainty 0%; harms 0%).*

https://www.conservationevidence.com/actions/1530

Provide artificial water sources

Three of five studies found that a minority of primates survived for between 10 months and seven years when provided with supplementary water, alongside other interventions. Two of five studies found that a majority of primates survived for between nine and ten months, when provided with supplementary water, alongside other interventions. *Assessment: unknown effectiveness — limited evidence (effectiveness 20%; certainty 10%; harms 0%).*

https://www.conservationevidence.com/actions/1531

No evidence found (no assessment)

We have captured no evidence for the following interventions:

- Provide salt licks for primates
- Provide supplementary food to primates through the establishment of prey populations.

7.12.3 Species reintroduction

Based on the collated evidence, what is the current assessment of the effectiveness of interventions for species reintroduction?	
Likely to be beneficial	• Reintroduce primates into habitat where the species is absent
Unknown effectiveness (limited evidence)	• Translocate (capture and release) wild primates from development sites to natural habitat elsewhere • Translocate (capture and release) wild primates from abundant population areas to non-inhabited environments • Allow primates to adapt to local habitat conditions for some time before introduction to the wild • Reintroduce primates in groups • Reintroduce primates as single/multiple individuals • Reintroduce primates into habitat where the species is present • Reintroduce primates into habitat with predators • Reintroduce primates into habitat without predators

Likely to be beneficial

Reintroduce primates into habitat where the species is absent

One of two studies found that primate populations increased after reintroduction into habitat where the species was absent, alongside other interventions. One study in Thailand found that lar gibbon populations declined post-reintroduction. One study in Indonesia found that a orangutan population persisted for at least four years after reintroduction. Eight of ten studies found that a majority of primates survived after reintroduction into habitat where the species was absent, alongside other interventions. Two studies in Malaysia and Vietnam found that a minority of primates survived after reintroduction into habitat where the species was absent, alongside other interventions. *Assessment: likely to be beneficial (effectiveness 60%; certainty 40%; harms 0%).*

https://www.conservationevidence.com/actions/1590

Unknown effectiveness (limited evidence)

Translocate (capture and release) wild primates from development sites to natural habitat elsewhere

Four studies found that the majority of primates survived following translocation from a development site to natural habitat, alongside other interventions. One study in French Guyana found that a minority of primates survived for at least 18 months. One study in India found that rhesus macaques remained at sites where they were released following translocation from a development site to natural habitat, alongside other interventions. *Assessment: unknown effectiveness — limited evidence (effectiveness 60%; certainty 30%; harms 10%).*

https://www.conservationevidence.com/actions/1558

Translocate (capture and release) wild primates from abundant population areas to non-inhabited environments

One study in Belize found that he majority of howler monkeys survived for at least 10 months after translocation from abundant population areas to an uninhabited site, along with other interventions. *Assessment: unknown effectiveness — limited evidence (effectiveness 50%; certainty 20%; harms 0%).*

https://www.conservationevidence.com/actions/1559

Allow primates to adapt to local habitat conditions for some time before introduction to the wild

Two of three studies found that primate populations declined despite allowing individuals to adapt to local habitat conditions before introduction into the wild, along with other interventions. One study in Belize found an increase in introduced howler monkey populations. Ten of 17 studies found that a majority of primates survived after allowing them to adapt to local habitat conditions before introduction into the wild, along with other interventions. Six studies found that a minority of primates survived and one study found that half of primates survived. One study found that a reintroduced chimpanzee repeatedly returned to human settlements after allowing it to adapt to local habitat conditions before introduction into the wild, along with other interventions. One study found that after allowing time to adapt to local habitat conditions, a pair of reintroduced Bornean agile gibbons had a similar diet to wild gibbons. *Assessment: unknown effectiveness — limited evidence (effectiveness 50%; certainty 10%; harms 0%).*

https://www.conservationevidence.com/actions/1564

Reintroduce primates in groups

Two of four studies found that populations of introduced primates declined after reintroduction in groups, alongside other interventions, while two studies recorded increases in populations. Two studies found that primate populations persisted for at least five to 55 years after reintroduction in groups, alongside other interventions. Seven of fourteen studies found that a majority of primates survived after reintroduction in groups, alongside other interventions. Seven of fourteen studies found that a minority of primates survived after reintroduction in groups, alongside other interventions. One study found that introduced primates had a similar diet to a wild population. *Assessment: unknown effectiveness — limited evidence (effectiveness 50%; certainty 20%; harms 0%).*

https://www.conservationevidence.com/actions/1567

Reintroduce primates as single/multiple individuals

Three of four studies found that populations of reintroduced primates declined after reintroduction as single/multiple individuals, alongside other interventions. One study in Tanzania found that the introduced chimpanzee population increased in size. Three of five studies found that a minority of primates survived after reintroduction as single/multiple individuals, alongside other interventions. One study found that a majority of primates survived and one study found that half of primates survived. Two of two studies in Brazil and Senegal found that abandoned primates were successfully reunited with their mothers after reintroduction as single/multiple individuals, alongside other interventions. *Assessment: unknown effectiveness — limited evidence (effectiveness 20%; certainty 10%; harms 0%).*

https://www.conservationevidence.com/actions/1589

Reintroduce primates into habitat where the species is present

One of two studies found that primate populations increased after reintroduction into habitat where the species was absent, alongside other interventions. One study in Malaysia found that an introduced orangutan population declined post-reintroduction. One study found that a primate population persisted for at least four years after reintroduction. Eight of ten studies found that a majority of primates survived after reintroduction into habitat where the species was absent, alongside other interventions. Two studies found that a minority of primates survived after reintroduction into habitat where the species was present, alongside other interventions. *Assessment: unknown effectiveness — limited evidence (effectiveness 50%; certainty 30%; harms 0%).*

https://www.conservationevidence.com/actions/1591

Reintroduce primates into habitat with predators

Eight of fourteen studies found that a majority of reintroduced primates survived after reintroduction into habitat with predators, alongside other interventions. Six studies found that a minority of primates survived. One study found that an introduced primate population increased after reintroduction into habitat with predators, alongside other interventions. *Assessment: unknown effectiveness — limited evidence (effectiveness 50%; certainty 10%; harms 0%).*

https://www.conservationevidence.com/actions/1593

Reintroduce primates into habitat without predators

One study in Tanzania found that a population of reintroduced chimpanzees increased over 16 years following reintroduction into habitat without predators. *Assessment: unknown effectiveness — limited evidence (effectiveness 50%; certainty 5%; harms 0%).*

https://www.conservationevidence.com/actions/1592

7.12.4 *Ex-situ* conservation

Based on the collated evidence, what is the current assessment of the effectiveness of interventions for *ex-situ* conservation?	
Unknown effectiveness (limited evidence)	• Captive breeding and reintroduction of primates into the wild: born and reared in cages • Captive breeding and reintroduction of primates into the wild: limited free-ranging experience • Captive breeding and reintroduction of primates into the wild: born and raised in a free-ranging environment • Rehabilitate injured/orphaned primates • Fostering appropriate behaviour to facilitate rehabilitation

Unknown effectiveness (limited evidence)

Captive breeding and reintroduction of primates into the wild: born and reared in cages

One study in Brazil found that the majority of reintroduced golden lion tamarins which were born and reared in cages, alongside other interventions, did not survive over seven years.

Two of two studies in Brazil and French Guiana found that more reintroduced primates that were born and reared in cages, alongside other interventions, died post-reintroduction compared to wild-born monkeys. *Assessment: unknown effectiveness — limited evidence (effectiveness 0%; certainty 15%; harms 0%).*

https://www.conservationevidence.com/actions/1594

Captive breeding and reintroduction of primates into the wild: limited free-ranging experience

One of three studies found that the majority of captive-bred primates, with limited free-ranging experience and which were reintroduced in the wild, alongside other interventions, had survived. One study in Madagascar found that a minority of captive-bred lemurs survived reintroduction over five years. One study found that reintroduced lemurs with limited free-ranging experience had a similar diet to wild primates. Reintroduction was undertaken alongside other interventions. *Assessment: unknown effectiveness — limited evidence (effectiveness 30%; certainty 10%; harms 0%).*

https://www.conservationevidence.com/actions/1595

Captive breeding and reintroduction of primates into the wild: born and raised in a free-ranging environment

One study in Brazil found that the majority of golden lion tamarins survived for at least four months after being raised in a free-ranging environment, alongside other interventions. One study found that the diet of lemurs that were born and raised in a free-ranging environment alongside other interventions, overlapped with that of wild primates. *Assessment: unknown effectiveness — limited evidence (effectiveness 40%; certainty 10%; harms 0%).*

https://www.conservationevidence.com/actions/1596

Rehabilitate injured/orphaned primates

Six of eight studies found that the majority of introduced primates survived after rehabilitation of injured or orphaned individuals, alongside other interventions. One study found that a minority of introduced primates survived, and one study found that half of primates survived. One of two studies found that an introduced chimpanzee population increased in size after rehabilitation of injured or orphaned individuals, alongside other interventions. One study found that an introduced rehabilitated or injured primate population declined. One review found that primates living in sanctuaries had a low reproduction rate. One study found that introduced primates had similar behaviour to wild primates after rehabilitation of injured or orphaned individuals, alongside other interventions. *Assessment: unknown effectiveness — limited evidence (effectiveness 50%; certainty 10%; harms 0%).*

https://www.conservationevidence.com/actions/1597

Fostering appropriate behaviour to facilitate rehabilitation

Three of five studies found that a minority of primates survived after they were fostered to encourage behaviour appropriate to facilitate rehabilitation, alongside other interventions. Two studies found that the majority of reintroduced primates fostered to facilitate rehabilitation along other interventions survived. Three studies found that despite fostering to encourage behaviour appropriate to facilitate rehabilitation, alongside other interventions, primates differed in their behaviour to wild primates. *Assessment: unknown effectiveness — limited evidence (effectiveness 10%; certainty 10%; harms 0%).*

https://www.conservationevidence.com/actions/1600

7.13 Livelihood; economic and other incentives

7.13.1 Provide benefits to local communities for sustainably managing their forest and its wildlife

Based on the collated evidence, what is the current assessment of the effectiveness of interventions for providing benefits to local communities for sustainably managing their forest and its wildlife?	
Unknown effectiveness (limited evidence)	• Provide monetary benefits to local communities for sustainably managing their forest and its wildlife (e.g. REDD, employment) • Provide non-monetary benefits to local communities for sustainably managing their forest and its wildlife (e.g. better education, infrastructure development)

Unknown effectiveness (limited evidence)

Provide monetary benefits to local communities for sustainably managing their forest and its wildlife (e.g. REDD, employment)

One before-and-after study in Belize found that howler monkey numbers increased after the provision of monetary benefits to local communities alongside other interventions. However, one before-and-after study in

Rwanda, Uganda and the Congo found that gorilla numbers decreased despite the implementation of development projects in nearby communities, alongside other interventions. One before-and-after study in Congo found that most chimpanzees reintroduced to an area where local communities received monetary benefits, alongside other interventions, survived over five years. *Assessment: unknown effectiveness — limited evidence (effectiveness 50%; certainty 25%; harms 0%).*

https://www.conservationevidence.com/actions/1509

Provide non-monetary benefits to local communities for sustainably managing their forest and its wildlife (e.g. better education, infrastructure development)

One before-and-after study India found that numbers of gibbons increased in areas were local communities were provided alternative income, alongside other interventions. One before-and-after study in Congo found that most chimpanzees reintroduced survived over seven years in areas where local communities were provided non-monetary benefits, alongside other interventions. *Assessment: unknown effectiveness — limited evidence (effectiveness 40%; certainty 10%; harms 0%).*

https://www.conservationevidence.com/actions/1510

7.13.2 Long-term presence of research/tourism project

Based on the collated evidence, what is the current assessment of the effectiveness of interventions for the long-term presence of research-/tourism project?	
Likely to be beneficial	• Run research project and ensure permanent human presence at site
Trade-off between benefit and harms	• Run tourism project and ensure permanent human presence at site
Unknown effectiveness (limited evidence)	• Permanent presence of staff/managers

Likely to be beneficial

Run research project and ensure permanent human presence at site

Three before-and-after studies, in Rwanda, Uganda, Congo and Belize found that numbers of gorillas and howler monkeys increased while populations were continuously monitored by researchers, alongside other interventions. One before-and-after study in Kenya found that troops of translocated baboons survived over 16 years post-translocation while being continuously monitored by researchers, alongside other interventions. One before-and-after study in the Congo found that most reintroduced chimpanzees survived over 3.5 years while being continuously monitored by researchers, alongside other interventions. However, one before-and-after study in Brazil found that most reintroduced tamarins did not survive over 7 years, despite being continuously monitored by researchers, alongside other interventions; but tamarins reproduced successfully. One review on gorillas in Uganda found that no individuals were killed while gorillas were continuously being monitored by researchers, alongside other interventions. *Assessment: likely to be beneficial (effectiveness 61%; certainty 40%; harms 0%).*

https://www.conservationevidence.com/actions/1511

Trade-off between benefit and harms

Run tourism project and ensure permanent human presence at site

Six studies, including four before-and-after studies, in Rwanda, Uganda, Congo and Belize found that numbers of gorillas and howler monkeys increased after local tourism projects were initiated, alongside other interventions. However, two before-and-after studies in Kenya and Madagascar found that numbers of colobus and mangabeys and two of three lemur species decreased after implementing tourism projects, alongside other interventions. One before-and-after study in China found that exposing macaques to intense tourism practices, especially through range restrictions to increase visibility for tourists, had increased stress levels and increased infant mortality, peaking at 100% in some years. *Assessment: trade-off between benefit and harms (effectiveness 40%; certainty 40%; harms 40%).*

https://www.conservationevidence.com/actions/1512

Unknown effectiveness (limited evidence)

Permanent presence of staff/managers

Two before-and-after studies in the Congo and Gabon found that most reintroduced chimpanzees and gorillas survived over a period of between nine months to five years while having permanent presence of reserve staff. One before-and-after study in Belize found that numbers of howler monkeys increased after permanent presence of reserve staff, alongside other interventions. However, one before-and-after study in Kenya found that numbers of colobus and mangabeys decreased despite permanent presence of reserve staff, alongside other interventions. *Assessment: unknown effectiveness — limited evidence (effectiveness 40%; certainty 30%; harms 0%).*

https://www.conservationevidence.com/actions/1517

8. SHRUBLAND AND HEATHLAND CONSERVATION

Philip A. Martin, Ricardo Rocha, Rebecca K. Smith & William J. Sutherland

Expert assessors

Andrew Bennet, La Trobe University, Australia
Brian van Wilgen, Stellenbosch University, South Africa
Rob Marrs, University of Liverpool, UK
Chris Diek, Royal Society for the Protection of Birds, UK
G. Matt Davies, Ohio State University, USA
David Le Maitre, CSIR, UK
Giles Groome, Consultant Ecologist, UK
Isabel Barrio, University of Iceland, Iceland
James Adler, Surrey Wildlife Trust, UK
Jon Keeley, US Geological Survey, Western Ecological Research Center and Department of Ecology and Evolutionary Biology, University of California, USA
Jonty Denton, Consultant Ecologist, UK
Penny Anderson, Penny Anderson Associates Limited, UK

Scope of assessment: for the conservation of shrubland and heathland habitats (not specific species within these habitats).

Assessed: 2017.

Effectiveness measure is the median % score for effectiveness.

Certainty measure is the median % certainty of evidence, determined by the quantity and quality of the evidence in the synopsis.

Harm measure is the median % score for negative side-effects on the shrubland and heathland habitats of concern.

 https://doi.org/10.11647/OBP.0267.08

This book is meant as a guide to the evidence available for different conservation interventions and as a starting point in assessing their effectiveness. The assessments are based on the available evidence for the target habitat for each intervention. The assessment may therefore refer to different habitat to the one(s) you are considering. Before making any decisions about implementing interventions it is vital that you read the more detailed accounts of the evidence in order to assess their relevance for your study species or system.

Full details of the evidence are available at**www. conservationevidence.com**

There may also be significant negative side-effects on the target habitats or other species or communities that have not been identified in this assessment.

A lack of evidence means that we have been unable to assess whether or not an intervention is effective or has any harmful impacts.

8.1 Threat: Residential and commercial development

Based on the collated evidence, what is the current assessment of the effectiveness of interventions for managing the impacts of residential and commercial development in shrublands and heathlands?	
No evidence found (no assessment)	• Remove residential or commercial development • Maintain/create habitat corridors in developed areas

No evidence found (no assessment)

We have captured no evidence for the following interventions:

- Remove residential or commercial development
- Maintain/create habitat corridors in developed areas.

8.2 Threat: Agriculture and aquaculture

Based on the collated evidence, what is the current assessment of the effectiveness of interventions for managing the impacts of agriculture and aquaculture in shrublands and heathlands?	
Beneficial	• Reduce number of livestock
Likely to be beneficial	• Use fences to exclude livestock from shrublands
Unknown effectiveness (limited evidence)	• Change type of livestock • Shorten the period in which livestock can graze

Beneficial

Reduce number of livestock

Two before-and-after trials in the UK and South Africa and one replicated, controlled study in the UK found that reducing or stopping grazing increased the abundance or cover of shrubs. Two site comparison studies in the UK found that cover of common heather declined in sites with high livestock density, but increased in sites with low livestock density. One site comparison study in the Netherlands found that dwarf shrub cover was higher in ungrazed sites. One replicated, randomized, before-and-after study in Spain found that reducing grazing increased the cover of western gorse. One randomized, controlled trial and one before-and-after trial in the USA found that stopping grazing did not increase shrub abundance. One site comparison study in France found that ungrazed sites had higher cover of ericaceous shrubs, but lower cover of non-ericaceous shrubs than grazed sites. One site comparison study in the UK found that reducing grazing had mixed effects on shrub cover. One replicated, randomized, controlled study in the UK found that reducing grazing increased vegetation height. However, one replicated, controlled, paired, site comparison study in the UK found that reducing grazing led to a reduction in the height of heather plants. Two site comparison studies in France and the Netherlands found that ungrazed sites had a lower number of plant species than grazed sites. One replicated, controlled, paired, site comparison study in Namibia and South Africa found that reducing livestock numbers increased plant cover and the number of plant species. One controlled study in Israel found that reducing grazing increased plant biomass. However, one randomized, site comparison on the island of Gomera, Spain found that reducing grazing did not increase plant cover and one replicated, controlled study in the UK found that the number of plant species did not change . One replicated, controlled study in the UK found no change in the cover of rush or herbaceous species as a result of a reduction in grazing. Two site comparison studies in France and the Netherlands found that grass cover and sedge cover were lower in ungrazed sites than in grazed sites. One randomized, controlled study in the USA found a mixed effect of reducing grazing on grass cover. *Assessment: Beneficial (effectiveness 65%, certainty 70%, harms 10%).*

https://www.conservationevidence.com/actions/1607

Likely to be beneficial

Use fences to exclude livestock from shrublands

Two replicated, controlled, randomized studies (one of which was also a before-and-after trial) and one controlled before-and-after trial in the UK found that using fences to exclude livestock increased shrub cover or abundance. Two replicated, controlled, randomized studies in Germany and the UK found that using fences increased shrub biomass or the biomass and height of individual heather plants. Two controlled studies (one of which was a before-and-after study) in Denmark and the UK found that heather presence or cover was higher in fenced areas that in areas that were not fenced. However, one site comparison study in the USA found that using fences led to decreased cover of woody plants. Three replicated, controlled studies (one of which was a before and after study) in the USA and the UK found that fencing either had a mixed effect on shrub cover or did not alter shrub cover. One randomized, replicated, controlled, paired study in the

UK found that using fences to exclude livestock did not alter the number of plant species, but did increase vegetation height and biomass. One controlled, before-and-after study in the UK found that fenced areas had lower species richness than unfenced areas. One randomized, replicated, controlled, before-and-after trial in the UK and one site comparison study in the USA found that using fences to exclude livestock led to a decline in grass cover. However, four controlled studies (one of which a before-and-after trial) in the USA, the UK, and Finland found that using fences did not alter cover of grass species. One site comparison study in the USA and one replicated, controlled study in the UK recorded an increase in grass cover. One controlled study in Finland found that using fences to exclude livestock did not alter the abundance of herb species and one site comparison in the USA found no difference in forb cover between fenced and unfenced areas. One replicated, controlled study in the USA found fencing had a mixed effect on herb cover. *Assessment: likely to be beneficial (effectiveness 51%; certainty 60%; harms 10%).*

https://www.conservationevidence.com/actions/1545

Unknown effectiveness (limited evidence)

Change type of livestock

Two replicated, before-and-after studies and one controlled study in Spain and the UK found changing the type of livestock led to mixed effects on shrub cover. However, in two of these studies changing the type of livestock reduced the cover of herbaceous species. One replicated, controlled, before-and-after study in the UK found that grazing with both cattle and sheep, as opposed to grazing with sheep, reduced cover of purple moor grass, but had no effect on four other plant species. *Assessment: unknown effectiveness (effectiveness 40%; certainty 29%; harms 5%).*

https://www.conservationevidence.com/actions/1608

Shorten the period during which livestock can graze

One replicated, controlled, before-and-after study in the UK found that shortening the period in which livestock can graze had mixed effects on heather, bilberry, crowberry, and grass cover. One replicated, randomized, controlled study in the UK found that grazing in only winter or summer did not affect the heather or grass height compared to year-round grazing. *Assessment: unknown effectiveness (effectiveness 32%; certainty 20%; harms 2%).*

https://www.conservationevidence.com/actions/1609

8.3 Threat: Energy production and mining

Based on the collated evidence, what is the current assessment of the effectiveness of interventions for managing the impacts of energy production and mining in shrublands and heathlands?	
No evidence found (no assessment)	• Maintain/create habitat corridors in areas of energy production or mining

No evidence found (no assessment)

We have captured no evidence for the following interventions:

- Maintain/create habitat corridors in areas of energy production or mining.

8.4 Threat: Biological resource use

Based on the collated evidence, what is the current assessment of the effectiveness of interventions for managing the impacts of biological resource use in shrublands and heathlands?	
No evidence found (no assessment)	• Legally protect plant species affected by gathering • Place signs to deter gathering of shrubland species • Reduce frequency of prescribed burning

No evidence found (no assessment)

We have captured no evidence for the following interventions:

- Legally protect plant species affected by gathering
- Place signs to deter gathering of shrubland species
- Reduce the frequency of prescribed burning.

8.5 Threat: Transportation and service corridors

Based on the collated evidence, what is the current assessment of the effectiveness of interventions for managing the impacts of transportation and service corridors in shrublands and heathlands?	
No evidence found (no assessment)	• Maintain habitat corridors over or under roads and other transportation corridors • Create buffer zones besides roads and other transportation corridors

No evidence found (no assessment)

We have captured no evidence for the following interventions:

- Maintain habitat corridors over or under roads and other transportation corridors
- Create buffer zones besides roads and other transportation corridors.

8.6 Threat: Human intrusions and disturbance

Based on the collated evidence, what is the current assessment of the effectiveness of interventions for managing the impacts of human intrusions and disturbance in shrublands and heathlands?	
Unknown effectiveness (limited evidence)	• Re-route paths to reduce habitat disturbance
No evidence found (no assessment)	• Use signs and access restrictions to reduce disturbance • Plant spiny shrubs to act as barriers to people

Unknown effectiveness (limited evidence)

Re-route paths to reduce habitat disturbance

One before-and-after trial in Australia found that closing paths did not alter shrub cover, but did increase the number of plant species in an alpine shrubland. *Assessment: unknown effectiveness (effectiveness 30%; certainty 10%; harms 0%).*

https://www.conservationevidence.com/actions/1619

No evidence found (no assessment)

We have captured no evidence for the following interventions:

- Use signs and access restrictions to reduce disturbance
- Plant spiny shrubs to act as barriers to people.

8.7 Threat: Natural system modifications

8.7.1 Modified fire regime

Based on the collated evidence, what is the current assessment of the effectiveness of interventions for managing the impacts of a modified fire regime in shrublands and heathlands?	
No evidence found (no assessment)	• Use prescribed burning to mimic natural fire cycle • Use prescribed burning to reduce the potential for large wild fires • Cut strips of vegetation to reduce the spread of fire

No evidence found (no assessment)

We have captured no evidence for the following interventions:

- Use prescribed burning to mimic natural fire cycle
- Use prescribed burning to reduce the potential for large wild fires
- Cut strips of vegetation to reduce the spread of fire.

8.7.2 Modified vegetation management

Based on the collated evidence, what is the current assessment of the effectiveness of interventions for managing the impacts of a modified vegetation management in shrublands and heathlands?	
Unknown effectiveness (limited evidence)	• Reinstate the use of traditional burning practices • Use cutting/mowing to mimic grazing • Increase number of livestock

Unknown effectiveness (limited evidence)

Reinstate the use of traditional burning practices

One before and after study in the UK found that prescribed burning initially decreased the cover of most plant species, but that their cover subsequently increased. A systematic review of five studies from the UK found that prescribed burning did not alter species diversity. A replicated, controlled study in the UK found that regeneration of heather was similar in cut and burned areas. A systematic review of five studies, from Europe found that prescribed burning did not alter grass cover relative to heather cover. *Assessment: unknown effectiveness (effectiveness 40%; certainty 30%; harms 12%).*

https://www.conservationevidence.com/actions/1625

Use cutting/mowing to mimic grazing

One systematic review of three studies in lowland heathland in North Western Europe found that mowing did not alter heather abundance relative to grass abundance. A site comparison in Italy found that mowing increased heather cover. Two replicated, randomized, before-and-after trials in Spain (one of which was controlled) found that using cutting to mimic grazing reduced heather cover. One replicated, randomized, controlled, before-and-after trial in Spain found that cutting increased the number of plant species. However, a replicated, randomized, before-and-after trial found that the number of plant species only increased in a minority of cases. One replicated, randomized, before-and-after trial in Spain found that cutting to mimic grazing increased grass cover. A site comparison in Italy found that mowing increased grass cover. One site comparison study in Italy found a reduction in tree cover. *Assessment: unknown effectiveness (effectiveness 30%; certainty 25%; harms 10%).*

https://www.conservationevidence.com/actions/1627

Increase number of livestock

Two site comparison studies in the UK found that cover of common heather declined in sites with a high density of livestock. One site comparison in the Netherlands found that dwarf shrub cover was lower in grazed areas than in ungrazed areas. One before-and-after study in Belgium found that grazing increased cover of heather. One site comparison in France found that areas grazed by cattle had higher cover of non-ericaceous shrubs, but lower cover of ericaceous shrubs. One before-and-after study in the Netherlands found that increasing the number of livestock resulted in an increase in the number of common heather and cross-leaved heath seedlings. One randomized, replicated, paired, controlled study in the USA found that increasing the number of livestock did not alter shrub cover. One replicated, site comparison study and one before-and-after study in the UK and Netherlands found that increasing grazing had mixed effects on shrub and heather cover. Three site comparisons in France, the Netherlands and Greece found that grazed areas had a higher number of plant species than ungrazed areas. One before-and-after study in Belgium found that the number of plant species did not change after the introduction of grazing. One replicated, before-and-after study in the Netherlands found a decrease in the number of plant species. One before-and-after study in the Netherlands found that increasing the number of livestock resulted in a decrease in vegetation height. One replicated, before-and-after trial in France found that grazing to control native woody species increased vegetation cover in one of five sites but did not increase vegetation cover in four of five sites. A systematic review of four studies in North Western Europe found that increased grazing intensity increased the cover of grass species, relative to heather species. One before-and-after study and two site comparisons in the Netherlands and France found areas with high livestock density had higher grass and sedge cover than ungrazed areas. One randomized, replicated, paired, controlled study in the USA found that increasing the number of livestock reduced grass and herb cover. One before-and-after study in Spain found that increasing the number of ponies in a heathland site reduced grass height. One replicated, site comparison in the UK and one replicated before-and-after study in the Netherlands found that increasing cattle had mixed effects on grass and herbaceous species. *Assessment: unknown effectiveness (effectiveness 30%; certainty 30%; harms 20%).*

https://www.conservationevidence.com/actions/1628

8.8 Threat: Invasive and other problematic species

8.8.1 Problematic tree species

Based on the collated evidence, what is the current assessment of the effectiveness of interventions for managing the impacts of invasive and other problematic tree species in shrublands and heathlands?	
Unknown effectiveness (limited evidence)	• Apply herbicide to trees • Cut trees • Cut trees and remove leaf litter • Cut trees and remove tree seedlings • Use prescribed burning to control trees • Use grazing to control trees • Cut trees and apply herbicide • Cut trees and use prescribed burning • Increase number of livestock and use prescribed burning to control trees
No evidence found (no assessment)	• Mow/cut shrubland to control trees • Cut trees and increase livestock numbers

Unknown effectiveness (limited evidence)

Apply herbicide to trees

One replicated, controlled, before-and-after study in South Africa found that using herbicide to control trees increased plant diversity but did not increase shrub cover. One randomized, replicated, controlled study in the

UK found that herbicide treatment of trees increased the abundance of common heather seedlings. *Assessment: unknown effectiveness (effectiveness 40%; certainty 35%; harms 10%).*

https://www.conservationevidence.com/actions/1629

Cut trees

One randomized, replicated, controlled study in the UK found that cutting birch trees increased density of heather seedlings but not that of mature common heather plants. One replicated, controlled study in South Africa found that cutting non-native trees increased herbaceous plant cover but did not increase cover of native woody plants. One site comparison study in South Africa found that cutting non-native Acacia trees reduced shrub and tree cover. *Assessment: unknown effectiveness (effectiveness 37%; certainty 30%; harms 3%).*

https://www.conservationevidence.com/actions/1630

Cut trees and remove leaf litter

One before-and-after trial in the Netherlands found that cutting trees and removing the litter layer increased the cover of two heather species and of three grass species. *Assessment: unknown effectiveness (effectiveness 45%; certainty 10%; harms 3%).*

https://www.conservationevidence.com/actions/1631

Cut trees and remove seedlings

A controlled, before-and-after study in South Africa found that cutting orange wattle trees and removing seedlings of the same species increased plant diversity and shrub cover. *Assessment: unknown effectiveness (effectiveness 62%; certainty 20%; harms 0%).*

https://www.conservationevidence.com/actions/1632

Use prescribed burning to control trees

One randomized, replicated, controlled, before-and-after trial in the USA found that burning to control trees did not change cover of two of three grass species. One randomized, controlled study in Italy found that prescribed burning to control trees reduced cover of common heather, increased cover of purple moor grass, and had mixed effects on the basal area of trees. *Assessment: unknown effectiveness (effectiveness 10%; certainty 20%; harms 22%).*

https://www.conservationevidence.com/actions/1721

Use grazing to control trees

One randomized, controlled, before-and-after study in Italy found that grazing to reduce tree cover reduced cover of common heather and the basal area of trees, but did not alter cover of purple moor grass. *Assessment: unknown effectiveness (effectiveness 20%; certainty 10%; harms 5%).*

https://www.conservationevidence.com/actions/1634

Cut trees and apply herbicide

One controlled study in the UK found that cutting trees and applying herbicide increased the abundance of heather seedlings. However, one replicated, controlled study in the UK found that cutting silver birch trees and applying herbicide did not alter cover of common heather when compared to cutting alone. Two controlled studies (one of which was a before-and-after study) in South Africa found that cutting of trees and applying herbicide did not increase shrub cover. Two controlled studies in South Africa found that cutting trees and applying herbicide increased the total number of plant species and plant diversity. One replicated, controlled study in the UK found that cutting and applying herbicide reduced cover of silver birch trees. *Assessment: unknown effectiveness (effectiveness 45%; certainty 35%; harms 3%).*

https://www.conservationevidence.com/actions/1636

Cut trees and use prescribed burning

One replicated, before-and-after trial in the USA found that cutting western juniper trees and using prescribed burning increased the cover of herbaceous plants. One replicated, randomized, controlled, before-and-after trial in the USA found that cutting western juniper trees and using prescribed burning increased cover of herbaceous plants but had no effect on the cover of most shrubs. One controlled study in South Africa found that cutting followed by prescribed burning reduced the cover of woody plants but did not alter herbaceous cover. *Assessment: unknown effectiveness (effectiveness 40%; certainty 35%; harms 5%).*

https://www.conservationevidence.com/actions/1637

Increase number of livestock and use prescribed burning to control trees

One randomized, controlled, before-and-after study in Italy found that using prescribed burning and grazing to reduce tree cover reduced the cover of common heather and the basal area of trees. However, it did not alter the cover of purple moor grass. *Assessment: unknown effectiveness (effectiveness 2%; certainty 12%; harms 12%).*

https://www.conservationevidence.com/actions/1722

No evidence found (no assessment)

We have captured no evidence for the following interventions:

- Cut/mow shrubland to control trees
- Cut trees and increase livestock numbers.

8.8.2 Problematic grass species

Based on the collated evidence, what is the current assessment of the effectiveness of interventions for managing the impacts of invasive and other problematic grass species in shrublands and heathlands?	
Unknown effectiveness (limited evidence)	• Cut/mow to control grass • Cut/mow to control grass and sow seed of shrubland plants • Rake to control grass • Cut/mow and rotovate to control grass • Apply herbicide and sow seeds of shrubland plants to control grass • Apply herbicide and remove plants to control grass • Use grazing to control grass • Use prescribed burning to control grass • Cut and use prescribed burning to control grass • Use herbicide and prescribed burning to control grass • Strip turf to control grass • Rotovate to control grass • Add mulch to control grass • Add mulch to control grass and sow seed • Cut/mow, rotovate and sow seed to control grass
Unlikely to be beneficial	• Use herbicide to control grass

Unknown effectiveness (limited evidence)

Cut/mow to control grass

One controlled study in the UK found that mowing increased the number of heathland plants in one of two sites. The same study found that the presence of a small minority of heathland plants increased, but the presence of non-heathland plants did not change. Three replicated, controlled studies in the UK and the USA found that cutting to control grass did not alter cover of common heather or shrub seedling abundance. One replicated, controlled study in the UK found that cutting to control purple moor grass reduced vegetation height, had mixed effects on purple moor grass cover and the number of plant species, and did not alter cover of common heather. Two randomized, controlled studies in the USA found that mowing did not increase the cover of native forb species. Both studies found that mowing reduced grass cover but in one of these studies grass cover recovered over time. One replicated, controlled study in the UK found that mowing did not alter the abundance of wavy hair grass relative to rotovating or cutting turf. *Assessment: unknown effectiveness (effectiveness 22%; certainty 35%; harms 5%).*

https://www.conservationevidence.com/actions/1638

Cut/mow to control grass and sow seed of shrubland plants

One randomized, replicated, controlled study in the USA found that the biomass of sagebrush plants in areas where grass was cut and seeds sown did not differ from areas where grass was not cut, but seeds were sown. One randomized controlled study in the USA found that cutting grass and sowing seeds increased shrub seedling abundance and reduced grass cover One randomized, replicated, controlled study in the USA found that sowing seeds and mowing did not change the cover of non-native plants or the number of native plant species. *Assessment: unknown effectiveness (effectiveness 31%; certainty 20%; harms 0%).*

https://www.conservationevidence.com/actions/1639

Rake to control grass

A randomized, replicated, controlled, paired study in the USA found that cover of both invasive and native grasses, as well as forbs was lower in areas that were raked than in areas that were not raked, but that the number of annual plants species did not differ. *Assessment: unknown effectiveness (effectiveness 30%; certainty 20%; harms 12%).*

https://www.conservationevidence.com/actions/1640

Cut/mow and rotovate to control grass

One controlled study in the UK found that mowing followed by rotovating increased the number of heathland plant species in one of two sites. The same study found that the presence of a minority of heathland and non-heathland species increased. *Assessment: unknown effectiveness (effectiveness 22%; certainty 15%; harms 7%).*

https://www.conservationevidence.com/actions/1641

Apply herbicide and sow seeds of shrubland plants to control grass

One randomized, controlled study in the USA found that areas where herbicide was sprayed and seeds of shrubland species were sown had more shrub seedlings than areas that were not sprayed or sown with seeds. One randomized, replicated, controlled study in the USA found that spraying with herbicide and sowing seeds of shrubland species did not increase the cover of native plant species, but did increase the number of native plant species. One of two studies in the USA found that spraying with herbicide and sowing seeds of shrubland species reduced non-native grass cover. One study in the USA found that applying herbicide and sowing seeds of shrubland species did not reduce the cover of non-native grasses. *Assessment: unknown effectiveness (effectiveness 35%; certainty 30%; harms 0%).*

https://www.conservationevidence.com/actions/1644

Apply herbicide and remove plants to control grass

One randomized, replicated, controlled, paired study in the USA found that areas sprayed with herbicide and weeded to control non-native grass cover had higher cover of native grasses and forbs than areas that were not sprayed or weeded, but not a higher number of native plant species. The same study found that spraying with herbicide and weeding reduced non-native grass cover. *Assessment: unknown effectiveness (effectiveness 42%; certainty 20%; harms 2%).*

https://www.conservationevidence.com/actions/1645

Use grazing to control grass

One replicated, controlled, before-and-after study in the Netherlands found that grazing to reduce grass cover had mixed effects on cover of common heather and cross-leaved heath. One replicated, controlled, before-and-after study in the Netherlands found that cover of wavy-hair grass increased and one before-and-after study in Spain found a reduction in grass height. *Assessment: unknown effectiveness (effectiveness 32%; certainty 17%; harms 10%).*

https://www.conservationevidence.com/actions/1646

Use precribed burning to control grass

One replicated controlled, paired, before-and-after study in the UK found that prescribed burning to reduce the cover of purple moor grass, did not reduce its cover but did reduce the cover of common heather. One randomized, replicated, controlled study in the UK found that prescribed burning initially reduced vegetation height, but this recovered over time. *Assessment: unknown effectiveness (effectiveness 0%; certainty 20%; harms 15%).*

https://www.conservationevidence.com/actions/1723

Cut and use prescribed burning to control grass

One randomized, replicated, controlled, paired, before-and-after study in the UK found that burning and cutting to reduce the cover of purple moor grass reduced cover of common heather but did not reduce cover of purple moor grass. *Assessment: unknown effectiveness (effectiveness 0%; certainty 10%; harms 10%).*

https://www.conservationevidence.com/actions/1724

Use herbicide and prescribed burning to control grass

One randomized, replicated, controlled, paired, before-and-after study in the UK found that burning and applying herbicide to reduce the cover of purple moor grass reduced cover of common heather but did not reduce cover of purple moor grass. *Assessment: unknown effectiveness (effectiveness 0%; certainty 10%; harms 20%).*

https://www.conservationevidence.com/actions/1725

Strip turf to control grass

One controlled study in the UK found that cutting and removing turf increased the number of heathland plants. The same study found that the presence of a small number of heathland plants increased, and that the presence of a small number of non-heathland plants decreased. One replicated, controlled study in the UK found that presence of heather was similar in areas where turf was cut and areas that were mown or rotovated. One replicated, controlled study in the UK found that the presence of wavy hair grass was similar in areas where turf was cut and those that were mown or rotovated. *Assessment: unknown effectiveness (effectiveness 32%; certainty 25%; harms 2%).*

https://www.conservationevidence.com/actions/1647

Rotovate to control grass

One replicated, controlled study in the UK found that rotovating did not alter the presence of heather compared to mowing or cutting. The same study found that wavy hair grass presence was not altered by rotovating, relative to areas that were mown or cut. *Assessment: unknown effectiveness (effectiveness 0%; certainty 5%; harms 0%).*

https://www.conservationevidence.com/actions/1648

Add mulch to control grass

One randomized, controlled study in the USA found that areas where mulch was used to control grass cover had a similar number of shrub seedlings to areas where mulch was not applied. The same study found that mulch application did not reduce grass cover. *Assessment: unknown effectiveness (effectiveness 0%; certainty 10%; harms 0%).*

https://www.conservationevidence.com/actions/1649

Add mulch to control grass and sow seed

One randomized, controlled study in the USA found that adding mulch, followed by seeding with shrub seeds, increased the seedling abundance of one of seven shrub species but did not reduce grass cover. *Assessment: unknown effectiveness (effectiveness 5%; certainty 7%; harms 0%).*

https://www.conservationevidence.com/actions/1650

Cut/mow, rotovate and sow seeds to control grass

One replicated, controlled study in the UK found that rotovating did not alter the presence of heather compared to mowing or cutting. The same study found that wavy hair grass presence was not altered by rotovating, relative to areas that were mown or cut. *Assessment: unknown effectiveness (effectiveness 50%; certainty 12%; harms 1%).*

https://www.conservationevidence.com/actions/1651

Unlikely to be beneficial

Use herbicide to control grass

Two randomized, controlled studies in the UK and the USA found that spraying with herbicide did not affect the number of shrub or heathland plant seedlings. One of these studies found that applying herbicide increased the abundance of one of four heathland plants, but reduced the abundance of one heathland species. However, one randomized, controlled study in the UK found that applying herbicide increased cover of heathland species. One randomized, replicated, controlled study in the UK reported no effect on the cover of common heather. One randomized, replicated study in the UK reported mixed effects of herbicide application on shrub cover. Two randomized, controlled studies in the USA and the UK found that herbicide application did not change the cover of forb species. However, one randomized, controlled, study in the USA found that herbicide application increased native forb cover. Four of five controlled studies (two of which were replicated) in the USA found that grass cover or non-native grass cover were lower in areas where herbicides were used to control grass than areas were herbicide was not used. Two randomized, replicated, controlled studies in the UK found that herbicide reduced cover of purple moor grass, but not cover of three grass/reed species. Two randomized, controlled studies in the UK found that herbicide application did not reduce grass cover. *Assessment: unlikely to be beneficial (effectiveness 32%; certainty 40%; harms 7%).*

https://www.conservationevidence.com/actions/1643

8.8.3 Bracken

Based on the collated evidence, what is the current assessment of the effectiveness of interventions for managing the impacts of bracken in shrublands and heathlands?	
Unknown effectiveness (limited evidence)	• Use herbicide to control bracken • Cut to control bracken • Cut and apply herbicide to control bracken • Cut bracken and rotovate • Use 'bracken bruiser' to control bracken • Use herbicide and remove leaf litter to control bracken
No evidence found (no assessment)	• Cut and burn bracken • Use herbicide and sow seed of shrubland plants to control bracken • Increase grazing intensity to control bracken • Use herbicide and increase livestock numbers to control bracken

Unknown effectiveness (limited evidence)

Use herbicide to control bracken

One controlled, before-and-after trial in the UK found that applying herbicide to control bracken increased the number of heather seedlings. However, two randomized, controlled studies in the UK found that spraying with herbicide did not increase heather cover. One randomized, controlled study in the UK found that applying herbicide to control bracken increased heather biomass. One replicated, randomized, controlled study in the UK found that the application of herbicide increased the number of plant species in a heathland site. However, one replicated, randomized, controlled study in the UK found that spraying bracken with herbicide had no effect on species richness or diversity. One randomized, controlled study in the UK found that applying herbicide to control bracken increased the cover of wavy hair-grass and sheep's fescue. One controlled study in the UK found that applying herbicide to control bracken increased the cover of gorse and the abundance of common cow-wheat. One controlled, before-and-after trial in the UK found that the application of herbicide reduced the abundance of bracken but increased the number of silver birch seedlings. Three randomized, controlled studies in the UK found that the application of herbicide reduced the biomass or cover of bracken. However, one controlled study in the UK found that applying herbicide did not change the abundance of bracken. *Assessment: unknown effectiveness (effectiveness 50%; certainty 35%; harms 10%).*

https://www.conservationevidence.com/actions/1652

Cut to control bracken

One randomized, controlled, before-and-after trial in Norway and one randomized, controlled study in the UK found that cutting bracken increased the cover or biomass of heather. However, two randomized, replicated, controlled studies in the UK found that cutting bracken did not increase heather cover or abundance of heather seedlings. One randomized, replicated, controlled study in the UK found that cutting to control bracken increased the species richness of heathland plant species. However, another randomized, replicated, controlled study in the UK found that cutting to control bracken did not alter species richness but did increase species diversity. One randomized, replicated, controlled study in the UK found that cutting bracken increased cover of wavy hair-grass and sheep's fescue. One controlled study in the UK found that cutting bracken did not increase the abundance of gorse or common cow-wheat. One randomized, controlled, before-and-after trial in Norway and two randomized, controlled studies in the UK found that cutting bracken reduced bracken cover or biomass. One randomized, replicated, controlled, paired study the UK found that cutting had mixed effects on bracken cover. However, one controlled study in the UK found that cutting bracken did not decrease the abundance of bracken. *Assessment: unknown effectiveness (effectiveness 50%; certainty 35%; harms 2%).*

https://www.conservationevidence.com/actions/1653

Cut and apply herbicide to control bracken

One randomized, controlled study in the UK found that cutting and applying herbicide to control bracken did not alter heather biomass. One randomized, controlled, before-and-after trial in Norway found that cutting and applying herbicide increased heather cover. One randomized, replicated, controlled, paired study in the UK found that cutting and using herbicide had no significant effect on the cover of seven plant species. One replicated, randomized, controlled study in the UK found that cutting bracken followed by applying herbicide increased plant species richness when compared with applying herbicide followed by cutting. Three randomized, controlled studies (one also a before-and-after trial, and one of which was a paired study) in the UK and Norway found that cutting and applying herbicide reduced bracken biomass or cover. *Assessment: unknown effectiveness (effectiveness 30%; certainty 30%; harms 4%).*

https://www.conservationevidence.com/actions/1654

Cut bracken and rotovate

One controlled study in the UK found that cutting followed by rotovating to control bracken did not increase total plant biomass or biomass of heather. *Assessment: unknown effectiveness (effectiveness 0%; certainty 10%; harms 0%).*

https://www.conservationevidence.com/actions/1656

Use 'bracken bruiser' to control bracken

One randomized, replicated, controlled, before-and-after, paired study in the UK found that bracken bruising increased bracken cover, though bracken cover also increased in areas where bracken bruising was not done .There was no effect on the number of plant species or plant diversity. *Assessment: unknown effectiveness (effectiveness 0%; certainty 10%; harms 7%).*

https://www.conservationevidence.com/actions/1726

Use herbicide and remove leaf litter to control bracken

One randomized, controlled study in the UK found that using herbicide and removing leaf litter did not increase total plant biomass after eight years. The same study found that for three of six years, heather biomass was higher in areas where herbicide was sprayed and leaf litter was removed than in areas that were sprayed with herbicide. *Assessment: unknown effectiveness (effectiveness 27%; certainty 12%; harms 2%).*

https://www.conservationevidence.com/actions/1660

No evidence found (no assessment)

We have captured no evidence for the following interventions:

- Cut and burn bracken
- Use herbicide and sow seed of shrubland plants to control bracken

- Increase grazing intensity to control bracken
- Use herbicide and increase livestock numbers to control bracken.

8.8.4 Problematic animals

Based on the collated evidence, what is the current assessment of the effectiveness of interventions for managing the impacts of problematic animals in shrublands and heathlands?	
Unknown effectiveness (limited evidence)	• Use fences to exclude large herbivores • Reduce numbers of large herbivores
No evidence found (no assessment)	• Use biological control to reduce the number of problematic invertebrates

Unknown effectiveness (limited evidence)

Use fences to exclude large herbivores

One controlled study in the USA found that using fences to exclude deer increased the height of shrubs, but not shrub cover. *Assessment: unknown effectiveness (effectiveness 7%; certainty 10%; harms 0%).*

https://www.conservationevidence.com/actions/1662

Reduce numbers of large herbivores

One before-and-after trial in the USA found that removing feral sheep, cattle and horses increased shrub cover and reduced grass cover. One replicated study in the UK found that reducing grazing pressure by red deer increased the cover and height of common heather. *Assessment: unknown effectiveness (effectiveness 70%; certainty 30%; harms 0%).*

https://www.conservationevidence.com/actions/1663

No evidence found (no assessment)

We have captured no evidence for the following interventions:

- Use biological control to reduce the number of problematic invertebrates.

8.9 Threat: Pollution

Based on the collated evidence, what is the current assessment of the effectiveness of interventions for managing the impacts of pollution in shrublands and heathlands?	
Unknown effectiveness (limited evidence)	• Mow shrubland to reduce impacts of pollutants • Burn shrublands to reduce impacts of pollutants
No evidence found (no assessment)	• Plant vegetation to act as a buffer to exclude vegetation • Reduce pesticide use on nearby agricultural/ forestry land • Reduce herbicide use on nearby agricultural/ forestry land • Reduce fertilizer use on nearby agricultural/ forestry land • Add lime to shrubland to reduce the impacts of sulphur dioxide pollution

Unknown effectiveness (limited evidence)

Mow shrubland to reduce impact of pollutants

One randomized, replicated, controlled study in the UK found that mowing to reduce the impact of nitrogen deposition did not alter shoot length of common heather or the number of purple moor grass seedlings. One controlled study in the UK found that mowing a heathland affected by nitrogen pollution did not alter the cover or shoot length of heather compared to areas where prescribed burning was used. *Assessment: unknown effectiveness (effectiveness 0%; certainty 17%; harms 0%).*

https://www.conservationevidence.com/actions/1669

Burn shrublands to reduce impacts of pollutants

One randomized, replicated, controlled study in the UK found that prescribed burning to reduce the impact of nitrogen deposition did not alter the shoot length of common heather or the number of purple moor grass seedlings compared to mowing. A controlled study in the UK found that burning to reduce the concentration of pollutants in a heathland affected by nitrogen pollution did not alter the cover or shoot length of heather relative to areas that were mowed. *Assessment: unknown effectiveness (effectiveness 0%; certainty 17%; harms 0%).*

https://www.conservationevidence.com/actions/1670

No evidence found (no assessment)

We have captured no evidence for the following interventions:

- Plant vegetation to act as a buffer to exclude vegetation
- Reduce pesticide use on nearby agricultural/forestry land
- Reduce herbicide use on nearby agricultural/forestry land
- Reduce fertilizer use on nearby agricultural/forestry land
- Add lime to shrubland to reduce the impacts of sulphur dioxide pollution.

8.10 Threat: Climate change and severe weather

Based on the collated evidence, what is the current assessment of the effectiveness of interventions for managing the impacts of climate change and severe weather in shrublands and heathlands?	
No evidence found (no assessment)	• Restore habitat in area predicted to have suitable habitat for shrubland species in the future • Improve connectivity between areas of shrubland to allow species movements and habitat shifts in response to climate change

No evidence found (no assessment)

We have captured no evidence for the following interventions:

- Restore habitat in area predicted to have suitable habitat for shrubland species in the future
- Improve connectivity between areas of shrubland to allow species movements and habitat shifts in response to climate change.

8.11 Threat: Habitat protection

Based on the collated evidence, what is the current assessment of the effectiveness of interventions for habitat protection in shrublands and heathlands?	
No evidence found (no assessment)	• Legally protect shrubland • Legally protect habitat around shrubland

No evidence found (no assessment)

We have captured no evidence for the following interventions:

- Legally protect shrubland
- Legally protect habitat around shrubland.

8.12 Habitat restoration and creation

8.12.1 General restoration

Based on the collated evidence, what is the current assessment of the effectiveness of interventions for general restoration of shrubland and heathland habitats?	
Likely to be beneficial	• Allow shrubland to regenerate without active management
No evidence found (no assessment)	• Restore/create connectivity between shrublands

Likely to be beneficial

Allow shrubland to regenerate without active management

Five before-and-after trials (two of which were replicated) in the USA, UK, and Norway, found that allowing shrubland to recover after fire without any active management increased shrub cover or biomass. One replicated, paired, site comparison in the USA found that sites that were allowed to recover without active restoration had similar shrub cover to unburned areas. One controlled, before-and-after trial in the USA found no increase in shrub cover. One before-and-after trial in Norway found an increase in heather height. One before-and-after trial in Spain found that there was an increase in seedlings for one of three shrub species. Two replicated, randomized, controlled, before-and-after trials in Spain and Portugal found that there was an increase in the cover of woody plant species. One before-and-after study in Spain found that cover of woody plants increased, but the number of woody plant species did not. One replicated, before-and-after study in South Africa found that the height of three protea species increased after recovery from fire. One before-and-after trial in South Africa found that there was an increase in vegetation cover, but not in the number of plant species. One before-and-after trial in South Africa found an increase in a minority of plant species. Two before-and-after trials in the USA and UK found that allowing shrubland to recover after fire without active management resulted in a decrease in grass cover or biomass. One controlled, before-and-after trial in the USA found an increase in the cover of a minority of grass species. One before-and-after study in Spain found that cover of herbaceous species declined. One replicated, before-and-after study in the UK found mixed effects on cover of wavy hair grass. One controlled, before-and-after trial in the USA found no increase in forb cover. One replicated, randomized, controlled before-and-after trial in Spain found that herb cover declined after allowing recovery of shrubland after fire. *Assessment: likely to be beneficial (effectiveness 62%; certainty 60%; harms 0%).*

https://www.conservationevidence.com/actions/1679

No evidence found (no assessment)

We have captured no evidence for the following interventions:

- Restore/create connectivity between shrublands.

8.12.2 Modify physical habitat

Based on the collated evidence, what is the current assessment of the effectiveness of interventions for restoring shrubland and heathland habitats by modifying the physical habitat?	
Likely to be beneficial	• Add topsoil
Unknown effectiveness (limited evidence)	• Disturb vegetation • Strip topsoil • Remove leaf litter • Add sulphur to soil

	• Use erosion blankets/mats to aid plant establishment • Add mulch and fertilizer to soil • Add manure to soil • Irrigate degraded shrublands
No evidence found (no assessment)	• Remove trees/crops to restore shrubland structure • Remove trees, leaf litter and topsoil • Add peat to soil • Burn leaf litter

Likely to be beneficial

Add topsoil

Two randomized, controlled studies in the UK found that the addition of topsoil increased the cover or abundance of heathland plant species. One replicated, site comparison in Spain found an increase in the abundance of woody plants. One randomized, controlled study in the UK found an increase in the number of seedlings for a majority of heathland plants. One controlled study in Namibia found that addition of topsoil increased plant cover and the number of plant species, but that these were lower than at a nearby undisturbed site. One randomized, controlled study in the UK found an increase in the cover of forbs but a reduction in the cover of grasses. *Assessment: likely to be beneficial (effectiveness 67%; certainty 45%; harms 0%).*

https://www.conservationevidence.com/actions/1686

Unknown effectiveness (limited evidence)

Disturb vegetation

One randomized, replicated, controlled study in the UK found that vegetation disturbance did not increase the abundance or species richness of specialist plants but increased the abundance of generalist plants. *Assessment: unknown effectiveness (effectiveness 0%; certainty 10%; harms 7%).*

https://www.conservationevidence.com/actions/1727

Strip topsoil

Two randomized, replicated, controlled studies in the UK found that removal of topsoil did not increase heather cover or cover of heathland species. However, one controlled study in the UK found an increase in heather cover. One randomized, replicated, controlled study in the UK found that removing topsoil increased the cover of both specialist and generalist plant species, but did not increase species richness. One randomized, replicated, paired, controlled study in the UK found that removal of topsoil increased cover of annual grasses but led to a decrease in the cover of perennial grasses. One controlled study in the UK found that removal of turf reduced cover of wavy hair grass. One controlled, before-and-after trial in the UK found that stripping surface layers of soil increased the cover of gorse and sheep's sorrel as well as the number of plant species. *Assessment: unknown effectiveness (effectiveness 30%; certainty 25%; harms 3%).*

https://www.conservationevidence.com/actions/1685

Remove leaf litter

One randomized, controlled study in the UK found that removing leaf litter did not alter the presence of heather. *Assessment: unknown effectiveness (effectiveness 0%; certainty 10%; harms 0%).*

https://www.conservationevidence.com/actions/1688

Add sulphur to soil

One randomized, replicated, controlled study in the UK found that adding sulphur to the soil of a former agricultural field did not increase the number of heather seedlings in five of six cases. *Assessment: unknown effectiveness (effectiveness 2%; certainty 10%; harms 0%).*

https://www.conservationevidence.com/actions/1691

Use erosion blankets/mats to aid plant establishment

One replicated, randomized, controlled study in the USA found that using an erosion control blanket increased the height of two shrub species. One replicated, randomized, controlled study in the USA did not find an increase in the number of shrub species, but one controlled study in China did find an increase in plant diversity following the use of erosion control blankets. The same study found an increase in plant biomass and cover. *Assessment: unknown effectiveness (effectiveness 30%; certainty 20%; harms 0%).*

https://www.conservationevidence.com/actions/1692

Add mulch and fertilizer to soil

One randomized, controlled study in the USA found that adding mulch and fertilizer did not increase the seedling abundance of seven shrub species. The same study also reported no change in grass cover. *Assessment: unknown effectiveness (effectiveness 0%; certainty 10%; harms 0%).*

https://www.conservationevidence.com/actions/1694

Add manure to soil

One replicated, randomized, controlled study in South Africa found that adding manure increased plant cover and the number of plant species. *Assessment: unknown effectiveness (effectiveness 30%; certainty 10%; harms 0%).*

https://www.conservationevidence.com/actions/1695

Irrigate degraded shrublands

One replicated, randomized, controlled study at two sites in USA found that temporary irrigation increased shrub cover. *Assessment: unknown effectiveness (effectiveness 30%; certainty 10%; harms 0%).*

https://www.conservationevidence.com/actions/1696

No evidence found (no assessment)

We have captured no evidence for the following interventions:

- Remove trees/crops to restore shrubland structure
- Remove trees, leaf litter and topsoil
- Add peat to soil
- Burn leaf litter.

8.12.3 Introduce vegetation or seeds

Based on the collated evidence, what is the current assessment of the effectiveness of interventions for restoring shrubland and heathland habitats by introducing vegetation or seeds?	
Beneficial	• Sow seeds
Unknown effectiveness (limited evidence)	• Plant individual plants • Sow seeds and plant individual plants

	• Spread clippings • Build bird perches to encourage colonization by plants • Plant turf

Beneficial

Sow seeds

Five of six studies (including three replicated, randomized, controlled studies, one site comparison study and one controlled study) in the UK, South Africa, and the USA found that sowing seeds of shrubland species increased shrub cover. One of six studies in the UK found no increase in shrub cover. One replicated site comparison in the USA found in sites where seed containing Wyoming big sagebrush was sown the abundance of the plant was higher than in sites where it was not sown. One replicated, randomized, controlled study in the USA found that shrub seedling abundance increased after seeds were sown. One study in the USA found very low germination of hackberry seeds when they were sown. One replicated, randomized, controlled study in the USA found that the community composition of shrublands where seeds were sown was similar to that found in undisturbed shrublands. One randomized, controlled study in the UK found an increase in the cover of heathland plants when seeds were sown. One replicated, randomized, controlled study in South Africa found that sowing seeds increased plant cover. One replicated, randomized, controlled study in the USA found that areas where seeds were sown did not differ significantly in native cover compared to areas where shrubland plants had been planted. One controlled study in the USA found higher plant diversity in areas where seeds were sown by hand than in areas where they were sown using a seed drill. Two of three studies (one of which was a replicated, randomized, controlled study) in the USA found that sowing seeds of shrubland species resulted in an increase in grass cover. One randomized, controlled study in the UK found no changes in the cover of grasses or forbs. *Assessment: beneficial (effectiveness 70%; certainty 60%; harms 0%).*

https://www.conservationevidence.com/actions/1698

Unknown effectiveness (limited evidence)

Plant individual plants

One replicated, randomized, controlled study in the USA found that planting California sagebrush plants did not increase the cover of native plant species compared to sowing of seeds or a combination of planting and sowing seeds. One replicated, randomized, controlled study in South Africa found that planting *Brownanthus pseudoschlichtianus* plants increased plant cover, but not the number of plant species. One study in the USA found that a majority of planted plants survived after one year. *Assessment: unknown effectiveness (effectiveness 40%; certainty 20%; harms 0%).*

https://www.conservationevidence.com/actions/1697

Sow seeds and plant individual plants

One replicated, controlled study in the USA found that planting California sagebrush and sowing of seeds did not increase cover of native plant species compared to sowing of seeds, or planting alone. *Assessment: unknown effectiveness (effectiveness 10%; certainty 10%; harms 0%).*

https://www.conservationevidence.com/actions/1700

Spread clippings

One randomized, controlled study in the UK found that the addition of shoots and seeds of heathland plants did not increase the abundance of mature plants for half of plant species. One randomized, controlled study in the UK found that the frequency of heather plants was not significantly different in areas where heather clippings had been spread and areas where they were not spread. One replicated, randomized, controlled study in the UK found an increase in the number of heather seedlings, but not of other heathland species. One randomized, controlled study in the UK found that the addition of shoots and seeds increased the number of seedlings for a minority of species. One replicated, randomized, controlled study in South Africa found that plant cover and the number of plant species did not differ significantly between areas where branches had been spread and those where branches had not been spread. *Assessment: unknown effectiveness (effectiveness 30%; certainty 32%; harms 0%).*

https://www.conservationevidence.com/actions/1701

Build bird perches to encourage colonization by plants

One replicated, controlled study in South Africa found that building artificial bird perches increased the number of seeds at two sites, but no shrubs became established at either of these sites. *Assessment: unknown effectiveness (effectiveness 10%; certainty 10%; harms 0%).*

https://www.conservationevidence.com/actions/1702

Plant turf

Two randomized, controlled studies in the UK found that planting turf from intact heathland sites increased the abundance or cover of heathland species. One of these studies also found that planting turf increased the seedling abundance for a majority of heathland plant species. One randomized, controlled study in the UK found that planting turf increased forb cover, and reduced grass cover. One randomized, replicated, controlled study in Iceland found that planting large turves from intact heathland sites increased the number of plant species, but smaller turves did not. A*ssessment: unknown effectiveness (effectiveness 62%; certainty 30%; harms 0%).*

https://www.conservationevidence.com/actions/1703

8.13 Actions to benefit introduced vegetation

Based on the collated evidence, what is the current assessment of the effectiveness of interventions to benefit introduced vegetation in shrubland heathland habitats?	
Unknown effectiveness (limited evidence)	• Add fertilizer to soil (alongside planting/seeding) • Add peat to soil (alongside planting/seeding) • Add mulch and fertilizer to soil (alongside planting/seeding) • Add gypsum to soil (alongside planting/seeding) • Add sulphur to soil (alongside planting/seeding) • Strip/disturb topsoil (alongside planting/seeding) • Add topsoil (alongside planting/seeding) • Plant seed balls • Plant/sow seeds of nurse plants alongside focal plants • Plant/seed under established vegetation • Plant shrubs in clusters • Add root associated bacteria/fungi to introduced plants

Unknown effectiveness (limited evidence)

Add fertilizer to soil (alongside planting/seeding)

A replicated, controlled study in Iceland found that adding fertilizer and sowing seeds increased cover of shrubs and trees in a majority of cases. The same study showed an increase in vegetation cover in two of three cases.

One controlled study in the USA found that adding fertilizer increased the biomass of four-wing saltbush in a majority of cases. *Assessment: unknown effectiveness (effectiveness 45%; certainty 25%; harms 0%).*

https://www.conservationevidence.com/actions/1704

Add peat to soil (alongside planting/seeding)

One replicated, randomized, controlled study in the UK found that adding peat to soil and sowing seed increased the cover of common heather in the majority of cases, compared to seeding alone. One replicated, randomized, controlled study in the UK found that adding peat to soil and sowing seed increased the density of heather seedlings, and led to larger heather plants than seeding alone, but that no seedlings survived after two years. *Assessment: unknown effectiveness (effectiveness 42%; certainty 20%; harms 0%).*

https://www.conservationevidence.com/actions/1705

Add mulch and fertilizer to soil (alongside planting/seeding)

A randomized, controlled study in the USA found that adding mulch and fertilizer, followed by sowing of seeds increased the abundance of seedlings for a minority of shrub species. The same study found that adding mulch and fertilizer, followed by sowing seeds had no significant effect on grass cover. *Assessment: unknown effectiveness (effectiveness 35%; certainty 15%; harms 0%).*

https://www.conservationevidence.com/actions/1707

Add gypsum to soil (alongside planting/seeding)

One randomized, controlled study in South Africa found that adding gypsum to soils and sowing seeds increased survival of seedlings for one of two species. *Assessment: unknown effectiveness (effectiveness 30%; certainty 10%; harms 0%).*

https://www.conservationevidence.com/actions/1708

Add sulphur to soil (alongside planting/seeding)

A randomized, replicated, controlled study in the UK found that adding sulphur to soil alongside sowing seeds did not increase heather cover in a majority of cases. One replicated, controlled study in the UK found that adding sulphur and spreading heathland clippings had mixed effects on cover of common heather, perennial rye-grass, and common bent. One randomized, controlled study in the UK found that adding sulphur to soil alongside planting of heather seedlings increased their survival, though after two years survival was very low. *Assessment: unknown effectiveness (effectiveness 20%; certainty 30%; harms 0%).*

https://www.conservationevidence.com/actions/1710

Strip/disturb topsoil (alongside planting/seeding)

Two replicated, controlled studies in the UK found that removal of topsoil and addition seed/clippings increased cover of heathland plants or cover of heather and gorse. One controlled study in the UK found that soil disturbance using a rotovator and spreading clippings of heathland plants (alongside mowing) increased the number of heathland plants. One replicated, controlled study in the UK found that stripping the surface layers of soil and adding seed reduced the cover of perennial rye-grass. One randomized, replicated, paired, and controlled study in the UK found that removal of topsoil and addition of the clippings of heathland plants did not alter the cover of annual grasses but led to a decrease in cover of perennial grasses. *Assessment: unknown effectiveness (effectiveness 60%; certainty 35%; harms 0%).*

https://www.conservationevidence.com/actions/1711

Add topsoil (alongside planting/seeding)

One randomized, replicated, paired, controlled study in the USA found that addition of topsoil alongside sowing of seed increased the biomass of grasses but reduced the biomass of forbs in comparison to addition of topsoil alone. *Assessment: unknown effectiveness (effectiveness 0%; certainty 10%; harms 0%).*

https://www.conservationevidence.com/actions/1857

Plant seed balls

A randomized, replicated, controlled study in the USA found that planting seed balls resulted in lower seedling numbers than sowing seed. *Assessment: unknown effectiveness (effectiveness 0%; certainty 10%; harms 0%).*

https://www.conservationevidence.com/actions/1712

Plant/sow seeds of nurse plants alongside focal plants

A randomized, replicated, controlled study in the UK found that sowing seeds of nurse plants and heathland plants did not increase the cover of common heather. One replicated, randomized, controlled study in the USA found that sowing seeds of nurse plants and California sagebrush seeds together reduced survival of shrubs in more than half of cases. The same study found that California sagebrush biomass was also reduced when its seeds were sown with those of nurse plants. *Assessment: unknown effectiveness (effectiveness 0%; certainty 20%; harms 10%).*

https://www.conservationevidence.com/actions/1713

Plant/seed under established vegetation

A replicated, randomized, controlled study in the USA found that sowing seed under established shrubs had mixed effects on blackbrush seedling emergence. *Assessment: unknown effectiveness (effectiveness 20%; certainty 10%; harms 0%).*

https://www.conservationevidence.com/actions/1714

Plant shrubs in clusters

A randomized, controlled study in South Africa found that when shrubs were planted in clumps more of them died than when they were planted alone. *Assessment: unknown effectiveness (effectiveness 0%; certainty 15%; harms 1%).*

https://www.conservationevidence.com/actions/1715

Add root associated bacteria/fungi to introduced plants

Two controlled studies (one of which was randomized) in Spain found that adding rhizobacteria to soil increased the biomass of shrubs. One of these studies also found an increase in shrub height. *Assessment: unknown effectiveness (effectiveness 60%; certainty 15%; harms 0%).*

https://www.conservationevidence.com/actions/1716

8.14 Education and awareness

Based on the collated evidence, what is the current assessment of the effectiveness of interventions for education and awareness of shrubland and heathland habitats?	
No evidence found (no assessment)	• Raise awareness amongst the general public • Provide education programmes about shrublands

No evidence found (no assessment)

We have captured no evidence for the following interventions:

- Raise awareness amongst the general public
- Provide education programmes about shrublands.

9. MANAGEMENT OF CAPTIVE ANIMALS

Coral S. Jonas, Lydia T. Timbrell, Fey Young, Silviu O. Petrovan, Andrew E. Bowkett & Rebecca K. Smith

Husbandry interventions for captive breeding amphibians

Expert assessors

Kay Bradfield, Perth Zoo, Australia
Jeff Dawson, Durrell Wildlife Conservation Trust, UK
Devin Edmonds, Association Mitsinjo, Madagascar
Jonathan Kolby, Honduras Amphibian Rescue and Conservation Center, Honduras
Stephanie Jayson, Veterinary Department, Zoological Society of London, UK
Daniel Nicholson, Queen Mary University of London, UK
Silviu O. Petrovan, Cambridge University, UK and Froglife Trust, UK
Jay Redbond, Wildfowl & Wetlands Trust, UK
Rebecca K. Smith, Cambridge University, UK
Benjamin Tapley, Herpetology Section, Zoological Society of London, UK

Scope of assessment: for husbandry interventions for captive breeding amphibians.
Assessed: 2017.

Promoting health and welfare in captive carnivores (felids, canids and ursids) through feeding practices

Expert assessors

Kathy Baker, Whitley Wildlife Conservation Trust, Newquay Zoo, UK
Marcus Clauss, University of Zurich, Switzerland
Ellen Dierenfeld, Independent comparative nutrition consultant, USA
Thomas Quirke, University College Cork, Republic of Ireland
Joanna Newbolt, Whitley Wildlife Conservation Trust, Paignton Zoo, and University of Plymouth, UK
Simon Marsh, Yorkshire Wildlife Wildlife Park, UK
Amy Plowman, Whitley Wildlife Conservation Trust, Paignton Zoo, UK
Katherine Whitehouse-Tedd, Nottingham Trent University, UK
Gwen Wirobski, University of Veterinary Medicine Vienna, Austria

Scope of assessment: for promoting health and welfare in captive carnivores (felids, canids and ursids) through feeding practices.
Assessed: 2018.

 https://doi.org/10.11647/OBP.0267.09

Promoting natural feeding behaviours in primates in captivity

Expert assessors

Francis Cabana, Wildlife Reserves Singapore, Singapore
Po-Han Chou, Taipei Zoo, Taiwan
Ellen Dierenfeld, Independent comparative nutrition consultant, USA
Mike Downman, Dartmoor Zoo, UK
Craig Gilchrist, Paignton Zoo, UK
Amy Plowman, Whitley Wildlife Conservation Trust, Paignton Zoo, UK

Scope of assessment: for promoting natural feeding behaviours in captive primates.
Assessed: 2017.
Effectiveness measure is the median % score for effectiveness.
Certainty measure is the median % certainty of evidence for effectiveness, determined by the quantity and quality of the evidence in the synopsis.
Harm measure is the median % score for negative side-effects on the species included.

This book is meant as a guide to the evidence available for different conservation interventions and as a starting point in assessing their effectiveness. The assessments are based on the available evidence for the target group of species for each intervention. The assessment may therefore refer to different species or habitat to the one(s) you are considering. Before making any decisions about implementing interventions it is vital that you read the more detailed accounts of the evidence in order to assess their relevance for your study species or system.

Full details of the evidence are available at**www.conservationevidence.com**

There may also be significant negative side-effects on the target groups or other species or communities that have not been identified in this assessment.

A lack of evidence means that we have been unable to assess whether or not an intervention is effective or has any harmful impacts.

9.1 *Ex-situ* conservation – breeding amphibians

9.1.1 Refining techniques using less threatened species

Based on the collated evidence, what is the current assessment of the effectiveness of interventions for refining techniques using less threatened species?	
Unknown effectiveness (limited evidence)	• Identify and breed a similar species to refine husbandry techniques prior to working with target species

Unknown effectiveness (limited evidence)

Identify and breed a similar species to refine husbandry techniques prior to working with target species

Two small, replicated interlinked studies in Brazil found that working with a less threatened surrogate species of frog first to establish husbandry interventions promoted successful breeding of a critically endangered species of frog. *Assessment: unknown effectiveness (effectiveness 68%; certainty 30%; harms 15%).*

https://www.conservationevidence.com/actions/1862

9.1.2 Changing environmental conditions/ microclimate

Based on the collated evidence, what is the current assessment of the effectiveness of interventions for changing environmental conditions/ microclimate?	
Unknown effectiveness (limited evidence)	• Vary enclosure temperature to simulate seasonal changes in the wild • Vary quality or quantity (UV% or gradients) of enclosure lighting to simulate seasonal changes in the wild • Provide artificial aquifers for species which breed in upwelling springs • Vary artificial rainfall to simulate seasonal changes in the wild
No evidence found (no assessment)	• Vary enclosure humidity to simulate seasonal changes in the wild using humidifiers, foggers/ misters or artificial rain • Vary duration of enclosure lighting to simulate seasonal changes in the wild • Simulate rainfall using sound recordings of rain and/or thunderstorms • Allow temperate amphibians to hibernate • Allow amphibians from highly seasonal environments to have a period of dormancy during a simulated drought period • Vary water flow/speed of artificial streams in enclosures for torrent breeding species

Unknown effectiveness (limited evidence)

Vary enclosure temperature to simulate seasonal changes in the wild

One small, replicated study in Italy found that one of six females bred following a drop in temperature from 20-24 to 17°C, and filling of an egg laying pond. One replicated, before-and-after study in 2006-2012 in Australia found that providing a pre-breeding cooling period, alongside allowing females to gain weight before the breeding period, along with separating sexes during the non-breeding period, providing mate choice for females and playing recorded mating calls, increased breeding success. *Assessment: unknown effectiveness (effectiveness 50%; certainty 35%; harms 0%).*

https://www.conservationevidence.com/actions/1864

Vary quality or quantity (UV% or gradients) of enclosure lighting to simulate seasonal changes in the wild

One replicated study in the UK found that there was no difference in clutch size between frogs given an ultraviolet (UV) boost compared with those that only received background levels. However, frogs given the UV boost had a significantly greater fungal load than frogs that were not UV-boosted. *Assessment: unknown effectiveness (effectiveness 0%; certainty 33%; harms 20%).*

https://www.conservationevidence.com/actions/1865

Provide artificial aquifers for species which breed in upwelling springs

One small study in the USA found that salamanders bred in an aquarium fitted with an artificial aquifer. *Assessment: unknown effectiveness (effectiveness 50%; certainty 15%; harms 0%).*

https://www.conservationevidence.com/actions/1871

Vary artificial rainfall to simulate seasonal changes in the wild

Two replicated, before-and-after studies in Germany and Austria found that simulating a wet and dry season, as well as being moved to an enclosure with more egg laying sites and flowing water in Austria, stimulated breeding and egg deposition. In Germany, no toadlets survived past 142 days old. *Assessment: unknown effectiveness (effectiveness 78%; certainty 33%; harms 0%).*

https://www.conservationevidence.com/actions/1872

No evidence found (no assessment)

We have captured no evidence for the following interventions:

- Vary enclosure humidity to simulate seasonal changes in the wild using humidifiers, foggers/misters or artificial rain

- Vary duration of enclosure lighting to simulate seasonal changes in the wild
- Simulate rainfall using sound recordings of rain and/or thunderstorms
- Allow temperate amphibians to hibernate
- Allow amphibians from highly seasonal environments to have a period of dormancy
- Vary water flow/speed of artificial streams in enclosures for torrent breeding species

9.1.3 Changing enclosure design for spawning or egg laying sites

Based on the collated evidence, what is the current assessment of the effectiveness of interventions for changing enclosure design for spawning or egg laying sites?	
Unknown effectiveness (limited evidence)	• Provide multiple egg laying sites within an enclosure • Provide natural substrate for species which do not breed in water (e.g. burrowing/tunnel breeders) • Provide particular plants as breeding areas or egg laying sites

Unknown effectiveness (limited evidence)

Provide multiple egg laying sites within an enclosure

One replicated study in Australia found that frogs only bred once moved into an indoor enclosure which had various types of organic substrate, allowed temporary flooding, and enabled sex ratios to be manipulated along with playing recorded mating calls. One small, replicated, before-and-after study in Fiji found that adding rotting logs and hollow bamboo pipes to an enclosure, as well as a variety of substrates, promoted egg laying in frogs. *Assessment: unknown effectiveness (effectiveness 50%; certainty 25%; harms 0%).*

https://www.conservationevidence.com/actions/1873

Provide natural substrate for species which do not breed in water (e.g. burrowing/tunnel breeders)

Two replicated studies in Australia and Fiji found that adding a variety of substrates to an enclosure, as well as rotting logs and hollow bamboo pipes in one case, promoted egg laying of frogs. The Australian study also temporarily flooded enclosures, manipulated sex ratios and played recorded mating calls. *Assessment: unknown effectiveness (effectiveness 50%; certainty 20%; harms 0%).*

https://www.conservationevidence.com/actions/1874

Provide particular plants as breeding areas or egg laying sites

One small, controlled study in the USA found that salamanders bred in an aquarium heavily planted with java moss and swamp-weed. *Assessment: unknown effectiveness (effectiveness 75%; certainty 20%; harms 0%).*

https://www.conservationevidence.com/actions/1875

9.1.4 Manipulate social conditions

Based on the collated evidence, what is the current assessment of the effectiveness of interventions for manipulating social conditions?	
Unknown effectiveness (limited evidence)	• Manipulate sex ratio within the enclosure • Separate sexes in non-breeding periods • Play recordings of breeding calls to simulate breeding season in the wild • Allow female mate choice
No evidence found (no assessment)	• Provide visual barriers for territorial species • Manipulate adult density within the enclosure

Unknown effectiveness (limited evidence)

Manipulate sex ratio within the enclosure

One replicated study in Australia found that frogs only bred once sex ratios were manipulated, along with playing recorded mating calls and moving frogs into an indoor enclosure which allowed temporary flooding, and had various types of organic substrate. *Assessment: unknown effectiveness (effectiveness 35%; certainty 15%; harms 0%).*

https://www.conservationevidence.com/actions/1879

Separate sexes in non-breeding periods

One replicated, before-and-after study in Australia found that clutch size of frogs increased when sexes were separated in the non-breeding periods, alongside providing female mate choice, playing recorded mating calls and allowing females to increase in weight before breeding. *Assessment: unknown effectiveness (effectiveness 65%; certainty 30%; harms 0%).*

https://www.conservationevidence.com/actions/1880

Play recordings of breeding calls to simulate breeding season in the wild

One replicated study in Australia found that frogs only bred when recorded mating calls were played, as well as manipulating the sex ratio after frogs were moved into an indoor enclosure that allowed temporary flooding and had various types of organic substrates. One replicated, before-and-after study in Australia found that clutch size of frogs increased when playing recorded mating calls, along with the sexes being separated in the non-breeding periods, providing female mate choice, and allowing females to increase in weight before breeding. *Assessment: unknown effectiveness (effectiveness 35%; certainty 28%; harms 0%).*

https://www.conservationevidence.com/actions/1881

Allow female mate choice

One replicated study in Australia found that frogs only bred after females carrying eggs were introduced to males, sex ratios were manipulated, recorded mating calls were played, and after being moved to an indoor enclosure which allowed temporary flooding and had various types of organic substrates.

One replicated, before-and-after study in Australia found that clutch size of frogs increased when female mate choice was provided, alongside playing recorded mating calls, sexes being separated in the non-breeding periods, and allowing females to increase in weight before breeding. *Assessment: unknown effectiveness (effectiveness 50%; certainty 20%; harms 0%).*

https://www.conservationevidence.com/actions/1882

No evidence found (no assessment)

We have captured no evidence for the following interventions:

- Provide visual barriers for territorial species
- Manipulate adult density within the enclosure.

9.1.5 Changing the diet of adults

Based on the collated evidence, what is the current assessment of the effectiveness of interventions for changing the diet of adults?	
Unknown effectiveness (limited evidence)	• Supplement diets with carotenoids (including for colouration) • Increase caloric intake of females in preparation for breeding
No evidence found (no assessment)	• Vary food provision to reflect seasonal availability in the wild • Formulate adult diet to reflect nutritional composition of wild foods • Supplement diets with vitamins/calcium fed to prey (e.g. prey gut loading) • Supplement diets with vitamins/calcium applied to food (e.g. dusting prey)

Unknown effectiveness (limited evidence)

Supplement diets with carotenoids (including for colouration)

One study in the USA found that adding carotenoids to fruit flies fed to frogs reduced the number of clutches, but increased the number of tadpoles and successful metamorphs. *Assessment: unknown effectiveness (effectiveness 70%; certainty 28%; harms 0%).*

https://www.conservationevidence.com/actions/1887

Increase caloric intake of females in preparation for breeding

One replicated, before-and-after study in Australia found that clutch size of frogs increased when females increased in weight before breeding, as well as having mate choice, recorded mating calls, and sexes being separated during the non-breeding periods. *Assessment: unknown effectiveness (effectiveness 60%; certainty 23%; harms 0%).*

https://www.conservationevidence.com/actions/1888

No evidence found (no assessment)

We have captured no evidence for the following interventions:

- Vary food provision to reflect seasonal availability in the wild
- Formulate adult diet to reflect nutritional composition of wild foods
- Supplement diets with vitamins/calcium fed to prey (e.g. prey gut loading)
- Supplement diets with vitamins/calcium applied to food (e.g. dusting prey).

9.1.6 Manipulate rearing conditions for young

Based on the collated evidence, what is the current assessment of the effectiveness of interventions for manipulating rearing conditions for the young	
Trade-off between benefit and harms	• Manipulate temperature of enclosure to improve development or survival to adulthood
Unknown effectiveness (limited evidence)	• Formulate larval diets to improve development or survival to adulthood • Manipulate larval density within the enclosure
No evidence found (no assessment)	• Leave infertile eggs at spawn site as food for egg-eating larvae • Manipulate humidity to improve development or survival to adulthood • Manipulate quality and quantity of enclosure lighting to improve development or survival to adulthood • Allow adults to attend their eggs

Trade-off between benefit and harms

Manipulate temperature of enclosure to improve development or survival to adulthood

One replicated study in Spain found that salamander larvae had higher survival rates when reared at lower temperatures. One replicated study in Germany found that the growth rate and development stage reached by harlequin toad tadpoles was faster at a higher constant temperature rather than a lower and varied water temperature. One replicated study in Australia found that frog tadpoles took longer to reach metamorphosis when reared at lower temperatures. One replicated, controlled study in Iran found that developing eggs reared within a temperature range of 12-25°C had higher survival rates, higher growth rates and lower abnormalities than those raised outside of that range. *Assessment: trade-offs between benefits and harms (effectiveness 80%; certainty 58%; harms 20%).*

https://www.conservationevidence.com/actions/1893

Unknown effectiveness (limited evidence)

Formulate larval diets to improve development or survival to adulthood

One randomized, replicated, controlled study in the USA found that tadpoles had higher body mass and reached a more advanced developmental stage when fed a control diet (rabbit chow and fish food) or freshwater algae, compared to those fed pine or oak pollen. One randomized, replicated study in Portugal found that tadpoles reared on a diet containing 46% protein had higher growth rates, survival and body weights at metamorphosis compared to diets containing less protein. *Assessment: unknown effectiveness (effectiveness 65%; certainty 35%; harms 0%).*

https://www.conservationevidence.com/actions/1889

Manipulate larval density within the enclosure

One randomized study in the USA found that decreasing larval density of salamanders increased larvae survival and body mass. *Assessment: unknown effectiveness (effectiveness 88%; certainty 28%; harms 0%).*

https://www.conservationevidence.com/actions/1894

No evidence found (no assessment)

We have captured no evidence for the following interventions:

- Leave infertile eggs at spawn site as food for egg-eating larvae
- Manipulate humidity to improve development or survival to adulthood
- Manipulate quality and quantity of enclosure lighting to improve development or survival to adulthood
- Allow adults to attend their eggs.

9.1.7 Artificial reproduction

Based on the collated evidence, what is the current assessment of the effectiveness of interventions for artificial reproduction?	
No evidence found (no assessment)	• Use artificial cloning from frozen or fresh tissue

No evidence found (no assessment)

We have captured no evidence for the following interventions:

- Use artificial cloning from frozen or fresh tissue

For summarised evidence for
Use hormone treatment to induce sperm and egg release
Use artificial fertilization in captive breeding
See Smith, R.K. and Sutherland, W.J. (2014) *Amphibian Conservation: Global Evidence for the Effects of Interventions*. Exeter, Pelagic Publishing.
Key messages and summaries are available here:

https://www.conservationevidence.com/actions/834

https://www.conservationevidence.com/actions/883

9.2 Promoting health and welfare in captive carnivores (felids, canids and ursids) through feeding practices

9.2.1 Diet and food type

Based on the collated evidence, what is the current assessment of the effectiveness of interventions for diet and food type?	
Likely to be beneficial	• Provide bones, hides or partial carcasses
Trade-off between benefit and harms	• Feed whole carcasses (with or without organs/ gastrointestinal tract)
Unknown effectiveness (limited evidence)	• Feed commercially prepared diets • Feed plant-derived protein • Supplement meat-based diets with prebiotic plant material to facilitate digestion • Supplement meat-based diet with amino acids
No evidence found (no assessment)	• Supplement meat-based diet with vitamins or minerals • Supplement meat-based diet with fatty acids • Increase variety of food items

Likely to be beneficial

Provide bones, hides or partial carcasses

One replicated, before-and-after study in the USA and one replicated, controlled study in Finland found that the provision of bones decreased the frequency of stereotypic behaviours in lions, tigers and Arctic foxes. Two replicated, before-and-after studies of felids and red foxes in the USA and Norway found that the provision of bones increased activity and manipulation time. *Assessment: likely to be beneficial (effectiveness 80%; certainty 60%; harms 0%).*

https://www.conservationevidence.com/actions/1902

Trade-off between benefit and harms

Feed whole carcasses (with or without organs/ gastrointestinal tract)

Two replicated, before-and-after studies in the USA found that feeding whole carcasses reduced pacing levels in lions, leopards, snow leopards and cougars. However, it increased pacing in tigers. One replicated, randomized, controlled study in Denmark found that when fed whole rabbit, cheetahs had lower blood protein urea, zinc and vitamin A levels compared to supplemented beef. One replicated before-and-after study in Denmark found that feeding whole rabbit showed lower levels of inflammatory bowel indicators in cheetahs. One replicated, randomized study and one controlled study in the USA found that when fed whole 1 to 3 day old chickens, ocelots had lower digestible energy and fat compared to a commercial diet and African wildcats had had lower organic matter digestibility compared to a ground-chicken diet. *Assessment: trade-offs between benefits and harms (effectiveness 80%; certainty 70%; harms 25%).*

https://www.conservationevidence.com/actions/1901

Unknown effectiveness (limited evidence)

Feed commercially prepared diets

One replicated, before-and-after study in the USA found that providing a commercial diet to maned wolves led to similar dry matter intake and digestibility despite having a lower protein content. One replicated, controlled study in South Africa found that cheetahs fed a commercial diet had a similar likelihood of developing gastritis as those fed horse meat, lower levels of blood protein urea but higher levels of creatine. One study in USA found that cheetahs fed a commercial meat diet or whole chicken carcasses had plasma a-tocopherol, retinol and taurine concentrations within the ranges recommended for domestic cats. *Assessment: unknown effectiveness (effectiveness 40%; certainty 35%; harms 50%).*

https://www.conservationevidence.com/actions/1900

Feed plant-derived protein

One replicated, randomized, controlled study and one replicated, controlled study in the USA found that a plant-derived protein diet increased digestible energy and dry matter digestibility but decreased mineral retention and plasma taurine levels in maned wolves compared to a (supplemented) animal-based protein diet. *Assessment: unknown effectiveness (effectiveness 10%; certainty 25%; harms 70%).*

https://www.conservationevidence.com/actions/1903

Supplement meat-based diets with prebiotic plant material to facilitate digestion

One replicated, before-and-after study in India found that providing Jerusalem artichoke as a supplement increased two types of gut microbiota, faecal scores and faecal moisture content in leopards. *Assessment: unknown effectiveness (effectiveness 50%; certainty 25%; harms 0%).*

https://www.conservationevidence.com/actions/1905

Supplement meat-based diet with amino acid

One replicated, before-and-after study in the USA found that supplementing an animal-protein diet with taurine, increased plasma taurine levels in maned wolves. *Assessment: unknown effectiveness (effectiveness 90%; certainty 25%; harms 0%).*

https://www.conservationevidence.com/actions/1908

No evidence found (no assessment)

We have captured no evidence for the following interventions:

- Supplement meat-based diet with vitamins or minerals
- Supplement meat-based diet with fatty acids
- Increase variety of food items.

9.2.2 Food presentation and enrichment

Based on the collated evidence, what is the current assessment of the effectiveness of interventions for food presentation and enrichment?	
Beneficial	• Hide food around enclosure
Likely to be beneficial	• Present food frozen in ice • Present food inside objects (e.g. Boomer balls)
Trade-off between benefit and harms	• Provide devices to simulate live prey, including sounds, lures, pulleys and bungees
Unknown effectiveness (limited evidence)	• Change location of food around enclosure • Scatter food around enclosure • Provide live vertebrate prey • Provide live invertebrate prey
No evidence found (no assessment)	• Present food in/on water • Use food as a reward in animal training

Beneficial

Hide food around enclosure

Four replicated, before-and-after studies in the USA, UK and Germany and one before-and-after study of a black bear, leopard cats, bush dogs, maned wolves and Malayan sun bears found that hiding food increased exploring and foraging behaviours. One replicated, before-and-after study and one before-and-after study in the USA found a decrease in stereotypical pacing in leopard cats and black bear. One before-and-after study in the USA found that hiding food reduced the time Canadian lynx spent sleeping during the day. *Assessment: beneficial (effectiveness 90%; certainty 70%; harms 10%).*

https://www.conservationevidence.com/actions/1915

Likely to be beneficial

Present food frozen in ice

Two replicated, before-and-after studies in the USA found that when presented with food frozen in ice, abnormal or stereotypic behaviours decreased and activity levels increased in bears and felids. One replicated, before-and-after study in the USA found that manipulation behaviours increased in lions, whereas a replicated study in the USA found that manipulation behaviours decreased in grizzly bears. *Assessment: likely to be beneficial (effectiveness 70%; certainty 52%; harms 10%).*

https://www.conservationevidence.com/actions/1923

Present food inside objects (e.g. Boomer balls)

Two before-and-after studies in Germany and India found that exploratory and foraging behaviours increased and stereotypic behaviours decreased in sloth bears and spectacled bears when presented with food inside objects. One before-and-after study in the USA found that exploring/foraging behaviours decreased in a sloth bear when presented with food inside objects. One replicated study in the USA found that grizzly bears spent a similar time manipulating food in a box and freely available food. *Assessment: likely to be beneficial (effectiveness 60%; certainty 70%; harms 10%).*

https://www.conservationevidence.com/actions/1924

Trade-off between benefit and harms

Provide devices to simulate live prey, including sounds, lures, pulleys and bungees

Two before-and-after studies in the USA and the UK found that activity levels and behavioural diversity increased in felids when presented with a lure or pulley system. One replicated, before-and-after study in the USA found that pacing behaviour decreased and walking increased in cougars, but pacing initially increased in tigers, when provided with a carcass on a bungee. *Assessment: trade-offs between benefits and harms (effectiveness 60%; certainty 50%; harms 25%).*

https://www.conservationevidence.com/actions/1927

Unknown effectiveness (limited evidence)

Change location of food around enclosure

One replicated, before-and-after study in Ireland found that altering the location of food decreased pacing behaviours in cheetahs. *Assessment: unknown effectiveness (effectiveness 90%; certainty 30%; harms 0%).*

https://www.conservationevidence.com/actions/1918

Scatter food around enclosure

One replicated, before-and-after study in Brazil found that scattered feeding increased locomotion in maned wolves. One replicated study in Brazil found that maned wolves spent more time in the section of their enclosure with scattered food than in a section with food on a tray. *Assessment: unknown effectiveness (effectiveness 70%; certainty 30%; harms 0%).*

https://www.conservationevidence.com/actions/1921

Provide live vertebrate prey

One small before-and-after study in the USA found that hunting behaviour increased and sleeping decreased when a fishing cat was provided with live fish. One replicated, before-and-after study in the USA found that there was no change in the occurrence of stereotypical behaviours in tigers when provided with live fish. *Assessment: unknown effectiveness (effectiveness 50%; certainty 30%; harms 0%).*

https://www.conservationevidence.com/actions/1925

Provide live invertebrate prey

One replicated study in the USA found that provision of live prey increased explorative behaviours in fennec foxes compared to other types of enrichment. *Assessment: unknown effectiveness (effectiveness 80%; certainty 20%; harms 0%).*

https://www.conservationevidence.com/actions/1926

No evidence found (no assessment)

We have captured no evidence for the following interventions:

- Present food in/on water
- Use food as a reward in animal training.

9.2.3 Feeding schedule

Based on the collated evidence, what is the current assessment of the effectiveness of interventions for feeding schedule?	
Trade-off between benefit and harms	• Provide food on a random temporal schedule

Unknown effectiveness (limited evidence)	• Allocate fast days
No evidence found (no assessment)	• Alter food abundance or type seasonally • Provide food during natural active periods • Use automated feeders • Alter feeding schedule according to visitor activity • Provide food during visitor experiences

Trade-off between benefit and harms

Provide food on a random temporal schedule

Three replicated, before-and-after studies and one replicated, controlled study found that an unpredictable feeding schedule reduced the frequency of stereotypic pacing behaviours in tigers and cheetahs. One replicated, before-and-after controlled study in the USA found that an unpredictable feeding schedule increased territorial behaviour in coyotes but did not affect travelling or foraging. One before-and-after study in Switzerland found that an unpredictable feeding schedule increased behavioural diversity in red foxes. *Assessment: trade-offs between benefits and harms (effectiveness 100%; certainty 80%; harms 20%).*

https://www.conservationevidence.com/actions/1904

Unknown effectiveness (limited evidence)

Allocate fast days

One replicated, before-and-after study in the UK found that large felids fed once every three days paced more frequently on non-feeding days. *Assessment: unknown effectiveness (effectiveness 6%; certainty 25%; harms 50%).*

https://www.conservationevidence.com/actions/1906

No evidence found (no assessment)

We have captured no evidence for the following interventions:

- Alter food abundance or type seasonally

- Provide food during natural active periods
- Use automated feeders
- Alter feeding schedule according to visitor activity
- Provide food during visitor experiences.

9.2.4 Social feeding

Based on the collated evidence, what is the current assessment of the effectiveness of interventions for social feeding?	
No evidence found (no assessment)	• Feed individuals separately • Feed individuals within a social group • Hand-feed

No evidence found (no assessment)

We have captured no evidence for the following interventions:

- Feed individuals separately
- Feed individuals within a social group
- Hand-feed.

9.3 Promoting natural feeding behaviours in primates in captivity

9.3.1 Food Presentation

Based on the collated evidence, what is the current assessment of the effectiveness of interventions for food presentation?	
Beneficial	• Scatter food throughout enclosure
Likely to be beneficial	• Hide food in containers (including boxes and bags) • Present food frozen in ice • Present food items whole instead of processed • Present feeds at different crowd levels • Maximise both vertical and horizontal presentation locations
Trade-off between benefit and harms	• Present food in puzzle feeders
Unknown effectiveness (limited evidence)	• Present food in water (including dishes and ponds) • Present food dipped in food colouring • Provide live vegetation in planters for foraging
No evidence found (no assessment)	• Present food which required the use (or modification) of tools • Paint gum solutions on rough bark • Add gum solutions to drilled hollow feeders

Beneficial

Scatter food throughout enclosure

Four studies, including one replicated study, in the USA, found that scattering food throughout enclosures increased overall activity, feeding and exploration and decreased abnormal behaviours and aggression. *Assessment: beneficial (effectiveness 80%; certainty 80%; harms 0%).*

https://www.conservationevidence.com/actions/1315

Likely to be beneficial

Hide food in containers (including boxes and bags)

Three studies including two before-and-after studies in the USA and Ireland found that the addition of food in boxes, baskets or tubes increased activity levels in lemurs and foraging levels in gibbons. *Assessment: likely to be beneficial (effectiveness 75%; certainty 50%; harms 0%).*

https://www.conservationevidence.com/actions/1316

Present food frozen in ice

Two studies in the USA and Ireland found that when frozen food was presented, feeding time increased and inactivity decreased. *Assessment: likely to be beneficial (effectiveness 60%; certainty 50%; harms 0%).*

https://www.conservationevidence.com/actions/1321

Present food items whole instead of processed

One before-and-after study in the USA found that when food items were presented whole instead of chopped, the amount of food consumed and feeding time increased in macaques. *Assessment: likely to be beneficial (effectiveness 80%; certainty 50%; harms 0%).*

https://www.conservationevidence.com/actions/1323

Present feeds at different crowd levels

One before-and-after study in the USA found that when smaller crowds were present foraging and object use in chimpanzees increased. *Assessment: likely to be beneficial (effectiveness 60%; certainty 40%; harms 0%).*

https://www.conservationevidence.com/actions/1324

Maximise both vertical and horizontal presentation locations

One controlled study in the UK and Madagascar found that less time was spent feeding on provisioned food in the indoor enclosure when food was hung in trees in an outdoor enclosure. One replicated, before-and-after study in the UK reported that when vertical and horizontal food locations were increased feeding time increased. *Assessment: likely to be beneficial (effectiveness 65%; certainty 50%; harms 0%).*

https://www.conservationevidence.com/actions/1328

Trade-off between benefit and harms

Present food in puzzle feeders

Three studies including two before-and-after studies in the USA and UK found that presenting food in puzzle feeders, increased foraging behaviour, time spent feeding and tool use but also aggression. *Assessment: trade-offs between benefits and harms (effectiveness 55%; certainty 80%; harms 60%).*

https://www.conservationevidence.com/actions/1318

Unknown effectiveness (limited evidence)

Present food in water (including dishes and ponds)

One replicated, before-and-after study in the USA found that when exposed to water filled troughs, rhesus monkeys were more active and increased their use of tools. *Assessment: unknown effectiveness (effectiveness 60%; certainty 30%; harms 0%).*

https://www.conservationevidence.com/actions/1320

Present food dipped in food colouring

One before-and-after study in the USA found that when food was presented after being dipped in food colouring, orangutans ate more and spent less time feeding. *Assessment: unknown effectiveness (effectiveness 50%; certainty 20%; harms 20%).*

https://www.conservationevidence.com/actions/1322

Provide live vegetation in planters for foraging

One replicated, before-and-after study in the USA reported that chimpanzees spent more time foraging when provided with planted rye grass and scattered sunflower seeds compared to browse and grass added to the enclosure with their normal diet. *Assessment: unknown effectiveness (effectiveness 80%; certainty 30%; harms 0%).*

https://www.conservationevidence.com/actions/1327

No evidence found (no assessment)

We have captured no evidence for the following interventions:

- Present food which required the use (or modification) of tools
- Paint gum solutions on rough bark
- Add gum solutions to drilled hollow feeders.

9.3.2 Diet manipulation

Based on the collated evidence, what is the current assessment of the effectiveness of interventions for diet manipulation?	
Likely to be beneficial	• Formulate diet to reflect nutritional composition of wild foods (including removal of domestic fruits) • Provide cut branches (browse) • Provide live invertebrates • Provide fresh produce
No evidence found (no assessment)	• Provide gum (including artificial gum) • Provide nectar (including artificial nectar) • Provide herbs or other plants for self-medication • Modify ingredients/nutrient composition seasonally (not daily) to reflect natural variability

Likely to be beneficial

Formulate diet to reflect nutritional composition of wild foods (including removal of domestic fruits)

Two replicated, before-and-after studies in the USA and UK found that when changing the diet of captive primates to reflect nutritional compositions of wild foods, there was a decrease in regurgitation and reingestion, aggression and self-directed behaviours. *Assessment: likely to be beneficial (effectiveness 70%; certainty 60%; harms 0%).*

https://www.conservationevidence.com/actions/1329

Provide cut branches (browse)

One replicated, before-and-after study in the Netherlands and Germany found that captive gorillas when presented with stinging nettles use the same processing skills as wild gorillas to forage. *Assessment: likely to be beneficial (effectiveness 70%; certainty 50%; harms 0%).*

https://www.conservationevidence.com/actions/1332

Provide live invertebrates

One before-and-after study in the UK found that providing live invertebrates to captive lorises increased foraging levels and reduced inactivity. *Assessment: likely to be beneficial (effectiveness 85%; certainty 50%; harms 0%).*

https://www.conservationevidence.com/actions/1333

Provide fresh produce

One replicated, before-and-after study in the USA found that when fresh produce was offered feeding time increased and inactivity decreased in rhesus macaques. *Assessment: likely to be beneficial (effectiveness 60%; certainty 40%; harms 1%).*

https://www.conservationevidence.com/actions/1335

No evidence found (no assessment)

We have captured no evidence for the following interventions:

- Provide gum (including artificial gum)
- Provide nectar (including artificial nectar)

- Provide herbs or other plants for self-medication
- Modify ingredients/nutrient composition seasonally (not daily) to reflect natural variability.

9.3.3 Feeding Schedule

Based on the collated evidence, what is the current assessment of the effectiveness of interventions for feeding schedule?	
Likely to be beneficial	• Change feeding times
Trade-off between benefit and harms	• Change the number of feeds per day
No evidence found (no assessment)	• Provide food at natural (wild) feeding times • Provide access to food at all times (day and night) • Use of automated feeders

Likely to be beneficial

Change feeding times

One controlled study in the USA found that changing feeding times decreased inactivity and abnormal behaviours in chimpanzees. *Assessment: likely to be beneficial (effectiveness 70%; certainty 50%; harms 0%).*

https://www.conservationevidence.com/actions/1338

Trade-off between benefit and harms

Change the number of feeds per day

Two before-and-after studies in Japan and the USA found that changing the number of feeds per day increased time spent feeding in chimpanzees but also increased hair eating in baboons. *Assessment: trade-offs between benefits and harms (effectiveness 70%; certainty 50%; harms 50%).*

https://www.conservationevidence.com/actions/1337

No evidence found (no assessment)

We have captured no evidence for the following interventions:

- Provide food at natural (wild) feeding times
- Provide access to food at all times (day and night)
- Use of automated feeders.

9.3.4 Social group manipulation

Based on the collated evidence, what is the current assessment of the effectiveness of interventions for social group manipulation?	
Trade-off between benefit and harms	• Feed individuals in social groups
No evidence found (no assessment)	• Feed individuals separately • Feed individuals in subgroups

Trade-off between benefit and harms

Feed individuals in social groups

One replicated, controlled study in the USA found that an enrichment task took less time to complete when monkeys were in social groups than when feeding alone. One before-and-after study in Italy found that in the presence of their groupmates monkeys ate more unfamiliar foods during the first encounter. *Assessment: trade-offs between benefits and harms (effectiveness 60%; certainty 50%; harms 25%).*

https://www.conservationevidence.com/actions/1343

No evidence found (no assessment)

We have captured no evidence for the following interventions:

- Feed individuals separately
- Feed individuals in subgroups.

10. SOME ASPECTS OF CONTROL OF FRESHWATER INVASIVE SPECIES

David Aldridge, Nancy Ockendon, Ricardo Rocha, Rebecca K. Smith & William J. Sutherland

Expert assessors

David Aldridge, University of Cambridge, UK
Olaf Booy, Animal and Plant Health Agency, UK
Manuel A. Duenas, Centre for Ecology & Hydrology, UK
Alison Dunn, University of Leeds, UK
Robert Francis, King's College London, UK
Belinda Gallardo, Pyrenean Institute of Ecology, Spain
Nancy Ockendon, University of Cambridge, UK
Trevor Renals, Environment Agency, UK
Emmanuelle Sarat, International Union for Conservation of Nature, France
Sonal Varia, The Centre for Agriculture and Bioscience International, UK
Alexandra Zieritz, University of Nottingham, UK
Ana L. Nunes, The Centre for Agriculture and Bioscience International, UK
Deborah Hofstra, National Institute of Water and Atmospheric Research, New Zealand
Jonathan Newman, Waterland Management Ltd, UK
Johan van Valkenburg, National Plant Protection Organization, The Netherlands
Ryan Wersal, Lonza Water Care, Alpharetta, Georgia, US
Ricardo Rocha, University of Cambridge, UK

Scope of assessment: for the control of 12 invasive freshwater species.

Assessed: American bullfrog and *Procambarus* spp. crayfish 2015; parrot's feather 2017; all other species 2016.

Effectiveness measure is the median % score for effectiveness.

Certainty measure is the median % certainty of evidence for effectiveness, determined by the quantity and quality of the evidence in the synopsis.

Harm measure is the median % score for negative side-effects to non-target native species. This was not assessed for some species in this chapter.

Potential impacts on non-target species should be considered carefully before implementing any control action.

 https://doi.org/10.11647/OBP.0267.10

This book is meant as a guide to the evidence available for different conservation interventions and as a starting point in assessing their effectiveness. The assessments are based on the available evidence for the target group of species for each intervention. The assessment may therefore refer to different species or habitat to the one(s) you are considering. Before making any decisions about implementing interventions it is vital that you read the more detailed accounts of the evidence in order to assess their relevance for your study species or system.

Full details of the evidence are available at**www. conservationevidence.com**

There may also be significant negative side-effects on the target groups or other species or communities that have not been identified in this assessment.

A lack of evidence means that we have been unable to assess whether or not an intervention is effective or has any harmful impacts.

10.1 Threat: Invasive plants

10.1.1 Parrot's feather *Myriophyllum aquaticum*

Based on the collated evidence, what is the current assessment of the effectiveness of interventions for controlling parrot's feather?	
Beneficial	• Chemical control using the herbicide 2,4-D
Likely to be beneficial	• Chemical control using the herbicide carfentrazone-ethyl • Chemical control using the herbicide triclopyr • Chemical control using the herbicide diquat • Chemical control using the herbicide endohall • Chemical control using other herbicides • Reduction of trade through legislation and codes of conduct
Trade-offs between benefit and harms	• Biological control using herbivores
Unknown effectiveness (limited evidence)	• Water level drawdown • Biological control using plant pathogens
No evidence found (no assessment)	• Mechanical harvesting or cutting • Mechanical excavation • Removal using water jets • Suction dredging and diver-assisted suction removal • Manual harvesting (hand-weeding) • Use of lightproof barriers • Dye application • Biological control using fungal-based herbicides • Use of salt

	• Decontamination / preventing further spread • Public education • Multiple integrated measures

Beneficial

Chemical control using the herbicide 2,4-D

Five laboratory studies (three replicated, controlled and two randomized, controlled) in the USA and Brazil and two replicated, randomized, field studies in Portugal reported that treatment with 2,4-D reduced growth, biomass or cover of parrot's feather. *Assessment: beneficial (effectiveness 80%; certainty 80%; harms 0%).*

https://www.conservationevidence.com/actions/1606

Likely to be beneficial

Chemical control using the herbicide carfentrazone-ethyl

Five laboratory studies (one replicated, controlled, before-and-after, three replicated, controlled and one randomized, controlled) in the USA reported that treatment with carfentrazone-ethyl reduced growth. *Assessment: likely to be beneficial (effectiveness 50%; certainty 40%; harms 5%).*

https://www.conservationevidence.com/actions/1676

Chemical control using the herbicide triclopyr

Three replicated, controlled laboratory studies in the USA and New Zealand reported that treatment with triclopyr reduced growth or that cover was lower than that of plants treated with glyphosate. One replicated, controlled field study and one replicated, before-and-after field study in New Zealand reported that cover was reduced after treatment with triclopyr but one of these studies reported that cover later increased to near pre-treatment levels. *Assessment: likely to be beneficial (effectiveness 60%; certainty 55%; harms 0%).*

https://www.conservationevidence.com/actions/1689

Chemical control using the herbicide diquat

Two replicated, controlled laboratory studies in the USA reported reduced growth after exposure to diquat. However, one replicated, randomized, controlled field study in Portugal reported no reduction in biomass following treatment with diquat. *Assessment: likely to be beneficial (effectiveness 60%; certainty 40%; harms 0%).*

https://www.conservationevidence.com/actions/1680

Chemical control using the herbicide endohall

Two replicated, controlled laboratory studies in the USA and New Zealand reported a reduction in biomass after treatment with endothall. However, one replicated, controlled field study in New Zealand found that cover declined after treatment with endothall but later cover increased close to pre-treatment levels. *Assessment: likely to be beneficial (effectiveness 50%; certainty 40%; harms 0%).*

https://www.conservationevidence.com/actions/1681

Chemical control using other herbicides

One replicated, randomized, controlled field study in Portugal and one replicated, controlled, laboratory study in the USA reported reduced growth or vegetation cover after treatment with glyphosate. Two replicated, randomized, controlled laboratory studies (one of which was randomized) in the USA have found that the herbicide imazapyr reduced growth. Four replicated, controlled (one of which was randomized) laboratory studies in the USA and New Zealand reported reduced growth after treatment with the herbicides imazamox, flumioxazin, dichlobenil and florpyrauxifen-benzyl. Two replicated, controlled (one of which was randomized) field studies in Portugal and New Zealand reported a decrease in cover after treatment with dichlobenil followed by recovery. One replicated, randomized, controlled field study in Portugal reported reduced biomass after treatment with gluphosinate-ammonium. Three replicated, controlled laboratory studies in New Zealand and the USA found no reduction in growth after treatment with clopyralid, copper chelate or fluridone. *Assessment: likely to be beneficial (effectiveness 50%; certainty 40%; harms 0%).*

https://www.conservationevidence.com/actions/1699

Reduction of trade through legislation and codes of conduct

One randomized, before-and-after trial in the Netherlands reported that the implementation of a code of conduct reduced the trade of invasive aquatic plants banned from sale. One study in the USA found that despite a state-wide trade ban on parrot's feather plants, these could still be purchased in some stores. *Assessment: likely to be beneficial (effectiveness 60%; certainty 45%; harms 0%).*

https://www.conservationevidence.com/actions/1604

Trade-off between benefit and harms

Biological control using herbivores

Two replicated, randomized studies in Argentina and the USA found that stocking with grass carp reduced the biomass or abundance of parrot's feather. However, one controlled laboratory study in Portugal found that grass carp did not reduce biomass or cover of parrot's feather. One field study in South Africa found that one *Lysathia* beetle species retarded the growth of parrot's feather. *Assessment: trade-offs between benefits and harms (effectiveness 50%; certainty 40%; harms 20%).*

https://www.conservationevidence.com/actions/1599

Unknown effectiveness (limited evidence)

Water level drawdown

One replicated, randomized, controlled laboratory study in the USA found that water removal to expose plants to drying during the summer led to lower survival of parrot's feather plants than water removal during winter. *Assessment: unknown effectiveness (effectiveness 60%; certainty 30%; harms 0%).*

https://www.conservationevidence.com/actions/1585

Biological control using plant pathogens

One study in South Africa found that exposure to a strain of the bacterium *Xanthomonas campestris* did not affect the survival of parrot's feather. *Assessment: unknown effectiveness (effectiveness 5%; certainty 10%; harms 0%).*

https://www.conservationevidence.com/actions/1601

No evidence found (no assessment)

We have captured no evidence for the following interventions:

- Mechanical harvesting or cutting
- Mechanical excavation

- Removal using water jets
- Suction dredging and diver-assisted suction removal
- Manual harvesting (hand-weeding)
- Use of lightproof barriers
- Dye application
- Biological control using fungal-based herbicides
- Use of salt
- Decontamination / preventing further spread
- Public education
- Multiple integrated measures

10.1.2 Floating pennywort *Hydrocotyle ranunculoides*

Based on the collated evidence, what is the current assessment of the effectiveness of interventions for controlling floating pennywort?	
Beneficial	• Chemical control using herbicides
Likely to be beneficial	• Flame treatment • Physical removal
Unknown effectiveness (limited evidence)	• Combination treatment using herbicides and physical removal
Unlikely to be beneficial	• Biological control using co-evolved, host-specific herbivores • Use of hydrogen peroxide
No evidence found (no assessment)	• Biological control using fungal-based herbicides • Biological control using native herbivores • Environmental control (e.g. shading, reduced flow, reduction of rooting depth, or dredging) • Excavation of banks • Public education • Use of liquid nitrogen

Beneficial

Chemical control using herbicides

A controlled, replicated field study in the UK found that the herbicide 2,4-D amine achieved almost 100% mortality of floating pennywort, compared with the herbicide glyphosate (applied without an adjuvant) which achieved negligible mortality. *Assessment: beneficial (effectiveness 80%; certainty 70%).*

https://www.conservationevidence.com/actions/1127

Likely to be beneficial

Flame treatment

A controlled, replicated study in the Netherlands found that floating pennywort plants were killed by a three second flame treatment with a three second repeat treatment 11 days later. *Assessment: likely to be beneficial (effectiveness 60%; certainty 50%).*

https://www.conservationevidence.com/actions/1131

Physical removal

Two studies, one in Western Australia and one in the UK, found physical removal did not completely eradicate floating pennywort. *Assessment: likely to be beneficial (effectiveness 40%; certainty 40%).*

https://www.conservationevidence.com/actions/1126

Unknown effectiveness (limited evidence)

Combination treatment using herbicides and physical removal

A before-and-after study in Western Australia found that a combination of cutting followed by a glyphosate chemical treatment, removed floating pennywort. *Assessment: unknown effectiveness (effectiveness 70%; certainty 35%).*

https://www.conservationevidence.com/actions/1128

Unlikely to be beneficial

Biological control using co-evolved, host-specific herbivores

A replicated laboratory and field study in South America found that the South American weevil fed on water pennywort but did not reduce the biomass. *Assessment: unlikely to be beneficial (effectiveness 20%; certainty 50%).*

https://www.conservationevidence.com/actions/1123

Use of hydrogen peroxide

A controlled, replicated study in the Netherlands found that hydrogen peroxide sprayed on potted floating pennywort plants at 30% concentration resulted in curling and transparency of the leaves but did not kill the plants. *Assessment: unlikely to be beneficial (effectiveness 10%; certainty 60%).*

https://www.conservationevidence.com/actions/1129

No evidence found (no assessment)

We have captured no evidence for the following interventions:

- Biological control using fungal-based herbicides
- Biological control using native herbivores
- Environmental control (e.g. shading, reduced flow, reduction of rooting depth, or dredging)
- Excavation of banks
- Public education
- Use of liquid nitrogen.

10.1.3 Water primrose *Ludwigia spp.*

Based on the collated evidence, what is the current assessment of the effectiveness of interventions for controlling water primrose?	
Likely to be beneficial	• Biological control using co-evolved, host specific herbivores • Chemical control using herbicides • Combination treatment using herbicides and physical removal

Unlikely to be beneficial	• Physical removal
No evidence found (no assessment)	• Biological control using fungal-based herbicides • Biological control using native herbivores • Environmental control (e.g. shading, altered flow, altered rooting depth, or dredging) • Excavation of banks • Public education • Use of a tarpaulin • Use of flame treatment • Use of hydrogen peroxide • Use of liquid nitrogen • Use of mats placed on the bottom of the water body

Likely to be beneficial

Biological control using co-evolved, host specific herbivores

A controlled, replicated study in China, found a flea beetle caused heavy feeding destruction to the prostrate water primrose. A before-and-after study in the USA found that the introduction of flea beetles to a pond significantly reduced the abundance of large-flower primrose-willow. *Assessment: likely to be beneficial (effectiveness 60%; certainty 50%).*

https://www.conservationevidence.com/actions/1135

Chemical control using herbicides

A controlled, replicated laboratory study in the USA found that the herbicide triclopyr TEA applied at concentrations of 0.25% killed 100% of young cultivated water primrose within two months. A before-and-after field study in the UK found that the herbicide glyphosate caused 97% mortality when mixed with a non-oil based sticking agent and 100% mortality when combined with TopFilm. A controlled, replicated, randomized study in Venezuela, found that use of the herbicide halosulfuron-methyl (Sempra) resulted in a significant reduction in water primrose coverage without apparent toxicity to rice plants. *Assessment: likely to be beneficial (effectiveness 80%; certainty 60%).*

https://www.conservationevidence.com/actions/1139

Combination treatment using herbicides and physical removal

A study in the USA found that application of glyphosate and a surface active agent called Cygnet-Plus followed by removal by mechanical means killed 75% of a long-standing population of water primrose. A study in Australia found that a combination of herbicide application, physical removal, and other actions such as promotion of native plants and mulching reduced the cover of Peruvian primrose-willow by 85–90%. *Assessment: likely to be beneficial (effectiveness 70%; certainty 55%).*

https://www.conservationevidence.com/actions/1140

Unlikely to be beneficial

Physical removal

A study in the USA found that hand pulling and raking water primrose failed to reduce its abundance at one site, whereas hand-pulling from the margins of a pond eradicated a smaller population of water primrose at a second site. *Assessment: unlikely to be beneficial (effectiveness 30%; certainty 50%).*

https://www.conservationevidence.com/actions/1138

No evidence found (no assessment)

We have captured no evidence for the following interventions:

- Biological control using fungal-based herbicides
- Biological control using native herbivores
- Environmental control (e.g. shading, reduced flow, reduction of rooting depth, or dredging)
- Excavation of banks
- Public education
- Use of a tarpaulin
- Use of flame treatment
- Use of hydrogen peroxide
- Use of liquid nitrogen
- Use of mats placed on the bottom of the waterbody.

10.1.4 Skunk cabbage *Lysichiton americanus*

Based on the collated evidence, what is the current assessment of the effectiveness of interventions for controlling skunk cabbage?	
Likely to be beneficial	• Chemical control using herbicides • Physical removal
No evidence found (no assessment)	• Biological control using co-evolved, host-specific herbivores • Biological control using fungal-based herbicides • Biological control using native herbivores • Combination treatment using herbicides and physical removal • Environmental control (e.g. shading, or promotion of native plants) • Public education • Use of a tarpaulin • Use of flame treatment • Use of hydrogen peroxide • Use of liquid nitrogen

Likely to be beneficial

Chemical control using herbicides

Two studies in the UK found that application of the chemical 2,4-D amine appeared to be successful in eradicating skunk cabbage stands. One of these studies also found glyphosate eradicated skunk cabbage. However, a study in the UK found that glyphosate did not eradicate skunk cabbage, but resulted in only limited reduced growth of plants. *Assessment: likely to be beneficial (effectiveness 60%; certainty 50%).*

https://www.conservationevidence.com/actions/1102

Physical removal

Two studies in Switzerland and the Netherlands, reported effective removal of recently established skunk cabbage plants using physical removal, one reporting removal of the entire stock within five years. A third study in

Germany reported that after four years of a twice yearly full removal programme, a large number of plants still needed to be removed each year. *Assessment: likely to be beneficial (effectiveness 65%; certainty 55%).*

https://www.conservationevidence.com/actions/1101

No evidence found (no assessment)

We have captured no evidence for the following interventions:

- Biological control using co-evolved, host-specific herbivores
- Biological control using fungal-based herbicides
- Biological control using native herbivores
- Combination treatment using herbicides and physical removal
- Environmental control (e.g. shading, or promotion of native plants)
- Public education
- Use of a tarpaulin
- Use of flame treatment
- Use of hydrogen peroxide
- Use of liquid nitrogen.

10.1.5 New Zealand pigmyweed *Crassula helmsii*

Based on the collated evidence, what is the current assessment of the effectiveness of interventions for controlling *Crassula helmsii*?	
Beneficial	• Chemical control using herbicides • Decontamination to prevent further spread
Likely to be beneficial	• Use lightproof barriers to control plants • Use salt water to kill plants
Unknown effectiveness (limited evidence)	• Use a combination of control measures
Unlikely to be beneficial	• Use dyes to reduce light levels • Use grazing to control plants • Use hot foam to control plants • Use hydrogen peroxide to control plants

No evidence found (no assessment)	• Alter environmental conditions to control plants (e.g. shading by succession, increasing turbidity, re-profiling or dredging) • Biological control using fungal-based herbicides • Biological control using herbivores • Bury plants • Dry out waterbodies • Physical control using manual/mechanical control or dredging • Plant other species to suppress growth • Public education • Surround with wire mesh • Use flame throwers • Use hot water • Use of liquid nitrogen

Beneficial

Chemical control using herbicides

Seven studies in the UK, including one replicated, controlled study, found that applying glyphosate reduced *Crassula helmsii*. Three out of four studies in the UK, including one controlled study, found that applying diquat or diquat alginate reduced or eradicated *C. helmsii*. One small trial found no effect of diquat on *C. helmsii* cover. One replicated, controlled study in the UK found dichlobenil reduced biomass of submerged *C. helmsii* but one small before-and-after study found no effect of dichlobenil on *C. helmsii*. A replicated, controlled study found that treatment with terbutryne partially reduced biomass of submerged *C. helmsii* and that asulam, 2,4-D amine and dalapon reduced emergent *C. helmsii*. *Assessment: beneficial (effectiveness 78%; certainty 75%).*

https://www.conservationevidence.com/actions/1279

Decontamination to prevent further spread

One controlled, replicated container trial in the UK found that submerging *Crassula helmsii* fragments in hot water led to higher mortality than drying out plants or a control. *Assessment: beneficial (effectiveness 80%; certainty 70%).*

https://www.conservationevidence.com/actions/1308

Likely to be beneficial

Use lightproof barriers to control plants

Five before-and-after studies in the UK found that covering with black sheeting or carpet eradicated or severely reduced cover of *Crassula helmsii*. *Assessment: likely to be beneficial (effectiveness 65%; certainty 50%).*

https://www.conservationevidence.com/actions/1294

Use salt water to kill plants

Two replicated, controlled container trials and two before-and-after field trials in the UK found that seawater eradicated *Crassula helmsii*. *Assessment: likely to be beneficial (effectiveness 80%; certainty 45%).*

https://www.conservationevidence.com/actions/1288

Unknown effectiveness (limited evidence)

Use a combination of control methods

One before-and-after study in the UK found that covering *Crassula helmsii* with carpet followed by treatment with glyphosate killed 80% of the plant. *Assessment: unknown effectiveness (effectiveness 75%; certainty 30%).*

https://www.conservationevidence.com/actions/1313

Unlikely to be beneficial

Use dyes to reduce light levels

One replicated, controlled study in the UK found that applying aquatic dye, along with other treatments, did not reduce cover of *Crassula helmsii*. *Assessment: unlikely to be beneficial (effectiveness 0%; certainty 53%).*

https://www.conservationevidence.com/actions/1293

Use grazing to control plants

One of two replicated, controlled studies in the UK found that excluding grazing reduce abundance and coverage of *Crassula helmsii*. The other study found that ungrazed areas had higher coverage of *C. helmsii* than grazed plots. *Assessment: unlikely to be beneficial (effectiveness 23%; certainty 43%).*

https://www.conservationevidence.com/actions/1301

Use hot foam to control plants

One replicated, controlled study in the UK found that treatment with hot foam, along with other treatments, did not control *Crassula helmsii*. A before-and-after study in the UK found that treatment with hot foam partially destroyed *C. helmsii*. *Assessment: unlikely to be beneficial (effectiveness 20%; certainty 50%).*

https://www.conservationevidence.com/actions/1286

Use hydrogen peroxide to control plants

One controlled tank trial in the UK found that hydrogen peroxide did not control *Crassula helmsii*. *Assessment: unlikely to be beneficial (effectiveness 0%; certainty 50%).*

https://www.conservationevidence.com/actions/1281

No evidence found (no assessment)

We have captured no evidence for the following interventions:

- Alter environmental conditions to control plants (e.g. shading by succession, increasing turbidity, re-profiling or dredging)
- Biological control using fungal-based herbicides
- Biological control using herbivores
- Bury plants
- Dry out waterbodies
- Physical control using manual/mechanical control or dredging
- Plant other species to suppress growth
- Public education
- Surround with wire mesh
- Use flame throwers
- Use hot water
- Use of liquid nitrogen.

10.2 Threat: Invasive molluscs

10.2.1 Asian clams

Based on the collated evidence, what is the current assessment of the effectiveness of interventions for controlling Asian clams?	
Beneficial	• Add chemicals to the water • Change salinity of the water • Mechanical removal
Likely to be beneficial	• Change temperature of water • Clean equipment • Use of gas-impermeable barriers
Unlikely to be beneficial	• Reduce oxygen in water
No evidence found (no assessment)	• Change pH of water • Drain the invaded waterbody • Exposure to disease-causing organisms • Exposure to parasites • Hand removal • Public awareness and education

Beneficial

Add chemicals to the water

Two replicated laboratory studies and one controlled, replicated field study found that chlorine, potassium and copper killed Asian clams. Increasing chemical concentration and water temperature killed more clams in less time. One controlled field trial achieved 80% and 100% mortality of Asian clams using encapsulated control agents (SB1000 and SB2000 respectively) in irrigation systems. *Assessment: beneficial (effectiveness 75%; certainty 70%).*

https://www.conservationevidence.com/actions/1118

Change salinity of water

A controlled, replicated laboratory study from the USA found that exposure to saline water killed all Asian clams. *Assessment: beneficial (effectiveness 65%; certainty 68%).*

https://www.conservationevidence.com/actions/1115

Mechanical removal

A controlled before-and-after study from North America found suction dredging of sediment reduced an Asian clam population by 96%, and these effects persisted for a year. A replicated, controlled, before-and-after field trial in Ireland showed that three types of dredges were effective at removing between 74% and >95% of the Asian clam biomass. *Assessment: beneficial (effectiveness 80%; certainty 78%).*

https://www.conservationevidence.com/actions/1120

Likely to be beneficial

Change temperature of water

A controlled laboratory study from the USA found that exposure to water at temperatures of 37°C and 36°C killed all Asian clams within 2 and 4 days, respectively. *Assessment: likely to be beneficial (effectiveness 60%; certainty 55%).*

https://www.conservationevidence.com/actions/1116

Clean equipment

A field study from Portugal found that mechanical removal, followed by regular cleaning and maintenance of industrial pipes at a power plant permanently removed an Asian clam population. A field study from Portugal found that adding a sand filter to a water treatment plant reduced an Asian clam population. *Assessment: likely to be beneficial (effectiveness 75%; certainty 50%).*

https://www.conservationevidence.com/actions/1119

Use of gas-impermeable barriers

One controlled study from North America found that placing gas impermeable fabric barriers on a lake bottom (several small and one large area) reduced populations of Asian clams. *Assessment: likely to be beneficial (effectiveness 78%; certainty 60%).*

https://www.conservationevidence.com/actions/1117

Unlikely to be beneficial

Reduce oxygen in water

A controlled laboratory study from the USA found that Asian clams were not susceptible to low oxygen levels in the water. *Assessment: unlikely to be beneficial (effectiveness 10%; certainty 50%).*

https://www.conservationevidence.com/actions/1113

No evidence found (no assessment)

We have captured no evidence for the following interventions:

- Change pH of water
- Drain the invaded waterbody
- Exposure to disease-causing organisms
- Exposure to parasites
- Hand removal
- Public awareness and education.

10.3 Threat: Invasive crustaceans

10.3.1 Ponto-Caspian gammarids

Based on the collated evidence, what is the current assessment of the effectiveness of interventions for controlling Ponto-Caspian gammarids?	
Likely to be beneficial	• Change salinity of the water • Change water temperature • Dewatering (drying out) habitat • Exposure to parasites
Unlikely to be beneficial	• Add chemicals to water • Change water pH • Control movement of gammarids
No evidence found (no assessment)	• Biological control using predatory fish • Cleaning equipment • Exchange ballast water • Exposure to disease-causing organisms

Likely to be beneficial

Change salinity of the water

One of two replicated studies, including one controlled study, in Canada and the UK found that increasing the salinity level of water killed the majority of invasive shrimp within five hours. One found that increased salinity did not kill invasive killer shrimp. *Assessment: likely to be beneficial (effectiveness 40%; certainty 50%).*

https://www.conservationevidence.com/actions/1091

Change water temperature

A controlled laboratory study from the UK found that heating water in excess of 40°C killed invasive killer shrimps. *Assessment: likely to be beneficial (effectiveness 80%; certainty 50%).*

https://www.conservationevidence.com/actions/1092

Dewatering (drying out) habitat

A replicated, controlled laboratory study from Poland found that lowering water levels in sand (dewatering) killed three species of invasive freshwater shrimp, although one species required water content levels of 4% and below before it was killed. *Assessment: likely to be beneficial (effectiveness 60%; certainty 50%).*

https://www.conservationevidence.com/actions/1094

Exposure to parasites

A replicated, controlled experimental study in Canada found that a parasitic mould reduced populations of freshwater invasive shrimp. *Assessment: likely to be beneficial (effectiveness 50%; certainty 50%).*

https://www.conservationevidence.com/actions/1089

Unlikely to be beneficial

Add chemicals to water

A controlled laboratory study from the UK found that four of nine substances added to freshwater killed invasive killer shrimp, but were impractical (iodine solution, acetic acid, Virkon S and sodium hypochlorite). Five substances did not kill invasive killer shrimp (methanol, citric acid, urea, hydrogen peroxide and sucrose). *Assessment: unlikely to be beneficial (effectiveness 35%; certainty 60%).*

https://www.conservationevidence.com/actions/1095

Change water pH

A controlled laboratory study from the UK found that lowering the pH of water did not kill invasive killer shrimp. *Assessment: unlikely to be beneficial (effectiveness 0%; certainty 50%).*

https://www.conservationevidence.com/actions/1093

Control movement of gammarids

Two replicated studies, including one controlled study, in the USA and UK found that movements of invasive freshwater shrimp slowed down or were stopped when shrimp were placed in water that had been exposed to predatory fish or was carbonated. *Assessment: likely to be beneficial (effectiveness 20%; certainty 40%).*

https://www.conservationevidence.com/actions/1088

No evidence found (no assessment)

We have captured no evidence for the following interventions:

- Biological control using predatory fish
- Cleaning equipment
- Exchange ballast water
- Exposure to disease-causing organisms.

10.3.2 *Procambarus* spp. crayfish

Based on the collated evidence, what is the current assessment of the effectiveness of interventions for controlling *Procambarus* spp. crayfish?	
Likely to be beneficial	• Add chemicals to the water • Sterilization of males • Trapping and removal • Trapping combined with encouragement of predators
Unknown effectiveness (limited evidence)	• Create barriers

Unlikely to be beneficial	• Encouraging predators
No evidence found (no assessment)	• Draining the waterway • Food source removal • Relocate vulnerable crayfish • Remove the crayfish by electrofishing

Likely to be beneficial

Add chemicals to the water

One replicated study in Italy found that natural pyrethrum at concentrations of 0.05 mg/l and above was effective at killing red swamp crayfish both in the laboratory and in a river, but not in drained burrows. *Assessment: likely to be beneficial (effectiveness 80%; certainty 50%; harms 0%).*

https://www.conservationevidence.com/actions/1036

Sterilization of males

One replicated laboratory study from Italy found that exposing male red swamp crayfish to X-rays reduced the number of offspring they produced. *Assessment: likely to be beneficial (effectiveness 50%; certainty 40%; harms 0%).*

https://www.conservationevidence.com/actions/1032

Trapping and removal

One controlled, replicated study from Italy found that food (tinned meat) was a more effective bait in trapping red swamp crayfish, than using pheromone treatments or no bait (control). Baiting with food increased trapping success compared to trapping without bait. *Assessment: likely to be beneficial (effectiveness 40%; certainty 60%; harms 0%).*

https://www.conservationevidence.com/actions/1029

Trapping combined with encouragement of predators

One before-and-after study in Switzerland and a replicated, paired site study from Italy found that a combination of trapping and predation was more effective at reducing red swamp crayfish populations than predation alone. *Assessment: likely to be beneficial (effectiveness 50%; certainty 50%; harms 0%).*

https://www.conservationevidence.com/actions/1031

Unknown effectiveness (limited evidence)

Create barriers

One before-and-after study from Italy found that the use of concrete dams across a stream was effective at containing spread of the population upstream. *Assessment: unknown effectiveness (effectiveness 30%; certainty 30%; harms 0%).*

https://www.conservationevidence.com/actions/1037

Unlikely to be beneficial

Encouraging predators

Two replicated, controlled studies in Italy found that eels fed on the red swamp crayfish and reduced population size. One replicated, controlled study found that pike predated red swamp crayfish. *Assessment: unlikely to be beneficial (effectiveness 30%; certainty 60%; harms 0%).*

https://www.conservationevidence.com/actions/1030

No evidence found (no assessment)

We have captured no evidence for the following interventions:

- Draining the waterway
- Food source removal
- Relocate vulnerable crayfish
- Remove the crayfish by electrofishing.

10.4 Threat: Invasive fish

10.4.1 Brown and black bullheads

Based on the collated evidence, what is the current assessment of the effectiveness of interventions for controlling brown and black bullheads?	
Beneficial	• Application of a biocide
Likely to be beneficial	• Netting
No evidence found (no assessment)	• Biological control of beneficial species • Biological control using native predators • Changing salinity • Changing pH • Draining invaded waterbodies • Electrofishing • Habitat manipulation • Increasing carbon dioxide concentrations • Public education • Trapping using sound or pheromonal lures • Using a combination of netting and electrofishing • UV radiation

Beneficial

Application of a biocide

Two studies in the UK and USA found that rotenone successfully eradicated black bullhead. *Assessment: beneficial (effectiveness 80%; certainty 80%).*

https://www.conservationevidence.com/actions/1050

Likely to be beneficial

Netting

A replicated study in a nature reserve in Belgium found that double fyke nets could be used to significantly reduce the population of large brown bullheads. *Assessment: likely to be beneficial (effectiveness 55%; certainty 55%).*

https://www.conservationevidence.com/actions/1051

No evidence found (no assessment)

We have captured no evidence for the following interventions:

- Biological control of beneficial species
- Biological control using native predators
- Changing salinity
- Changing pH
- Draining invaded waterbodies
- Electrofishing
- Habitat manipulation
- Increasing carbon dioxide concentrations
- Public education
- Trapping using sound or pheromonal lures
- Using a combination of netting and electrofishing
- UV radiation.

10.4.2 Ponto-Caspian gobies

Based on the collated evidence, what is the current assessment of the effectiveness of interventions for controlling Ponto-Caspian gobies?	
Beneficial	• Changing salinity
Likely to be beneficial	• Use of barriers to prevent migration
No evidence found (no assessment)	• Application of a biocide • Biological control of beneficial species • Biological control using native predators • Changing pH • Draining invaded waterbodies • Electrofishing • Habitat manipulation • Increasing carbon dioxide concentrations • Netting • Public education • Trapping using visual, sound and pheromonal lures • Using a combination of netting and electrofishing • UV radiation

Beneficial

Changing salinity

A replicated controlled laboratory study in Canada found 100% mortality of round gobies within 48 hours of exposure to water of 30% salinity. *Assessment: beneficial (effectiveness 90%; certainty 75%).*

https://www.conservationevidence.com/actions/1072

Likely to be beneficial

Use of barriers to prevent migration

A controlled, replicated field study in the USA found that an electrical barrier prevented movement of round gobies across it, and that increasing electrical pulse duration and voltage increased the effectiveness of the barrier. *Assessment: likely to be beneficial (effectiveness 50%; certainty 45%).*

https://www.conservationevidence.com/actions/1074

No evidence found (no assessment)

We have captured no evidence for the following interventions:

- Application of a biocide
- Biological control of beneficial species
- Biological control using native predators
- Changing pH
- Draining invaded waterbodies
- Electrofishing
- Habitat manipulation
- Increasing carbon dioxide concentrations
- Netting
- Public education
- Trapping using visual, sound and pheromonal lures
- Using a combination of netting and electrofishing
- UV radiation.

10.5 Threat: Invasive reptiles

10.5.1 Red-eared terrapin *Trachemys scripta*

Based on the collated evidence, what is the current assessment of the effectiveness of interventions for controlling red-eared terrapin?	
Likely to be beneficial	• Direct removal of adults
Unlikely to be beneficial	• Application of a biocide
No evidence found (no assessment)	• Biological control using native predators • Draining invaded waterbodies • Public education • Search and removal using sniffer dogs

Likely to be beneficial

Direct removal of adults

Two studies, a replicated study from Spain using Aranzadi turtle traps, and an un-replicated study in the British Virgin Islands using sein netting, successfully captured but did not eradicate red-eared terrapin populations. *Assessment: likely to be beneficial (effectiveness 40%; certainty 50%).*

https://www.conservationevidence.com/actions/1055

Unlikely to be beneficial

Application of a biocide

A replicated, controlled laboratory study in the USA, found that application of glyphosate to the eggs of red-eared terrapins reduced hatching success to 73% but only at the highest experimental concentration of glyphosate and a surface active agent. *Assessment: unlikely to be beneficial (effectiveness 15%; certainty 50%).*

https://www.conservationevidence.com/actions/1059

No evidence found (no assessment)

We have captured no evidence for the following interventions:

- Biological control using native predators
- Draining invaded waterbodies
- Public education
- Search and removal using sniffer dogs.

10.6 Threat: Invasive amphibians

10.6.1 American bullfrog *Lithobates catesbeiana*

Based on the collated evidence, what is the current assessment of the effectiveness of interventions for controlling American bullfrogs?	
Likely to be beneficial	• Biological control using native predators • Direct removal of adults • Direct removal of juveniles
Unknown effectiveness (limited evidence)	• Application of a biocide
No evidence found (no assessment)	• Biological control of co-occurring beneficial species • Collection of egg clutches • Draining ponds • Fencing • Habitat modification • Pond destruction • Public education

Likely to be beneficial

Biological control using native predators

One replicated, controlled study conducted in northeast Belgium found the introduction of the northern pike led to a strong decline in bullfrog tadpole numbers. *Assessment: likely to be beneficial (effectiveness 70%; certainty 40%; harms 0%).*

https://www.conservationevidence.com/actions/1039

Direct removal of adults

One replicated study in Belgium found catchability of adult bullfrogs in small shallow ponds using a double fyke net to be very low. One small study in the USA found that adult bullfrogs can be captured overnight in a single trap floating on the water surface. One replicated, controlled study in the USA found that bullfrog populations rapidly rebounded following intensive removal of the adults. One study in France found a significant reduction in the number of recorded adults and juveniles following the shooting of metamorphosed individuals before reproduction, when carried out as part of a combination treatment. *Assessment: likely to be beneficial (effectiveness 50%; certainty 70%; harms 0%).*

https://www.conservationevidence.com/actions/1045

Direct removal of juveniles

One replicated study in Belgium found double fyke nets were effective in catching bullfrog tadpoles in small shallow ponds. One study in France found a significant reduction in the number of recorded adults and juveniles following the removal of juveniles by trapping, when carried out as part of a combination treatment. *Assessment: likely to be beneficial (effectiveness 70%; certainty 60%; harms 0%).*

https://www.conservationevidence.com/actions/1046

Unknown effectiveness (limited evidence)

Application of a biocide

One replicated, controlled study in the USA reported a number of chemicals killed American bullfrogs, including caffeine (10% solution), chloroxylenol (5% solution), and a combined treatment of Permethrin (4.6% solution) and Rotenone (1% solution). *Assessment: unknown effectiveness (effectiveness 50%; certainty 20%; harms 0%).*

https://www.conservationevidence.com/actions/1048

No evidence found (no assessment)

We have captured no evidence for the following interventions:

- Biological control of co-occurring beneficial species
- Collection of egg clutches
- Draining ponds
- Fencing
- Habitat modification
- Pond destruction
- Public education.

11. SOME ASPECTS OF ENHANCING NATURAL PEST CONTROL

Hugh L. Wright, Joscelyne E. Ashpole, Lynn V. Dicks, James Hutchison, Caitlin G. McCormack & William J. Sutherland

Expert assessors
Barbara Smith, Game and Wildlife Conservation Trust, UK
Tony Harding, Rothamsted Research, UK
Anthony Goggin, Linking Environment and Farming (LEAF), UK
Felix Wackers, BioBest/University of Lancaster, Belgium/UK
Melvyn Fidgett, Syngenta, UK
Michael Garratt, University of Reading, UK
Michelle Fountain, East Malling Research, UK
Phillip Effingham, Greentech Consultants, UK
Stephanie Williamson, Pesticides Action Network, UK
Toby Bruce, Rothamsted Research, UK
Andrew Wilby, University of Lancaster, UK
Eve Veromann, Estonian University of Life Sciences, Estonia
Mattias Jonsson, Swedish University of Agricultural Sciences, Sweden
Vicky Kindemba, Buglife, UK
Steve Sait, University of Leeds, UK

Scope of assessment: 22 of 92 possible actions to enhance natural regulation of pests (including animals, plants, fungi, bacteria and viruses) in agricultural systems across the world.

Assessed: 2014.

Effectiveness measure is the median % score.

Certainty measure is the median % certainty of evidence, determined by the quantity and quality of the evidence in the synopsis.

Harm measure is the median % score for negative side-effects for the farmer such as reduced yield and profits or increased costs.

 https://doi.org/10.11647/OBP.0267.11

This book is meant as a guide to the evidence available for different conservation interventions and as a starting point in assessing their effectiveness. The assessments are based on the available evidence for the target group of species for each intervention. The assessment may therefore refer to different species or habitat to the one(s) you are considering. Before making any decisions about implementing interventions it is vital that you read the more detailed accounts of the evidence in order to assess their relevance for your study species or system.

Full details of the evidence are available at**www.conservationevidence.com**

There may also be significant negative side-effects on the target groups or other species or communities that have not been identified in this assessment.

A lack of evidence means that we have been unable to assess whether or not an intervention is effective or has any harmful impacts.

11.1 Reducing agricultural pollution

Based on the collated evidence, what is the current assessment of the effectiveness of interventions that reduce agricultural pollution for enhancing natural pest regulation?	
Unknown effectiveness (limited evidence)	• Alter the timing of insecticide use • Delay herbicide use • Incorporate parasitism rates when setting thresholds for insecticide use • Use pesticides only when pests or crop damage reach threshold levels
Evidence not assessed	• Convert to organic farming

Unknown effectiveness (limited evidence)

Alter the timing of insecticide use

- *Natural enemies*: One controlled study from the UK reported more natural enemies when insecticides were sprayed earlier rather than later in the growing season.
- *Pests*: Two of four studies from Mozambique, the UK and the USA found fewer pests or less disease damage when insecticides were applied early rather than late. Effects on a disease-carrying pest varied with insecticide type. Two studies (including one randomized, replicated, controlled test) found no effect on pests or pest damage.

- *Yield*: Four studies (including one randomized, replicated, controlled test) from Mozambique, the Philippines, the UK and the USA measured yields. Two studies found mixed effects and one study found no effect on yield when insecticides were applied early. One study found higher yields when insecticides were applied at times of suspected crop susceptibility.
- *Profit and costs:* One controlled study from the Philippines found higher profits and similar costs when insecticides were only applied at times of suspected crop susceptibility.
- *Crops studied:* aubergine, barley, maize, pear, stringbean.
- *Assessment: unknown effectiveness (effectiveness 40%; certainty 28%; harms 13%).*

http://www.conservationevidence.com/actions/723

Delay herbicide use

- *Natural enemies:* Two randomized, replicated, controlled trials from Australia and Denmark found more natural enemies when herbicide treatments were delayed. One of the studies found some but not all natural enemy groups benefited and fewer groups benefitted early in the season.
- *Weeds:* One randomized, replicated, controlled study found more weeds when herbicide treatments were delayed.
- *Insect pests and damage:* One of two randomized, replicated, controlled studies from Canada and Denmark found more insect pests, but only for some pest groups, and one study found fewer pests in one of two experiments and for one of two crop varieties. One study found lower crop damage in some but not all varieties and study years.
- *Yield:* One randomized, replicated, controlled study found lower yields.
- *Crops studied:* beet and oilseed.
- *Assessment: unknown effectiveness (effectiveness 20%; certainty 25%; harms 50%).*

http://www.conservationevidence.com/actions/774

Incorporate parasitism rates when setting thresholds for insecticide use

- *Pest damage:* One controlled study from New Zealand found using parasitism rates to inform spraying decisions resulted in acceptable levels of crop damage from pests. Effects on natural enemy populations were not monitored.
- *The crop studied* was tomato.
- *Assessment: unknown effectiveness (effectiveness 50%; certainty 10%; harms 5%).*

http://www.conservationevidence.com/actions/726

Use pesticides only when pests or crop damage reach threshold levels

- *Natural enemies*: One randomized, replicated, controlled study from Finland found that threshold-based spraying regimes increased numbers of natural enemies in two of three years but effects lasted for as little as three weeks.
- *Pests and disease*: Two of four studies from France, Malaysia and the USA reported that pests were satisfactorily controlled. One randomized, replicated, controlled study found pest numbers were similar under threshold-based and conventional spraying regimes and one study reported that pest control was inadequate. A randomized, replicated, controlled study found mixed effects on disease severity.
- *Crop damage*: Four of five randomized, replicated, controlled studies from New Zealand, the Philippines and the USA found similar crop damage under threshold-based and conventional, preventative spraying regimes, but one study found damage increased. Another study found slightly less crop damage compared to unsprayed controls.
- *Yield:* Two of four randomized, replicated, controlled studies found similar yields under threshold-based and conventional spraying regimes. Two studies found mixed effects depending on site, year, pest stage/type or control treatment.

- *Profit:* Two of three randomized, replicated, controlled studies found similar profits using threshold-based and conventional spraying regimes. One study found effects varied between sites and years.
- *Costs:* Nine studies found fewer pesticide applications were needed and three studies found or predicted lower production costs.
- *Crops studied:* barley, broccoli, cabbages, cauliflower, celery, cocoa, cotton, grape, peanut, potato, rice, tomato, and wheat.
- *Assessment: unknown effectiveness (effectiveness 39%; certainty 30%; harms 20%).*

http://www.conservationevidence.com/actions/750

Evidence not assessed

Convert to organic farming

- *Parasitism and mortality (caused by natural enemies)*: One of five studies (three replicated, controlled tests and two also randomized) from Europe, North America, Asia and Australasia found that organic farming increased parasitism or natural enemy-induced mortality of pests. Two studies found mixed effects of organic farming and two randomized, replicated, controlled studies found no effect.
- *Natural enemies:* Eight of 12 studies (including six randomized, replicated, controlled tests) from Europe, North America Asia and Australasia found more natural enemies under organic farming, although seven of these found effects varied over time or between natural enemy species or groups and/or crops or management practices. Three studies (one randomized, replicated, controlled) found no or inconsistent effects on natural enemies and one study found a negative effect.
- *Pests and diseases:* One of eight studies (including five randomized, replicated, controlled tests) found that organic farming reduced pests or disease, but two studies found more pests. Three studies found mixed effects and two studies found no effect.
- *Crop damage:* One of seven studies (including five randomized, replicated, controlled tests) found less crop damage in organic fields but two studies found more. One study found a mixed response and three studies found no or inconsistent effects.

- *Weed seed predation and weed abundance:* One randomized, replicated, controlled study from the USA found mixed effects of organic farming on weed seed predation by natural enemies. Two of three randomized, replicated, controlled studies from the USA found more weeds in organically farmed fields, but in one of these studies this effect varied between crops and years. One study found no effect.
- *Yield and profit:* Six randomized, replicated, controlled studies measured yields and found one positive effect, one negative effect and one mixed effect, plus no or inconsistent effects in three studies. One study found net profit increased if produce received a premium, but otherwise profit decreased. Another study found a negative or no effect on profit.
- *Crops studied:* apple, barley, beans, cabbage, carrot, gourd, maize, mixed vegetables, pea, pepper, safflower, soybean, tomato and wheat.

http://www.conservationevidence.com/actions/717

11.2 All farming systems

Based on the collated evidence, what is the current assessment of the effectiveness of interventions on all farming systems for enhancing natural pest regulation?	
Likely to be beneficial	• Grow non-crop plants that produce chemicals that attract natural enemies • Use chemicals to attract natural enemies
Trade-offs between benefit and harms	• Leave part of the crop or pasture unharvested or uncut
Unknown effectiveness (limited evidence)	• Plant new hedges • Use alley cropping
Evidence not assessed	• Use mass-emergence devices to increase natural enemy populations

Likely to be beneficial

Grow non-crop plants that produce chemicals that attract natural enemies

- *Natural enemies:* Four studies from China, Germany, India and Kenya tested the effects of growing plants that produce chemicals that attract natural enemies. Three (including one replicated, randomized, controlled trail) found higher numbers of natural enemies in plots with plants that produce attractive chemicals, and one also found that the plant used attracted natural enemies in lab studies. One found no effect on parasitism but the plant used was found not to be attractive to natural enemies in lab studies.

- *Pests:* All four studies found a decrease in either pest population or pest damage in plots with plants that produce chemicals that attract natural enemies.
- *Yield:* One replicated, randomized, controlled study found an increase in crop yield in plots with plants that produce attractive chemicals.
- *Crops studied:* sorghum, safflower, orange and lettuce.
- *Assessment: likely to be beneficial (effectiveness 68%; certainty 40%; harms 0%).*

http://www.conservationevidence.com/actions/724

Use chemicals to attract natural enemies

- *Parasitism and predation (by natural enemies):* One review and two of five studies from Asia, Europe and North America found that attractive chemicals increased parasitism. Two studies, including one randomized, replicated, controlled trial, found greater parasitism for some but not all chemicals, crops, sites or years and one study found no effect. One study showed that parasites found pests more rapidly. One study found lower egg predation by natural predators.
- *Natural enemies:* Five of 13 studies from Africa, Asia, Australasia, Europe and North America found more natural enemies while eight (including seven randomized, replicated, controlled trials) found positive effects varied between enemy groups, sites or study dates. Four of 13 studies (including a meta-analysis) found more natural enemies with some but not all test chemicals. Two of four studies (including a review) found higher chemical doses attracted more enemies, but one study found lower doses were more effective and one found no effect.
- *Pests:* Three of nine studies (seven randomized, replicated, controlled) from Asia, Australasia, Europe and North America found fewer pests, although the effect occurred only in the egg stage in one study. Two studies found more pests and four found no effect.
- *Crop damage:* One study found reduced damage with some chemicals but not others, and one study found no effect.
- *Yield:* One study found higher wheat yields.

- *Crops studied:* apple, banana, bean, broccoli, Chinese cabbage, cotton, cowpea, cranberry, grape, grapefruit, hop, maize, oilseed, orange, tomato, turnip and wheat.
- *Assessment: likely to be beneficial (effectiveness 40%; certainty 50%; harms 15%).*

http://www.conservationevidence.com/actions/754

Trade-off between benefit and harms

Leave part of the crop or pasture unharvested or uncut

- *Natural enemies:* We found eight studies from Australia, Germany, Hungary, New Zealand, Switzerland and the USA that tested leaving part of the crop or pasture unharvested or unmown. Three (including one replicated, controlled trial) found an increase in abundance of predatory insects or spiders in the crop field or pasture that was partly uncut, while four (including three replicated, controlled trials), found more predators in the unharvested or unmown area itself. Two studies (one replicated and controlled) found that the ratio of predators to pests was higher in partially cut plots and one replicated, controlled study found the same result in the uncut area. Two replicated, controlled studies found differing effects between species or groups of natural enemies.
- *Predation and parasitism:* One replicated, controlled study from Australia found an increase in predation and parasitism rates of pest eggs in unharvested strips.
- *Pests:* Two studies (including one replicated, controlled study) found a decrease in pest numbers in partially cut plots, one of them only for one species out of two. Two studies (one replicated, the other controlled) found an increase in pest numbers in partially cut plots, and two studies (including one replicated, controlled study) found more pests in uncut areas.
- *Crops studied:* alfalfa and meadow pastures.
- *Assessment: trade-offs between benefits and harms (effectiveness 45%; certainty 50%; harms 25%).*

http://www.conservationevidence.com/actions/725

Unknown effectiveness (limited evidence)

Plant new hedges

- *Natural enemies:* One randomized, replicated, controlled study from China compared plots with and without hedges and found no effect on spiders in crops. One of two studies from France and China found more natural enemies in a hedge than in adjacent crops while one study found this effect varied between crop types, hedge species and years. Two randomized, replicated, controlled studies from France and Kenya found natural enemy abundance in hedges was affected by the type of hedge shrub/ tree planted and one also found this effect varied between natural enemy groups.
- *Pests:* One randomized, replicated, controlled study from Kenya compared fallow plots with and without hedges and found effects varied between nematode (roundworm) groups.
- *Crops studied:* barley, beans, maize and wheat.
- *Assessment: unknown effectiveness (effectiveness 20%; certainty 19%; harms 20%).*

http://www.conservationevidence.com/actions/752

Use alley cropping

- *Parasitism, infection and predation:* Two of four studies from Kenya and the USA (including three randomized, replicated, controlled trials) found that effects of alley cropping on parasitism varied between study sites, sampling dates, pest life stages or the width of crop alleys. Two studies found no effect on parasitism. One study found mixed effects on fungal infections in pests and one study found lower egg predation.
- *Natural enemies:* One randomized, replicated, controlled study from Kenya found more wasps and spiders but fewer ladybirds. Some natural enemy groups were affected by the types of trees used in hedges.
- *Pests and crop damage:* Two of four replicated, controlled studies (two also randomized) from Kenya, the Philippines and the UK found more pests in alley cropped plots. One study found fewer pests and one study found effects varied with pest group and between years.

One study found more pest damage to crops but another study found no effect.

- *Weeds:* One randomized, replicated, controlled study from the Philippines found mixed effects on weeds, with more grasses in alley cropped than conventional fields under some soil conditions.
- *Yield:* One controlled study from the USA found lower yield and one study from the Philippines reported similar or lower yields.
- *Costs and profit:* One study from the USA found lower costs but also lower profit in alley cropped plots.
- *Crops studied:* alfalfa, barley, cowpea, maize, pea, rice and wheat.
- *Assessment: unknown effectiveness (effectiveness 15%; certainty 35%; harms 50%).*

http://www.conservationevidence.com/actions/718

Evidence not assessed

Use mass-emergence devices to increase natural enemy populations

- *Parasitism:* One randomized, replicated, controlled study in Switzerland found higher parasitism at one site but no effect at another site when mass-emergence devices were used in urban areas.
- *Pest damage:* The same study found no effect on pest damage to horse chestnut trees.

http://www.conservationevidence.com/actions/775

11.3 Arable farming

Based on the collated evidence, what is the current assessment of the effectiveness of interventions on arable farming systems for enhancing natural pest regulation?	
Beneficial	• Combine trap and repellent crops in a push-pull system
Trade-offs between benefit and harms	• Use crop rotation in potato farming systems
Unlikely to be beneficial	• Create beetle banks
Likely to be ineffective or harmful	• Incorporate plant remains into the soil that produce weed-controlling chemicals

Beneficial

Combine trap and repellent crops in a push-pull system

- *Parasitism:* Two randomized, replicated, controlled studies from Kenya found that push-pull cropping systems increased parasitism of stem borer larvae. One of the studies found no effect on egg parasitism.
- *Natural enemies:* Two randomized, replicated, controlled studies from Kenya and South Africa found push-pull systems had more natural predators, both in overall totals and the abundance of different predator groups.

- *Pests:* Two of three studies (two randomized, replicated, controlled) in Ethiopia, Kenya and South Africa found fewer pests. One study found no effect on pest infestation, but pests were scarce throughout. Two replicated, controlled studies (one also randomized) found fewer witchweeds.
- *Crop damage*: Two of three replicated, controlled studies (one randomized) found less pest damage, but one study (where pest numbers were low) found effects varied between years and types of damage symptom.
- *Yield:* Four of five replicated, controlled studies (two also randomized) found higher yields and one found no effect.
- *Profit and cost:* Two studies in Kenya and a review found greater economic benefits. One study found higher production costs in the first year, but equal or lower costs in the following five years.
- *Crops studied:* maize and beans.
- *Assessment: beneficial (effectiveness 70%; certainty 68%; harms 5%).*

http://www.conservationevidence.com/actions/753

Trade-off between benefit and harms

Use crop rotation in potato farming systems

- *Pests:* Nine studies from Canada and the USA and one review investigated the effect of crop rotation on pest or pathogen populations in potato. Three studies (including two replicated studies of which one randomized and one controlled) and a review found crop rotation reduced pest populations and crop diseases in at least one year or at least one site. One paired study found pest populations increased in crop rotation. Four studies (including one replicated, randomized, controlled trial) found increases and decreases in pest populations depending on rotation crops used and other treatments. One replicated, randomized, controlled study found no effect.
- *Yield:* Three out of five studies (all replicated, controlled, two also randomized) from Canada and the USA, found that crop rotation increased crop yield in some years or with certain rotation crops. The two other studies (both replicated, one also randomized and

one replicated) found yield increases and decreases depending on rotation crops used.

- *Profit:* One replicated, controlled study found that crop rotation increased profit.
- *Insecticides:* Two studies (one replicated, controlled) found that fewer insecticide treatments were needed on rotated plots.
- *Crops studied:* alfalfa, barley, broccoli, brown mustard, buckwheat, cotton, lupins, maize, oats, pearl millet, peas, potato, rye, sorghum, soybean, sugar beet, timothy grass, wheat and yellow sweet clover.
- *Assessment: trade-offs between benefits and harms (effectiveness 50%; certainty 50%; harms 25%).*

http://www.conservationevidence.com/actions/719

Unlikely to be beneficial

Create beetle banks

- *Natural enemies in fields:* Six studies from Canada, the UK and USA (three replicated, controlled, of which two were also randomized) examined the effects on predator numbers in adjacent crops. A review found that predators increased in adjacent crops, but one study found effects varied with time and another found no effect. Two studies found small or slow movements of predators from banks to crops. One study found greater beetle activity in fields but this did not improve pest predation.
- *Natural enemies on banks:* Four studies and a review found more invertebrate predators on beetle banks than in surrounding crops, but one of these found that effects varied with time.
- Eight studies from the UK and USA (including two randomized, replicated, controlled trials and two reviews) compared numbers of predatory invertebrates on beetle banks with other refuge habitats. Two studies found more natural enemies on beetle banks, but one of these found only seasonal effects. One review found similar or higher numbers of predators on beetle banks and four studies found similar or lower numbers.
- *Pests:* A replicated, randomized, controlled study and a review found the largest pest reductions in areas closest to a beetle bank or on the

beetle bank itself. One review found fewer pests in fields with than without a beetle bank.

- *Economics:* One replicated, randomized, controlled trial and a review showed that beetle banks could make economic savings if they prevented pests from reaching a spray threshold or causing 5% yield loss.
- *Beetle bank design:* Two studies from the UK found certain grass species held higher numbers of predatory invertebrates than others.
- *Crops studied:* barley, field bean, maize, oats, pasture, pea, radish, rapeseed, soybean and wheat.
- *Assessment: unlikely to be beneficial (effectiveness 25%; certainty 60%; harms 10%).*

http://www.conservationevidence.com/actions/729

Likely to be ineffective or harmful

Incorporate plant remains into the soil that produce weed-controlling chemicals

- *Weeds:* Six studies (including six randomized, replicated, controlled tests) from Asia, Europe and North America examined the effect of allelopathic plant residues on weeds by comparing amended soils with weeded controls. Three studies found a reduction in weed growth, and three found effects varied between years, weed groups, or type of weeding method in controls.
- Four studies from Asia and North America examined the effect on weeds by comparing amended soils with unweeded controls. Two studies found a reduction in weed growth, but one found that residues applied too far in advance of crop planting had the reverse effect.
- Two studies found that effects varied between trials, weed species or the type of residue used.
- *Weed control:* Two studies, including one randomized, replicated, controlled laboratory study, found that the decrease in weeds did not last beyond a few days or weeks after residue incorporation.

- *Pests:* One randomized, replicated, controlled study in the Philippines found mixed effects on pests.
- *Crop growth:* Two of three studies found that crop growth was inhibited by allelopathic residues, but these effects could be minimized by changing the timing of application. One study found effects varied between years.
- *Yield:* Three randomized, replicated, controlled studies compared crop yields in amended plots with weeded controls and found positive, negative and mixed effects. Three studies compared amended plots with unweeded controls, two found positive effects on yield and one found mixed effects (depending on crop type).
- *Profit:* One study found that amending soils increased profit compared to unweeded controls, but not compared to weeded controls.
- *Crops studied:* beans, cotton, maize, rice and wheat.
- *Assessment: likely to be ineffective or harmful (effectiveness 39%; certainty 47%; harms 30%).*

http://www.conservationevidence.com/actions/728

11.4 Perennial farming

Based on the collated evidence, what is the current assessment of the effectiveness of interventions on perennial farming systems for enhancing natural pest regulation?	
Likely to be beneficial	• Exclude ants that protect pests
Unknown effectiveness (limited evidence)	• Allow natural regeneration of ground cover beneath perennial crops • Isolate colonies of beneficial ants

Likely to be beneficial

Exclude ants that protect pests

- *Parasitism:* One of two replicated, controlled studies (one also randomized) from Japan and the USA found greater parasitism of pests by natural enemies when ants were excluded from trees. The other study found greater parasitism at one site but no effect at another.
- *Natural enemies:* Five studies (including four randomized, replicated, controlled trials) from Japan, Switzerland and the USA found effects varied between natural enemy species and groups, sampling dates, sites, crop varieties and ground cover types beneath trees.
- *Pests:* Three of seven studies (including four randomized, replicated, controlled trials) found fewer pests and another found fewer pests at times of peak abundance only. One study found mixed effects depending on date and other actions taken simultaneously (predator attractant and ground cover treatments). One study found no effect.

- *Damage and tree growth:* One study found no effect on damage to tree foliage but one study found greater tree growth.
- *Ants:* Six studies found that glue or pesticide barriers reduced ant numbers in tree or vine canopies. One study found that citrus oil barriers had no effect.
- *Crops studied:* cherimoyas, cherry, grape, grapefruit, orange, pecan and satsuma mandarin.
- *Assessment: likely to be beneficial (effectiveness 40%; certainty 50%; harms 12%).*

http://www.conservationevidence.com/actions/886

Unknown effectiveness (limited evidence)

Allow natural regeneration of ground cover beneath perennial crops

- *Natural enemies on crop trees and vines:* Five studies (including one replicated, randomized, controlled test) from Australia, China, Italy and Portugal compared natural and bare ground covers by measuring numbers of natural enemies in fruit tree or vine canopies. Three found effects varied between groups of natural enemies, two found no difference. Two studies from Australia and France compared natural to sown ground cover and found no effect on enemies in crop canopies.
- *Natural enemies on the ground:* Five studies (including three replicated, randomized, controlled trials) from Australia, Canada, China, France, and Spain compared natural and bare ground covers by measuring natural enemies on the ground. Two studies found more natural enemies in natural ground cover, but in one the effects were only short-term for most natural enemy groups. Three studies found mixed effects, with higher numbers of some natural enemy groups but not others. Two studies compared natural and sown ground covers, one study found more natural enemies and one found no effect.
- *Pests and crop damage:* Four studies (three controlled, one also replicated and randomized) from Italy, Australia and China measured pests and crop damage in regenerated and bare ground

covers. Two studies found fewer pests, whilst two studies found effects on pests and crop damage varied for different pest or disease groups. One study found more pests in natural than in sown ground covers.

- *Crops studied:* apple, grape, lemon, olive and pear.
- *Assessment: unknown effectiveness (effectiveness 35%; certainty 29%; harms 20%).*

http://www.conservationevidence.com/actions/720

Isolate colonies of beneficial ants

- *Natural enemies:* One replicated, controlled study from Australia found predatory ants occupied more cashew trees when colonies were kept isolated.
- *Pest damage and yield:* The same study found lower pest damage to cashews and higher yields.
- *The crop studied* was cashew.
- *Assessment: unknown effectiveness (effectiveness 60%; certainty 19%; harms 0%).*

http://www.conservationevidence.com/actions/773

11.5 Livestock farming and pasture

Based on the collated evidence, what is the current assessment of the effectiveness of interventions on livestock and pasture farming systems for enhancing natural pest regulation?	
Likely to be beneficial	• Grow plants that compete with damaging weeds
Unknown effectiveness (limited evidence)	• Delay mowing or first grazing date on pasture or grassland
Likely to be ineffective or harmful	• Use grazing instead of cutting for pasture or grassland management • Use mixed pasture

Likely to be beneficial

Grow plants that compete with damaging weeds

- *Weed weight and cover*: Nine studies from Australia, Slovakia, the UK and the USA tested the effects of planting species to compete with weeds. All (including four replicated, randomized, controlled trials) found reduced weed plant weight or ground cover, although two found this only in some years or conditions.
- *Weed reproduction and survival:* Five studies (including three replicated, randomized, controlled trials) also found that competition reduced weed reproduction, survival or both. One of these found an effect only in one year only.

- *Crops studied*: clovers, fescues, ryegrass, other grasses and turnip.
- *Assessment: likely to be beneficial (effectiveness 70%; certainty 60%; harms 5%).*

http://www.conservationevidence.com/actions/722

Unknown effectiveness (limited evidence)

Delay mowing or first grazing date on pasture or grassland

- *Natural enemy abundance:* One replicated, randomized, controlled study found fewer predatory spiders with delayed cutting. Three studies from the UK (two of them replicated, randomized and controlled) found no change in insect predator numbers and one replicated study from Sweden found mixed effects between different predator groups.
- *Natural enemy diversity:* One replicated study from Sweden found a decrease in ant diversity with delayed cutting and one replicated, randomized, controlled study from the UK found no effect on spider and beetle diversity.
- *Pests:* One of two replicated, randomized, controlled studies from the UK and USA found more pest insects in late-cut plots and one found no effect.
- *Insects in general:* Four replicated, randomized, controlled studies measured the abundance of insect groups without classifying them as pests or natural enemies. One UK study found lower numbers in late-cut plots, while two found effects varied between groups. Two studies from the UK and USA found no effect on insect numbers.
- *Crops studied:* barley, bird's-foot trefoil, clovers, fescues, rapeseed, ryegrass, other grasses and wheat.
- *Assessment: unknown effectiveness (effectiveness 5%; certainty 20%; harms 15%).*

http://www.conservationevidence.com/actions/727

Likely to be ineffective or harmful

Use grazing instead of cutting for pasture or grassland management

- *Natural enemies:* Two studies (one before-and-after and one replicated trial) from Australia and the UK found grazing instead of cutting had mixed effects on natural enemies, with some species and groups affected on some dates but not others. One replicated study from New Zealand found no effect.
- *Pests and diseases:* One of five studies (including three replicated trials) from Australia, New Zealand, the UK and the USA found more pests, and two studies found effects varied between pest groups and sampling dates. Two studies found no effect on pests. One study found no effect on disease when grazing was used in addition to cutting.
- *Pasture damage and plant survival:* One randomized study found more ryegrass shoots were attacked by pests. One study found lower survival of alfalfa plants but another found no effect.
- *Yield:* One of four randomized, replicated studies (one also controlled) found lower yields and two found no effect. One study found lower ryegrass and higher clover yields, but no difference between clover varieties. Another randomized study found more ryegrass shoots.
- *Crops studied:* alfalfa, cock's-foot, perennial ryegrass, other grasses and white clover.
- *Assessment: likely to be ineffective or harmful (effectiveness 10%; certainty 45%; harms 40%).*

http://www.conservationevidence.com/actions/885

Use mixed pasture

- *Weeds:* Two of two studies (randomized and replicated and one also controlled) from the USA found weeds were negatively affected by mixed compared to monoculture pasture.

- *Pests:* Five studies from North America measured pests including four randomized, replicated, controlled tests. One study found fewer pests and two studies found negative or mixed effects depending on different pests groups or pasture mixes. One study found no effect ad another found more pests, although the effect was potentially inseparable from grazing treatments.
- *Crop mortality:* One randomized, replicated study from the USA found no effect on forage crop mortality caused by nematodes.
- *Yield:* Two of five studies (including two randomized, replicated, controlled tests) from North America found increased forage crop yields and two studies found mixed effects depending on the crop type and year. One study found no effect.
- *Crops studied:* alfalfa, bird's-foot trefoil, chicory, cicer milkvetch, clovers, fescues, oats, plantain, ryegrass, other grasses, other legumes, rapeseed and turnip.
- *Assessment: likely to be ineffective or harmful (effectiveness 35%; certainty 45%; harms 20%).*

http://www.conservationevidence.com/actions/721

12. ENHANCING SOIL FERTILITY

Georgina Key, Mike Whitfield, Lynn V. Dicks, William J. Sutherland & Richard D. Bardgett

Expert assessors

Martin Collison, Collison and Associates Limited, UK
Julia Cooper, Newcastle University, UK
Thanasis Dedousis, PepsiCo Europe
Richard Heathcote, Heineken, S&N UK Ltd
Shamal Mohammed, Agriculture and Horticulture Development Board, Cranfield University, UK
Andrew Molyneux, Huntapac Produce Ltd, UK
Wim van der Putten, Netherlands Institute of Ecology
Brendan Roth, Department for Environment, Food & Rural Affairs, UK
Franciska de Vries, University of Manchester, UK

Scope of assessment: actions to enhance soil fertility for agricultural systems across the world.

Assessed: 2014.

Effectiveness measure is the median % score.

Certainty measure is the median % certainty of evidence, determined by the quantity and quality of the evidence in the synopsis.

Harm measure is the median % score for negative side-effects for the farmer such as reduced yield, crop quality or profits, or increased costs.

 https://doi.org/10.11647/OBP.0267.12

This book is meant as a guide to the evidence available for different conservation interventions and as a starting point in assessing their effectiveness. The assessments are based on the available evidence for the target group of species for each intervention. The assessment may therefore refer to different species or habitat to the one(s) you are considering. Before making any decisions about implementing interventions it is vital that you read the more detailed accounts of the evidence in order to assess their relevance for your study species or system.

Full details of the evidence are available at**www. conservationevidence.com**

There may also be significant negative side-effects on the target groups or other species or communities that have not been identified in this assessment.

A lack of evidence means that we have been unable to assess whether or not an intervention is effective or has any harmful impacts.

12.1 Reducing agricultural pollution

Based on the collated evidence, what is the current assessment of the effectiveness of interventions to reduce agricultural pollution for enhancing soil fertility?	
Unknown effectiveness (limited evidence)	• Change the timing of manure application
Likely to be ineffective or harmful	• Reduce fertilizer, pesticide or herbicide use generally

Unknown effectiveness (limited evidence)

Change the timing of manure application

- One controlled, randomized, replicated, site comparison study from the UK found less nitrate was lost from the soil when manure application was delayed from autumn until December or January.
- *Soil types covered*: sandy loam.
- *Assessment: unknown effectiveness (effectiveness 50%; certainty 33%; harms 24%).*

http://www.conservationevidence.com/actions/893

Likely to be ineffective or harmful

Reduce fertilizer, pesticide or herbicide use generally

- *Biodiversity*: Two site comparison studies from Italy and Pakistan (one also replicated) found a higher diversity of soil invertebrates and microorganisms in low chemical-input systems.
- *Nutrient loss*: One study from Canada found lower nutrient levels and yields in low-input systems.
- *Soil types covered*: course sandy, loam, sandy loam, and silt.
- *Assessment: likely to be ineffective or harmful (effectiveness 26%; certainty 40%; harms 48%).*

http://www.conservationevidence.com/actions/904

12.2 All farming systems

Based on the collated evidence, what is the current assessment of the effectiveness of interventions on all farming systems for enhancing soil fertility?	
Likely to be beneficial	• Control traffic and traffic timing
Trade-off between benefit and harms	• Change tillage practices • Convert to organic farming • Plant new hedges
Unknown effectiveness (limited evidence)	• Change the timing of ploughing

Likely to be beneficial

Control traffic and traffic timing

- *Biodiversity*: One randomised, replicated study from Poland found higher numbers and bacterial activity under controlled traffic. One replicated site comparison study from Denmark found higher microbial biomass when farm traffic was not controlled.
- *Erosion*: Five trials from Europe and Australia (including three replicated trials, one controlled before-and-after trial, and one review) found a higher number of pores in the soil, less compaction, reduced runoff and increased water filtration into soil under controlled traffic. One controlled, replicated trial in India found increased soil crack width when traffic was not controlled.

- *Yield*: One replicated trial from Australia found increased yield under controlled traffic.
- *Soil types covered*: clay, loamy silt, sandy loam, silty, silty clay, silt-loam.
- *Assessment: likely to be beneficial (effectiveness 55%; certainty 62%; harms 18%).*

http://www.conservationevidence.com/actions/899

Trade-off between benefit and harms

Change tillage practices

- *Biodiversity loss*: Nine studies from Canada, Europe, Mexico, or the USA measured effects of reduced tillage on soil animals or microbes. Of these, six (including three replicated trials (two also randomized and one also controlled) found more microbes, more species of earthworm, or higher microbe activity under reduced tillage. One replicated trial found increased numbers of soil animals and earthworms under reduced tillage. Two (including one controlled, replicated trial), found no effect of reduced tillage on earthworm activity or microbe activity.
- *Compaction*: Five studies from Australia, Canada, and Europe measured the effect of controlled traffic and reduced tillage on compacted soils. Of these, two (including one before-and-after trial and one replicated trial) found reduced compaction and subsequent effects (reduced water runoff, for example) under controlled traffic, and one also found that crop yields increased under no-tillage. Three replicated trials, including one site comparison study, found higher compaction under reduced tillage.
- *Drought*: Three replicated trials from Europe and India (one randomized) found the size of soil cracks decreased, and ability of soil to absorb water and soil water content increased with conventional tillage and sub-soiling.
- *Erosion*: Ten replicated trials from Brazil, Europe, India, Nigeria and the USA, and one review showed mixed results of tillage on soil erosion. Seven trials (one also controlled and randomized) showed

reduced soil loss and runoff under reduced tillage compared to conventional ploughing. One trial showed no differences between tillage systems, but demonstrated that across-slope cultivation reduced soil loss compared to up-and-downslope cultivation. Two trials, showed that no-tillage increased soil loss in the absence of crop cover.

- *Soil organic carbon*: Twelve studies from Australia, Canada, China, Europe, Japan and the USA compared the effect of no-tillage and conventionally tilled systems on soil organic carbon. All (including two randomized, five replicated, two randomized, replicated, and one controlled, randomized, replicated) found higher soil organic carbon in soils under a no-tillage or reduced tillage system compared to conventionally tilled soil. One review showed that no-tillage with cover cropping plus manure application increased soil organic carbon. One randomized, replicated trial from Spain found greater soil organic carbon in conventionally tilled soil.
- *Soil organic matter*: Twelve studies from Canada, China, Europe, Morocco, and the USA measured effects of reduced tillage on soil organic matter content and nutrient retention. Of these, six studies (including three replicated, two site comparisons (one also replicated) and one controlled) found maintained or increased soil organic matter and improved soil structure under reduced tillage. Four trials (including two replicated and two site comparison studies) found higher nutrient retention under reduced tillage. One controlled, replicated trial from the USA found less carbon and nitrate in no-till compared to conventionally tilled soil, but conventionally tilled soil lost more carbon and nitrate.
- *Soil types covered*: anthrosol, calcareous silt-loam, chalky, clay, clay-loam, fine sandy loam, loam, loamy clay, loam/sandy loam, loam silt-loam, loamy silt, non-chalky clay, sandy, sandy clay-loam, sandy loam, sandy silt-loam, silt-loam, silty, silty clay, silty clay-loam, silty loam.
- *Assessment: trade-offs between benefit and harms (effectiveness 61%; certainty 72%; harms 46%).*

http://www.conservationevidence.com/actions/906

Convert to organic farming

- *Biodiversity*: Four studies in Asia, Europe, and the USA (including two site comparison studies and three replicated trials) found higher numbers, diversity, functional diversity (see background) or activity of soil organisms under organic management.
- *Soil organic carbon*: Two replicated trials in Italy and the USA showed that organically managed orchards had higher soil carbon levels compared to conventionally managed orchards. One randomised, replicated trial in the USA found soil carbon was lower under organic management compared to alley cropping.
- *Soil organic matter*: One replicated trial in Canada found that soil nutrients were lower in organically managed soils.
- *Yield*: One replicated trial in Canada found lower yields in organically managed soils. Two replicated trials in the USA (one also randomised) found that fruit was of a higher quality and more resistant to disease, though smaller or that organic management had mixed effects on yield.
- *Soil types covered*: clay, clay-loam, fine sandy loam, loam, sandy loam, sandy clay-loam, silt, silty clay, silt-loam.
- *Assessment: trade-offs between benefit and harms (effectiveness 55%; certainty 52%; harms 64%).*

http://www.conservationevidence.com/actions/895

Plant new hedges

- Five studies in Slovakia, Kenya and Thailand measured the effects of planting grass or shrub hedgerows on soil animals and soil fertility. All five found hedgerows to maintain or improve soil fertility and soil animal activity. Of these, three replicated studies found reduced soil erosion and higher soil organic matter levels. Another replicated trial found a higher diversity of soil animals near to the hedgerows. One of the replicated studies and one review found that adding woody species to the hedgerows improved many factors contributing to soil fertility.
- *Soil types covered*: alluvial, clay, sandy loam.

- *Assessment: trade-offs between benefit and harms (effectiveness 49%; certainty 45%; harms 20%).*

http://www.conservationevidence.com/actions/744

Unknown effectiveness (limited evidence)

Change the timing of ploughing

- *Nutrient loss*: Two replicated site comparison studies from Denmark and Norway (one also randomised) found reduced erosion soil loss and nitrate leaching when ploughing was delayed until spring.
- *Soil types covered*: Sandy, sandy loam, silty clay loam.
- *Assessment: unknown effectiveness (effectiveness 46%; certainty 38%; harms 33%).*

http://www.conservationevidence.com/actions/712

12.3 Arable farming

Based on the collated evidence, what is the current assessment of the effectiveness of interventions on arable farming systems for enhancing soil fertility?	
Beneficial	• Amend the soil using a mix of organic and inorganic amendments • Grow cover crops when the field is empty • Use crop rotation
Likely to be beneficial	• Amend the soil with formulated chemical compounds • Grow cover crops beneath the main crop (living mulches) or between crop rows
Trade-off between benefit and harms	• Add mulch to crops • Amend the soil with fresh plant material or crop remains • Amend the soil with manures and agricultural composts • Amend the soil with municipal wastes or their composts • Incorporate leys into crop rotation • Retain crop residues
Unknown effectiveness (limited evidence)	• Amend the soil with bacteria or fungi • Amend the soil with composts not otherwise specified • Amend the soil with crops grown as green manures • Amend the soil with non-chemical minerals and mineral wastes • Amend the soil with organic processing wastes or their composts • Encourage foraging waterfowl • Use alley cropping

Beneficial

Amend the soil using a mix of organic and inorganic amendments

- *Biodiversity*: Five controlled trials from China and India (four also randomized and replicated), and one study from Japan found higher microbial biomass and activity in soils with a mix of manure and inorganic fertilizers. Manure alone also increased microbial biomass. One trial found increased microbial diversity.
- *Erosion*: One controlled, replicated trial from India found that mixed amendments were more effective at reducing the size of cracks in dry soil than inorganic fertilizers alone or no fertilizer.
- *Soil organic carbon loss*: Four controlled, randomized, replicated trials and one controlled trial all from China and India found more organic carbon in soils with mixed fertilizers. Manure alone also increased organic carbon. One trial also found more carbon in soil amended with inorganic fertilizers and lime.
- *Soil organic matter loss*: Three randomized, replicated trials from China and India (two also controlled), found more nutrients in soils with manure and inorganic fertilizers. One controlled, randomized, replicated trial from China found inconsistent effects of using mixed manure and inorganic fertilizers.
- *Yield*: Two randomized, replicated trials from China (one also controlled) found increased maize or rice and wheat yields in soils with mixed manure and inorganic fertilizer amendments. One study found lower yields of rice and wheat under mixed fertilizers.
- *Soil types covered*: clay, clay-loam, sandy loam, silt clay-loam, silty loam.
- *Assessment: beneficial (effectiveness 69%; certainty 64%; harms 15%).*

http://www.conservationevidence.com/actions/902

Grow cover crops when the field is empty

- *Biodiversity*: One controlled, randomized, replicated experiment in Martinique found that growing cover crops resulted in more diverse nematode communities. One replicated trial from the USA found greater microbial biomass under ryegrass compared to a ryegrass/vetch cover crop mix.

- *Soil structure:* Three randomized, replicated studies from Denmark, Turkey and the UK found that growing cover crops improved soil structure and nutrient retention. One trial found higher soil porosity, interconnectivity and one lower resistance in soil under cover crops, and one found reduced nitrate leaching.
- *Soil organic carbon:* One replicated study from Denmark and one review based mainly in Japan found increased soil carbon levels under cover crops. One study also found soil carbon levels increased further when legumes were included in cover crops.
- *Soil organic matter:* Two controlled, randomized, replicated studies from Australia and the USA found increased carbon and nitrogen levels under cover crops, with one showing that they increased regardless of whether those crops were legumes or not. Two studies from Europe (including one controlled, replicated trial) found no marked effect on soil organic matter levels.
- *Yield:* One replicated trial from the USA found higher tomato yield from soils which had been under a ryegrass cover crop.
- *Soil types covered*: clay, loam, sandy clay, sandy loam, silty clay, silty loam.
- *Assessment: beneficial (effectiveness 75%; certainty 67%; harms 16%).*

http://www.conservationevidence.com/actions/898

Use crop rotation

- *Biodiversity:* Three randomized, replicated trials from Canada and Zambia measured the effect of including legumes in crop rotations and found the number of microbes and diversity of different soil animals increased.
- *Erosion:* One randomized, replicated trial from Canada found that including forage crops in crop rotations reduced rainwater runoff and soil loss, and one replicated trial from Syria showed that including legumes in rotation increased water infiltration (movement of water into the soil).
- *Soil organic carbon*: Three studies from Australia, Canada, and Denmark (including one controlled replicated trial and one replicated

site comparison study), found increased soil organic carbon under crop rotation, particularly when some legumes were included.

- *Soil organic matter:* Two of four replicated trials from Canada and Syria (one also controlled and randomized) found increased soil organic matter, particularly when legumes were included in the rotation. One study found lower soil organic matter levels when longer crop rotations were used. One randomized, replicated study found no effect on soil particle size.
- *Soil types covered*: clay, clay-loam, fine clay, loam, loam/silt loam, sandy clay, sandy loam, silty loam.
- *Assessment: beneficial (effectiveness 66%; certainty 75%; harms 8%).*

http://www.conservationevidence.com/actions/857

Likely to be beneficial

Amend the soil with formulated chemical compounds

- *Nutrient loss*: Three of five replicated trials from New Zealand and the UK measured the effect of applying nitrification inhibitors to the soil and three found reduced nitrate losses and nitrous oxide emissions, although one of these found that the method of application influenced its effect. One trial found no effect on nitrate loss. One trial found reduced nutrient and soil loss when aluminium sulphate was applied to the soil.
- *Soil organic matter*: Four of five studies (including two controlled, randomised and replicated and one randomised and replicated) in Australia, China, India, Syria and the UK testing the effects of adding chemical compounds to the soil showed an increase in soil organic matter or carbon when nitrogen or phosphorus fertilizer was applied. One site comparison study showed that a slow-release fertilizer resulted in higher nutrient retention. One study found higher carbon levels when NPK fertilizers were applied with straw, than when applied alone, and one replicated study from France found higher soil carbon when manure rather than chemical compounds were applied.
- *Yield*: One replicated experiment from India showed that maize and wheat yield increased with increased fertilizer application.

- *Soil types covered*: clay, fine loamy, gravelly sandy loam, loam, sandy loam, silty, silty clay, silt-loam.
- *Assessment: likely to be beneficial (effectiveness 64%; certainty 46%; harms 19%).*

http://www.conservationevidence.com/actions/909

Grow cover crops beneath the main crop (living mulches) or between crop rows

- *Biodiversity:* One randomized, replicated study from Spain found that cover crops increased bacterial numbers and activity.
- *Erosion:* Two studies from France and the USA showed reduced erosion under cover crops. One controlled study showed that soil stability was highest under a grass cover, and one randomized replicated study found that cover crops reduced soil loss.
- *Soil organic matter:* Two controlled trials from India and South Africa (one also randomized and replicated) found that soil organic matter increased under cover crops, and one trial from Germany found no effect on soil organic matter levels.
- *Soil types covered*: gravelly sandy loam, sandy loam, sandy, silty loam.
- *Assessment: likely to be beneficial (effectiveness 65%; certainty 54%; harms 19%).*

http://www.conservationevidence.com/actions/897

Trade-off between benefit and harms

Add mulch to crops

- *Biodiversity*: Three replicated trials from Canada, Poland and Spain (including one also controlled, one also randomised and one also controlled and randomised) showed that adding mulch to crops (whether shredded paper, municipal compost or straw) increased soil animal and fungal numbers, diversity and activity. Of these, one trial also showed that mulch improved soil structure and increased soil organic matter.
- *Nutrient loss*: One replicated study from Nigeria found higher nutrient levels in continually cropped soil.

- *Erosion*: Five studies from India, France, Nigeria and the UK (including one controlled, randomised, replicated trial, one randomised, replicated trial, two replicated (one also controlled), and one controlled trial) found that mulches increased soil stability, and reduced soil erosion and runoff. One trial found that some mulches are more effective than others.
- *Drought*: Two replicated trials from India found that adding mulch to crops increased soil moisture.
- *Yield*: Two replicated trials from India found that yields increased when either a live mulch or vegetation barrier combined with mulch was used.
- *Soil types covered*: clay, fine loam, gravelly sandy loam, sandy, sandy clay, sandy loam, sandy silt-loam, silty, silty loam.
- *Assessment: trade-offs between benefit and harms (effectiveness 60%; certainty 64%; harms 23%).*

http://www.conservationevidence.com/actions/887

Amend the soil with fresh plant material or crop remains

- *Biodiversity*: One randomized, replicated experiment from Belgium found increased microbial biomass when crop remains and straw were added.
- *Compaction*: One before-and-after trial from the UK found that incorporating straw residues by discing (reduced tillage) did not improve anaerobic soils (low oxygen levels) in compacted soils.
- *Erosion*: Two randomized, replicated studies from Canada and India measured the effect of incorporating straw on erosion. One found straw addition reduced soil loss, and one found mixed effects depending on soil type.
- *Nutrient loss*: Two replicated studies from Belgium and the UK (one also controlled and one also randomized) reported higher soil nitrogen levels when compost or straw was applied, but mixed results when processed wastes were added.
- *Soil organic carbon*: Three randomized, replicated studies (two also controlled) from China and India, and one controlled before-and-after site comparison study from Denmark found higher carbon levels when plant material was added. One found higher carbon

levels when straw was applied along with NPK fertilizers. One also found larger soil aggregates.

- *Soil types covered*: clay, clay-loam, loam/sandy loam, loamy sand, sandy, sandy clay-loam, sandy loam, silt-loam, silty, silty clay.
- *Assessment: trade-offs between benefit and harms (effectiveness 53%; certainty 53%; harms 34%).*

http://www.conservationevidence.com/actions/910

Amend the soil with manures and agricultural composts

- *Biodiversity loss*: Three controlled, replicated studies from the UK and USA found higher microbial biomass when manure or compost was applied, and higher microbial respiration when poultry manure was applied.
- *Erosion*: One controlled, randomized, replicated study from India found lower soil loss and water runoff with manure application in combination with other treatments.
- *Nutrient management*: Two randomized, replicated studies from Canada and the UK (one also controlled) found lower nitrate loss or larger soil aggregates (which hold more nutrients) when manure was applied, compared to broiler (poultry) litter, slurry or synthetic fertilizers. One study found that treatment in winter was more effective than in autumn and that farmyard manure was more effective than broiler (poultry) litter or slurry in reducing nutrient loss. One controlled, replicated study from Spain found higher nitrate leaching.
- *Soil organic carbon*: Three studies (including two controlled, replicated studies and a review) from India, Japan and the UK found higher carbon levels when manures were applied.
- *Soil organic matter*: One controlled, randomized, replicated study from Turkey found higher organic matter, larger soil aggregations and a positive effect on soil physical properties when manure and compost were applied. One study from Germany found no effect of manure on organic matter levels.
- *Yield*: Four controlled, replicated studies (including four also randomized) from India, Spain and Turkey found higher crop yields

when manures or compost were applied. One study found higher yields when manure were applied in combination with cover crops.

- *Soil types covered*: clay-loam, loam, loamy, sandy loam, sandy clay-loam, silty loam, sandy silt-loam.
- *Assessment: trade-offs between benefit and harms (effectiveness 70%; certainty 59%; harms 26%).*

http://www.conservationevidence.com/actions/911

Amend the soil with municipal wastes or their composts

- *Erosion*: Two controlled, replicated trials in Spain and the UK measured the effect of adding wastes to the soil. One trial found that adding municipal compost to semi-arid soils greatly reduced soil loss and water runoff. One found mixed results of adding composts and wastes.
- *Soil types covered*: coarse loamy, sandy loam.
- *Assessment: trade-offs between benefit and harms (effectiveness 45%; certainty 44%; harms 54%).*

http://www.conservationevidence.com/actions/890

Incorporate leys into crop rotation

- *Nutrient loss*: One replicated study from Denmark showed that reducing the extent of grass pasture in leys reduced the undesirable uptake of nitrogen by grasses, therefore requiring lower rates of fertilizer for subsequent crops.
- *Soil types covered*: sandy loam.
- *Assessment: trade-offs between benefit and harms (effectiveness 46%; certainty 45%; harms 36%).*

http://www.conservationevidence.com/actions/900

Retain crop residues

- *Biodiversity:* One replicated study from Mexico found higher microbial biomass when crop residues were retained.
- *Erosion*: One review found reduced water runoff, increased water storage and reduced soil erosion. One replicated site comparison from Canada found mixed effects on soil physical properties,

including penetration resistance and the size of soil aggregates. One replicated study from the USA found that tillage can have mixed results on soil erosion when crop remains are removed.

- *Soil organic matter:* One randomized, replicated trial from Australia found higher soil organic carbon and nitrogen when residues were retained, but only when fertilizer was also applied.
- *Yield:* One randomized, replicated trial from Australia found higher yields when residues were retained in combination with fertilizer application and no-tillage.
- *Soil types covered:* clay, loam, sandy loam, silt-loam.
- *Assessment: trade-offs between benefit and harms (effectiveness 63%; certainty 54%; harms 29%).*

http://www.conservationevidence.com/actions/907

Unknown effectiveness (limited evidence)

Amend the soil with bacteria or fungi

- *Biodiversity*: One randomised, replicated trial from India showed that adding soil bacteria and arbuscular mycorrhizal fungi resulted in higher microbial diversity.
- *Soil organic matter*: One controlled, randomised, replicated trial from Turkey found increased soil organic matter content in soil under mycorrhizal-inoculated compost applications.
- *Yield*: Two randomised, replicated trials (including one also controlled) from India and Turkey found higher crop yields.
- *Soil types covered*: clay-loam, sandy loam.
- *Assessment: unknown effectiveness (effectiveness 40%; certainty 31%; harms 17%).*

http://www.conservationevidence.com/actions/888

Amend the soil with composts not otherwise specified

- *Soil organic matter*: One controlled, randomised, replicated trial in Italy found that applying a high rate of compost increased soil organic matter levels, microbial biomass and fruit yield.

- *Soil types covered*: silty clay.
- *Assessment: unknown effectiveness (effectiveness 54%; certainty 29%; harms 19%).*

http://www.conservationevidence.com/actions/889

Amend the soil with crops grown as green manures

- *Soil organic matter*: Two controlled, randomized, replicated studies from India and Pakistan found higher soil organic carbon, and one found increased grain yields when green manures were grown.
- *Soil types covered*: clay-loam.
- *Assessment: unknown effectiveness (effectiveness 53%; certainty 36%; harms 16%).*

http://www.conservationevidence.com/actions/908

Amend the soil with non-chemical minerals and mineral wastes

- *Nutrient loss*: Two replicated studies from Australia and New Zealand measured the effects of adding minerals and mineral wastes to the soil. Both found reduced nutrient loss and one study found reduced erosion.
- *Soil types covered*: sandy clay, silt-loam.
- *Assessment: unknown effectiveness (effectiveness 35%; certainty 37%; harms 23%).*

http://www.conservationevidence.com/actions/892

Amend the soil with organic processing wastes or their composts

- *Nutrient loss*: Two controlled, replicated trials from Spain and the UK (one also randomized) measured the effect of adding composts to soil. One trial found applying high rates of cotton gin compost and poultry manure improved soil structure and reduced soil loss, but increased nutrient loss. One trial found improved nutrient retention and increased barley *Hordeum vulgare* yield when molasses were added.

- *Soil types covered*: sandy clay, sandy loam, silty clay.
- *Assessment: unknown effectiveness (effectiveness 58%; certainty 35%; harms 20%).*

http://www.conservationevidence.com/actions/891

Encourage foraging waterfowl

- *Soil organic matter:* One controlled, replicated experiment from the USA found increased straw decomposition when ducks were allowed to forage.
- *Soil types covered:* silty clay.
- *Assessment: unknown effectiveness (effectiveness 14%; certainty 34%; harms 20%).*

http://www.conservationevidence.com/actions/711

Use alley cropping

- *Biodiversity*: A controlled, randomized, replicated study from Canada found that intercropping with trees resulted in a higher diversity of arbuscular mycorrhizal fungi.
- *Soil types covered*: sandy loam.
- *Assessment: unknown effectiveness (effectiveness 36%; certainty 23%; harms 19%).*

http://www.conservationevidence.com/actions/903

12.4 Livestock and pasture farming

Based on the collated evidence, what is the current assessment of the effectiveness of interventions on livestock and pasture farming systems for enhancing soil fertility?	
Likely to be beneficial	• Reduce grazing intensity
Trade-off between benefit and harms	• Restore or create low input grasslands

Likely to be beneficial

Reduce grazing intensity

- *Compaction*: One replicated study from Australia found compacted soils recovered when sheep were excluded for 2.5 years.
- *Erosion*: Two replicated studies from New Zealand, and Syria (one also controlled) measured the effect of grazing animals on soil nutrient and sediment loss. Of these, one trial found increased soil carbon and nitrogen when grazing animals were excluded. One trial found higher soil phosphate levels, and less sediment erosion when grazing time in forage crops was reduced.
- *Soil types covered*: clay, clay-loam, loamy, silt-loam.
- *Assessment: likely to be beneficial (effectiveness 51%; certainty 58%; harms 14%).*

http://www.conservationevidence.com/actions/901

Trade-off between benefit and harms

Restore or create low input grasslands

- *Biodiversity*: One randomized, replicated trial in the Netherlands and one controlled trial from France found that restoring grasslands increased the diversity of soil animals. One trial also found higher microbial biomass, activity and carbon under grassland.
- *Soil types covered*: sandy loam, silty.
- *Assessment: trade-offs between benefit and harms (effectiveness 53%; certainty 59%; harms 32%).*

http://www.conservationevidence.com/actions/905

13. SUBTIDAL BENTHIC INVERTEBRATE CONSERVATION

Anaelle J. Lemasson, Laura R. Pettit, Rebecca K. Smith & William J. Sutherland

Expert assessors
Silviu Petrovan, University of Cambridge, United Kingdom
Anaelle Lemasson, JNCC, United Kingdom
Ann Thornton, University of Cambridge, United Kingdom
Lucy Shuff, Gardline Limited, United Kingdom
Christopher Barrett, CEFAS, United Kingdom

Scope of assessment: for native wild subtidal species across the world.

Assessed: 2020.

Effectiveness measure is the median % score for effectiveness.

Certainty measure is the median % certainty of evidence for effectiveness, determined by the quantity and quality of the evidence in the synopsis.

Harm measure is the median % score for negative side-effects to the group of species of concern.

https://doi.org/10.11647/OBP.0267.13

This book is meant as a guide to the evidence available for different conservation interventions and as a starting point in assessing their effectiveness. The assessments are based on the available evidence for the target group of species for each intervention. The assessment may therefore refer to different species or habitat to the one(s) you are considering. Before making any decisions about implementing interventions it is vital that you read the more detailed accounts of the evidence in order to assess their relevance for your study species or system.

Full details of the evidence are available atwww.conservationevidence.com

There may also be significant negative side-effects on the target groups or other species or communities that have not been identified in this assessment.

A lack of evidence means that we have been unable to assess whether or not an intervention is effective or has any harmful impacts.

13.1 Threat: Energy production and mining

13.1.1 Oil and gas drilling

Based on the collated evidence, what is the current assessment of the effectiveness of interventions for oil and gas drilling?	
No evidence found (no assessment)	• Bury drill cuttings in the seabed rather than leaving them on the seabed surface • Cease or prohibit oil and gas drilling • Cease or prohibit the deposit of drill cuttings on the seabed • Dispose of drill cuttings on land rather than on the seabed • Limit the thickness of drill cuttings • Recycle or repurpose fluids used in the drilling process • Remove drill cuttings after decommissioning • Set limits for change in sediment particle size during aggregate extraction • Use water-based muds instead of oil-based muds (drilling fluids) in the drilling process

No evidence found (no assessment)

We have captured no evidence for the following interventions:

- Bury drill cuttings in the seabed rather than leaving them on the seabed surface
- Cease or prohibit oil and gas drilling

- Cease or prohibit the deposit of drill cuttings on the seabed
- Dispose of drill cuttings on land rather than on the seabed
- Limit the thickness of drill cuttings
- Recycle or repurpose fluids used in the drilling process
- Remove drill cuttings after decommissioning
- Set limits for change in sediment particle size during aggregate extraction
- Use water-based muds instead of oil-based muds (drilling fluids) in the drilling process.

13.1.2 General

Based on the collated evidence, what is the current assessment of the effectiveness of interventions for energy production and mining?	
No evidence found (no assessment)	• Bury pipelines instead of surface laying and rock dumping • Leave pipelines and infrastructure in place following decommissioning • Limit the amount of stabilisation material used • Remove pipelines and infrastructure following decommissioning • Set limits for change in sediment particle size during rock dumping • Use stabilisation material that can be more easily recovered at decommissioning stage

No evidence found (no assessment)

We have captured no evidence for the following interventions:

- Bury pipelines instead of surface laying and rock dumping
- Leave pipelines and infrastructure in place following decommissioning
- Limit the amount of stabilisation material used
- Remove pipelines and infrastructure following decommissioning
- Set limits for change in sediment particle size during rock dumping

- Use stabilisation material that can be more easily recovered at decommissioning stage.

13.1.3 Mining, quarrying, and aggregate extraction

Based on the collated evidence, what is the current assessment of the effectiveness of interventions for mining, quarrying, and aggregate extraction?	
Beneficial	• Cease or prohibit aggregate extraction
Unknown effectiveness	• Cease or prohibit marine mining • Extract aggregates from a vessel that is moving rather than static • Leave mining waste (tailings) in place following cessation of disposal operations
No evidence found (no assessment)	• Cease or prohibit mining waste (tailings) disposal at sea • Limit, cease, or prohibit sediment discard during aggregate extraction • Remove discarded sediment material from the seabed following cessation of aggregate extraction

Beneficial

Cease or prohibit aggregate extraction

Seven studies examined the effects of ceasing or prohibiting aggregate extraction on subtidal benthic invertebrate populations. One study was in the English Channel (France), one in the Mediterranean Sea (Italy), one a global study, and four in the North Sea (UK, Belgium).

COMMUNITY RESPONSE (6 STUDIES)

Overall community composition (4 studies): One global systematic review found that it took nine months to several decades for overall invertebrate community composition to recover after ceasing aggregate extraction. One before-and-after, site comparison study in the Mediterranean Sea and one of two site comparison studies in the North Sea found that after ceasing aggregate extraction overall invertebrate community composition became more similar to pre-extraction and/or natural site communities.

Overall richness/diversity (5 studies): Two before-and-after, site comparison studies in the English Channel and the Mediterranean Sea and one of two site comparison studies in the North Sea found that after ceasing aggregate extraction, overall invertebrate species richness and/or diversity became more similar to that of pre-extraction and/or natural sites. The other site comparison found that species richness did not change over time and remained different to that of natural sites. One replicated, site comparison study in the North Sea found that 21 months after ceasing aggregate extraction, invertebrate species richness was similar to that of natural sites.

Worm community composition (1 study): One before-and-after study in the North Sea found that after ceasing aggregate extraction, nematode worm community composition remained different to the pre-extraction community.

Worm richness/diversity (1 study): One before-and-after study in the North Sea found that after ceasing aggregate extraction, nematode worm species richness remained different to pre-extraction richness.

POPULATION RESPONSE (6 STUDIES)

Overall abundance (5 studies): Two before-and-after, site comparison studies in the English Channel and the Mediterranean Sea and one of two site comparison studies in the North Sea found that after ceasing aggregate extraction overall invertebrate abundance and/or biomass became more similar to that of pre-extraction and/or natural sites. The other site comparison found that abundance and biomass did not change over time and remained different to that of natural sites. One replicated, site comparison study in the North Sea found that 21 months after ceasing aggregate extraction, invertebrate abundance was similar to that of natural sites.

Worm abundance (1 study): One before-and-after study in the North Sea found that after ceasing aggregate extraction, nematode worm abundance remained different to pre-extraction abundance.

Assessment: beneficial (effectiveness 65%; certainty 60%; harms 0%).

https://www.conservationevidence.com/actions/2070

Unknown effectiveness

Cease or prohibit marine mining

One study examined the effects of ceasing or prohibiting mining on subtidal benthic invertebrate populations. The study was in the Bering Sea (USA).

COMMUNITY RESPONSE (1 STUDY)

Overall community composition (1 study): One site comparison study in the Bering Sea found that following cessation of gold mining, overall invertebrate community composition became similar to that of an unmined site.

Overall richness/diversity (1 study): One site comparison study in the Bering Sea found that following cessation of gold mining, overall invertebrate richness and diversity became similar to that of an unmined site.

POPULATION RESPONSE (1 STUDY)

Overall abundance (1 study): One site comparison study in the Bering Sea found that following cessation of gold mining, overall invertebrate abundance and biomass became similar to that of an unmined site.

Assessment: unknown effectiveness (effectiveness 70%; certainty 31%; harms 0%).

https://www.conservationevidence.com/actions/2075

Extract aggregates from a vessel that is moving rather than static

One study examined the effects of dredging from a vessel that is moving rather than static on subtidal benthic invertebrate populations. The study was in the English Channel (UK).

COMMUNITY RESPONSE (1 STUDY)

Overall species richness/diversity (1 study): One site comparison study in the English Channel found that a site where aggregate extraction was undertaken using a moving trailer suction hopper dredger had similar invertebrate species richness and lower diversity compared to a site where extraction occurred using a static suction hopper dredger.

POPULATION RESPONSE (1 STUDY)

Overall abundance (1 study): One site comparison study in the English Channel found that a site where aggregate extraction was undertaken using a moving trailer suction hopper dredger had higher abundance of invertebrates compared to a site where extraction occurred using a static suction hopper dredger.

Assessment: unknown effectiveness (effectiveness 35%; certainty 20%; harms 18%).

https://www.conservationevidence.com/actions/2071

Leave mining waste (tailings) in place following cessation of disposal operations

One study examined the effects of leaving mining waste (tailings) in place following cessation of disposal operations on subtidal benthic invertebrate populations. The study was in Auke Bay (USA).

COMMUNITY RESPONSE (1 STUDY)

Overall community composition (1 study): One replicated, paired, controlled study in Auke Bay found that plots where mine tailings were left in place had similar invertebrate community composition as plots where tailings had been removed, but both had different communities to plots of natural sediment.

Overall richness/diversity (1 study): One replicated, paired, controlled study in Auke Bay found that plots where mine tailings were left in place had similar invertebrate species richness as plots where tailings had been removed, but both had lower richness compared to plots of natural sediment.

POPULATION RESPONSE (1 STUDY)

Overall abundance (1 study): One replicated, paired, controlled study in Auke Bay found that plots where mine tailings were left in place had similar invertebrate overall abundance and biomass as plots where tailings had been removed. While plots with and without tailings had similar abundances to plots of natural sediment, their biomasses were higher.

Assessment: unknown effectiveness (effectiveness 15%; certainty 25%; harms 5%).

https://www.conservationevidence.com/actions/2077

No evidence found (no assessment)

We have captured no evidence for the following interventions:

- Cease or prohibit mining waste (tailings) disposal at sea
- Limit, cease, or prohibit sediment discard during aggregate extraction
- Remove discarded sediment material from the seabed following cessation of aggregate extraction.

13.1.4 Renewable energy

Based on the collated evidence, what is the current assessment of the effectiveness of interventions for renewable energy?	
No evidence found (no assessment)	• Co-locate aquaculture systems with other activities and other infrastructures (such as wind farms) to maximise use of marine space • Limit the number and/or extent of, or prohibit additional, renewable energy installations in an area

No evidence found (no assessment)

We have captured no evidence for the following interventions:

- Co-locate aquaculture systems with other activities and other infrastructures (such as wind farms) to maximise use of marine space
- Limit the number and/or extent of, or prohibit additional, renewable energy installations in an area.

13.2 Threat: Transportation and service corridors

13.2.1 Utility and service lines

Based on the collated evidence, what is the current assessment of the effectiveness of interventions for utility and service lines?	
No evidence found (no assessment)	• Bury cables and pipelines in the seabed rather than laying them on the seabed • Leave utility and service lines in place after decommissioning • Remove utility and service lines after decommissioning • Set limits on the area that can be covered by utility and service lines at one location • Use a different technique when laying and burying cables and pipelines • Use cables and pipelines of smaller width

No evidence found (no assessment)

We have captured no evidence for the following interventions:

- Bury cables and pipelines in the seabed rather than laying them on the seabed
- Leave utility and service lines in place after decommissioning
- Remove utility and service lines after decommissioning
- Set limits on the area that can be covered by utility and service lines at one location

- Use a different technique when laying and burying cables and pipelines
- Use cables and pipelines of smaller width.

13.2.2 Shipping lanes

Based on the collated evidence, what is the current assessment of the effectiveness of interventions for shipping lanes?	
Unknown effectiveness	• Cease or prohibit shipping
No evidence found (no assessment)	• Divert shipping routes • Limit, cease or prohibit anchoring from ships/boats/vessels • Limit, cease or prohibit recreational boating • Periodically move and relocate moorings • Provide additional moorings to reduce anchoring • Reduce ships/boats/vessels speed limits • Set limits on hull depth • Use a different type of anchor • Use moorings which reduce or avoid contact with the seabed (eco- moorings)

Unknown effectiveness

Cease or prohibit shipping

Three studies examined the effects of ceasing or prohibiting shipping on subtidal benthic invertebrate populations. All studies were in the North Sea (Belgium, Germany, Netherlands).

COMMUNITY RESPONSE (2 STUDIES)

Overall community composition (1 study): One site comparison study in the North Sea found that areas closed to shipping developed different overall invertebrate community compositions compared to areas where shipping occurred.

Overall species richness/diversity (1 study): One site comparison study in the North Sea found that areas closed to shipping did not develop different overall invertebrate species richness and diversity compared to areas where shipping occurred.

POPULATION RESPONSE (2 STUDIES)

Overall abundance (2 studies): Two site comparison studies (one before-and-after) in the North Sea found that areas closed to shipping had similar overall invertebrate abundance and biomass compared to areas where shipping occurred.

Overall abundance (2 studies): Two site comparison studies (one before-and-after) in the North Sea found that areas closed to shipping had similar overall invertebrate abundance and biomass compared to areas where shipping occurred.

OTHER (2 STUDIES)

Overall community energy flow (1 study): One before-after, site comparison study in the North Sea found that after closing an area to shipping, invertebrate community energy flow did not change, but it increased in nearby areas where shipping occurred.

Species energy flow (1 study): One before-and-after, site comparison study in the North Sea found that closing an area to shipping had mixed effects on species-level energy flow.

Assessment: unknown effectiveness (effectiveness 35%; certainty 30%; harms 0%).

https://www.conservationevidence.com/actions/2086

No evidence found (no assessment)

We have captured no evidence for the following interventions:

- Divert shipping routes
- Limit, cease or prohibit anchoring from ships/boats/vessels
- Limit, cease or prohibit recreational boating
- Periodically move and relocate moorings
- Provide additional moorings to reduce anchoring
- Reduce ships/boats/vessels speed limits
- Set limits on hull depth
- Use a different type of anchor
- Use moorings which reduce or avoid contact with the seabed (eco- moorings).

13.3 Threat: Biological resource use

13.3.1 Spatial and Temporal Management

Based on the collated evidence, what is the current assessment of the effectiveness of interventions for spatial and temporal management?	
Beneficial	• Cease or prohibit all towed (mobile) fishing gear
Likely to be beneficial	• Cease or prohibit all types of fishing • Cease or prohibit bottom trawling • Cease or prohibit dredging
Unknown effectiveness	• Cease or prohibit commercial fishing • Establish temporary fisheries closures
No evidence found (no assessment)	• Cease or prohibit midwater/semi-pelagic trawling • Cease or prohibit static fishing gear

Beneficial

Cease or prohibit all towed (mobile) fishing gear

Eight studies examined the effects of ceasing or prohibiting all towed fishing gear on subtidal benthic invertebrate populations. One study was in the Limfjord (Denmark), two in the English Channel (UK), three in Georges Bank in the North Atlantic Ocean (USA and Canada), one in the Ria Formosa lagoon (Portugal), and one in the Irish Sea (Isle of Man).

COMMUNITY RESPONSE (4 STUDIES)

Overall community composition (3 studies): Two of three replicated, site comparison studies in the Limfjord and the English Channel, found that areas excluding towed fishing gear for either an unspecified amount of time or two to 23 years had different overall invertebrate community composition compared to areas where towed-fishing occurred and one found that ceasing towed-gear fishing for nine years had mixed effects.

Overall species richness/diversity (3 studies): Two replicated, site comparison studies in the English Channel reported that areas excluding towed fishing gear for either an unspecified amount of time or two to 23 years had different or greater invertebrate species richness and diversity to areas where towed-fishing occurred. One site comparison study in Georges Bank found no difference in invertebrate species richness between an area closed to mobile fishing gear for 10 to 14 years and a fished area.

POPULATION RESPONSE (3 STUDIES)

Overall abundance (3 studies): Two site comparison studies (one replicated) in the English Channel and Georges Bank found that sites excluding towed gear for either two to 23 years or 10 to 14 years had greater overall invertebrate biomass compared to sites where towed-gear fishing occurred, but one also found that abundance was similar in both areas. One replicated, controlled, before-and-after study in the Ria Formosa lagoon found that ceasing towed gear for 10 months led to increases in the cover of mobile but not sessile

Mollusc abundance (2 studies): Two site comparison studies (one replicated) in the Irish Sea and the English Channel found that areas closed to towed fishing gear for either two to 23 years or 14 years had more scallops compared to adjacent fished areas.

Mollusc condition (1 study): One site comparison study the Irish Sea found that an area closed to towed fishing gear for 14 years had higher proportions of older and larger scallops compared to an adjacent fished area.

Starfish abundance (1 study): One replicated, site comparison study in Georges Bank found more starfish in areas closed to towed fishing gear for five to nine years compared to adjacent fished areas.

Starfish condition (1 study): One replicated, site comparison study in Georges Bank found that starfish arm length was similar in areas closed to towed fishing gear for five to nine years and adjacent fished areas.

OTHER (1 STUDY)

Overall community biological production (1 study): One before-and-after, site comparison study in Georges Bank found an increase in the biological production from invertebrate in sites closed to towed fishing gear for approximately five years compared to adjacent fished sites.

Assessment: beneficial (effectiveness 70%; certainty 70%; harms 0%).

https://www.conservationevidence.com/actions/2102

Likely to be beneficial

Cease or prohibit all types of fishing

Five studies examined the effects of ceasing or prohibiting all types of fishing on subtidal benthic invertebrate populations. All studies were in the North Sea (Belgium, Germany, Netherlands, UK).

COMMUNITY RESPONSE (2 STUDIES)

Overall community composition (2 studies): Two site comparison studies (one before-and-after) in the North Sea found that areas closed to all fishing developed different overall invertebrate community compositions compared to fished areas.

Overall species richness/diversity (2 studies): One of two site comparison studies (one before-and-after) in the North Sea found that areas closed to all fishing did not develop different overall invertebrate species richness and diversity compared to fished areas after three years, but the other found higher species richness in the closed areas after 20 years.

POPULATION RESPONSE (3 STUDIES)

Overall abundance (2 studies): Two site comparison studies (one before-and-after) in the North Sea found that areas closed to all fishing had similar overall invertebrate abundance and biomass compared to fished areas after three and five years.

Crustacean abundance (1 study): One before-and-after, site comparison study in the North Sea found that closing a site to all fishing led to similar numbers of lobster compared to a fished site after 20 months.

Crustacean condition (1 study): One before-and-after, site comparison study in the North Sea found that closing a site to all fishing led to larger sizes of lobster compared to a fished site after 20 months.

OTHER (1 STUDY)

Overall community energy flow (1 study): One before-after, site comparison study in the North Sea found that, during the 12–14 months after closing an area to all fishing, the invertebrate community structure (measured as energy flow) at sites within the closed area did not change, but that it increased in nearby fished sites.

Species energy flow (1 study): One before-and-after, site comparison study in the North Sea found that closing an area to all fishing for 12–14 months had mixed effects on species-level energy flow.

Assessment: likely to be beneficial (effectiveness 55%; certainty 50%; harms 0%).

https://www.conservationevidence.com/actions/2096

Cease or prohibit bottom trawling

Four studies examined the effects of ceasing or prohibiting bottom trawling on subtidal benthic invertebrate populations. Two studies were in the Bering Sea (USA), one in the North Sea, and one in the Mediterranean Sea (Italy).

COMMUNITY RESPONSE (3 STUDIES)

Overall community composition (2 studies): Two site comparison studies (one before-and-after, one replicated) in the North Sea and the Mediterranean Sea found that in areas prohibiting trawling for either 15 or 20 years, overall invertebrate community composition was different to that of trawled areas.

Overall species richness/diversity (3 studies): Two of three site comparison studies (one paired, one before-and-after, one replicated) in the Bering Sea, the North Sea, and the Mediterranean Sea found that invertebrate diversity was higher in sites closed to trawling compared to trawled sites after either 37 or 15 years, but the other found no differences after 20 years.

POPULATION RESPONSE (3 STUDIES)

Overall abundance (2 studies): One of two site comparison studies (one paired, one replicated) in the Bering Sea and the Mediterranean Sea found that total invertebrate abundance was higher in sites closed to trawling compared to trawled sites after 37 years, but the other found no differences after 20 years. Both found no differences in total invertebrate biomass.

Unwanted catch overall abundance (1 study): One replicated, before-and-after, site comparison study in the Bering Sea found that during the three years after closing areas to all bottom trawling, unwanted catch of crabs appeared to have decreased, while no changes appeared to have occurred in nearby trawled areas.

Assessment: likely to be beneficial (effectiveness 55%; certainty 50%; harms 5%).

https://www.conservationevidence.com/actions/2099

Cease or prohibit dredging

Four studies examined the effects of ceasing or prohibiting dredging on subtidal benthic invertebrate populations. One study was in the North Atlantic Ocean (Portugal), one in the South Atlantic Ocean (Argentina), one in the English Channel and one in the Irish Sea (UK).

COMMUNITY RESPONSE (3 STUDIES)

Overall community composition (3 studies): One of three site comparison studies (one replicated, one before-and-after) in Atlantic Ocean and the Irish Sea found that after ceasing dredging, overall invertebrate community composition was different to that in dredged areas. The other two found that communities remained similar in dredged and non-dredged areas.

Overall richness/diversity (3 studies): One of three site comparison studies (one replicated, one before-and-after) in Atlantic Ocean and the Irish Sea found that after ceasing dredging, large (macro-) invertebrate diversity was higher but small (meio-) invertebrate diversity was lower compared to dredged areas. The other two found that overall diversity remained similar in dredged and non-dredged areas.

POPULATION RESPONSE (3 STUDIES)

Overall abundance (3 studies): One of three site comparison studies (one replicated, one before-and-after) in Atlantic Ocean and the Irish Sea found that four years after ceasing dredging, large (macro-) and small (meio-) invertebrate abundance and/or biomass appeared higher to that in dredged areas. The other two found that abundance and/or biomass remained similar in dredged and non-dredged areas after either two or six years.

Tunicate abundance (1 study): One replicated, site comparison study in the English Channel found that a year after ceasing dredging in three areas, abundance of ascidians/sea squirts (tunicates) was similar to that in dredged areas.

Bryozoan abundance (1 study): One replicated, site comparison study in the English Channel found that a year after ceasing dredging in three areas, abundance of bryozoan was higher than in dredged areas.

Crustacean abundance (1 study): One replicated, site comparison study in the English Channel found that a year after ceasing dredging in three areas, abundance of spider crabs was higher than in dredged areas, but abundance of edible crab was similar.

Cnidarian abundance (1 study): One replicated, site comparison study in the English Channel found that a year after ceasing dredging in three areas, abundance of sea fans was higher than in dredged areas.

Sponge abundance (1 study): One replicated, site comparison study in the English Channel found that a year after ceasing dredging in three areas, abundance of sponges was higher than in dredged areas.

Assessment: likely to be beneficial (effectiveness 75%; certainty 55%; harms 0%).

https://www.conservationevidence.com/actions/2101

Unknown effectiveness

Cease or prohibit commercial fishing

Three studies examined the effects of ceasing or prohibiting commercial fishing on subtidal benthic invertebrate populations. Two studies were in the Tasman Sea (New Zealand), the third on Gorges Bank in the North Atlantic Ocean (USA).

COMMUNITY RESPONSE (2 STUDIES)

Overall community composition (1 study): One site comparison study in the Tasman Sea found that an area closed to commercial trawling and dredging for 28 years had different overall invertebrate communities than an area subject to commercial fishing.

Overall species richness/diversity (1 study): One site comparison study on Georges Bank found no difference in invertebrate species richness between an area closed to commercial fishing for 10 to 14 years and a fished area.

POPULATION RESPONSE (3 STUDIES)

Overall abundance (2 studies): Two site comparison studies in the Tasman Sea and on Georges Bank found that areas prohibiting commercial fishing for 10 to 14 years and 28 years had greater overall invertebrate abundance compared to areas where commercial fishing occurred. One of the studies also found higher biomass, while the other found similar biomass in closed and fished areas.

Crustacean abundance (1 study): One replicated, site comparison study in the Tasman Sea found that in commercial fishing exclusion zones lobster abundance was not different to adjacent fished areas after up to two years.

OTHER (1 STUDY)

Overall community biological production (1 study): One site comparison study in the Tasman Sea found that an area closed to commercial trawling and dredging for 28 years had greater biological production from invertebrates than an area where commercial fishing occurred.

Assessment: unknown effectiveness (effectiveness 55%; certainty 34%; harms 0%).

https://www.conservationevidence.com/actions/2097

Establish temporary fisheries closures

Six studies examined the effects of establishing temporary fisheries closures on subtidal benthic invertebrates. One study was in the English Channel (UK), one in the D'Entrecasteaux Channel (Australia), one in the North Pacific Ocean (USA), two in the Mozambique Channel (Madagascar), and one in the North Sea (UK).

COMMUNITY RESPONSE (2 STUDIES)

Overall species richness/diversity (1 study): One replicated, site comparison study in the English Channel found that sites seasonally closed to towed-gear fishing did not have greater invertebrate species richness than sites where towed-fishing occurred year-round.

Mollusc community composition (1 study): One replicated, before-and after study in the D'Entrecasteaux Channel found that temporarily reopening an area previously closed to all fishing for 12 years only to recreational fishing led to changes in scallop species community composition over four fishing seasons.

POPULATION RESPONSE (1 STUDY)

Overall abundance (1 study): One replicated, site comparison study in the English Channel found that sites seasonally closed to towed-gear fishing did not have a greater invertebrate biomass than sites where towed-fishing occurred year-round.

Crustacean abundance (1 study): One before-and-after, site comparison study in the North Sea found that reopening a site to fishing following a temporary 20-month closure led to lower total abundance but similar marketable abundance of European lobsters compared to a continuously-fished site after a month.

Mollusc abundance (5 studies): One replicated, site comparison study English Channel found that sites seasonally closed to towed gear did not have higher abundance of great scallops than sites where towed-fishing occurred year-round. Two before-and after, site comparison studies (one replicated) in the Mozambique Channel found that temporarily closing an area to reef octopus fishing did not increase octopus abundance/biomass compared to before closure and to continuously fished areas. Two replicated, before-and after studies in the D'Entrecasteaux Channel and the North Pacific Ocean found that temporarily reopening an area previously closed to all fishing to recreational fishing only led to a decline in scallop abundance after four fishing seasons and in red abalone after three years.

Mollusc condition (3 studies): One replicated, before-and after study in the North Pacific Ocean found that temporarily reopening an area previously closed to fishing led to a decline in the size of red abalone after three years. Two before-and after, site comparison studies (one replicated) in the Mozambique Channel found that temporarily closing an area to reef octopus fishing increased the weight of octopus compared to before closure and to continuously fished areas, but one also found that this effect did not last once fishing resumed.

Assessment: unknown effectiveness (effectiveness 30%; certainty 36%; harms 10%).

https://www.conservationevidence.com/actions/2098

No evidence found (no assessment)

We have captured no evidence for the following interventions:

- Cease or prohibit midwater/semi-pelagic trawling
- Cease or prohibit static fishing gear.

13.3.2 Effort and Capacity Reduction

Based on the collated evidence, what is the current assessment of the effectiveness of interventions for effort and capacity reduction?	
Unknown effectiveness	• Establish territorial user rights for fisheries • Install physical barriers to prevent trawling
No evidence found (no assessment)	• Eliminate fisheries subsidies that encourage overfishing • Introduce catch shares • Limit the density of traps • Limit the number of fishing days • Limit the number of fishing vessels • Limit the number of traps per fishing vessels • Purchase fishing permits and/or vessels from fishers • Set commercial catch quotas • Set commercial catch quotas and habitat credits systems • Set habitat credits systems

Unknown effectiveness

Establish territorial user rights for fisheries

One study examined the effects of establishing territorial user rights for fisheries on subtidal benthic invertebrate populations. The study was in the South Pacific Ocean (Chile).

COMMUNITY RESPONSE (0 STUDIES)

POPULATION RESPONSE (1 STUDY)

Mollusc reproductive success (1 study): One site comparison study in South Pacific Ocean found that an area with territorial user rights for fisheries had larger-sized and more numerous egg capsules, and more larvae of the Chilean abalone up to 21 months after establishing fishing restrictions compared to an open-access area.

Assessment: unknown effectiveness (effectiveness 65%; certainty 10%; harms 0%).

https://www.conservationevidence.com/actions/2104

Install physical barriers to prevent trawling

One study examined the effects of installing physical barriers to prevent trawling on subtidal benthic invertebrate populations. The study was in the Bay of Biscay (Spain).

COMMUNITY RESPONSE (1 STUDY)

Overall community composition (1 study): One before-and-after study in the Bay of Biscay found that one to four years after installing artificial reefs as physical barriers to prevent trawling invertebrate community composition changed.

POPULATION RESPONSE (1 STUDY)

Overall abundance (1 study): One before-and-after study in the Bay of Biscay found that one to four years after installing artificial reefs as physical barriers to prevent trawling overall invertebrate biomass increased.

Echinoderm abundance (1 study): One before-and-after study in the Bay of Biscay found that one to four years after installing artificial reefs as physical barriers to prevent trawling the biomass of sea urchins and starfish increased.

Molluscs abundance (1 study): One before-and-after study in the Bay of Biscay found that one to four years after installing artificial reefs as physical barriers to prevent trawling the biomass of gastropods (sea snails), of one species of cuttlefish, and of two species of octopus increased.

Assessment: unknown effectiveness (effectiveness 75%; certainty 32%; harms 0%).

https://www.conservationevidence.com/actions/2112

No evidence found (no assessment)

We have captured no evidence for the following interventions:

- Eliminate fisheries subsidies that encourage overfishing
- Introduce catch shares
- Limit the density of traps

- Limit the number of fishing days
- Limit the number of fishing vessels
- Limit the number of traps per fishing vessels
- Purchase fishing permits and/or vessels from fishers
- Set commercial catch quotas
- Set commercial catch quotas and habitat credits systems
- Set habitat credits systems.

13.3.3 Reduce Unwanted catch, Discards and Impacts on seabed communities

Based on the collated evidence, what is the current assessment of the effectiveness of interventions for reduce unwanted catch, discards and impacts on seabed communities?	
Likely to be beneficial	• Fit one or more mesh escape panels/windows to trawl nets • Fit one or more soft, semi-rigid, or rigid grids or frames to trawl nets • Modify the design of dredges • Modify the position of traps • Use a larger codend mesh size on trawl nets • Use a midwater/semi-pelagic trawl instead of bottom/demersal trawl
Unknown effectiveness	• Fit a funnel (such as a sievenet) or other escape devices on shrimp/prawn trawl nets • Fit one or more mesh escape panels/windows and one or more soft, rigid or semi-rigid grids or frames to trawl nets • Fit one or more mesh escape panels/windows to trawl nets and use a square mesh instead of a diamond mesh codend • Fit one or more soft, semi-rigid, or rigid grids or frames and increase the mesh size of pots and traps • Fit one or more soft, semi-rigid, or rigid grids or frames on pots and traps

	• Fit one or more soft, semi-rigid, or rigid grids or frames to trawl nets and use square mesh instead of a diamond mesh at the codend • Hand harvest instead of using a dredge • Increase the mesh size of pots and traps • Modify the design of traps • Modify the design/attachments of a shrimp/prawn W-trawl net • Reduce the number or modify the arrangement of tickler chains/chain mats on trawl nets • Use a larger mesh size on trammel nets • Use a pulse trawl instead of a beam trawl • Use a smaller beam trawl • Use a square mesh instead of a diamond mesh codend on trawl nets • Use an otter trawl instead of a beam trawl • Use an otter trawl instead of a dredge • Use different bait species in traps • Use traps instead of fishing nets
No evidence found (no assessment)	• Fit one or more mesh escape panels/windows on pots and traps • Limit the maximum weight and/or size of bobbins on the footrope • Modify harvest methods of macroalgae • Modify trawl doors to reduce sediment penetration • Outfit trawls with a raised footrope • Release live unwanted catch first before handling commercial species • Set unwanted catch quotas • Use alternative means of getting mussel seeds rather than dredging from natural mussel beds • Use hook and line fishing instead of other fishing methods • Use lower water pressure during hydraulic dredging • Use more than one net on otter trawls

Likely to be beneficial

Fit one or more mesh escape panels/windows to trawl nets

Seven studies examined the effects of adding one or more mesh escape panels/windows to trawl nets on subtidal benthic invertebrate populations. Six were in the North Sea (Belgium, Netherlands, UK), two in the Thames estuary (UK), one in the English Channel (UK), and one in the Gulf of Carpentaria (Australia).

COMMUNITY RESPONSE (0 STUDIES)

POPULATION RESPONSE (7 STUDIES)

Overall survival (1 study): One replicated, paired, controlled study in the English Channel and the North Sea found that fitting nets with either one of seven designs of square mesh escape panels (varying mesh size and twine type) led to higher survival rates of invertebrates that escaped the nets compared to unmodified nets.

Unwanted catch overall abundance (7 studies): Three of seven replicated, paired, controlled studies in the North Sea, the Thames estuary, the English Channel and the Gulf of Carpentaria found that trawl nets fitted with one or more mesh escape panels/windows/zones reduced the unwanted catch of invertebrates compared to unmodified nets. Two found mixed effects of fitting escape panels on the unwanted catch of invertebrates and fish depending on the panel design. Two found that trawl nets fitted with escape panels caught similar amounts of unwanted invertebrates and fish compared to unmodified nets.

OTHERS (7 STUDIES)

Commercially targeted catch abundance (7 studies): Three of seven replicated, paired, controlled studies in the North Sea, the Thames estuary, the English Channel and the Gulf of Carpentaria, found that trawl nets fitted with one or more mesh escape panels/windows/zones caught similar amounts of all or most commercial species to unmodified nets. Three found mixed effects of fitting escape panels on the catch of all or most commercial species depending on the species and/or panel design. One found that trawl nets fitted with escape panels reduced the catch of commercial species compared to unmodified nets.

Assessment: likely to be beneficial (effectiveness 60%; certainty 50%; harms 5%).

https://www.conservationevidence.com/actions/2132

Fit one or more soft, semi-rigid, or rigid grids or frames to trawl nets

Two studies examined the effects of fitting one or more soft, semi-rigid, or rigid grids or frames to trawl nets on subtidal benthic invertebrate populations. The studies were in the Gulf of Carpentaria and Spencer Gulf (Australia).

COMMUNITY RESPONSE (0 STUDIES)

POPULATION RESPONSE (2 STUDIES)

Unwanted catch abundance (1 study): Two replicated, paired, controlled studies in the Gulf of Carpentaria and in Spencer Gulf found that nets fitted with a 'downward'-oriented grid but not an 'upward'-oriented grid reduced the weight of small unwanted catch and that both grid orientations caught fewer unwanted large sponges, and that nets fitted with two sizes of grids reduced the number and biomass of unwanted blue swimmer crabs and giant cuttlefish caught, compared to unmodified nets.

OTHER (2 STUDIES)

Commercial catch abundance (2 studies): Two replicated, paired, controlled studies in the Gulf of Carpentaria and Spencer Gulf found that nets fitted with a 'downward'-oriented grid or a small grid reduced the catch of commercially targeted prawns, compared to unmodified nets, but those fitted with an 'upward'-oriented grid or a large grid caught similar amounts to unmodified nets.

Assessment: likely to be beneficial (effectiveness 60%; certainty 40%; harms 0%).

https://www.conservationevidence.com/actions/2133

Modify the design of dredges

Six studies examined the effects of modifying the design of dredges on subtidal benthic invertebrate populations. Four were in the North Atlantic Ocean (Portugal) and two were in the Irish Sea (Isle of Man).

COMMUNITY RESPONSE (1 STUDY)

Unwanted catch overall composition (1 study): One replicated, controlled, study in the Irish Sea found that a new design of scallop dredge caught a similar species composition of unwanted catch to a traditional dredge.

POPULATION RESPONSE (3 STUDIES)

Overall abundance (2 studies): One of two controlled studies in the North Atlantic Ocean and in the Irish Sea found that a new dredge design damaged or killed fewer invertebrates left in the sediment tracks following dredging. The other found no difference in total invertebrate abundance or biomass living in or on the sediment tracks following fishing with two dredge designs.

Unwanted catch overall abundance (2 studies): Two controlled studies (one replicated) in the North Atlantic Ocean and the Irish Sea found that a modified or a new design of bivalve dredge caught less unwanted catch compared to traditional unmodified dredges.

Unwanted catch condition (6 studies): Six controlled studies (one replicated and paired, four replicated) in the North Atlantic Ocean and the Irish Sea found that new or modified bivalve dredges damaged or killed similar proportions of unwanted catch (retained and/or escaped) compared to traditional or unmodified designs, three of which also found that they did not reduce the proportion of damaged or dead unwanted crabs (retained and/or escaped).

OTHER (1 study)

Commercial catch abundance (1 study): One replicated, controlled, study in the Irish Sea found that a new dredge design caught a similar amount of commercially targeted queen scallops compared to a traditional dredge.

Assessment: likely to be beneficial (effectiveness 40%; certainty 42%; harms 19%).

https://www.conservationevidence.com/actions/2119

Modify the position of traps

Two studies examined the effects of modifying the position of traps on subtidal benthic invertebrate populations. One study was in the Varangerfjord (Norway), the other in the North Atlantic Ocean (Spain).

COMMUNITY RESPONSE (1 STUDY)

Unwanted catch species richness/diversity (1 study): One replicated, controlled study in the North Atlantic found that semi-floating traps caught fewer unwanted catch species compared to standard bottom traps.

POPULATION RESPONSE (2 STUDIES)

Unwanted catch abundance (2 studies): Two replicated, controlled studies in the Varangerfjord and the North Atlantic found that floating or semi-floating traps caught fewer unwanted invertebrates compared to standard bottom traps.

OTHER (2 STUDIES)

Commercial catch abundance (2 studies): Two replicated, controlled studies in the Varangerfjord and the North Atlantic found that floating or semi-floating traps caught similar amounts (abundance and biomass) of commercially targeted species as standard bottom traps.

Assessment: likely to be beneficial (effectiveness 65%; certainty 40%; harms 0%).

https://www.conservationevidence.com/actions/2144

Use a larger codend mesh size on trawl nets

One study examined the effects of using a larger codend mesh size on trawl nets on unwanted catch of subtidal benthic invertebrate populations. The study was in the Gulf of Mexico (Mexico).

COMMUNITY RESPONSE (1 STUDY)

Unwanted catch species richness/diversity (1 study): One replicated, paired, controlled study in the Gulf of Mexico found that trawl nets fitted with a larger mesh codend caught fewer combined species of non-commercial unwanted invertebrates and fish compared to a traditional codend.

POPULATION RESPONSE (1 STUDY)

Unwanted catch abundance (1 study): One replicated, paired, controlled study in the Gulf of Mexico found that trawl nets fitted with a larger mesh codend caught lower combined biomass and abundance of non-commercial unwanted invertebrates and fish compared to a traditional codend.

OTHER (1 STUDY)

Commercial catch abundance (1 study): One replicated, paired, controlled study in the Gulf of Mexico found that trawl nets fitted with a larger mesh codend caught less biomass and abundance of commercially targeted shrimps compared to a traditional codend, but that the biomass ratios of commercially targeted to discard species was similar for both.

Assessment: likely to be beneficial (effectiveness 65%; certainty 42%; harms 0%).

https://www.conservationevidence.com/actions/2135

Use a midwater/semi-pelagic trawl instead of bottom/demersal trawl

One study examined the effects of using a semi-pelagic trawl instead of a demersal trawl on subtidal benthic invertebrates. The study was in the Indian Ocean (Australia).

COMMUNITY RESPONSE (0 STUDIES)

POPULATION RESPONSE (1 STUDY)

Overall abundance (1 study): One replicated, controlled, study in the Indian Ocean found that fishing with a semi-pelagic trawl did not reduce the abundance of large sessile invertebrates, which was similar to non-trawled plots, but a demersal trawl did.

OTHER (1 STUDY)

Commercial catch abundance (1 study): One replicated, controlled, study in the Indian Ocean found that fishing with a semi-pelagic trawl reduced the abundance of retained commercially targeted fish compared to fishing with a demersal trawl.

Assessment: likely to be beneficial (effectiveness 70%; certainty 41%; harms 0%).

https://www.conservationevidence.com/actions/2118

Unknown effectiveness

Fit a funnel (such as a sievenet) or other escape devices on shrimp/prawn trawl nets

One study examined the effects of fitting a funnel, sievenet, or other escape devices on trawl nets on marine subtidal invertebrate. The study was in the North Sea (UK).

COMMUNITY RESPONSE (0 STUDIES)

POPULATION RESPONSE (1 STUDY)

Unwanted catch abundance (1 study): One replicated, paired, controlled study in the North Sea found that trawl nets fitted with a sievenet appeared to catch fewer unwanted catch of non-commercial invertebrates compared to unmodified nets.

Assessment: unknown effectiveness (effectiveness 65%; certainty 20%; harms 0%).

https://www.conservationevidence.com/actions/2131

Fit one or more mesh escape panels/windows and one or more soft, rigid or semi-rigid grids or frames to trawl nets

One study examined the effects on subtidal benthic invertebrate populations of fitting one or more mesh escape panels/windows and one or more soft, rigid or semi-rigid grids or frames to trawl nets. The study was in the Gulf of Carpentaria (Australia).

COMMUNITY RESPONSE (0 STUDIES)

POPULATION RESPONSE (1 STUDY)

Unwanted catch abundance (1 study): One replicated, paired, controlled study in Gulf of Carpentaria found that trawl nets fitted with an escape window and a grid reduced the total weight of small unwanted catch and caught fewer unwanted large sponges, compared to unmodified nets.

OTHER (1 STUDY)

Commercial catch abundance (1 study): One replicated, paired, controlled study in Carpentaria found that trawl nets fitted with an escape window and a grid reduced the catch of commercially targeted prawns, compared to unmodified nets.

Assessment: unknown effectiveness (effectiveness 65%; certainty 32%; harms 0%).

https://www.conservationevidence.com/actions/2134

Fit one or more mesh escape panels/windows to trawl nets and use a square mesh instead of a diamond mesh codend

One study examined the effects of fitting one or more mesh escape panels to trawl nets and using a square mesh instead of a diamond mesh codend on subtidal benthic invertebrate populations. The study was in the English Channel (UK).

COMMUNITY RESPONSE (0 STUDIES)

POPULATION RESPONSE (1 STUDY)

Unwanted catch abundance (1 study): One replicated, paired, controlled study in the English Channel found that trawl nets fitted with two large square mesh release panels and a square mesh codend caught fewer unwanted catch of non-commercial invertebrates compared to standard trawl nets.

OTHER (1 STUDY)

Commercial catch abundance (1 study): One replicated, paired, controlled study in the English Channel found that trawl nets fitted with two large square mesh release panels and a square mesh codend caught fewer commercial shellfish, and fewer but more valuable commercially important fish, compared to standard trawl nets.

Assessment: unknown effectiveness (effectiveness 65%; certainty 30%; harms 0%).

https://www.conservationevidence.com/actions/2138

Fit one or more soft, semi-rigid, or rigid grids or frames and increase the mesh size of pots and traps

One study examined the effects of fitting one or more soft, semi-rigid, or rigid grids or frames and increasing the mesh size of pots and traps on subtidal benthic invertebrates. The study took place in the Corindi River system (Australia).

COMMUNITY RESPONSE (0 STUDIES)

POPULATION RESPONSE (1 STUDY)

Unwanted catch abundance (1 study): One replicated, controlled study in the Corindi River system found that traps fitted with escape frames and designed with larger mesh appeared to reduce the proportion of unwanted undersized mud crabs caught, compared to conventional traps without escape frames and smaller mesh.

Assessment: unknown effectiveness (effectiveness 50%; certainty 20%; harms 10%).

https://www.conservationevidence.com/actions/2149

Fit one or more soft, semi-rigid, or rigid grids or frames on pots and traps

One study examined the effects of fitting one or more soft, semi-rigid, or rigid grids or frames on pots and traps on subtidal benthic invertebrates. The study took place in the Corindi River system (Australia).

COMMUNITY RESPONSE (0 STUDIES)

POPULATION RESPONSE (1 STUDY)

Unwanted catch abundance (1 study): One replicated, controlled study in the Corindi River system found that traps fitted with escape frames appeared to reduce the proportion of unwanted undersized mud crabs caught, compared to conventional traps without escape frames.

Assessment: unknown effectiveness (effectiveness 55%; certainty 20%; harms 0%).

https://www.conservationevidence.com/actions/2146

Fit one or more soft, semi-rigid, or rigid grids or frames to trawl nets and use square mesh instead of a diamond mesh at the codend

One study examined the effects of fitting one or more soft, semi-rigid, or rigid grids or frames to trawl nets and using a square mesh codend on subtidal benthic invertebrates. The study was in the Gulf of St Vincent (Australia).

COMMUNITY RESPONSE (0 STUDIES)

POPULATION RESPONSE (1 STUDY)

Unwanted catch abundance (1 study): One replicated, paired, controlled study in Gulf of St Vincent found that trawl nets fitted with a rigid U-shaped grid and a square-oriented mesh codend reduced the catch rates of three dominant groups of unwanted invertebrate catch species, compared to unmodified nets.

OTHER (1 STUDY)

Commercial catch abundance (1 study): One replicated, paired, controlled study in the Gulf of St Vincent found that trawl nets fitted with a rigid U-shaped grid and a square-oriented mesh codend reduced the catch rates of the commercially targeted western king prawn, due to reduced catch of less valuable smaller-sized prawns, compared to unmodified nets.

Assessment: unknown effectiveness (effectiveness 70%; certainty 20%; harms 0%).

https://www.conservationevidence.com/actions/2137

Hand harvest instead of using a dredge

Two studies examined the effects of hand harvesting instead of using a dredge on subtidal benthic invertebrate populations. Both were in San Matías Gulf, South Atlantic Ocean (Argentina).

COMMUNITY RESPONSE (2 STUDIES)

Unwanted catch community composition (1 study): One replicated, controlled study in San Matías Gulf found that, when harvesting mussels, the community composition of the unwanted catch was similar by hand harvesting and by using a dredge.

Unwanted catch richness/diversity (1 study): One replicated, controlled study in San Matías Gulf found that, when harvesting mussels, hand harvesting caught fewer species of unwanted catch compared to using a dredge.

POPULATION RESPONSE (2 STUDIES)

Unwanted catch abundance (1 study): One replicated, controlled study in San Matías Gulf found that, when harvesting mussels, hand harvesting caught fewer unwanted sea urchins and brittle stars compared to using a dredge.

Unwanted catch condition (1 study): One replicated, controlled study in San Matías Gulf found that, when harvesting mussels, the damage caused to unwanted sea urchins and brittle stars was similar by hand harvesting and by using a dredge.

OTHER 1 STUDY)

Commercial catch abundance (1 study): One replicated, controlled study in San Matías Gulf found that more commercially targeted mussels were caught by hand harvesting than by using a dredge.

Assessment: unknown effectiveness (effectiveness 45%; certainty 18%; harms 10%).

https://www.conservationevidence.com/actions/2121

Increase the mesh size of pots and traps

One study examined the effects of increasing the mesh size of pots and traps on subtidal benthic invertebrates. The study took place in the Corindi River system (Australia).

COMMUNITY RESPONSE (0 STUDIES)

POPULATION RESPONSE (1 STUDY)

Unwanted catch abundance (1 study): One replicated, controlled study in the Corindi River system found that traps designed with larger mesh appeared to reduce the proportion of unwanted undersized mud crabs caught, compared to conventional traps of smaller mesh.

Assessment: unknown effectiveness (effectiveness 61%; certainty 29%; harms 0%).

https://www.conservationevidence.com/actions/2148

Modify the design of traps

Two studies examined the effects of modifying the design of traps on subtidal benthic invertebrates. One study took place in the Mediterranean Sea (Spain), and one in the South Pacific Ocean (New Zealand).

COMMUNITY RESPONSE (0 STUDIES)

POPULATION RESPONSE (2 STUDIES)

Unwanted catch abundance (2 studies): Two replicated, controlled studies in the Mediterranean Sea and the South Pacific Ocean found that the amount of combined unwanted catch of invertebrates and fish varied with the type of trap design used and the area.

OTHER (1 STUDY)

Commercial catch abundance (1 study): One replicated, controlled study in the Mediterranean Sea found that plastic traps caught some legal-size commercially targeted lobsters while collapsible traps caught none.

Assessment: unknown effectiveness (effectiveness 40%; certainty 21%; harms 20%).

https://www.conservationevidence.com/actions/2143

Modify the design/attachments of a shrimp/prawn W-trawl net

One study examined the effects of modifying the design/attachments of a W-trawl net used in shrimp/prawn fisheries on unwanted catch of subtidal benthic invertebrate. The study was in Moreton Bay (Australia).

COMMUNITY RESPONSE (0 STUDIES)

POPULATION RESPONSE (1 STUDY)

Unwanted catch overall abundance (1 study): One replicated, paired, controlled study in Moreton Bay found that four designs of W-trawl nets used in shrimp/prawn fisheries caught less non-commercial unwanted catch of crustaceans compared to a traditional Florida Flyer trawl net.

OTHERS (1 STUDY)

Commercial catch abundance (1 study): One replicated, paired, controlled study in Moreton Bay found that four designs of W-trawl nets used in shrimp/

prawn fisheries caught lower amounts of the commercially targeted prawn species compared to a traditional Florida Flyer trawl net.
Assessment: unknown effectiveness (effectiveness 61%; certainty 24%; harms 0%).

https://www.conservationevidence.com/actions/2139

Reduce the number or modify the arrangement of tickler chains/chain mats on trawl nets

Three studies examined the effects of reducing the number or modifying the arrangement of tickler chains/chain mats on subtidal benthic invertebrates. All studies were in the North Sea (Germany and Netherlands).
COMMUNITY RESPONSE (0 STUDIES)
POPULATION RESPONSE (3 STUDIES)
Overall abundance (1 study): One replicated, paired, controlled study in the North Sea found that using a beam trawl with a chain mat caused lower mortality of benthic invertebrates in the trawl tracks compared to using a beam trawl with tickler chains.
Unwanted catch abundance (2 studies): One of two replicated, paired, controlled studies in the North Sea found that all three modified parallel tickler chain arrangements reduced the combined amount of non-commercial unwanted invertebrate and fish catch compared to unmodified trawl nets, but the other found that none of three modified parabolic tickler chain arrangements reduced it.
OTHER (2 STUDIES)
Commercial catch abundance (2 studies): One of two replicated, paired, controlled studies in the North Sea found that three modified parabolic tickler chain arrangements caught similar amounts of commercial species to unmodified nets, but the other found that three modified parallel tickler chain arrangements caught lower amounts.
Assessment: unknown effectiveness (effectiveness 43%; certainty 32%; harms 10%).

https://www.conservationevidence.com/actions/2140

Use a larger mesh size on trammel nets

One study examined the effects of using a larger mesh size on trammel nets on subtidal benthic invertebrates. The study was in the North Atlantic Ocean (Portugal).
COMMUNITY RESPONSE (1 STUDY)
Unwanted catch community composition (1 study): One replicated, controlled, study in the North Atlantic Ocean found that using larger mesh sizes in the

inner and/or outer panels of trammel nets did not affect the community composition of unwanted catch of non-commercial invertebrates (discard).
POPULATION RESPONSE (1 STUDY)
Unwanted catch abundance (1 study): One replicated, controlled, study in the North Atlantic Ocean found that using larger mesh sizes in the inner and/or outer panels of trammel nets did not reduce the abundance of unwanted catch of non-commercial invertebrates (discard).
Assessment: unknown effectiveness (effectiveness 20%; certainty 36%; harms 0%).

https://www.conservationevidence.com/actions/2141

Use a pulse trawl instead of a beam trawl

One study examined the effects of using a pulse trawl instead of a beam trawl on subtidal benthic invertebrates. The study was in the North Sea (Netherlands).
COMMUNITY RESPONSE (0 STUDIES)
POPULATION RESPONSE (1 STUDY)
Unwanted catch abundance (1 study): One replicated, controlled, study in the North Sea found that pulse trawls caught less unwanted invertebrate catch compared to traditional beam trawls, but the effects varied with species.
OTHER (1 STUDY)
Commercial catch abundance (1 study): One replicated, controlled, study in the North Sea found that pulse trawls reduced the volume of commercial catch by 19% compared to beam trawls.
Assessment: unknown effectiveness (effectiveness 41%; certainty 34%; harms 15%).

https://www.conservationevidence.com/actions/2126

Use a smaller beam trawl

One study examined the effects of using a smaller beam trawl on subtidal benthic invertebrates. The study was in the North Sea (Germany and Netherlands).
COMMUNITY RESPONSE (0 STUDIES)
POPULATION RESPONSE (1 STUDY)
Overall abundance (1 study): One replicated, paired, controlled study in the North Sea found that a smaller beam trawl caused similar mortality of invertebrates in the trawl tracks compared to a larger beam trawl.
Assessment: unknown effectiveness (effectiveness 30%; certainty 35%; harms 0%).

https://www.conservationevidence.com/actions/2127

Use a square mesh instead of a diamond mesh codend on trawl nets

One study examined the effects of using a square mesh instead of a diamond mesh codend on trawl nets on unwanted catch of subtidal benthic invertebrate populations. The study was in the English Channel (UK).

COMMUNITY RESPONSE (0 STUDIES)

POPULATION RESPONSE (1 STUDY)

Unwanted catch abundance (1 study): One replicated, paired, controlled study in the English Channel found that a trawl net with a square mesh codend caught less non-commercial unwanted invertebrates in one of two areas, and similar amounts in the other area, compared to a standard diamond mesh codend.

OTHER (1 STUDY)

Commercial catch abundance (1 study): One replicated, paired, controlled study in the English Channel found that a trawl net with a square mesh codend caught similar amounts of commercially targeted fish species in two areas, and that in one of two areas it caught more commercially important shellfish, compared to a standard diamond mesh codend.

Assessment: unknown effectiveness (effectiveness 50%; certainty 20%; harms 10%).

https://www.conservationevidence.com/actions/2136

Use an otter trawl instead of a beam trawl

One study examined the effects of using an otter trawl instead of a beam trawl on subtidal benthic invertebrates. The study was in the North Sea (Germany and Netherlands).

COMMUNITY RESPONSE (0 STUDIES)

POPULATION RESPONSE (1 STUDY)

Overall abundance (1 study): One replicated, paired, controlled study in the North Sea found that otter trawls caused similar mortality of invertebrates in the trawl tracks compared to beam trawls in sandy areas but lower mortality in silty areas.

Assessment: unknown effectiveness (effectiveness 50%; certainty 34%; harms 10%).

https://www.conservationevidence.com/actions/2125

Use an otter trawl instead of a dredge

One study examined the effects of using an otter trawl instead of a dredge on subtidal benthic invertebrates. The study was in the Irish Sea (Isle of Man).

COMMUNITY RESPONSE (1 STUDY)

Unwanted catch overall composition (1 study): One replicated, controlled, study in the Irish Sea found that an otter trawl caught a different species composition of unwanted invertebrate and fish species (combined) compared to two scallop dredges.

POPULATION RESPONSE (1 STUDY)

Overall abundance (1 study): One replicated, controlled, study in the Irish Sea found no difference in total invertebrate abundance and biomass living in or on the sediment of the trawl tracks following fishing with either an otter trawl or two scallop dredges.

Unwanted catch overall abundance (1 study): One replicated, controlled, study in the Irish Sea found that an otter trawl caught fewer unwanted invertebrates and fish (combined) compared to two scallop dredges.

OTHER (1 STUDY)

Commercial catch abundance (1 study): One replicated, controlled, study in the Irish Sea found that an otter trawl caught similar number of commercially targeted queen scallops compared to two scallop dredges.

Assessment: unknown effectiveness (effectiveness 45%; certainty 35%; harms 0%).

https://www.conservationevidence.com/actions/2123

Use different bait species in traps

One study examined the effects of using different bait species in traps on subtidal benthic invertebrates. The study took place in the South Pacific Ocean (New Zealand).

COMMUNITY RESPONSE (0 STUDIES)

POPULATION RESPONSE (1 STUDY)

Unwanted catch abundance (1 study): One replicated, controlled study in the South Pacific Ocean found that the type of bait used in fishing pots did not change the amount of unwanted invertebrates caught.

Assessment: unknown effectiveness (effectiveness 1%; certainty 37%; harms 0%).

https://www.conservationevidence.com/actions/2145

Use traps instead of fishing nets

One study examined the effects of using traps instead of fishing nets on subtidal benthic invertebrates. The study took place in the Mediterranean Sea (Spain).

COMMUNITY RESPONSE (0 STUDIES)

POPULATION RESPONSE (1 STUDY)

Unwanted catch abundance (1 study): One replicated, controlled study in the Mediterranean Sea found that the combined amount of unwanted catch of invertebrates and fish appeared lower using plastic traps than trammel nets, but higher using collapsible traps.

OTHER (1 STUDY)

Commercial catch abundance (1 study): One replicated, controlled study in the Mediterranean Sea found that the catch of commercially targeted lobsters was lower using traps than in trammel nets.

Assessment: unknown effectiveness (effectiveness 40%; certainty 32%; harms 20%).

https://www.conservationevidence.com/actions/2142

No evidence found (no assessment)

We have captured no evidence for the following interventions:

- Fit one or more mesh escape panels/windows on pots and traps
- Limit the maximum weight and/or size of bobbins on the footrope
- Modify harvest methods of macroalgae
- Modify trawl doors to reduce sediment penetration
- Outfit trawls with a raised footrope
- Release live unwanted catch first before handling commercial species
- Set unwanted catch quotas
- Use alternative means of getting mussel seeds rather than dredging from natural mussel beds
- Use hook and line fishing instead of other fishing methods
- Use lower water pressure during hydraulic dredging
- Use more than one net on otter trawls.

13.4 Threat: Human intrusions and disturbances

13.4.1 Recreational Activities

<table>
<tr><th colspan="2">Based on the collated evidence, what is the current assessment of the effectiveness of interventions for recreational activities?</th></tr>
<tr><td>No evidence found (no assessment)</td><td>• Limit, cease or prohibit access for recreational purposes
• Limit, cease or prohibit recreational diving
• Limit, cease or prohibit recreational fishing and/or harvesting</td></tr>
</table>

No evidence found (no assessment)

We have captured no evidence for the following interventions:

- Limit, cease or prohibit access for recreational purposes
- Limit, cease or prohibit recreational diving
- Limit, cease or prohibit recreational fishing and/or harvesting.

13.5 Threat: Invasive and other problematic species, genes and diseases

13.5.1 Aquaculture

Based on the collated evidence, what is the current assessment of the effectiveness of interventions for aquaculture?	
No evidence found (no assessment)	• Implement quarantine to avoid accidental introduction of disease, non-native or problem species • Implement regular inspections to avoid accidental introduction of disease or non-native or problem species • Import spat and/or eggs to aquaculture facilities rather than juveniles and adults to reduce the risk of introducing hitchhiking species • Prevent the attachment of biofouling organisms/species in aquaculture • Reduce and/or eradicate aquaculture escapees in the wild • Remove biofouling organisms/species in aquaculture • Source spat and juveniles from areas or hatcheries not infested with diseases or non-native or problematic species • Use native species instead of non-native species in aquaculture systems • Use sterile individuals in aquaculture systems using non-native species

No evidence found (no assessment)

We have captured no evidence for the following interventions:

- Implement quarantine to avoid accidental introduction of disease, non-native or problem species
- Implement regular inspections to avoid accidental introduction of disease or non-native or problem species
- Import spat and/or eggs to aquaculture facilities rather than juveniles and adults to reduce the risk of introducing hitchhiking species
- Prevent the attachment of biofouling organisms/species in aquaculture
- Reduce and/or eradicate aquaculture escapees in the wild
- Remove biofouling organisms/species in aquaculture
- Source spat and juveniles from areas or hatcheries not infested with diseases or non-native or problematic species
- Use native species instead of non-native species in aquaculture systems
- Use sterile individuals in aquaculture systems using non-native species.

13.5.2 Shipping, transportation and anthropogenic structures

Based on the collated evidence, what is the current assessment of the effectiveness of interventions for Shipping, transportation and anthropogenic structures?	
No evidence found (no assessment)	• Clean anthropogenic platforms, structures or equipment • Clean the hull, anchor and chain of commercial and recreational vessels • Limit, cease or prohibit ballast water exchange in specific areas • Treat ballast water before exchange • Use antifouling coatings on the surfaces of vessels and anthropogenic structures

No evidence found (no assessment)

We have captured no evidence for the following interventions:

- Clean anthropogenic platforms, structures or equipment
- Clean the hull, anchor and chain of commercial and recreational vessels
- Limit, cease or prohibit ballast water exchange in specific areas
- Treat ballast water before exchange
- Use antifouling coatings on the surfaces of vessels and anthropogenic structures.

13.5.3 Other

Based on the collated evidence, what is the current assessment of the effectiveness of interventions for sources of non-native, invasive or other problematic species?	
Unknown effectiveness	• Remove or capture non-native, invasive or other problematic species
No evidence found (no assessment)	• Limit, cease or prohibit the sale and/or transportation of commercial non-native species • Use biocides or other chemicals to control non-native, invasive or other problematic species • Use biological control to manage non-native, invasive or other problematic species populations • Use of non-native, invasive or other problematic species from populations established in the wild for recreational or commercial purposes

Unknown effectiveness

Remove or capture non-native, invasive or other problematic species

One study examined the effects of removing or capturing non-native, invasive or other problematic species on subtidal benthic invertebrates. The study was in the South Atlantic Ocean (Brazil).

COMMUNITY RESPONSE (0 STUDIES)

POPULATION RESPONSE (1 STUDY)

Cnidarian abundance (1 study): One replicated, controlled, before-and-after study in the southwest Atlantic found that, regardless of the method used, removing invasive corals reduced the cover of native zoanthids.

Sponge abundance (1 study): One replicated, controlled, before-and-after study in the southwest Atlantic found that the effect of removing invasive corals on the cover of native sponges varied with the removal method used.

Assessment: unknown effectiveness (effectiveness 40%; certainty 23%; harms 22%).

https://www.conservationevidence.com/actions/2173

No evidence found (no assessment)

We have captured no evidence for the following interventions:

- Limit, cease or prohibit the sale and/or transportation of commercial non-native species
- Use biocides or other chemicals to control non-native, invasive or other problematic species
- Use biological control to manage non-native, invasive or other problematic species populations
- Use of non-native, invasive or other problematic species from populations established in the wild for recreational or commercial purposes.

13.6 Threat: Pollution

13.6.1 General

Based on the collated evidence, what is the current assessment of the effectiveness of interventions for pollution?	
Unknown effectiveness	• Add chemicals or minerals to sediments to remove or neutralise pollutants
No evidence found (no assessment)	• Establish pollution emergency plans • Transplant/translocate 'bioremediating' species

Unknown effectiveness

Add chemicals or minerals to sediments to remove or neutralise pollutants

Two studies examined the effects of adding chemicals or minerals to sediments to remove or neutralise pollutants on subtidal benthic invertebrate populations. Both studies evaluated the use of coal ash in Hiroshima Bay (Japan).

COMMUNITY RESPONSE (1 STUDY)

Overall richness/diversity (1 study): One controlled, before-and-after study in Hiroshima Bay found that adding coal ash increased invertebrate species richness in winter but not summer compared to untreated sites.

POPULATION RESPONSE (2 STUDIES)

Overall abundance (2 studies): One controlled, before-and-after study in Hiroshima Bay found that adding coal ash increased invertebrate abundance in winter but not summer compared to untreated sites. One controlled study in Hiroshima Bay found that one of two types of coal ash increased combined invertebrate and fish abundance, but not biomass.

Assessment: unknown effectiveness (effectiveness 40%; certainty 20%; harms 0%).

https://www.conservationevidence.com/actions/2176

No evidence found (no assessment)

We have captured no evidence for the following interventions:

- Establish pollution emergency plans
- Transplant/translocate 'bioremediating' species.

13.6.2 Domestic and urban wastewater

Based on the collated evidence, what is the current assessment of the effectiveness of interventions for domestic and urban wastewater?	
Likely to be beneficial	• Limit, cease or prohibit the dumping of sewage sludge
Unknown effectiveness	• Set or improve minimum sewage treatment standards
No evidence found (no assessment)	• Limit the amount of storm wastewater overflow • Limit, cease or prohibit the dumping of untreated sewage

Likely to be beneficial

Limit, cease or prohibit the dumping of sewage sludge

Two studies examined the effects of ceasing or prohibiting the dumping of sewage sludge on subtidal benthic invertebrate populations. One study was in the New York Bight (USA), one in the North Sea (UK).

COMMUNITY RESPONSE (2 STUDIES)

Overall community composition (2 studies): One before-and-after, site comparison study in the New York Bight found that after ceasing sewage sludge dumping, overall invertebrate community composition became more similar to less disturbed sites. One replicated, site comparison study in the North Sea found that overall invertebrate community composition changed but remained different to that of natural sites.

POPULATION RESPONSE (2 STUDIES)

Overall abundance (1 study): One replicated, site comparison study in the North Sea found that after ceasing sewage sludge dumping, overall invertebrate abundance became similar to that of natural sites.

Worm abundance (1 study): One before-and-after, site comparison study in the New York Bight found that after ceasing sewage sludge dumping, abundance of pollution-indicator polychaete worms decreased and became similar to that of natural sites.

Assessment: likely to be beneficial (effectiveness 70%; certainty 47%; harms 0%).

https://www.conservationevidence.com/actions/2179

Unknown effectiveness

Set or improve minimum sewage treatment standards

One study examined the effects of improving minimum sewage treatment standards on subtidal benthic invertebrates. The study was in the Bay of Biscay (Spain).

COMMUNITY RESPONSE (1 STUDY)

Overall community composition (1 study): One before-and-after, site comparison study in the Bay of Biscay found that after introducing a secondary treatment of sewage wastewaters, invertebrate community composition at an impacted site did not significantly change compared to unimpacted sites.

Overall richness/diversity (1 study): One before-and-after, site comparison study in the Bay of Biscay found that after introducing a secondary treatment of sewage wastewaters, invertebrate richness and diversity at an impacted site did not significantly change compared to unimpacted sites.

POPULATION RESPONSE (1 STUDY)

Overall abundance (1 study): One before-and-after, site comparison study in the Bay of Biscay found that after introducing a secondary treatment of sewage wastewaters, total cover of invertebrates significantly increased at an impacted site at 8 m but not 3 m depth, compared to unimpacted sites.

Assessment: unknown effectiveness (effectiveness 23%; certainty 28%; harms 0%).

https://www.conservationevidence.com/actions/2180

No evidence found (no assessment)

We have captured no evidence for the following interventions:

- Limit the amount of storm wastewater overflow
- Limit, cease or prohibit the dumping of untreated sewage.

13.6.3 Industrial and military effluents

Based on the collated evidence, what is the current assessment of the effectiveness of interventions for industrial and military effluents?	
Unknown effectiveness	• Remove or clean-up oil pollution following a spill
No evidence found (no assessment)	• Set regulatory ban on marine burial of nuclear waste • Use double hulls to prevent oil spills

Unknown effectiveness

Remove or clean-up oil pollution following a spill

One study examined the effects of removing and cleaning-up oil pollution following a spill on subtidal benthic invertebrates. The study was in the Baltic Proper (Sweden).

COMMUNITY RESPONSE (0 STUDIES)

POPULATION RESPONSE (1 STUDY)

Mollusc condition (1 study): One replicated, controlled, before-and-after study in the Baltic Proper found that after cleaning-up spilled oil using high pressure hot water, crude oil content increased in mussels and did not naturally decrease over time, and was higher than in mussels from an uncleaned contaminated and a non-contaminated site.

Assessment: unknown effectiveness (effectiveness 20%; certainty 26%; harms 30%).

https://www.conservationevidence.com/actions/2183

No evidence found (no assessment)

We have captured no evidence for the following interventions:

- Set regulatory ban on marine burial of nuclear waste
- Use double hulls to prevent oil spills.

13.6.4 Aquaculture effluents

Based on the collated evidence, what is the current assessment of the effectiveness of interventions for aquaculture effluents?	
Unknown effectiveness	• Cease or prohibit aquaculture activity • Leave a fallow period during fish/shellfish farming
No evidence found (no assessment)	• Improve fish food and pellets to reduce aquaculture waste production • Locate aquaculture systems in areas with fast currents • Locate aquaculture systems in already impacted areas • Locate aquaculture systems in vegetated areas • Locate artificial reefs near aquaculture systems (and vice versa) to act as biofilters • Moor aquaculture cages so they move in response to changing current direction • Reduce aquaculture stocking densities • Reduce the amount of antibiotics used in aquaculture systems • Reduce the amount of pesticides used in aquaculture systems • Use other bioremediation methods in aquaculture • Use species from more than one level of a food web in aquaculture systems

Unknown effectiveness

Cease or prohibit aquaculture activity

Two studies examined the effects of ceasing or prohibiting aquaculture activity on subtidal benthic invertebrate populations. Both studies were in the Mediterranean Sea (Italy and Spain).

COMMUNITY RESPONSE (2 STUDIES)

Overall community composition (1 study): One before-and-after, site comparison study in the Mediterranean Sea found that after ceasing aquaculture activity invertebrate community composition remained different to that of an unfarmed site.

Worm community composition (1 study): One before-and-after, site comparison study in the Mediterranean Sea found that after ceasing aquaculture activity worm community composition community composition remained different to that of an unfarmed site.

POPULATION RESPONSE (2 STUDIES)

Overall abundance (1 study): One before-and-after, site comparison study in the Mediterranean Sea found that after ceasing aquaculture activity overall invertebrate abundance was similar to an unfarmed site.

Worm abundance (1 study): One before-and-after, site comparison study in the Mediterranean Sea found that after ceasing aquaculture activity abundance of health-indicating worms increased, and abundance of pollution-indicating worms decreased.

Assessment: unknown effectiveness (effectiveness 25%; certainty 20%; harms 0%).

https://www.conservationevidence.com/actions/2185

Leave a fallow period during fish/shellfish farming

Three studies examined the effects of leaving a fallow period during fish farming on subtidal benthic invertebrate populations. Two studies were in the Tasman Sea (Australia), and one in the North Pacific Ocean (USA).

COMMUNITY RESPONSE (3 STUDIES)

Overall community composition (2 study): Two replicated, before-and-after, site comparison study in the Tasman Sea found that after a fallow period invertebrate community composition became similar to that occurring before the fish were added but remained different to communities at sites without fish farms.

Worm community composition (1 study): One replicated, before-and-after, site comparison study in the North Pacific Ocean found that after a fallow period polychaete worm community composition changed but remained different to communities at sites without fish farms.

Worm richness/diversity (1 study): One replicated, before-and-after, site comparison study in the North Pacific Ocean found that after a fallow period polychaete worm diversity did not change and remained lower compared to sites without fish farms.

POPULATION RESPONSE (2 STUDIES)

Worm abundance (2 studies): Two replicated, before-and-after, site comparison studies in the Tasman Sea and the North Pacific Ocean found that following a fallow period, abundances of pollution-indicator polychaete worms decreased, but remained higher compared to sites without fish farms.

Assessment: unknown effectiveness (effectiveness 40%; certainty 35%; harms 0%).

https://www.conservationevidence.com/actions/2191

No evidence found (no assessment)

We have captured no evidence for the following interventions:

- Improve fish food and pellets to reduce aquaculture waste production
- Locate aquaculture systems in areas with fast currents
- Locate aquaculture systems in already impacted areas
- Locate aquaculture systems in vegetated areas
- Locate artificial reefs near aquaculture systems (and vice versa) to act as biofilters
- Moor aquaculture cages so they move in response to changing current direction
- Reduce aquaculture stocking densities
- Reduce the amount of antibiotics used in aquaculture systems
- Reduce the amount of pesticides used in aquaculture systems
- Use other bioremediation methods in aquaculture
- Use species from more than one level of a food web in aquaculture systems.

13.6.5 Agricultural and forestry effluents

Based on the collated evidence, what is the current assessment of the effectiveness of interventions for agricultural and forestry effluents?	
No evidence found (no assessment)	• Create artificial wetlands to reduce the amount of pollutants reaching the sea • Establish aquaculture to extract the nutrients from run-offs • Regulate the use, dosage and disposal of agrichemicals • Treat wastewater from intensive livestock holdings

No evidence found (no assessment)

We have captured no evidence for the following interventions:

- Create artificial wetlands to reduce the amount of pollutants reaching the sea
- Establish aquaculture to extract the nutrients from run-offs
- Regulate the use, dosage and disposal of agrichemicals
- Treat wastewater from intensive livestock holdings.

13.6.6 Garbage and solid waste

Based on the collated evidence, what is the current assessment of the effectiveness of interventions for garbage and solid waste?	
No evidence found (no assessment)	• Bury electricity cables to reduce electromagnetic fields • Install stormwater traps or grids • Limit, cease or prohibit discharge of solid waste overboard from vessels • Recover lost fishing gear • Remove litter from the marine environment • Use biodegradable panels in fishing pots

No evidence found (no assessment)

We have captured no evidence for the following interventions:

- Bury electricity cables to reduce electromagnetic fields
- Install stormwater traps or grids
- Limit, cease or prohibit discharge of solid waste overboard from vessels
- Recover lost fishing gear
- Remove litter from the marine environment
- Use biodegradable panels in fishing pots.

13.6.7 Excess energy

Based on the collated evidence, what is the current assessment of the effectiveness of interventions for excess energy?	
No evidence found (no assessment)	• Limit, cease or prohibit industrial and urban lighting at night • Limit, cease or prohibit the discharge of cooling effluents from power stations • Limit, cease or prohibit the use of sonars • Reduce underwater noise (other than sonar)

No evidence found (no assessment)

We have captured no evidence for the following interventions:

- Limit, cease or prohibit industrial and urban lighting at night
- Limit, cease or prohibit the discharge of cooling effluents from power stations
- Limit, cease or prohibit the use of sonars
- Reduce underwater noise (other than sonar).

13.6.8 Other pollution

Based on the collated evidence, what is the current assessment of the effectiveness of interventions for other pollution?	
Likely to be beneficial	• Restrict the use of tributyltin or other toxic antifouling coatings
Unknown effectiveness	• Remove and clean-up shoreline waste disposal sites
No evidence found (no assessment)	• Limit, cease or prohibit the discharge of waste effluents overboard from vessels • Use non-toxic antifouling coatings on surfaces

Likely to be beneficial

Restrict the use of tributyltin or other toxic antifouling coatings

Four studies examined the effects of restricting the use of tributyltin as an antifouling coating on subtidal benthic invertebrate populations. One study was located in the English Channel (UK), two in the River Crouch estuary (UK), and one in Otsuchi Bay (Japan).

COMMUNITY RESPONSE (2 STUDIES)

Overall community composition (1 study): One replicated, before-and-after study in the River Crouch estuary found that after restricting the use of tributyltin, invertebrate community composition changed, but that changes varied with locations.

Overall richness/diversity (1 study): One replicated, before-and-after study in the River Crouch estuary found that after restricting the use of tributyltin, overall invertebrate species richness and diversity increased.

POPULATION RESPONSE (2 STUDIES)

Molluscs condition (1 study): One replicated, before-and-after study in the English Channel found that after restricting the use of tributyltin, there was a decrease in its concentration in dogwhelks and the penis length of female dogwhelks.

Crustacean condition (1 study): One study in Otsuchi Bay found that after restricting the use of tributyltin its concentration decreased in skeleton shrimps.
Assessment: likely to be beneficial (effectiveness 69%; certainty 45%; harms 0%).

https://www.conservationevidence.com/actions/2214

Unknown effectiveness

Remove and clean-up shoreline waste disposal sites

One study examined the effects of removing and cleaning-up shoreline waste disposal sites on subtidal benthic invertebrates. The study was in the Southern Ocean (Antarctica).

COMMUNITY RESPONSE (1 STUDY)

Overall community composition (1 study): One replicated, controlled, before-and-after study in the Southern Ocean found that after removing and cleaning-up a disused waste disposal site, invertebrate community composition changed, and no further negative impacts were detected, but communities remained different to natural sites.

Overall richness/diversity (1 study): One replicated, controlled, before-and-after study in the Southern Ocean found that after removing and cleaning-up a disused waste disposal site, invertebrate species richness did not change over time and remained different to that of natural sites, but no further negative impacts were detected.

Assessment: unknown effectiveness (effectiveness 49%; certainty 35%; harms 0%).

https://www.conservationevidence.com/actions/2215

No evidence found (no assessment)

We have captured no evidence for the following interventions:

- Limit, cease or prohibit the discharge of waste effluents overboard from vessels
- Use non-toxic antifouling coatings on surfaces.

13.7 Threat: Climate change and severe weather

Based on the collated evidence, what is the current assessment of the effectiveness of interventions for climate change and severe weather?	
No evidence found (no assessment)	• Create a Marine Protected Area or set levels of legal protection where natural climate refugia occur to further promote the persistence and recovery of species facing climate change • Limit, cease or prohibit the degradation and/or removal of carbon sequestering species and/or habitats • Manage climate-driven range extensions of problematic species • Promote natural carbon sequestration species and/or habitats • Restore habitats and/or habitat-forming (biogenic) species following extreme events • Transplant captive-bred or hatchery-reared individuals of habitat-forming (biogenic) species that are resistant to climate change • Transplant/release climate change-resistant captive-bred or hatchery-reared individuals to re-establish or boost native populations

No evidence found (no assessment)

We have captured no evidence for the following interventions:

- Create a Marine Protected Area or set levels of legal protection where natural climate refugia occur to further promote the persistence and recovery of species facing climate change
- Limit, cease or prohibit the degradation and/or removal of carbon sequestering species and/or habitats
- Manage climate-driven range extensions of problematic species
- Promote natural carbon sequestration species and/or habitats
- Restore habitats and/or habitat-forming (biogenic) species following extreme events
- Transplant captive-bred or hatchery-reared individuals of habitat-forming (biogenic) species that are resistant to climate change
- Transplant/release climate change-resistant captive-bred or hatchery-reared individuals to re-establish or boost native populations.

13.8 Habitat protection

Based on the collated evidence, what is the current assessment of the effectiveness of interventions for habitat protection?	
Likely to be beneficial	• Designate a Marine Protected Area and introduce some fishing restrictions (types unspecified) • Designate a Marine Protected Area and prohibit all types of fishing • Designate a Marine Protected Area and prohibit the harvesting of sea urchins • Designate a Marine Protected Area with a zonation system of activity restrictions
Unknown effectiveness	• Designate a Marine Protected Area and install physical barriers to prevent trawling • Designate a Marine Protected Area and only allow hook and line fishing • Designate a Marine Protected Area and prohibit all towed (mobile) fishing gear • Designate a Marine Protected Area and prohibit aquaculture activity • Designate a Marine Protected Area and prohibit bottom trawling • Designate a Marine Protected Area and prohibit commercial fishing • Designate a Marine Protected Area and prohibit dredging • Designate a Marine Protected Area and prohibit the harvesting of conch • Establish community-based fisheries management

No evidence found (no assessment)	• Designate a Marine Protected Area and limit the density of traps • Designate a Marine Protected Area and limit the number of fishing vessels • Designate a Marine Protected Area and prohibit static fishing gear • Designate a Marine Protected Area and prohibit the harvesting of scallops • Designate a Marine Protected Area and set a no-anchoring zone • Designate a Marine Protected Area without setting management measures, usage restrictions, or enforcement • Designate a Particularly Sensitive Sea Area (PSSA) to regulate impactful maritime activities • Engage with stakeholders when designing Marine Protected Areas

Likely to be beneficial

Designate a Marine Protected Area and introduce some fishing restrictions (types unspecified)

Four studies examined the effects of introducing unspecified types of fishing restrictions in marine protected areas on subtidal benthic invertebrate populations. Two studies were in the Indian Ocean (Seychelles), one was a global systematic review, and one was in the Mediterranean Sea (Italy).

COMMUNITY RESPONSE (2 STUDIES)

Overall community composition (2 studies): One of two site comparison studies (one replicated) in the Indian Ocean and the Mediterranean Sea found that a marine protected area with unspecified fishing restrictions (year of designation unspecified) had a different combined invertebrate and algae community composition, while the other (time since designation unspecified) found similar compositions compared to fished areas.

POPULATION RESPONSE (3 STUDIES)

Overall abundance (1 study): One replicated, site comparison study in the Mediterranean Sea found that a marine protected area with unspecified fishing restrictions had similar invertebrate abundance compared to unprotected fished areas (time since designation unspecified).

Bryozoan abundance (1 study): One site comparison study in the Indian Ocean found that a marine protected area with unspecified fishing restrictions (year of designation unspecified) had similar abundance of bryozoans compared to fished areas.

Crustacean abundance (1 study): One global systematic review found that marine protected areas with unspecified fishing restrictions had more lobsters compared to fished areas.

Echinoderm abundance (2 studies): One of two site comparison studies (one replicated) in the Indian Ocean found that marine protected areas with unspecified fishing restrictions had more sea cucumbers after more than 20 years but the other found fewer sea lilies (year of designation unspecified) compared to fished areas.

Hydrozoan abundance (1 study): One site comparison study in the Indian Ocean found that a marine protected area with unspecified fishing restrictions (year of designation unspecified) had more hydrozoans compared to fished areas.

Mollusc abundance (1 study): One global systematic review found that marine protected areas with unspecified fishing restrictions had more scallops compared to fished areas.

Sponge abundance (1 study): One site comparison study in the Indian Ocean found that a marine protected area with unspecified fishing restrictions (year of designation unspecified) had more sponges compared to fished areas.

Tunicate abundance (1 study): One site comparison study in the Indian Ocean found that a marine protected area closed to fishing with unspecified fishing restrictions (year of designation unspecified) had similar abundance of ascidians/sea squirts (tunicates) compared to fished areas.

Assessment: likely to be beneficial (effectiveness 50%; certainty 40%; harms 0%).

https://www.conservationevidence.com/actions/2239

Designate a Marine Protected Area and prohibit all types of fishing

Thirty studies examined the effects of prohibiting all types of fishing in marine protected areas on subtidal benthic invertebrate populations. Four studies were systematic reviews of marine reserves (New Zealand and across the world). Two studies were in the North Atlantic Ocean (Bahamas). Five were in the South Pacific Ocean (New Zealand, French Polynesia). Three were in the North Pacific Ocean (USA). Seven were in the Tasman Sea (New Zealand, Australia). One was in the Florida Keys (USA). One was in the Coral Sea

(Australia). Three were in the Mediterranean Sea (Italy, Spain). One was in the Bristol Channel and the Irish Sea (UK). Two were in the Firth of Clyde (UK). One was in the Foveaux Straight (New Zealand).

COMMUNITY RESPONSE (5 STUDIES)

Overall community composition (3 studies): Three site comparison studies (one replicated and paired, one replicated, one paired) in the Mediterranean Sea, the Tasman Sea, and the Firth of Clyde found that marine protected areas that had been prohibiting all fishing for five to 16 years depending on the study, had similar combined algae, invertebrate and fish community composition, similar combined mollusc and echinoderm community composition, and similar overall community composition of large invertebrates but different composition of small sessile invertebrates, compared to fished areas.

Overall species richness/diversity (5 studies): One global systematic review, and three site comparison studies (one replicated and paired, one replicated, one paired) in the Mediterranean Sea, the Tasman Sea, and the Firth of Clyde found that marine protected areas that had been prohibiting all fishing for five to 16 years depending on the study, had similar overall invertebrate species richness/diversity, similar combined algae, invertebrate and fish species richness, and similar combined mollusc and echinoderm species richness, compared to fished areas. One site comparison study in the Tasman Sea found inside a marine protected area prohibiting all mobile fishing that macroinvertebrate species richness remained stable over the 15 years after its designation and enforcement, but decreased at fished sites.

POPULATION RESPONSE (2 STUDIES)

Overall abundance (4 studies): Two systematic reviews of marine protected areas across the world prohibiting all fishing found that they had greater overall invertebrate abundance and biomass compared to fished areas. Two site comparison studies (one before-and-after, one replicated) in the Tasman Sea found that inside marine protected areas prohibiting all fishing, overall invertebrate abundance did not change over the 15 years after their designation and enforcement and that it did not change in fished areas either, and that all areas had similar combined mollusc and echinoderm abundance after 16 years.

Overall condition (1 study): One global systematic review found that in marine protected areas prohibiting all fishing, invertebrates were bigger compared to fished areas.

Crustacean abundance (17 studies): Two reviews (one global and systematic, one of New Zealand areas) found that marine protected areas prohibiting all fishing had more lobsters compared to marine protected areas only partially prohibiting fishing and unrestricted fished areas. Eleven of 15 site comparison studies (including replicated, randomized, paired, before-and-after) in the North Atlantic Ocean, the Bristol Channel and the Irish Sea, the Firth of Clyde, the Mediterranean Sea, the North Pacific Ocean, the Florida Keys, the South Pacific Ocean, the Tasman Sea, and the Coral Sea found that inside marine protected areas prohibiting all fishing, the abundances and/or biomasses of lobsters and mud crabs were higher compared to areas where seasonal or unrestricted fishing was allowed, after four to 33 years depending on the study. Four found that they had mixed effects on the abundances of lobster, and crab species, after one to seven years depending on the study. Two found that they had similar abundance of lobsters compared to fished areas after either five to seven years or after approximately 30 years.

Crustacean reproductive success (4 studies): Two site comparison studies (one replicated, randomized) in the Florida Keys and the Firth of Clyde found that marine protected areas prohibiting all fishing and harvesting had similar population sex ratios of lobsters compared to where seasonal fishing or all fishing was allowed, after four to seven years depending on the study. Two replicated, site comparison studies (one randomized) in the Tasman Sea and the Mediterranean Sea found that marine protected areas prohibiting all fishing had greater lobster egg production potential compared to commercial fishing exclusion zones and fully fished areas, after either 15 years or 21 to 25 years. One site comparison study in the Firth of Clyde found that marine protected areas prohibiting all fishing had more female lobsters with eggs than fished areas, after four to seven years.

Crustacean condition (8 studies): One review of studies in New Zealand, and five of seven site comparison studies (four replicated, one replicated and randomized) in the North Atlantic Ocean, the Bristol Channel and the Irish Sea, the Firth of Clyde, the Florida Keys, the South Pacific Ocean, the Coral Sea, and the Tasman Sea, found that marine protected areas prohibiting all fishing had bigger lobsters and crabs compared to seasonally fished or fully fished areas, after four to seven years depending on the study. Three found mixed effects on lobsters and crabs depending on species, sex, and locations, after one to seven years depending on the study.

Crustacean population structure (2 studies): Two replicated site comparison studies (one randomized) in the Tasman Sea and the Mediterranean Sea found that marine protected areas prohibiting all fishing had different population size structures of lobsters compared to commercial fishing exclusion zones (only for females) and compared to fished areas, after either 15 years or 21 to 25 years.

Echinoderm abundance (3 studies): Two of three site comparison studies (two replicated, one paired) in the North Pacific Ocean, the South Pacific Ocean, and the North Pacific Ocean, found that marine protected areas prohibiting all fishing had similar abundance of Kina sea urchins after more than 10 years, and sea cucumbers after eight years to fished areas, and a third found higher abundance of red sea urchins after approximately 30 years. One also found that the effects on abundance of red sea urchins depended on the age of the protected area and the size of the urchins.

Echinoderm condition (1 study): One paired, site comparison study in the South Pacific Ocean found that marine protected areas that had been prohibiting all fishing for over 10 years had heavier Kina sea urchins compared to fished areas.

Mollusc abundance (10 studies): Four of 10 site comparison studies (including replicated before-and-after, and site comparison) in the North Atlantic Ocean, the North Pacific Ocean the South Pacific Ocean, the Tasman Sea, and the Foveaux Straight found that inside a marine reserve prohibiting all fishing, abundances/biomass of giant clams, adult queen conch, Cook's turban snails, rock scallops and green abalone were higher compared to a fished area, after eight to 36 years depending on the study. Six found similar abundances of scallop species, pink abalone, juvenile queen conch, and top shell species, after five to 36 years depending on the study. Three found lower abundances of star limpets after 23 to 25 years and blacklip abalone after 15 to 16 years. One found that the effects of marine protected areas prohibiting all fishing on the abundance of mussel species compared to a commercial fishing exclusion zone varied with the age and location of the protected areas.

Mollusc reproductive success (1 study): One site comparison study in the North Atlantic Ocean found that inside a marine protected area that had been prohibiting all fishing for 33 to 36 years, abundance of queen conch larvae was higher compared to an unprotected fished area.

Mollusc condition (1 study): One site comparison study in the North Pacific Ocean found that in marine protected areas that had been prohibiting all fishing pink abalone were bigger five to 23 years after their designation, compared to fished site.

Assessment: likely to be beneficial (effectiveness 59%; certainty 60%; harms 0%).

https://www.conservationevidence.com/actions/2224

Designate a Marine Protected Area and prohibit the harvesting of sea urchins

Two studies examined the effects of prohibiting the harvest of sea urchins in marine protected areas on their populations and/or other subtidal benthic invertebrates. Both studies were in the North Pacific Ocean (USA).

COMMUNITY RESPONSE (0 STUDIES)

POPULATION RESPONSE (2 STUDIES)

Echinoderm abundance (1 study): One replicated, site comparison study in the North Pacific Ocean found that marine protected areas prohibiting the harvest of red sea urchins had higher adult sea urchin biomass six to 33 years after their designations, compared to harvested areas.

Echinoderm reproductive success (1 study): One replicated, site comparison study in the North Pacific Ocean found that marine protected areas prohibiting the harvest of red sea urchins had higher urchin population reproductive biomasses, but similar reproductive indices six to 33 years after their designations, compared to harvested areas.

Echinoderm condition (1 study): One replicated, site comparison study in the North Pacific Ocean found that marine protected areas prohibiting the harvest of red sea urchins had bigger adult sea urchins six to 33 years after their designations, compared to harvested areas.

Mollusc abundance (1 study): One replicated, site comparison study in the North Pacific Ocean found that marine protected areas prohibiting the harvest of red sea urchins (year of designation unspecified) had more juvenile red abalone and juvenile flat abalone compared to harvested areas, and that juvenile abalone abundance was positively related to sea urchin abundance.

Assessment: likely to be beneficial (effectiveness 70%; certainty 41%; harms 0%).

https://www.conservationevidence.com/actions/2238

Designate a Marine Protected Area with a zonation system of activity restrictions

Thirteen studies examined the effects of designating a marine protected area with a zonation system of activity restrictions on subtidal benthic invertebrate populations. Four studies were in the Caribbean Sea (Belize, Mexico), three in the Mediterranean Sea (Italy), one in the Central Pacific Ocean (Ecuador), three in the Bristol Channel and the Irish Sea (UK), one in the Indian Ocean (Australia), and one in the North Atlantic Ocean (Portugal).

COMMUNITY RESPONSE (2 STUDIES)

Overall community composition (1 study): One site comparison study in the Mediterranean Sea found that inside a marine protected area with a zonation system, the combined invertebrate and algae species community composition was different at a site prohibiting all fishing compared to sites where some fishing occurs, after six years.

Overall species richness/diversity (1 study): One site comparison study in the North Atlantic Ocean found that inside a marine protected area with a zonation system, sites prohibiting nearly all fishing had similar invertebrate species richness to sites where fishing was mostly allowed, after two years.

POPULATION RESPONSE (13 STUDIES)

Overall abundance (1 study): One site comparison study in the North Atlantic Ocean found that inside a marine protected area with a zonation system, abundances of specific invertebrate groups varied between sites prohibiting nearly all fishing and sites where fishing was mostly allowed, after two years.

Crustacean abundance (7 studies): Three of seven site comparison studies (two replicated) in the Caribbean Sea, the Central Pacific Ocean, and in the Bristol Channel and the Irish Sea found that inside a marine protected area with a zonation system, abundance and/or biomass of spiny lobsters increased in a zone closed to all/commercial fishing and were greater than in a zone where fewer fishing restrictions occurred, after four to 20 years depending on the study. One found that sites closed to all fishing had higher abundances of spiny lobsters and slipper lobsters after eight to ten years compared to fished sites. Two found that sites closed to all fishing for six to seven years had more European lobsters than sites where potting was allowed. And one found that abundances of European lobsters, velvet crabs, brown crabs and spider crabs, after one to four years, varied with the levels of protection.

Crustacean condition (4 studies): Three of five site comparison studies (one replicated) in the Bristol Channel and the Irish Sea, and in the Caribbean Sea found that, inside a marine protected area with a zonation system, sites prohibiting all fishing for seven years or commercial fishing (duration unspecified) had bigger lobsters compared to fished areas. One found that the sizes of lobsters, velvet crabs, brown crabs and spider crabs varied with the levels of protection, and one study found that the size of spiny lobsters decreased similarly in an area prohibiting all fishing and in an area with fewer restrictions 14 to 20 years after designation of the protected area. Two studies undertaken in the same area found conflicting effects of prohibiting all fishing for six to seven years on disease and injury of lobsters.

Echinoderm abundance (2 studies): One of two site comparison studies in the Mediterranean Sea found that inside a marine protected area with a zonation system, at a site prohibiting all fishing for 17 to 18 years, abundances of two species of sea urchins were higher than at sites allowing the recreational fishing of purple sea urchins. The other one found similar abundance of purple sea urchins inside fully protected sites, sites where some restricted urchin harvest occurs, and unprotected fished sites outside the protected area after five years.

Echinoderm condition (2 studies): Two site comparison studies in the Mediterranean Sea found that inside a marine protected area with a zonation system, sites prohibiting all fishing had bigger sea urchins compared to sites where some restricted urchin harvest occurs and compared to unprotected fished sites outside the protected area, after either four years or 17 to 18 years.

Mollusc abundance (3 studies): One replicated, randomized, controlled study in the Indian Ocean found that inside a marine protected area with a zonation system, abundance of blacklip abalone was higher in sites that had been prohibiting all fishing for five years compared to those prohibiting commercial fishing only. Two site comparison studies in the Caribbean Sea found that inside marine protected areas with a zonation system, abundances of adult queen conch increased over time in a zone closed to all fishing and were greater than in zones with fewer restrictions, but abundances of juvenile conch did not differ or vary differently between zones, after either five to eight years or 14 to 20 years.

Mollusc condition (1 study): One site comparison study in the Caribbean Sea found that inside a marine protected area with a zonation system, the size of queen conch decreased similarly in the area prohibiting all fishing and in the area with fewer restrictions, after 14 to 20 years.

Sponge abundance (1 study): One site comparison study in the Mediterranean Sea found that inside a marine protected area with a zonation system, the cover of sponges Cliona was higher at a site prohibiting all fishing for six years compared to sites where some fishing occurred.

BEHAVIOUR (1 STUDY)

Crustacean behaviour (1 study): One site comparison study in the Caribbean Sea found that, inside a marine protected area with a zonation system (year of designation unspecified), 80% of the lobster population occurring in the unfished area remained in the protected unfished area, and thus remained protected.

Assessment: likely to be beneficial (effectiveness 60%; certainty 60%; harms 10%).

https://www.conservationevidence.com/actions/2230

Unknown effectiveness

Designate a Marine Protected Area and install physical barriers to prevent trawling

One study examined the effects of installing physical barriers to prevent trawling in a protected area on subtidal benthic invertebrate populations. The study was in the South China Sea (Hong Kong).

COMMUNITY RESPONSE (1 STUDY)

Worm community composition (1 study): One replicated, site comparison study in the South China Sea found that sites in a protected area where physical barriers were installed to prevent trawling had a different community composition of nematode worms compared to nearby unprotected fished sites, after up to two years.

Worm species richness/diversity (1 study): One replicated, site comparison study in the South China Sea found that sites in a protected area where physical barriers were installed to prevent trawling had similar diversity and species richness of nematode worms to nearby unprotected fished sites, after up to two years.

POPULATION RESPONSE (1 STUDY)

Overall abundance (1 study): One replicated, site comparison study in the South China Sea found that sites in a protected area where physical barriers were installed to prevent trawling had fewer small invertebrates compared to nearby unprotected fished sites, after up to two years.

Worm abundance (1 study): One replicated, site comparison study in the South China Sea found that sites in a protected area where physical barriers were installed to prevent trawling had fewer nematode worms compared to nearby unprotected fished sites, after up to two years.

Assessment: unknown effectiveness (effectiveness 30%; certainty 30%; harms 10%).

https://www.conservationevidence.com/actions/2227

Designate a Marine Protected Area and only allow hook and line fishing

One study examined the effects of allowing only hook and line fishing in marine protected areas on subtidal benthic invertebrate populations. The study was in the Skagerrak (Norway).

COMMUNITY RESPONSE (0 STUDIES)

POPULATION RESPONSE (1 STUDY)

Crustacean abundance (1 study): One replicated, controlled, before-and-after study in the Skagerrak found that sites inside a protected area only allowing hook and line fishing had greater increases in lobster abundance over the four years after the area was designated compared to unprotected fully fished sites.

Crustacean condition (1 study): One replicated, controlled, before-and-after study in the Skagerrak found that sites inside a protected area only allowing hook and line fishing had greater increases in lobster size over the four years after the area was designated compared to unprotected fully fished sites.

Assessment: unknown effectiveness (effectiveness 75%; certainty 32%; harms 0%).

https://www.conservationevidence.com/actions/2233

Designate a Marine Protected Area and prohibit all towed (mobile) fishing gear

Two studies examined the effects of prohibiting all towed gear in marine protected areas on subtidal benthic invertebrate populations. One study was in the Bristol Channel and the Irish Sea (UK), the other in the English Channel (UK).

COMMUNITY RESPONSE (1 STUDY)

Overall community composition (1 study): One before-and-after, site comparison study in the English Channel found that, over the three years after closing a marine protected area to all towed gears, the community composition of reef-indicative invertebrate species became different to that of unprotected fished sites.

Overall diversity/species richness (1 study): One before-and-after, site comparison study in the English Channel found that, over the three years after closing a marine protected area to all towed gears, the number of reef-indicative invertebrate species remained similar to unprotected fished sites.

POPULATION RESPONSE (2 STUDIES)

Overall abundance (1 study): One before-and-after, site comparison study in the English Channel found that, over the three years after closing a marine protected area to all towed gears, the abundance of reef-indicative invertebrate species became greater than at unprotected fished sites.

Crustacean abundance (1 study): One replicated, site comparison study in the Bristol Channel and the Irish Sea found that a marine protected area closed to all towed gear (only allowing potting) for 33 to 36 years had mixed effects on the abundances of lobsters and crabs depending on species.

Crustacean condition (1 study): One replicated, site comparison study in the Bristol Channel and the Irish Sea found that a marine protected area closed to all towed gear (only allowing potting) for 33 to 36 years had mixed effects on the sizes of lobsters and crabs depending on species.
Assessment: unknown effectiveness (effectiveness 45%; certainty 23%; harms 10%).

https://www.conservationevidence.com/actions/2229

Designate a Marine Protected Area and prohibit aquaculture activity

One study examined the effects of prohibiting aquaculture activity in a protected area on subtidal benthic invertebrate populations. The study was in Tapong Bay lagoon (Taiwan).

COMMUNITY RESPONSE (0 STUDIES)

POPULATION RESPONSE (1 STUDY)

Crustacean abundance (1 study): One before-and-after study in Tapong Bay lagoon found that two and a half years after removing oyster aquaculture in a marine protected area, the biomasses of amphipods and shrimps had decreased, and that the biomass of crabs had not changed.

Mollusc abundance (1 study): One before-and-after study in Tapong Bay lagoon found that two and a half years after removing oyster aquaculture in a marine protected area, the biomasses of gastropods and bivalves had decreased.

Worm abundance (1 study): One before-and-after study in Tapong Bay lagoon found that two and a half years after removing oyster aquaculture in a marine protected area, the biomass of polychaete worms had stayed the same.
Assessment: unknown effectiveness (effectiveness 40%; certainty 10%; harms 20%).

https://www.conservationevidence.com/actions/2240

Designate a Marine Protected Area and prohibit bottom trawling

Three studies examined the effects of prohibiting bottom trawling in marine protected areas on subtidal benthic invertebrates. Two studies were in the South Pacific Ocean (Australia) and one in the Coral Sea (Australia).

COMMUNITY RESPONSE (3 STUDIES)

Overall community composition (2 studies): One of two replicated, site comparison studies in the South Pacific Ocean found that seamounts within a protected area closed to trawling had different invertebrate community composition compared to trawled seamounts and to never-trawled seamounts after four to nine years. The second study found that seamounts within a protected area closed to trawling had different invertebrate community composition compared to shallow unprotected seamounts (heavily trawled) after two years, but not compared to deep unprotected seamounts (lightly trawled).

Overall diversity/species richness (3 studies): One of two replicated, site comparison studies in the South Pacific Ocean found that seamounts within a protected area closed to trawling had similar invertebrate species richness and diversity to trawled seamounts and never-trawled seamounts after four to nine years. The second study found that seamounts within a protected area closed to trawling had more invertebrate species compared to shallow unprotected seamounts (heavily trawled) after two years, but not compared to deep unprotected seamounts (lightly trawled). One randomized, replicated, site comparison study in the Coral Sea found similar combined invertebrate and fish species richness in areas closed to trawling and adjacent fished areas, after seven to eight years.

POPULATION RESPONSE (3 STUDIES)

Overall abundance (3 studies): One of two replicated, site comparison studies in the South Pacific Ocean found that seamounts within a protected area closed to trawling had lower invertebrate biomass compared to trawled seamounts and never-trawled seamounts after four to nine years. The second study found that seamounts within a protected area closed to trawling had higher invertebrate biomass compared to shallow unprotected seamounts (heavily trawled) after two years, but not compared to deep unprotected seamounts (lightly trawled). One randomized, replicated, site comparison study in the Coral Sea found similar invertebrate and fish biomass in areas closed to trawling and adjacent fished areas, after seven to eight years.

Assessment: unknown effectiveness (effectiveness 50%; certainty 22%; harms 0%).

https://www.conservationevidence.com/actions/2226

Designate a Marine Protected Area and prohibit commercial fishing

Three studies examined the effects of prohibiting commercial fishing in marine protected areas on subtidal benthic invertebrates. Two studies were in the South Pacific Ocean (New Zealand), and one in the Caribbean Sea (Mexico).

COMMUNITY RESPONSE (0 STUDIES)

POPULATION RESPONSE (3 STUDIES)

Crustacean abundance (2 studies): Two replicated studies (one before-and-after, one site comparison) in the South Pacific Ocean found that after implementing a marine park prohibiting commercial fishing but allowing the recreational harvest of lobsters, lobster abundance inside the park did not increase over the 12 years after implementation, and abundance was similar inside the park and outside where fishing occurred.

Crustacean condition (3 studies): One replicated, before-and-after study in the South Pacific Ocean found that over the 12 years after implementing a marine park prohibiting commercial fishing but allowing the recreational harvest of lobsters, the biomass of legal-size lobsters inside the park did not increase. One of two site comparison studies (one replicated) in the South Pacific Ocean and the Caribbean Sea found bigger lobsters in an area closed to commercial fishing for an unspecified amount of time compared to a fished area. The second study found that 10 years after implementing a marine park prohibiting commercial fishing but allowing the recreational harvest of lobsters, lobster size was similar inside the park and outside where fishing occurred.

BEHAVIOUR (1 STUDY)

Crustacean behaviour (1 study): One site comparison study in the Caribbean Sea found that 80% of the lobster population occurring in a protected area (year of designation unspecified) where commercial fishing was prohibited remained in the unfished area, and thus remained protected.

Assessment: unknown effectiveness (effectiveness 35%; certainty 30%; harms 0%).

https://www.conservationevidence.com/actions/2225

Designate a Marine Protected Area and prohibit dredging

One study examined the effects of prohibiting dredging in marine protected areas on subtidal benthic invertebrates. The study was in the Firth of Lorn (UK).

COMMUNITY RESPONSE (1 STUDY)

Overall community composition (1 study): One paired, replicated, site comparison study in the Firth of Lorn found that sites inside a protected area that had been prohibiting dredging for approximately 2.5 years had different combined invertebrate and fish community composition compared to unprotected dredged sites.

POPULATION RESPONSE (1 STUDY)

Overall abundance (1 study): One paired, replicated, site comparison study in the Firth of Lorn found that sites inside a protected area that had been prohibiting dredging for approximately 2.5 years typically had greater combined cover of bryozoans and hydroids (combined) compared to unprotected dredged sites.

Assessment: unknown effectiveness (effectiveness 60%; certainty 20%; harms 0%).

https://www.conservationevidence.com/actions/2228

Designate a Marine Protected Area and prohibit the harvesting of conch

One study examined the effects of prohibiting the harvesting of conch in marine protected areas on their populations and/or other subtidal benthic invertebrates. The study was in the North Atlantic Ocean (British Overseas Territories).

COMMUNITY RESPONSE (0 STUDIES)

POPULATION RESPONSE (1 STUDY)

Mollusc abundance (1 study): One site comparison study in the North Atlantic Ocean found that a marine protected area prohibiting the commercial harvest of conch had more conch after five years compared to a fished area.

Mollusc condition (1 study): One site comparison study in the North Atlantic Ocean found that a marine protected area prohibiting the commercial harvest of conch had smaller adult conch after five years compared to a fished area.

Assessment: unknown effectiveness (effectiveness 50%; certainty 22%; harms 12%).

https://www.conservationevidence.com/actions/2237

Establish community-based fisheries management

One study examined the effects of establishing community-based fisheries management on subtidal benthic invertebrate populations. The study was in the Foveaux Straight (New Zealand).

COMMUNITY RESPONSE (0 STUDIES)

POPULATION RESPONSE (1 STUDY)

Mollusc abundance (1 study): One replicated, site comparison study in the Foveaux Straight found that a customary fisheries area where management was community-based had more New Zealand scallops compared to a protected area prohibiting all fishing and an area allowing recreational harvest.

Mollusc condition (1 study): One replicated, site comparison study in the Foveaux Straight found that a customary fisheries area where management

was community-based, tended to have smaller New Zealand scallops compared to a protected area prohibiting all fishing and an area allowing recreational harvest.
Assessment: unknown effectiveness (effectiveness 45%; certainty 21%; harms 0%).

https://www.conservationevidence.com/actions/2242

No evidence found (no assessment)

We have captured no evidence for the following interventions:

- Designate a Marine Protected Area and limit the density of traps
- Designate a Marine Protected Area and limit the number of fishing vessels
- Designate a Marine Protected Area and prohibit static fishing gear
- Designate a Marine Protected Area and prohibit the harvesting of scallops
- Designate a Marine Protected Area and set a no-anchoring zone
- Designate a Marine Protected Area without setting management measures, usage restrictions, or enforcement
- Designate a Particularly Sensitive Sea Area (PSSA) to regulate impactful maritime activities
- Engage with stakeholders when designing Marine Protected Areas.

13.9 Habitat restoration and creation

13.9.1 Natural habitat restoration

Based on the collated evidence, what is the current assessment of the effectiveness of interventions for natural habitat restoration?	
Likely to be beneficial	• Restore biogenic habitats (other methods) - Restore oyster reefs • Translocate habitat-forming (biogenic) species - Translocate reef-forming corals
Unknown effectiveness	• Install a pump on or above the seabed in docks, ports, harbour, or other coastal areas to increase oxygen concentration • Refill disused borrow pits • Restore biogenic habitats (other methods) - Restore mussel beds • Restore biogenic habitats (other methods) - Restore seagrass beds/meadows • Restore coastal lagoons • Translocate habitat-forming (biogenic) species - Translocate reef- or bed-forming molluscs
No evidence found (no assessment)	• Transplant captive-bred or hatchery-reared habitat-forming (biogenic) species

Likely to be beneficial

Restore biogenic habitats (other methods) - Restore oyster reefs

Eight studies examined the effects of restoring oyster reefs (not by transplanting or translocating oysters) on oysters and oyster reef-associated subtidal benthic invertebrates. Two were in the Gulf of Mexico (USA), one was a global review, four were in the North Pacific Ocean (USA), and one was in the Mission-Aransas estuary (USA).

COMMUNITY RESPONSE (3 STUDIES)

Overall community composition (2 studies): One of two replicated, controlled studies in the Gulf of Mexico and the Mission-Aransas estuary found that after restoring eastern oyster reefs, the community composition of combined mobile decapod invertebrates and fish was similar on all types of restoration material used, but the other found that composition varied with the material used.

Overall species richness/diversity (3 studies): One replicated, site comparison study in the Gulf of Mexico found that diversity of reef-associated invertebrates was similar in reefs restored by laying rocks regardless of age, in young reefs restored by laying oyster shells, and in natural reefs, but lower in old shell-restored reefs. One replicated, controlled study in the Gulf of Mexico found that diversity of reef-associated invertebrates was higher in all restored reefs than on unrestored sediment, but that diversity varied between the restoration materials used. One replicated, controlled study in the Mission-Aransas estuary found that diversity of fish, crabs and shrimps varied with the restoration material used.

POPULATION RESPONSE (7 STUDIES)

Overall abundance (2 studies): One replicated, site comparison study in the Gulf of Mexico found that the effect of restoring eastern oyster reefs on the abundance of reef-associated invertebrates depended on the material used for restoration and the age of the reef. One replicated, controlled study in the Gulf of Mexico found that abundance of combined reef-associated mobile decapod invertebrate and fish was similar on all restored reefs regardless of the restoration material used, and higher than on unrestored sediment.

Crustacean abundance (1 study): One replicated, controlled study in the Mission-Aransas estuary found that after restoring eastern oyster reefs, crab abundance, but not biomass, and shrimp biomass, but not abundance, varied with the restoration material used.

Oyster abundance (6 studies): One replicated, site comparison study in the Gulf of Mexico found that oyster reefs restored by laying rocks had similar oyster abundance to natural reefs, and higher than reefs restored by laying oyster shells. One replicated, controlled study in the Mission-Aransas estuary found that oyster cover and abundance varied with the restoration material used. One replicated, controlled study in the Gulf of Mexico found that oyster spat abundance was similar on all types of restoration material used, and higher than on unrestored sediment. Three replicated, controlled studies in the North Pacific Ocean found that restoring oyster reefs by placing lines of clam shells below Mean Lower Low Water (MLLW) led to higher cover of clam shells by oysters than when placing the lines above MLLW, that for those placed below MLLW, keeping them there led to similar cover compared to moving them above MLLW halfway through the study, and that placing the lines on cobbly seabed led to similar cover compared to placing them on muddy seabed.

Oyster reproductive success (3 studies): Three replicated, controlled studies in the North Pacific Ocean found that restoring oyster reefs by placing lines of clam shells below Mean Lower Low Water (MLLW) led to higher recruitment of oyster spat on clam shells than by lacing lines above MLLW, that recruitment was higher on lines placed on cobbly seabed than on muddy seabed, and that recruitment was similar on lines placed near or far from the nearest adult oyster populations.

Oyster survival (5 studies): One global systematic review found that two of nine restoration techniques (restoring oyster reef by transplanting juveniles, and by creating no-harvest sanctuaries) assessed resulted in over 85% survival of restored oysters. Four replicated, controlled studies in the North Pacific Ocean found that restoring oyster reefs by placing lines of clam shells below Mean Lower Low Water (MLLW) led to similar survival of oysters than when placing the lines above MLLW, but that for those placed below MLLW, moving them above MLLW halfway through the study led to higher survival than keeping then below, that survival was similar on lines placed on cobbly seabed or muddy seabed, and that survival was similar on lines placed near or far from the nearest adult oyster populations.

Oyster condition (5 studies): One replicated, controlled study in the Gulf of Mexico found that the effect of restoring eastern oyster reefs on average spat size varied with the restoration material used. One replicated, controlled study in the North Pacific Ocean found that restoring oyster reefs by placing lines

of clam shells below Mean Lower Low Water (MLLW) led to similar growth of oysters on the shells than placing lines above MLLW. Four replicated, controlled studies in the North Pacific Ocean found that restoring oyster reefs by placing lines of clam shells below Mean Lower Low Water (MLLW) led to higher cover of clam shells by non-native species than placing lines above MLLW, but that for those placed below MLLW, moving them above MLLW halfway through the study led to lower cover than keeping then below, that cover was similar on lines placed on cobbly seabed or muddy seabed, and that cover of clam shells by non-native species was higher on lines placed near compared to far from the nearest adult oyster populations.
Assessment: likely to be beneficial (effectiveness 60%; certainty 50%; harms 0%).

https://www.conservationevidence.com/actions/2248

Translocate habitat-forming (biogenic) species - Translocate reef-forming corals

Two studies examined the effects of translocating habitat-forming corals on associated subtidal benthic invertebrate populations. One was in Tayabas Bay (Philippines) and one in the South China Sea (Philippines).
COMMUNITY RESPONSE (2 STUDIES)
Overall community composition (1 study): One replicated, controlled, before-and-after study in the South China Sea found that following coral translocation associated invertebrate communities did not change and remained similar to plots without translocated corals.
Overall richness/diversity (2 studies): One replicated, controlled, before-and-after study in the South China Sea found that following coral translocation richness of associated invertebrates increased but also increased in plots without corals, likely due to spill-over. One replicated, controlled study in Tayabas Bay found that richness of associated invertebrates was higher in plots with translocated corals than in plots without.
POPULATION RESPONSE (1 STUDY)
Overall abundance (1 study): One replicated, controlled, before-and-after study in the South China Sea found that following coral translocation abundance of associated invertebrates increased and became higher than in plots without translocated corals.
Assessment: likely to be beneficial (effectiveness 65%; certainty 43%; harms 0%).

https://www.conservationevidence.com/actions/2246

Unknown effectiveness

Install a pump on or above the seabed in docks, ports, harbour, or other coastal areas to increase oxygen concentration

One study examined the effects of installing a pump on or above the seabed in docks, ports, harbour, or other coastal areas to increase oxygen concentration on subtidal benthic invertebrate populations. The study was in Osaka Bay (Japan).

COMMUNITY RESPONSE (1 STUDY)

Overall richness/diversity (1 study): One before-and-after study in Osaka Bay found that installing a pump on the seabed of a port to mix seawater and increase oxygen concentration led to an increase in combined invertebrate and fish species richness.

POPULATION RESPONSE (1 STUDY)

Overall abundance (1 study): One before-and-after study in Osaka Bay found that installing a pump on the seabed of a port to mix seawater and increase oxygen concentration led to an increase in combined invertebrates and fish abundance.

Assessment: unknown effectiveness (effectiveness 75%; certainty 20%; harms 0%).

https://www.conservationevidence.com/actions/2252

Refill disused borrow pits

One study examined the effects of refilling disused borrow pits on subtidal benthic invertebrate populations. The study was in Barnegat Bay estuary (USA).

COMMUNITY RESPONSE (1 STUDY)

Overall richness/diversity (1 study): One before-and-after, site comparison study in Barnegat Bay estuary found that overall invertebrate species richness and diversity increased at a disused borrow pit after being refilled with sediments but remained lower than at a natural non-dredged site.

POPULATION RESPONSE (1 STUDY)

Overall abundance (1 study): One before-and-after, site comparison study in Barnegat Bay estuary found that overall invertebrate abundance increased at a disused borrow pit after being refilled with sediments but remained lower than at a natural non-dredged site.

Assessment: unknown effectiveness (effectiveness 60%; certainty 32%; harms 0%).

https://www.conservationevidence.com/actions/2251

Restore biogenic habitats (other methods) - Restore mussel beds

Two studies examined the effects of restoring mussel beds (not by transplanting or translocating mussels) on mussels and mussel bed-associated subtidal benthic invertebrates. Both were in Strangford Lough (UK).

COMMUNITY RESPONSE (2 STUDIES)

Overall community composition (2 studies): One replicated, controlled study in Strangford Lough found that after restoring beds of horse mussels by adding scallop shells to the seabed, overall invertebrate community composition in restored plots was different to that of unrestored plots. One replicated, controlled study in the same area found that after restoring beds of horse mussels by adding scallop shells to the seabed and translocating horse mussels, overall invertebrate community composition in plots restored with shells and mussels was different to plots restored without mussels (shells only), and both were different to unrestored plots and to nearby natural horse mussel beds.

Overall species richness/diversity (2 studies): One replicated, controlled study in Strangford Lough found that after restoring beds of horse mussels by adding scallop shells to the seabed, overall invertebrate species diversity was lower in restored plots compared to unrestored plots, but species richness was similar. One replicated, controlled study in the same area found that after restoring beds of horse mussels by adding scallop shells to the seabed and translocating horse mussels, species richness and diversity were higher in restored plots with mussels and shells compared to plots with shells only, and similar to nearby natural horse mussel beds.

POPULATION RESPONSE (1 STUDY)

Overall abundance (1 study): One replicated, controlled study in Strangford Lough found that after restoring beds of horse mussels by adding scallop shells to the seabed, overall invertebrate abundance was higher in restored plots compared to unrestored plots.

Assessment: unknown effectiveness (effectiveness 55%; certainty 30%; harms 5%).

https://www.conservationevidence.com/actions/2247

Restore biogenic habitats (other methods) - Restore seagrass beds/meadows

Three studies examined the effects of restoring seagrass beds (not by transplanting or translocating seagrass) on seagrass bed-associated subtidal benthic invertebrates. One was in the North Atlantic Ocean (USA), one in the Indian Ocean (Kenya), and one in the Florida Keys (USA).

COMMUNITY RESPONSE (2 STUDIES)

Overall community composition (1 study): One randomized, replicated, controlled study in the Florida Keys found that restoring seagrass beds by fertilizing the seabed had no effect on overall invertebrate community composition, but adding sand led to communities different from both unrestored and natural sites.

Overall species richness/diversity (2 studies): One randomized, replicated, controlled study in the Florida Keys found that after restoring seagrass beds by fertilizing the seabed and adding sand, overall invertebrate species richness was similar at restored, unrestored, and natural sites. One replicated, controlled study in the Indian Ocean found that transplanting plastic seagrass mimics into bare sites, previously-restored seagrass sites, and natural seagrass sites, resulted in similar invertebrate diversity on mimic leaves and in the surrounding sediment, and similar species richness on mimic leaves at all restored sites as on natural seagrass leaves.

POPULATION RESPONSE (3 STUDIES)

Overall abundance (3 studies): One replicated, randomized, controlled, before-and-after study in the North Atlantic Ocean found that after restoring seagrass beds, the abundance of mobile invertebrates had increased and was higher in restored than unrestored plots, but the abundance of sessile invertebrates had not increased. One replicated, controlled study in the Indian Ocean found that transplanting plastic seagrass mimics into bare sites, previously-restored seagrass sites, and natural seagrass sites, resulted in similar abundance of invertebrate in the surrounding sediment across sites, and resulted in different abundance of invertebrates on mimic leaves between sites although all had lower abundances than on natural seagrass leaves. One randomized, replicated, controlled study in the Florida Keys found that after restoring seagrass beds by fertilizing the seabed or adding sand, overall invertebrate abundance was not different at restored sites compared to both unrestored and natural sites.

Assessment: unknown effectiveness (effectiveness 40%; certainty 30%; harms 0%).

https://www.conservationevidence.com/actions/2249

Restore coastal lagoons

Three studies examined the effects restoring coastal lagoons on subtidal benthic invertebrate populations. One study was in the Chilika lagoon (India), and two in East Harbor lagoon (USA).

COMMUNITY RESPONSE (3 STUDIES)

Crustacean richness/diversity (1 study): One before-and-after study in Chilika lagoon found that following hydrological restoration total crustacean species richness decreased, but changes varied with species groups (decreases in prawn and crab species; increases in lobster species). The lagoon also hosted new species not found before.

Mollusc richness/diversity (2 studies): Two studies in East Harbor lagoon found that following hydrological restoration molluscs recolonised the lagoon and their species richness increased in the first three years but later decreased over the following six.

POPULATION RESPONSE (3 STUDIES)

Crustacean abundance (1 study): One before-and-after study in Chilika lagoon found that following hydrological restoration abundances of prawns and crabs increased.

Mollusc abundance (2 studies): Two studies in East Harbor lagoon found that following hydrological restoration molluscs recolonised the lagoon and their total abundance increased in the first three years, but later decreased over the following six.

Assessment: unknown effectiveness (effectiveness 40%; certainty 28%; harms 20%).

https://www.conservationevidence.com/actions/2250

Translocate habitat-forming (biogenic) species - Translocate reef- or bed-forming molluscs

Two studies examined the effects of translocating habitat-forming molluscs on associated subtidal benthic invertebrate populations. Both were in Strangford Lough (UK).

COMMUNITY RESPONSE (2 STUDIES)

Overall community composition (2 studies): One replicated, site comparison study in Strangford Lough found that plots with translocated mussels had different associated invertebrate communities to plots without mussels, but also to natural mussel beds. One replicated, controlled study in Strangford Lough found that translocating mussels onto scallop shells or directly onto the seabed led to similar associated invertebrate communities.

Overall richness/diversity (2 studies): One replicated, site comparison study in Strangford Lough found that plots with translocated mussels had higher richness and diversity of associated invertebrates to plots without mussels, and similar to natural mussel beds. One replicated, controlled study in Strangford Lough found that translocating mussels onto scallop shells or directly onto the seabed led to similar richness and diversity of associated invertebrates.

POPULATION RESPONSE (2 STUDIES)

Overall abundance (2 studies): One replicated, site comparison study in Strangford Lough presented unclear abundance results. One replicated, controlled study in Strangford Lough found that translocating mussels onto scallop shells or directly onto the seabed led to higher abundance of associated invertebrates in one of two comparisons.

Assessment: unknown effectiveness (effectiveness 65%; certainty 35%; harms 0%).

https://www.conservationevidence.com/actions/2245

No evidence found (no assessment)

We have captured no evidence for the following interventions:

- Transplant captive-bred or hatchery-reared habitat-forming (biogenic) species.

13.9.2 Habitat enhancement

Based on the collated evidence, what is the current assessment of the effectiveness of interventions for habitat enhancement?	
Beneficial	• Provide artificial shelters
Unknown effectiveness	• Landscape or artificially enhance the seabed (natural habitats)
No evidence found (no assessment)	• Use green engineering techniques on artificial structures - Cover subsea cables with artificial reefs • Use green engineering techniques on artificial structures - Cover subsea cables with materials that encourage the accumulation of natural sediments • Use green engineering techniques on artificial structures - Modify rock dump to make it more similar to natural substrate

Beneficial

Provide artificial shelters

Five studies examined the effects of providing artificial shelters on subtidal benthic invertebrates. Three studies were in the Caribbean Sea (Mexico); one in Florida Bay and one in the Florida Keys (USA).

COMMUNITY RESPONSE (0 STUDIES)

POPULATION RESPONSE (2 STUDIES)

Lobster abundance (2 studies): Two replicated, controlled, before-and-after studies in the Caribbean Sea found that abundance of lobsters either increased in plots with artificial shelters but not in plots without, or increased in all plots but more so in plots with artificial shelters than those without.

Lobster condition (1 study): One replicated, controlled, before-and-after study in the Caribbean Sea found that lobsters in plots with artificial shelters were bigger than in plots without.

BEHAVIOUR (3 STUDIES)

Use (3 studies): Three replicated studies (two controlled) in Florida Bay, the Florida Keys, and the Caribbean Sea, found that artificial shelters were occupied by lobsters and molluscs, that occupancy by lobsters varied with artificial shelter designs, that lobsters occupied artificial shelters more than natural ones (crevices), and that lobsters occupying artificial shelters were larger, had greater nutritional condition, and had similar sex ratio and survival rate, compared to lobsters occupying natural shelters.

Assessment: beneficial (effectiveness 70%; certainty 63%; harms 0%).

https://www.conservationevidence.com/actions/2257

Unknown effectiveness

Landscape or artificially enhance the seabed (natural habitats)

Three studies examined the effects of landscaping or artificially enhancing the seabed on subtidal benthic invertebrates. One study was in the North Sea (UK), one in the Westerschelde estuary (Netherlands), and one in the Persian Gulf (Kuwait).

COMMUNITY RESPONSE (3 STUDIES)

Overall community composition (2 studies): One controlled, before-and after study in the North Sea found that following addition of gravels, invertebrate community composition became more similar to natural seabed communities. One before-and-after, site comparison study in the Westerschelde estuary found no change in invertebrate community composition following addition of sedimentary dredge material.

Overall richness/diversity (3 studies): One controlled, before-and after study in the North Sea and one site comparison study in the Persian Gulf found that invertebrate species richness increased following addition of gravels or coral and limestone rubbles, and one also found that richness became similar to natural seabed. One before-and-after, site comparison study in the Westerschelde estuary found no change in species richness following addition of sedimentary dredged material.

POPULATION RESPONSE (3 STUDIES)

Overall abundance (3 studies): One controlled, before-and after study in the North Sea and one site comparison study in the Persian Gulf found that invertebrate abundance and biomass increased following addition of gravels or coral and limestone rubbles, and one also found that abundance became similar to natural seabed. One before-and-after, site comparison study in the Westerschelde estuary found no change in invertebrate abundance and biomass following addition of sedimentary dredge material.

Assessment: unknown effectiveness (effectiveness 50%; certainty 35%; harms 5%).

https://www.conservationevidence.com/actions/2253

No evidence found (no assessment)

We have captured no evidence for the following interventions:

- Use green engineering techniques on artificial structures - Cover subsea cables with artificial reefs
- Use green engineering techniques on artificial structures - Cover subsea cables with materials that encourage the accumulation of natural sediments
- Use green engineering techniques on artificial structures - Modify rock dump to make it more similar to natural substrate.

13.9.3 Artificial habitat creation

Based on the collated evidence, what is the current assessment of the effectiveness of interventions for artificial habitat creation?	
Likely to be beneficial	• Create artificial reefs • Create artificial reefs of different 3-D structure and material used
Unknown effectiveness	• Locate artificial reefs near aquaculture systems to benefit from nutrient run-offs • Repurpose obsolete offshore structures to act as artificial reefs
No evidence found (no assessment)	• Place anthropogenic installations (e.g. windfarms) in an area such that they create artificial habitat and reduce the level of fishing activity

Likely to be beneficial

Create artificial reefs

Twelve studies examined the effects of creating artificial reefs on subtidal benthic invertebrate populations. Three studies were in the Mediterranean Sea (Italy); three were in the North Atlantic Ocean (USA, Portugal, France); one in the Firth of Lorn (UK); two in the North Pacific Ocean (USA); one in the English Channel (UK), one in the Gulf of Mexico (USA); and one in the Yellow Sea (China).

COMMUNITY RESPONSE (8 STUDIES)

Overall community composition (3 studies): Two site comparison studies (one replicated) in the English Channel and North Atlantic Ocean found that invertebrate communities growing on artificial reefs were different to that of natural reefs. One replicated study the North Pacific Ocean found that invertebrate community composition changed over time on an artificial reef. Overall richness/diversity (6 studies): Two site comparison studies (one replicated) in the Mediterranean Sea and North Atlantic Ocean found that invertebrate species richness and/or diversity on the artificial reef or in the sediments inside and adjacent to the reef area were lower compared to on natural reefs or in nearby natural sediments. One replicated, site comparison study in the Gulf of Mexico found that artificial breakwaters had more species of nekton compared to adjacent mudflats. One site comparison study in English Channel recorded 263 taxa on the artificial reef, including at least nine not recorded on nearby natural reefs but excluding at least 39 recorded on natural reefs. One replicated study in the North Pacific Ocean found a 49% increase in species richness over five years on an artificial reef. One study in the North Atlantic Ocean found that artificial reefs hosted at least five species of large mobile invertebrates.

Mollusc richness/diversity (1 study): One replicated, site comparison study in the Mediterranean Sea found that mollusc species richness and diversity were lower on artificial reefs compared to natural reefs.

Worm community composition (1 study): One replicated, site comparison study in the North Pacific Ocean found that polychaete worm community composition was similar at one of two artificial reefs compared to a natural reef.

Worm richness/diversity (1 study): One replicated, site comparison study in the North Pacific Ocean found that polychaete worm species richness and diversity were similar at one of two artificial reefs compared to a natural reef, but lower at the second artificial reef.

POPULATION RESPONSE (12 STUDIES)

Overall abundance (10 studies): One of two site comparison studies (one replicated) in the Mediterranean Sea found that abundance of invertebrates in the sediment was lower at the reef sites than in nearby natural sediments, but increased in the sediments directly adjacent to the reefs, while the other study found that abundance was similar in the sediments inside and directly adjacent to the artificial reef area, but lower than in nearby natural sediments. Of five site comparison studies (four replicated) in the North Pacific Ocean, the North Atlantic Ocean, the Gulf of Mexico and the Yellow Sea, one found that invertebrate biomass was higher on the artificial reef than in adjacent natural sediments, two that invertebrate abundance and biomass and nekton abundance were similar on artificial reefs and natural habitats (reef; mudflat), and two found mixed effects on abundances of invertebrates. One site comparison study in the English Channel reported that the abundances of some species were lower on the artificial reef compared to natural reefs. One replicated study in the North Pacific Ocean reported an 86% increase in invertebrate abundance growing on an artificial reef over five years. One study in the North Atlantic Ocean found that two of five species at one artificial reef, and three of seven at another, were recorded during >50% of dives.

Overall condition (1 study): One replicated, site comparison study in the Yellow Sea found mixed effects of creating an artificial reef on the sizes of mobile invertebrates.

Mollusc abundance (1 study): One replicated, site comparison study in the Mediterranean Sea found that mollusc abundance was lower on artificial reefs compared to natural reefs.
Crustacean abundance (1 study): One replicated, site comparison in the Firth of Lorn found that abundances of edible crabs and velvet swimming crabs were typically higher on artificial than natural reefs.
OTHER (1 STUDY)
Biological production (1 study): One site comparison study in North Atlantic Ocean found that secondary production was higher from invertebrates growing on an artificial reef than from invertebrates in adjacent natural sediments.
Assessment: likely to be beneficial (effectiveness 55%; certainty 60%; harms 0%).

https://www.conservationevidence.com/actions/2258

Create artificial reefs of different 3-D structure and material used

Eight studies examined the effects of creating artificial reefs of different typology on subtidal benthic invertebrate populations. One study was in the English Channel (UK), three in the Mediterranean Sea (Israel, Italy), one in the North Atlantic Ocean (USA), one in the Firth of Lorn (UK), one in the North Pacific Ocean (USA), and one in the Gulf of Mexico (USA).
COMMUNITY RESPONSE (6 STUDIES)
Overall community composition (3 studies): One controlled study in the English Channel found that artificial reef modules made of scrap tyres developed a similar sessile invertebrate community composition as traditional artificial concrete modules. Two controlled studies (one replicated) in the Mediterranean Sea found that pyramids reefs made of "sea-friendly" concrete developed different invertebrate community compositions compared to reefs of either traditional concrete plinth-pole structures or bundles of traditional concrete tubes.
Overall richness/diversity (5 studies): Four controlled studies (three replicated) in the Mediterranean Sea, the North Pacific Ocean, and the Gulf of Mexico found no differences in overall invertebrate richness/diversity or combined mobile invertebrate and fish richness between reef structure and/or material. One controlled study in the Mediterranean Sea found that invertebrate species richness was lower on "sea-friendly" pyramid reefs compared to bundle reefs of traditional concrete.
POPULATION RESPONSE (7 STUDIES)
Overall abundance (5 studies): Four controlled studies (three replicated) in the English Channel, the Mediterranean Sea, the North Pacific Ocean, and

the Gulf of Mexico found no differences in overall invertebrate abundances or combined mobile invertebrate and fish abundance between reef structure and/or material. One controlled study in the Mediterranean Sea found that "sea-friendly" concrete pyramids had lower abundance compared to plinth-pole structures after two years, but higher after three.

Crustacean abundance (2 studies): One replicated, controlled study in the North Atlantic Ocean found that artificial reefs made of limestone boulders, gravel concrete aggregate, or tyre-concrete aggregate had similar abundance of spiny lobsters. One replicated, controlled study in the Firth of Lorn found that the complexity of artificial reef modules had mixed effects on the abundance of edible crab and velvet swimming crab.

Mollusc abundance (1 study): One replicated, controlled study in the Gulf of Mexico found that breakwaters made of bags of oyster shells recruited more oysters and ribbed mussels compared to "ReefBall" breakwaters.

Assessment: likely to be beneficial (effectiveness 43%; certainty 40%; harms 10%).

https://www.conservationevidence.com/actions/2259

Unknown effectiveness

Locate artificial reefs near aquaculture systems to benefit from nutrient run-offs

Two studies examined the effects of locating artificial reefs near aquaculture systems to benefit from nutrient run-offs on subtidal benthic invertebrate populations. One study was in the Gulf of Aqaba (Israel and Jordan), and one in the Mediterranean Sea (Spain).

COMMUNITY RESPONSE (1 STUDY)

Overall community composition (1 study): One controlled study in the Mediterranean Sea found that an artificial reef located under aquaculture cages had similar invertebrate community composition to artificial reefs located at sites without aquaculture cages.

POPULATION RESPONSE (1 STUDY)

Overall abundance (1 study): One controlled study in the Gulf of Aqaba found that an artificial reef located at an aquaculture site had similar invertebrate biomass growing on it compared to an artificial reef located at a site without aquaculture cages.

Assessment: unknown effectiveness (effectiveness 35%; certainty 24%; harms 0%).

https://www.conservationevidence.com/actions/2260

Repurpose obsolete offshore structures to act as artificial reefs

One study examined the effects of repurposing obsolete offshore structures on subtidal benthic invertebrates. The study was of a sunken oil rig in the Mediterranean Sea (Italy).

COMMUNITY RESPONSE (1 STUDY)

Overall species richness/diversity (1 study): One study in the Mediterranean Sea recorded at least 53 invertebrate species having colonised a sunken oil rig after 30 years. Species included 14 species of molluscs, 14 species of worms, and 11 species of crustaceans.

POPULATION RESPONSE (0 STUDIES)

Assessment: unknown effectiveness (effectiveness 65%; certainty 26%; harms 0%).

https://www.conservationevidence.com/actions/2262

No evidence found (no assessment)

We have captured no evidence for the following interventions:

- Place anthropogenic installations (e.g. windfarms) in an area such that they create artificial habitat and reduce the level of fishing activity.

13.9.4 Other habitat restoration and creation interventions

Based on the collated evidence, what is the current assessment of the effectiveness of interventions for other habitat restoration and creation interventions?	
Unknown effectiveness	• Offset habitat loss from human activity by restoring or creating habitats elsewhere • Remove and relocate habitat-forming (biogenic) species before onset of impactful activities
No evidence found (no assessment)	• Pay monetary compensation for habitat damage remediation

Unknown effectiveness

Offset habitat loss from human activity by restoring or creating habitats elsewhere

Two studies examined the effects of offsetting habitat loss from human activity by restoring or creating habitats elsewhere on subtidal benthic invertebrate populations. One study was in the Delaware Bay (USA), the other in the Persian Gulf (Kuwait).

COMMUNITY RESPONSE (1 STUDY)

Overall richness/diversity (1 study): One study in the Persian Gulf found that an area of low ecological value restored to offset habitat lost to land reclamation was colonized by over 198 invertebrate species.

POPULATION RESPONSE (0 STUDIES)

OTHER (1 STUDY)

Biological production (1 study): One study in Delaware Bay found that an artificial reef built to offset lost soft-sediment habitat had higher annual secondary production/unit area from sessile invertebrates, but lower total annual secondary production, compared to habitat similar to that lost.

Assessment: unknown effectiveness (effectiveness 45%; certainty 20%; harms 0%).

https://www.conservationevidence.com/actions/2265

Remove and relocate habitat-forming (biogenic) species before onset of impactful activities

One study examined the effects of removing and relocating habitat-forming species before onset of impactful activities on subtidal benthic invertebrates. The study was in the Fal Estuary (UK).

COMMUNITY RESPONSE (1 STUDY)

Overall community composition (1 study): One replicated, paired, controlled study in the Fal Estuary found that invertebrate community composition was different in plots where maërl bed habitat had been removed and relayed compared to undisturbed maërl after five weeks, but similar after 44 weeks.

Overall species richness/diversity (1 study): One replicated, paired, controlled study in the Fal Estuary found that invertebrate species richness was lower in plots where maërl bed habitat had been removed and relayed compared to undisturbed maërl after five weeks, but similar after 44 weeks.

POPULATION RESPONSE (1 STUDY)

Overall abundance (1 study): One replicated, paired, controlled study in the Fal Estuary found that invertebrate abundance was different in plots where maërl bed habitat had been removed and relayed compared to undisturbed maërl after five weeks, but similar after 44 weeks.

Assessment: unknown effectiveness (effectiveness 55%; certainty 20%; harms 0%).

https://www.conservationevidence.com/actions/2264

No evidence found (no assessment)

We have captured no evidence for the following interventions:

- Pay monetary compensation for habitat damage remediation.

13.10 Species management

Based on the collated evidence, what is the current assessment of the effectiveness of interventions for species management?	
Likely to be beneficial	• Translocate species - Translocate molluscs • Transplant/release captive-bred or hatchery-reared species - Transplant/release crustaceans • Transplant/release captive-bred or hatchery-reared species - Transplant/release molluscs
Unknown effectiveness	• Cease or prohibit the harvesting of scallops • Tag species to prevent illegal fishing or harvesting • Translocate species - Translocate crustaceans • Translocate species - Translocate worms • Transplant/release captive-bred or hatchery-reared species in predator exclusion cages
No evidence found (no assessment)	• Cease or prohibit the harvest of conch • Cease or prohibit the harvest of sea urchins • Establish size limitations for the capture of recreational species • Provide artificial shelters following release • Remove and relocate invertebrate species before onset of impactful activities • Set recreational catch quotas

Likely to be beneficial

Translocate species - Translocate molluscs

Nine studies examined the effects of translocating mollusc species on their wild populations. Two examined scallops in the North Atlantic Ocean (USA) and one examined scallops in the Tasman Sea and South Pacific Ocean (New Zealand). One study examined conch in the Florida Keys (USA). One examined clams in the North Atlantic Ocean (Portugal). One examined abalone in the North Pacific Ocean (USA). One examined mussels in Strangford Lough (UK). Two examined mussels in the Gulf of Corinth (Greece).

COMMUNITY RESPONSE (0 STUDIES)

POPULATION RESPONSE (8 STUDIES)

Mollusc abundance (3 studies): One replicated, controlled, before-and-after study in the North Atlantic Ocean found that translocating bay scallops increased larval recruitment into the adult population compared to before translocation. One before-and-after study in the North Pacific Ocean found that following translocation of adult pink abalone to existing patchy populations, total abalone abundance (translocated and resident) decreased to similar levels as before translocation. One replicated, site comparison study in Strangford Lough found that after translocating horse mussels, the abundance of young mussels was higher in site with translocated mussels compared to both sites without translocated mussels and natural mussel reefs.

Mollusc reproductive success (1 study): One replicated, controlled, before-and-after study in the North Atlantic Ocean found that translocating bay scallops did not increase larval production compared to before translocation.

Mollusc survival (5 studies): Three replicated studies (one before-and-after and two site comparisons) in the North Atlantic Ocean and in the Tasman Sea and South Pacific Ocean, found that following translocation, scallops and clams survived. Survival of translocated New Zealand scallops was higher in areas closed to commercial fishing compared to fished areas. Two studies in the Gulf of Corinth found that Mediterranean fan mussels survived when translocated to a deep site, and had similar survival compared to naturally-occurring mussels, but did not survive when translocated to a shallow site.

Mollusc condition (2 studies): One replicated, site comparison study in the North Atlantic Ocean found that following translocation, clams had similar condition indices to clams in the source site. One study in the Gulf of Corinth found that translocated Mediterranean fan mussels had similar size-specific growth-rates compared to naturally-occurring mussels.

BEHAVIOUR (1 STUDY)

Mollusc behaviour (1 study): One replicated study in the Florida Keys found that translocating non-reproductive adult queen conch to aggregations of reproductive conch did not have adverse effects on the movement patterns of non-translocated resident conch, and all conch displayed similar total distance travelled, movement rates, migration patterns, home-range sizes, and sociability.

Assessment: likely to be beneficial (effectiveness 60%; certainty 60%; harms 10%).

https://www.conservationevidence.com/actions/2270

Transplant/release captive-bred or hatchery-reared species - Transplant/release crustaceans

Five studies examined the effects of transplanting or releasing hatchery-reared crustacean species on their wild populations. Four examined lobsters in the North Sea (Germany, Norway, UK), and one examined prawns in the Swan-Canning Estuary (Australia).

COMMUNITY RESPONSE (0 STUDIES)

POPULATION RESPONSE (5 STUDIES)

Crustacean abundance (1 study): One study in the Swan-Canning Estuary found that after releasing hatchery-reared prawn larvae into the wild, the abundance of egg-bearing female prawns increased.

Crustacean reproductive success (3 studies): Two studies (one controlled) in the North Sea found that after their release, recaptured hatchery-reared female lobsters carried eggs, and the number, size and developmental stage of eggs were similar to that of wild females. One study in the Swan-Canning Estuary found that after releasing hatchery-reared prawn larvae into the wild the overall population fecundity (egg production/area) increased.

Crustacean survival (2 studies): Two studies in the North Sea found that 50–84% and 32–39% of hatchery-reared lobsters survived in the wild after release, up to eight and up to five years, respectively.

Crustacean condition (4 studies): Two studies in the North Sea found that hatchery-reared lobsters grew in the wild after release. One controlled study in the North Sea found that after release into the wild, hatchery-reared female lobsters had similar growth rates as wild females. One study in the North Sea found that after releasing hatchery-reared lobsters, no recaptured lobsters displayed signs of "Black Spot" disease, and 95% had developed a crusher-claw. One study in the Swan-Canning Estuary found that after releasing hatchery-reared prawn larvae into the wild, the size of egg-bearing female prawns increased.

BEHAVIOUR (1 STUDY)

Crustacean movement (1 study): One controlled study in the North Sea found that after release into the wild, hatchery-reared female lobsters had similar movement patterns as wild females.

Assessment: likely to be beneficial (effectiveness 70%; certainty 45%; harms 0%).

https://www.conservationevidence.com/actions/2266

Transplant/release captive-bred or hatchery-reared species - Transplant/release molluscs

Eight studies examined the effects of transplanting or releasing hatchery-reared mollusc species on their wild populations. One examined abalone in the North Pacific Ocean (Canada), one examined clams off the Strait of Singapore (Singapore), one examined oysters in the North Atlantic Ocean (USA), and four examined scallops in the North Atlantic Ocean and Gulf of Mexico (USA).

COMMUNITY RESPONSE (0 STUDIES)

POPULATION RESPONSE (8 STUDIES)

Mollusc abundance (2 studies): One replicated, before-and-after study in the North Atlantic Ocean found that after transplanting hatchery-reared scallops, abundance of juvenile scallops typically increased, but not that of adult scallops. Two replicated, randomized, controlled studies in the North Atlantic Ocean, found that after releasing hatchery-reared oyster larvae, more spat initially settled using a direct technique compared to a traditional remote technique, and equal number of spat settled on cleaned and natural oyster shells.

Mollusc reproductive success (1 study): One replicated, before-and-after study in the North Atlantic Ocean found that after transplanting hatchery-reared scallops, larval recruitment increased across all areas studied.

Mollusc survival (5 studies): One replicated study in the Strait of Singapore found that, after transplantation in the field, aquarium-reared clams had a high survival rate. One replicated, controlled study in the North Atlantic Ocean found that after transplanting hatchery-reared scallops, the number of transplanted scallops surviving decreased regardless of the methods used, and maximum mortalities was reported to be 0–1.5%. One replicated, controlled study in the North Pacific Ocean found that transplanting hatchery-reared abalone into the wild reduced survivorship compared to non-transplanted hatchery-reared abalone kept in tanks. Two replicated, randomized, controlled studies in the North Atlantic Ocean found that after releasing hatchery-reared oyster larvae, 61% of the settled spat survived the winter, and settled spat survived equally on cleaned and natural oyster shells.

Mollusc condition (3 studies): Two replicated studies in the Strait of Singapore and the North Atlantic Ocean found after transplantation in the wild, aquarium-reared clams and hatchery-reared scallops increased in weight and/or grew. Scallops grew in both free-planted plots and suspended bags but grew more in free-planted plots. One replicated, before-and-after study in the Gulf of Mexico found that after transplanting hatchery-reared scallops, wild populations had not become genetically more similar to hatchery-reared scallops. One replicated, controlled study in the North Atlantic Ocean found that after transplanting hatchery-reared scallops, free-planted scallops developed less shell biofouling than suspended scallops.

Assessment: likely to be beneficial (effectiveness 45%; certainty 40%; harms 15%).

https://www.conservationevidence.com/actions/2267

Unknown effectiveness

Cease or prohibit the harvesting of scallops

Three studies examined the effects of ceasing or prohibiting the harvesting of scallops on their populations. One study was in the South Atlantic Ocean (Argentina), one in the English Channel (UK) and one in the Irish Sea (UK).

COMMUNITY RESPONSE (0 STUDIES)

POPULATION RESPONSE (3 STUDIES)

Scallop abundance (3 studies): Two of three site comparison studies (one replicated, one before-and-after) in the South Atlantic Ocean, the English Channel, and the Irish Sea found that in areas where scallop harvesting had stopped scallop abundance was similar, and one found that scallop biomass was higher, compared to harvested areas.

Assessment: unknown effectiveness (effectiveness 45%; certainty 33%; harms 0%).

https://www.conservationevidence.com/actions/2277

Tag species to prevent illegal fishing or harvesting

One study examined the effects of tagging species to prevent illegal fishing or harvesting on subtidal benthic invertebrates. The study examined the effects on the Californian abalone fishery (USA).

COMMUNITY RESPONSE (0 STUDIES)

POPULATION RESPONSE (0 STUDIES)
BEHAVIOURS (1 STUDY)
Behaviour-change (1 study): One before-and-after study in California found no significant reduction in non-compliance with daily quotas of abalones after introducing tagging regulations.
OTHER (1 STUDY)
Illegal catch (1 study): One before-and-after study in California found no significant reduction in illegal takes of abalones after introducing tagging regulations.
Assessment: unknown effectiveness (effectiveness 20%; certainty 22%; harms 0%).

https://www.conservationevidence.com/actions/2275

Translocate species - Translocate crustaceans

One study examined the effects of translocating crustacean species on their wild populations. The study took place in the Tasman Sea (Australia).
COMMUNITY RESPONSE (0 STUDIES)
POPULATION RESPONSE (1 STUDY)
Crustacean survival (1 study): One study in the Tasman Sea found that following translocation survival of southern rock lobsters was similar to that of resident lobsters.
Assessment: unknown effectiveness (effectiveness 75%; certainty 24%; harms 0%).

https://www.conservationevidence.com/actions/2269

Translocate species - Translocate worms

One study examined the effects of translocating worm species on their wild populations. The study was in Scottish Lochs (UK).
COMMUNITY RESPONSE (0 STUDIES)
POPULATION RESPONSE (1 STUDY)
Worm survival (1 study): One replicated, controlled study in Scottish Lochs found that no reef-forming red tube worm survived when translocated to a new Loch, but survival was high when worms were translocated back to its source Loch.
Worm condition (1 study): One replicated, controlled study in Scottish Lochs found that no reef-forming red tube worm survived and so no growth was recorded when translocated to a new loch, worms translocated back to its source Loch grew.
Assessment: unknown effectiveness (effectiveness 10%; certainty 25%; harms 15%).

https://www.conservationevidence.com/actions/2271

Transplant/release captive-bred or hatchery-reared species in predator exclusion cages

One study examined the effects of transplanting or releasing hatchery-reared species in predator exclusion cages on their wild populations. The study was in the North Pacific Ocean (Canada).

COMMUNITY RESPONSE (0 STUDIES)

POPULATION RESPONSE (1 STUDY)

Survival (1 study): One replicated, controlled study the North Pacific Ocean found that hatchery-reared abalone transplanted in predator exclusion cages had similar survivorship following release compared to those transplanted directly onto the seabed.

Assessment: unknown effectiveness (effectiveness 35%; certainty 26%; harms 0%).

https://www.conservationevidence.com/actions/2268

No evidence found (no assessment)

We have captured no evidence for the following interventions:

- Cease or prohibit the harvest of conch
- Cease or prohibit the harvest of sea urchins
- Establish size limitations for the capture of recreational species
- Provide artificial shelters following release
- Remove and relocate invertebrate species before onset of impactful activities
- Set recreational catch quotas.

13.11 Education and awareness

Based on the collated evidence, what is the current assessment of the effectiveness of interventions for education and awareness?	
Unknown effectiveness	• Provide educational or other training programmes about the marine environment to improve behaviours towards marine invertebrates
No evidence found (no assessment)	• Organise educational marine wildlife tours to improve behaviours towards marine invertebrates

Unknown effectiveness

Provide educational or other training programmes about the marine environment to improve behaviours towards marine invertebrates

One study examined the effects of providing educational or other training programmes about the marine environment on subtidal benthic invertebrate populations. The study took place in Hong Kong.

COMMUNITY RESPONSE (0 STUDIES)

POPULATION RESPONSE (0 STUDIES)

BEHAVIOUR (1 STUDY)

Behaviour change (1 study): One replicated, before-and-after survey study in Hong Kong found that a conservation education programme on the Asian horseshoe crab in secondary schools significantly increased the students' behaviour towards Asian horseshoe crab conservation.

Assessment: unknown effectiveness (effectiveness 65%; certainty 21%; harms 0%).

https://www.conservationevidence.com/actions/2281

No evidence found (no assessment)

We have captured no evidence for the following interventions:

- Organise educational marine wildlife tours to improve behaviours towards marine invertebrates.

14. MARINE AND FRESHWATER MAMMAL CONSERVATION

Anna Berthinussen, Rebecca K. Smith & William J. Sutherland

Expert assessors
Jay Barlow, NOAA / UC San Diego, United States
Andrew Bladon, University of Cambridge, United Kingdom
Elizabeth Campbell, ProDelphinus, Peru
Vera da Silva, Projeto Boto, Brazil
José Martins da Silva-Jr, Spinner Dolphin project, Brazil
Nico de Bruyn, University of Pretoria, South Africa
Frances Gulland, University of California, Davis, United States
Lucy Keith-Diagne, African Aquatic Conservation Fund, Senegal
Nachiket Kelkar, Ashoka Trust for Research in Ecology & the Environment, India
Giuseppe Notarbartolo di Sciara, Tethys Research Institute, Italy
Silviu Petrovan, University of Cambridge, United Kingdom
Nikki Taylor, JNCC, United Kingdom
Paul Thompson, Lighthouse Field Station, Aberdeen, United Kingdom
Razvan Zaharia, SEOPMM Oceanic Club, Romania

Scope of assessment: for native wild marine species across the world.

Assessed: 2021.

Effectiveness measure is the median % score for effectiveness.

Certainty measure is the median % certainty of evidence for effectiveness, determined by the quantity and quality of the evidence in the synopsis.

Harm measure is the median % score for negative side-effects to the group of species of concern.

 https://doi.org/10.11647/OBP.0267.14

This book is meant as a guide to the evidence available for different conservation interventions and as a starting point in assessing their effectiveness. The assessments are based on the available evidence for the target group of species for each intervention. The assessment may therefore refer to different species or habitat to the one(s) you are considering. Before making any decisions about implementing interventions it is vital that you read the more detailed accounts of the evidence in order to assess their relevance for your study species or system.

Full details of the evidence are available at
www.conservationevidence.com

There may also be significant negative side-effects on the target groups or other species or communities that have not been identified in this assessment.

A lack of evidence means that we have been unable to assess whether or not an intervention is effective or has any harmful impacts.

14.1 Threat: Aquaculture and agriculture

Based on the collated evidence, what is the current assessment of the effectiveness of interventions for aquaculture and agriculture?	
Trade-off between benefit and harms	• Use acoustic devices at aquaculture systems
Unknown effectiveness	• Replace or repair damaged anti-predator nets around aquaculture systems
Likely to be ineffective or harmful	• Translocate mammals away from aquaculture systems to reduce human-wildlife conflict
No evidence found (no assessment)	• Introduce and enforce legislation to prevent intentional killing of mammals at aquaculture systems • Introduce and enforce regulations to prevent the use of harmful deterrents on mammals at aquaculture systems • Minimize food waste at aquaculture systems • Modify anti-predator nets around aquaculture systems • Modify aquaculture gear

Trade-off between benefit and harms

Use acoustic devices at aquaculture systems

Six studies evaluated the effects on marine and freshwater mammals of using acoustic devices at aquaculture systems. Four studies were in the North

Atlantic Ocean (USA, UK), one was in the Reloncaví fjord (Chile) and one in the Mediterranean Sea (Italy).

COMMUNITY RESPONSE (0 STUDIES)

POPULATION RESPONSE (0 STUDIES)

BEHAVIOUR (0 STUDIES)

OTHER (6 STUDIES)

Human-wildlife conflict (6 studies): Four of six studies (including five before-and-after and/or site comparison studies and one controlled study) in the North Atlantic Ocean, the Reloncaví fjord and the Mediterranean Sea found that using acoustic devices at salmon farms reduced predation on caged salmon by grey seals, harbour seals and South American sea lions, or reduced the number of harbour seals approaching a fish cage. The two other studies found that using acoustic devices did not reduce harbour seal predation at salmon farms, or reduce the presence, approach distances, groups sizes or time spent around fin-fish farms by common bottlenose dolphins.

Assessment: trade-off between benefit and harms (effectiveness 45%; certainty 45%; harms 45%).

https://www.conservationevidence.com/actions/2775

Unknown effectiveness

Replace or repair damaged anti-predator nets around aquaculture systems

One study evaluated the effects on marine mammals of replacing anti-predator nets around aquaculture systems. The study was in the North Atlantic Ocean (USA).

COMMUNITY RESPONSE (0 STUDIES)

POPULATION RESPONSE (0 STUDIES)

BEHAVIOUR (0 STUDIES)

OTHER (1 STUDY)

Human-wildlife conflict (1 study): One site comparison study in the North Atlantic Ocean found that replacing anti-predator nets more frequently at salmon farms resulted in fewer salmon losses to harbour seal predation.

Assessment: unknown effectiveness (effectiveness 50%; certainty 28%; harms 10%).

https://www.conservationevidence.com/actions/2774

Likely to be ineffective or harmful

Translocate mammals away from aquaculture systems to reduce human-wildlife conflict

Two studies evaluated the effects of translocating mammals away from aquaculture systems to reduce human-wildlife conflict. Both studies were in the Tasman Sea and one was also in the Southern Ocean (Tasmania).

COMMUNITY RESPONSE (0 STUDIES)

POPULATION RESPONSE (0 STUDIES)

BEHAVIOUR (0 STUDIES)

OTHER (2 STUDIES)

Human-wildlife conflict (2 studies): Two studies (including one site comparison study) in the Tasman Sea (one also in the Southern Ocean) found that more than half or nearly all of Australian and New Zealand fur seals translocated away from salmon farms returned.

Assessment: likely to be ineffective or harmful (effectiveness 10%; certainty 40%; harms 40%).

https://www.conservationevidence.com/actions/2776

No evidence found (no assessment)

We have captured no evidence for the following interventions:

- Introduce and enforce legislation to prevent intentional killing of mammals at aquaculture systems
- Introduce and enforce regulations to prevent the use of harmful deterrents on mammals at aquaculture systems
- Minimize food waste at aquaculture systems
- Modify anti-predator nets around aquaculture systems
- Modify aquaculture gear.

14.2 Threat: Energy production and mining

14.2.1 Renewable energy

Based on the collated evidence, what is the current assessment of the effectiveness of interventions for renewable energy?	
No evidence found (no assessment)	• Modify design of underwater turbines • Use acoustic devices at renewable energy sites • Use real-time automated tools at renewable energy sites to detect marine and freshwater mammals and allow operations to be stopped or modified

No evidence found (no assessment)

We have captured no evidence for the following interventions:

- Modify design of underwater turbines
- Use acoustic devices at renewable energy sites
- Use real-time automated tools at renewable energy sites to detect marine and freshwater mammals and allow operations to be stopped or modified.

14.2.2 Power plants

Based on the collated evidence, what is the current assessment of the effectiveness of interventions for power plants?	
No evidence found (no assessment)	• Install diversion or return systems on cooling water intake structures • Reduce capacity of cooling water intake structures • Use acoustic devices at cooling water intake structures • Use cooling towers instead of once-through cooling systems

No evidence found (no assessment)

We have captured no evidence for the following interventions:

- Install diversion or return systems on cooling water intake structures
- Reduce capacity of cooling water intake structures
- Use acoustic devices at cooling water intake structures
- Use cooling towers instead of once-through cooling systems.

14.3 Threat: Transportation and service corridors

14.3.1 Shipping lanes

Based on the collated evidence, what is the current assessment of the effectiveness of interventions for shipping lanes?	
Likely to be beneficial	• Set and enforce vessel speed limits
No evidence found (no assessment)	• Develop and implement regulations for operating vessels around mammals • Divert shipping routes • Limit vessel traffic in shallow rivers • Modify vessels to reduce risk of physical injury to mammals • Provide educational materials at marinas and ports to encourage vessel operators to carry out safe practices around mammals (e.g. signs, leaflets) • Provide training to vessel operators on mammal behaviour and appropriate avoidance techniques • Reduce shipping along inland waterways • Use acoustic devices on moving vessels • Use observers on board vessels to detect mammals and allow vessel course or speed to be altered • Use real-time automated tools on board vessels to detect mammals and allow vessel course or speed to be altered • Use remote tools to detect mammals in an area and allow vessel course or speed to be altered

Likely to be beneficial

Set and enforce vessel speed limits

Two studies evaluated the effects on marine and freshwater mammals of setting and enforcing vessel speed limits. One study was in the Indian River estuarine system (USA) and the other in the North Atlantic Ocean (USA).

COMMUNITY RESPONSE (0 STUDIES)

POPULATION RESPONSE (2 STUDIES)

Survival (2 studies): One before-and-after study in the Indian River estuarine system found similar numbers of manatee deaths before and after vessel speed limits were set in 'zones', but fewer deaths were recorded after speed limits were set and enforced in all areas. One before-and-after study in the North Atlantic Ocean found that setting vessel speed limits during specific periods in key habitats resulted in fewer North Atlantic right whale deaths caused by collisions.

BEHAVIOUR (0 STUDIES)

Assessment: likely to be beneficial (effectiveness 80%; certainty 60%; harms 0%).

https://www.conservationevidence.com/actions/2777

No evidence found (no assessment)

We have captured no evidence for the following interventions:

- Develop and implement regulations for operating vessels around mammals
- Divert shipping routes
- Limit vessel traffic in shallow rivers
- Modify vessels to reduce risk of physical injury to mammals
- Provide educational materials at marinas and ports to encourage vessel operators to carry out safe practices around mammals (e.g. signs, leaflets)
- Provide training to vessel operators on mammal behaviour and appropriate avoidance techniques
- Reduce shipping along inland waterways
- Use acoustic devices on moving vessels
- Use observers on board vessels to detect mammals and allow vessel course or speed to be altered
- Use real-time automated tools on board vessels to detect mammals and allow vessel course or speed to be altered

- Use remote tools to detect mammals in an area and allow vessel course or speed to be altered.

14.3.2 Flight paths

Based on the collated evidence, what is the current assessment of the effectiveness of interventions for flight paths?	
No evidence found (no assessment)	• Introduce regulations for flying aircraft over marine and freshwater mammals

No evidence found (no assessment)

We have captured no evidence for the following interventions:

- Introduce regulations for flying aircraft over marine and freshwater mammals.

14.4 Threat: Biological resource use

14.4.1 Reduce hunting and persecution

Based on the collated evidence, what is the current assessment of the effectiveness of interventions for reduce hunting and persecution?	
Beneficial	• Prohibit or restrict hunting of marine and freshwater mammal species
No evidence found (no assessment)	• Educate local communities and fishers on mammal protection laws to reduce killing of marine and freshwater mammals • Enforce legislation to prevent the trafficking and trade of marine and freshwater mammal products • Inform local communities and fishers about the negative impacts of hunting to reduce the killing of marine and freshwater mammals • Introduce alternative food sources to replace marine and freshwater mammal meat • Introduce alternative income sources to reduce marine and freshwater mammal exploitation and trade • Introduce alternative sources of bait to replace the use of marine and freshwater mammals • Introduce alternative treatments to reduce the use of marine and freshwater mammals in traditional medicine • Introduce and enforce legislation to prevent intentional killing of mammals at wild fisheries • Introduce and enforce regulations for sustainable hunting of marine and freshwater mammals for traditional subsistence and handicrafts • Introduce and enforce regulations to prevent the use of harmful deterrents on mammals at wild fisheries • Restrict capture of marine and freshwater mammals for research or aquariums and zoos

Beneficial

Prohibit or restrict hunting of marine and freshwater mammal species

Five studies evaluated the effects of prohibiting hunting of marine mammal species. One study was in each of the Kattegat and Skagerrak seas (Denmark and Sweden), the North Atlantic Ocean, North Pacific Ocean and the Southern Hemisphere, the South Pacific Ocean (Australia), the North Atlantic Ocean (Greenland) and the Southern Ocean (Australia).

COMMUNITY RESPONSE (0 STUDIES)

POPULATION RESPONSE (5 STUDIES)

Abundance (5 studies): Four of five studies (including three before-and-after studies) in the Kattegat and Skagerrak Seas, the South Pacific Ocean, the North Atlantic Ocean and the Southern Ocean found that after hunting was prohibited, the abundance of harbour seals and humpback whales increased over 7–30 years. The other study found that numbers of mature male sperm whales did not differ significantly before or 31 years after hunting was prohibited. One review in the North Atlantic Ocean, North Pacific Ocean and the Southern Hemisphere found significant increase rates for 10 of 12 baleen whale populations during 7–21 years after legislation to prohibit hunting was introduced.

BEHAVIOUR (0 STUDIES)

Assessment: beneficial (effectiveness 84%; certainty 70%; harms 0%).

https://www.conservationevidence.com/actions/2780

No evidence found (no assessment)

We have captured no evidence for the following interventions:

- Educate local communities and fishers on mammal protection laws to reduce killing of marine and freshwater mammals
- Enforce legislation to prevent the trafficking and trade of marine and freshwater mammal products
- Inform local communities and fishers about the negative impacts of hunting to reduce the killing of marine and freshwater mammals
- Introduce alternative food sources to replace marine and freshwater mammal meat

- Introduce alternative income sources to reduce marine and freshwater mammal exploitation and trade
- Introduce alternative sources of bait to replace the use of marine and freshwater mammals
- Introduce alternative treatments to reduce the use of marine and freshwater mammals in traditional medicine
- Introduce and enforce legislation to prevent intentional killing of mammals at wild fisheries
- Introduce and enforce regulations for sustainable hunting of marine and freshwater mammals for traditional subsistence and handicrafts
- Introduce and enforce regulations to prevent the use of harmful deterrents on mammals at wild fisheries
- Restrict capture of marine and freshwater mammals for research or aquariums and zoos.

14.4.2 Reduce unwanted catch ('bycatch') of mammals and improve survival of released or escaped mammals

Based on the collated evidence, what is the current assessment of the effectiveness of interventions for reduce unwanted catch ('bycatch') of mammals and improve survival of released or escaped mammals?	
Likely to be beneficial	• Increase visual detectability of fishing gear for mammals • Install exclusion and/or escape devices for mammals on fishing nets • Modify fishing pots and traps to exclude mammals • Use acoustically reflective fishing gear materials • Use catch and hook protection devices on fishing gear
Trade-off between benefit and harms	• Use acoustic devices on fishing gear • Use acoustic devices on fishing vessels • Use acoustic devices on moorings

Based on the collated evidence, what is the current assessment of the effectiveness of interventions for reduce unwanted catch ('bycatch') of mammals and improve survival of released or escaped mammals?	
Unknown effectiveness	• Deploy fishing gear at different depths • Establish handling and release protocols for mammals captured by fisheries • Establish 'move-on rules' for fishing vessels if mammals are encountered • Install barriers at wild fisheries • Play predator calls to deter mammals from fishing gear • Switch off artificial lighting at wild fisheries • Use a larger mesh size for fishing trap-nets • Use acoustic decoys to divert mammals away from fishing gear • Use an electric current to deter mammals from fishing gear • Use noise aversive conditioning to deter mammals from fishing gear • Use stiffened materials or increase tension of fishing gear • Use 'mammal-safe' nets to capture and release mammals trapped in fishing structures
Unlikely to be beneficial	• Attach acoustically reflective objects to fishing gear
No evidence found (no assessment)	• Deploy fishing gear at times when mammals are less active • Educate the public to reduce consumer demand for fisheries that threaten mammals • Enforce legislation to control illegal fishing using gear or methods that are harmful to mammals • Finance low interest loans to convert to fishing gear that reduces unwanted catch of mammals • Introduce fishing gear exchange programmes to encourage fishers to use gear that reduces unwanted catch of mammals • Introduce legislation to prohibit or restrict the use of fishing gear types or methods that are harmful to mammals

Based on the collated evidence, what is the current assessment of the effectiveness of interventions for reduce unwanted catch ('bycatch') of mammals and improve survival of released or escaped mammals?	
	• Involve fishers in designing and trialling new fishing gear types to encourage uptake of gear that reduces unwanted catch of mammals • Limit size of trawl net openings • Limit the length of fishing gear in an area • Limit the number of fishing vessels or fishing days in an area • Promote fish and seafood certification (e.g. ecolabels) to reduce consumer demand for fisheries that threaten mammals • Provide training and tools for safe release of mammals captured by fisheries • Reduce duration of time fishing gear is in the water • Retain buoys and lines at the sea floor or river bed when not hauling • Retain offal on fishing vessels instead of discarding overboard • Use a smaller mesh size for fishing nets • Use bindings to keep trawl nets closed until they have sunk below the water surface • Use different bait species for fishing that are less attractive to mammals • Use passive listening devices to detect mammals and prompt fishing vessels to move away • Use sinking lines instead of floating lines • Use weakened fishing gear

Likely to be beneficial

Increase visual detectability of fishing gear for mammals

Two studies evaluated the effects on marine mammals of increasing the visual detectability of fishing gear for mammals. One study was in the Gulf of St. Lawrence (Canada) and one was in Cape Cod Bay (USA).

COMMUNITY RESPONSE (0 STUDIES)

POPULATION RESPONSE (0 STUDIES)

BEHAVIOUR (2 STUDIES)

Behaviour change (2 studies): One study in the Gulf of St. Lawrence found that minke whales approached white ropes more slowly and changed their bearing more when approaching black ropes compared to ropes of other colours. One study in Cape Cod Bay found that simulated ropes painted red or orange were detected by North Atlantic right whales at greater distances than green but not black ropes, and more whales collided with green ropes than the other three rope colours.

Assessment: likely to be beneficial (effectiveness 55%; certainty 45%; harms 5%).

https://www.conservationevidence.com/actions/2805

Install exclusion and/or escape devices for mammals on fishing nets

Seven studies evaluated the effects on marine mammals of installing exclusion and/or escape devices on fishing nets. Four studies were in the Indian Ocean (Australia, Tasmania) and/or Tasman Sea (Tasmania) and three studies were in the South Atlantic Ocean (South Georgia).

COMMUNITY RESPONSE (0 STUDIES)

POPULATION RESPONSE (2 STUDIES)

Survival (2 studies): One study in the Indian Ocean found that less than one third of common bottlenose dolphins exited escape hatches on trawl nets alive. One replicated study in the Tasman Sea and Indian Ocean found that fewer fur seals died in exclusion devices with large escape openings than in those with small openings.

BEHAVIOUR (0 STUDIES)

OTHER (5 STUDIES)

Reduction in entanglements/unwanted catch (5 studies): Three studies (including two controlled studies) in the South Atlantic Ocean and Indian Ocean found that installing exclusion and/or escape devices on trawl nets reduced the number of trapped or entangled Antarctic fur seals and common bottlenose dolphins. One before-and-after study in the Indian Ocean found that installing exclusion and escape devices on trawl nets reduced common bottlenose dolphin entanglements for three of four fishing vessels. One study in the South Atlantic Ocean found that modifying an exclusion and escape device by enlarging and relocating the escape panel resulted in fewer Antarctic fur seal entanglements.

Assessment: likely to be beneficial (effectiveness 60%; certainty 60%; harms 15%).

https://www.conservationevidence.com/actions/2823

Modify fishing pots and traps to exclude mammals

Six studies evaluated the effects on marine mammals of modifying fishing pots and traps to exclude mammals. Two studies were in the North Sea (UK, Sweden) and one study was in each of the Indian River Lagoon (USA), the Gulf of Finland (Finland), the Bothnian Sea (Finland), the Indian Ocean (Australia) and the Baltic Sea (Sweden).

COMMUNITY RESPONSE (0 STUDIES)

POPULATION RESPONSE (0 STUDIES)

BEHAVIOUR (0 STUDIES)

OTHER (6 STUDIES)

Reduction in entanglements/unwanted catch (2 studies): Two controlled studies (including one replicated study) in the Indian Ocean, and the Baltic Sea and North Sea found that installing steel rods on lobster pots or metal frames on fishing pots reduced the number of Australian sea lion pups or grey seals and harbour seals that entered or became trapped in pots.

Human wildlife conflict (4 studies): Two controlled studies (including one replicated study) in the Bothnian Sea and the North Sea found that installing wire grids or steel bars on fishing trap-nets or bag-nets, along with strengthened netting or other modifications to prevent seal access, reduced damage to salmon catches by seals. One controlled study in the Indian River Lagoon found that one of two methods of securing crab pot doors with a bungee cord reduced the number of common bottlenose dolphin interactions. One controlled study in the Gulf of Finland found that installing wire grids on trap-nets, along with strengthened netting, resulted in higher catches of undamaged salmon but not whitefish, likely due to reduced seal predation.

Assessment: likely to be beneficial (effectiveness 70%; certainty 56%; harms 10%).

https://www.conservationevidence.com/actions/2822

Use acoustically reflective fishing gear materials

Five studies evaluated the effects on marine mammals of using acoustically reflective fishing gear materials. Two studies were in the Bay of Fundy (Canada) and one study was in each of the Fortune Channel (Canada), the North Sea (Denmark) and the South Atlantic Ocean (Argentina).

COMMUNITY RESPONSE (0 STUDIES)

POPULATION RESPONSE (0 STUDIES)

BEHAVIOUR (2 STUDIES)

Behaviour change (2 studies): One controlled study in the Fortune Channel found that harbour porpoises approached nets made from acoustically reflective material (barium sulfate) and conventional nets to similar distances and for similar durations, but porpoises used fewer echolocation clicks at barium sulfate nets. One controlled study in the Bay of Fundy found that harbour porpoise echolocation activity was similar at barium sulfate and conventional nets.

OTHER (3 STUDIES)

Reduction in entanglements/unwanted catch (3 studies): Two of three controlled studies (including two replicated studies) in the North Sea, the Bay of Fundy and the South Atlantic Ocean found that fishing nets made from acoustically reflective materials (iron-oxide or barium sulfate) had fewer entanglements of harbour porpoises than conventional fishing nets. The other study found that nets made from barium sulfate did not reduce the number of dolphin entanglements.

Assessment: likely to be beneficial (effectiveness 40%; certainty 45%; harms 0%).

https://www.conservationevidence.com/actions/2807

Use catch and hook protection devices on fishing gear

Five studies evaluated the effects on marine mammals of using catch and hook protection devices on fishing gear. Two studies were in the South Pacific Ocean (Chile, Australia and Fiji), two were in the Indian Ocean (Seychelles, Madagascar) and one was in the Southwest Atlantic Ocean.

COMMUNITY RESPONSE (0 STUDIES)

POPULATION RESPONSE (0 STUDIES)

BEHAVIOUR (0 STUDIES)

OTHER (5 STUDIES)

Reduction in entanglements/unwanted catch (1 study): One study in the South Pacific Ocean found that using cage or chain devices on fishing hooks resulted in fewer unwanted catches of toothed whales.

Human-wildlife conflict (5 studies): Two of four studies (including three controlled and one before-and-after study) in the South Pacific Ocean, the Southwest Atlantic Ocean and the Indian Ocean found that net sleeves or cage and chain devices on fishing hooks reduced damage to fish catches by sperm whales, killer whales and toothed whales. The two other studies found that attaching 'umbrella' or 'spider' devices on fishing hooks did not reduce predation and/or damage to fish catches by sperm whales or toothed whales. One controlled study in the Indian Ocean found that attaching catch protection devices made from streamers to fishing lines reduced Indo-Pacific bottlenose and spinner dolphin predation on fish bait, but only during the first two trials.

Assessment: likely to be beneficial (effectiveness 43%; certainty 42%; harms 10%).

https://www.conservationevidence.com/actions/2821

Trade-off between benefit and harms

Use acoustic devices on fishing gear

Thirty-three studies evaluated the effects on marine mammals of using acoustic devices on fishing gear. Eight studies were in the North Atlantic Ocean (Canada, USA, UK), four studies were in each of the North Pacific Ocean (USA) and the North Sea (Germany, Denmark, UK), three studies were in the Mediterranean Sea (Spain, Italy), two studies were in each of the Fortune Channel (Canada), the South Atlantic Ocean (Argentina, Brazil) and the Baltic Sea (Denmark, Germany, Sweden), and one study was in each of Moreton Bay (Australia), the Black Sea (Turkey), the Celtic Sea (UK), the South Pacific Ocean (Peru), the Rainbow Channel (Australia), the UK (water body not stated), the Great Belt (Denmark), Omura Bay (Japan), and the Indian Ocean (Australia).

COMMUNITY RESPONSE (0 STUDIES)

POPULATION RESPONSE (0 STUDIES)

BEHAVIOUR (16 STUDIES)

Behaviour change (16 studies): Twelve of 16 controlled studies (including three replicated studies) in the North Atlantic Ocean, the Fortune Channel, the South Atlantic Ocean, Moreton Bay, the Mediterranean Sea, the Celtic Sea, the Rainbow Channel, a coastal site in the UK, the Great Belt, the North Sea, Omura Bay and the Indian Ocean found that using acoustic devices on fishing nets, float lines or simulated fishing nets resulted in harbour porpoises, common bottlenose dolphins, tuxuci dolphins, finless porpoises and seals approaching nets or lines less closely, having fewer encounters or interactions with nets, or activity and sightings were reduced in the surrounding area. The other four studies found that using acoustic devices on trawl nets, float lines or simulated fishing nets did not have a significant effect on the behaviour of common bottlenose dolphins, harbour porpoises, Indo-Pacific humpback dolphins or dugongs.

OTHER (19 STUDIES)

Reduction in entanglements/unwanted catch (14 studies): Nine studies (including seven controlled studies and two before-and after studies) in the North Atlantic Ocean, the North Sea, the South Atlantic Ocean, the North Pacific Ocean, the Black Sea, and the South Pacific Ocean found that using acoustic devices on cod traps or fishing nets resulted in fewer collisions of humpback whales or entanglements of harbour porpoises, Franciscana dolphins, beaked whales and small cetaceans. Three studies (including two controlled studies and one before-and-after study) in the North Pacific Ocean found that using acoustic devices on fishing nets resulted in fewer entanglements of some species but not others. One controlled study in the North Atlantic Ocean found that fishing nets with a 'complete' set of acoustic devices had fewer entanglements of harbour porpoises, but those with an 'incomplete' set did not. One replicated, controlled study in the North Sea and Baltic Sea found that using acoustic devices on fishing nets reduced harbour porpoise entanglements in one fishing area but not the other.

Human-wildlife conflict (6 studies): Five of six studies (including six controlled studies, one of which was replicated) in the Baltic Sea, the Mediterranean Sea, the North Pacific Ocean, a coastal site in the UK and the North Sea found that using acoustic devices reduced damage to fish catches and/or fishing nets caused by common bottlenose dolphins and seals. The other study found that acoustic devices did not reduce damage to swordfish catches by California sea lions.

Assessment: trade-off between benefit and harms (effectiveness 70%; certainty 65%; harms 20%).

https://www.conservationevidence.com/actions/2808

Use acoustic devices on fishing vessels

Five studies evaluated the effects on marine mammals of using acoustic devices on vessels. One study was in each of the Shannon Estuary (Ireland), the Rainbow Channel (Australia), Keppel Bay (Australia), the North Atlantic Ocean (Azores) and the Indian Ocean (Crozet Islands).

COMMUNITY RESPONSE (0 STUDIES)

POPULATION RESPONSE (0 STUDIES)

BEHAVIOUR (4 STUDIES)

Behaviour change (4 studies): One controlled study in the Shannon Estuary found that common bottlenose dolphins avoided a boat more frequently when acoustic devices of two types were deployed alongside it. One controlled study in the Indian Ocean found that killer whales were recorded further from a fishing vessel when an acoustic device was used during hauls, but distances decreased after the first exposure to the device. Two before-and-after studies in the Rainbow Channel and Keppel Bay found that an acoustic device deployed alongside a vessel reduced surfacing and echolocation rates and time spent foraging or socializing of Indo-Pacific humpback dolphins and Australian snubfin dolphins but there was no effect on 8–10 other types of behaviour (e.g. vocalizing, diving, travelling etc.).

OTHER (1 STUDY)

Human-wildlife conflict (1 study): One randomized, controlled study in the North Atlantic Ocean found that using acoustic devices of two types did not reduce predation of squid catches by Risso's dolphins.

Assessment: trade-off between benefit and harms (effectiveness 40%; certainty 50%; harms 20%).

https://www.conservationevidence.com/actions/2815

Use acoustic devices on moorings

Eight studies evaluated the effects on marine and freshwater mammals of using acoustic devices on moorings. Two studies were in the South Pacific Ocean and one study was in each of the Puntledge River (Canada), the Bay of Fundy (Canada), the Shannon Estuary (Ireland), the Rivers Conon and Esk (UK), the Kyle of Sutherland estuary (UK) and the North Atlantic Ocean (UK).

COMMUNITY RESPONSE (0 STUDIES)

POPULATION RESPONSE (0 STUDIES)

BEHAVIOUR (7 STUDIES)

Behaviour change (7 studies): Two of four controlled studies in the South Pacific Ocean, the Kyle of Sutherland estuary and the North Atlantic Ocean found that deploying acoustic devices on moorings reduced numbers of grey and harbour seals, and the activity of harbour porpoises, short-beaked common dolphins and common bottlenose dolphins. The two other studies found that using an acoustic device on a mooring did not have a significant effect on the number, direction of movement, speed, or dive durations of migrating humpback whales. One controlled study in the Bay of Fundy found that using an acoustic device on a mooring reduced harbour porpoise echolocation activity, but the probability of porpoises approaching within 125 m of the device increased over 10–11 days. One controlled study in the Shannon Estuary found that one of two types of acoustic device reduced the activity of common bottlenose dolphins. One replicated, controlled study in the Rivers Conon and Esk found that using acoustic devices reduced the

number of grey and harbour seals upstream of the device but did not reduce seal numbers overall.
OTHER (1 STUDY)
Human-wildlife conflict (1 study): One randomized controlled study in the Puntledge River found that deploying an acoustic device on a mooring reduced the number of harbour seals feeding on migrating juvenile salmon.
Assessment: trade-off between benefit and harms (effectiveness 42%; certainty 50%; harms 20%).

https://www.conservationevidence.com/actions/2816

Unknown effectiveness

Deploy fishing gear at different depths

One study evaluated the effects on marine mammals of deploying fishing gear at different depths. The study was in the Arafura Sea (Australia).
COMMUNITY RESPONSE (0 STUDIES)
POPULATION RESPONSE (0 STUDIES)
BEHAVIOUR (0 STUDIES)
OTHER (1 STUDY)
Reduction in entanglements/unwanted catch (1 study): One controlled study in the Arafura Sea found that fishing nets deployed 4.5 m below the water surface had fewer entanglements of dolphins than surface nets.
Assessment: unknown effectiveness (effectiveness 50%; certainty 35%; harms 15%).

https://www.conservationevidence.com/actions/2793

Establish handling and release protocols for mammals captured by fisheries

One study evaluated the effects on marine mammals of establishing handling and release protocols for mammals captured by wild fisheries. The study was in the Great Australian Bight (Australia).
COMMUNITY RESPONSE (0 STUDIES)
POPULATION RESPONSE (1 STUDY)
Survival (1 study): One before-and-after study in the Great Australian Bight found that introducing a code of conduct for releasing dolphins trapped in nets, along with avoiding dolphins during fishing, resulted in lower mortality of short-beaked common dolphins.
BEHAVIOUR (0 STUDIES)

Assessment: unknown effectiveness (effectiveness 67%; certainty 27%; harms 0%).

https://www.conservationevidence.com/actions/2829

Establish 'move-on rules' for fishing vessels if mammals are encountered

One study evaluated the effects on marine mammals of establishing move-on rules for fishing vessels if mammals are encountered. The study was in the Great Australian Bight (Australia).

COMMUNITY RESPONSE (0 STUDIES)

POPULATION RESPONSE (1 STUDY)

Survival (1 study): One before-and-after study in the Great Australian Bight found that introducing measures to delay or relocate fishing if dolphins were encountered, along with releasing trapped dolphins, resulted in fewer short-beaked common dolphins being encircled and killed.

BEHAVIOUR (0 STUDIES)

Assessment: unknown effectiveness (effectiveness 70%; certainty 27%; harms 0%).

https://www.conservationevidence.com/actions/2790

Install barriers at wild fisheries

One study evaluated the effects on freshwater mammals of installing a barrier at a wild fishery. The study was in the Puntledge River (Canada).

COMMUNITY RESPONSE (0 STUDIES)

POPULATION RESPONSE (0 STUDIES)

BEHAVIOUR (0 STUDIES)

OTHER (1 STUDY)

Human-wildlife conflict (1 study): One randomized, controlled study in the Puntledge River found that installing a 'cork line' barrier did not deter harbour seals from feeding on salmon released from a hatchery.

Assessment: unknown effectiveness (effectiveness 0%; certainty 20%; harms 0%).

https://www.conservationevidence.com/actions/2824

Play predator calls to deter mammals from fishing gear

One study evaluated the effects of playing predator calls to deter mammals from fishing gear. The study was in the South Atlantic Ocean (Africa).

COMMUNITY RESPONSE (0 STUDIES)

POPULATION RESPONSE (0 STUDIES)

BEHAVIOUR (0 STUDIES)

OTHER (1 STUDY)

Human-wildlife conflict (1 study): One study in the South Atlantic Ocean found that playing killer whale vocalisations did not deter Cape fur seals from feeding on fish catches in a purse-seine net or trawl net.
Assessment: unknown effectiveness (effectiveness 0%; certainty 20%; harms 10%).

https://www.conservationevidence.com/actions/2817

Switch off artificial lighting at wild fisheries

One study evaluated the effects on freshwater mammals of switching off artificial lights at a wild fishery. The study was in the Puntledge River (Canada).
COMMUNITY RESPONSE (0 STUDIES)
POPULATION RESPONSE (0 STUDIES)
BEHAVIOUR (0 STUDIES)
OTHER (1 STUDY)
Human-wildlife conflict (1 study): One randomized, controlled study in the Puntledge River found that switching off artificial lights on a bridge did not deter harbour seals from feeding on salmon released from a hatchery.
Assessment: unknown effectiveness (effectiveness 5%; certainty 25%; harms 0%).

https://www.conservationevidence.com/actions/2825

Use a larger mesh size for fishing trap-nets

One study evaluated the effects on freshwater mammals of using a larger mesh size for fishing trap-nets. The study was in the River Indal (Sweden).
COMMUNITY RESPONSE (0 STUDIES)
POPULATION RESPONSE (0 STUDIES)
BEHAVIOUR (0 STUDIES)
OTHER (1 STUDY)
Human-wildlife conflict (1 study): One controlled study in the River Indal found that a fishing trap-net with a larger mesh size in the first two sections had fewer grey seals feeding around it and less damage caused by seals.
Assessment: unknown effectiveness (effectiveness 50%; certainty 30%; harms 15%).

https://www.conservationevidence.com/actions/2803

Use acoustic decoys to divert mammals away from fishing gear

One study evaluated the effects on marine mammals of using acoustic decoys to divert mammals away from fishing gear. The study was in the Gulf of Alaska (USA).
COMMUNITY RESPONSE (0 STUDIES)

POPULATION RESPONSE (0 STUDIES)

BEHAVIOUR (1 STUDY)

Behaviour change (1 study): One study in the Gulf of Alaska found that increasing the distance between an acoustic decoy device and fishing lines resulted in fewer sperm whales at the lines, but sperm whale presence and time of arrival did not differ.

Assessment: unknown effectiveness (effectiveness 37%; certainty 30%; harms 0%).

https://www.conservationevidence.com/actions/2820

Use an electric current to deter mammals from fishing gear

One study evaluated the effects of using an electric current to deter mammals from fishing gear. The study was in the Fraser River (Canada).

COMMUNITY RESPONSE (0 STUDIES)

POPULATION RESPONSE (0 STUDIES)

BEHAVIOUR (0 STUDIES)

OTHER (1 STUDY)

Human-wildlife conflict (1 study): One controlled study in the Fraser River found that using an electric current on a fishing net reduced Pacific harbour seal predation on salmon catches.

Assessment: unknown effectiveness (effectiveness 50%; certainty 30%; harms 20%).

https://www.conservationevidence.com/actions/2818

Use noise aversive conditioning to deter mammals from fishing gear

One study evaluated the effects on marine mammals of using noise aversive conditioning to deter mammals from fishing gear. The study was in the North Pacific Ocean (USA).

COMMUNITY RESPONSE (0 STUDIES)

POPULATION RESPONSE (0 STUDIES)

BEHAVIOUR (0 STUDIES)

OTHER (1 STUDY)

Human-wildlife conflict (1 study): One study in the North Pacific Ocean found that noise aversive conditioning did not reduce bait foraging behaviour by California sea lions.

Assessment: unknown effectiveness (effectiveness 20%; certainty 30%; harms 10%).

https://www.conservationevidence.com/actions/2819

Use stiffened materials or increase tension of fishing gear

One study evaluated the effects on marine mammals of using stiffened materials in fishing nets. The study was in the South Atlantic Ocean (Argentina).

COMMUNITY RESPONSE (0 STUDIES)

POPULATION RESPONSE (0 STUDIES)

BEHAVIOUR (0 STUDIES)

OTHER (1 STUDY)

Reduction in entanglements/unwanted catch (1 study): One controlled study in the South Atlantic Ocean found that using stiffened fishing nets did not reduce the number of Franciscana dolphin entanglements.

Assessment: unknown effectiveness (effectiveness 0%; certainty 28%; harms 10%).

https://www.conservationevidence.com/actions/2801

Use 'mammal-safe' nets to capture and release mammals trapped in fishing structures

One study evaluated the effects on marine mammals of using 'mammal-safe' nets to capture and release mammals trapped in fishing structures. The study was in the Bay of Fundy (Canada).

COMMUNITY RESPONSE (0 STUDIES)

POPULATION RESPONSE (1 STUDY)

Survival (1 study): One controlled study in the Bay of Fundy found that using 'marine mammal nets' with a larger mesh size to release harbour porpoises from herring weirs resulted in lower porpoise mortality compared to using conventional herring nets.

BEHAVIOUR (0 STUDIES)

Assessment: unknown effectiveness (effectiveness 65%; certainty 32%; harms 0%).

https://www.conservationevidence.com/actions/2828

Unlikely to be beneficial

Attach acoustically reflective objects to fishing gear

Two studies evaluated the effects on marine mammals of attaching acoustically reflective objects to fishing gear. One study was in the Timor Sea and Arafura Sea (Australia) and one was in the Gulf of Alaska (USA).

COMMUNITY RESPONSE (0 STUDIES)

POPULATION RESPONSE (0 STUDIES)

BEHAVIOUR (0 STUDIES)

OTHER (2 STUDIES)

Reduction in entanglements/unwanted catch (1 study): One controlled study in the Timor Sea and Arafura Sea found that attaching metallic bead chains to fishing nets did not reduce the number of dolphin entanglements.

Human-wildlife conflict (1 study): One controlled study in the Gulf of Alaska found that attaching acrylic beads next to fishing hooks did not reduce predation on fish catches by sperm whales.

Assessment: unlikely to be beneficial (effectiveness 10%; certainty 41%; harms 15%).

https://www.conservationevidence.com/actions/2806

No evidence found (no assessment)

We have captured no evidence for the following interventions:

- Deploy fishing gear at times when mammals are less active
- Educate the public to reduce consumer demand for fisheries that threaten mammals
- Enforce legislation to control illegal fishing using gear or methods that are harmful to mammals
- Finance low interest loans to convert to fishing gear that reduces unwanted catch of mammals
- Introduce fishing gear exchange programmes to encourage fishers to use gear that reduces unwanted catch of mammals
- Introduce legislation to prohibit or restrict the use of fishing gear types or methods that are harmful to mammals
- Involve fishers in designing and trialling new fishing gear types to encourage uptake of gear that reduces unwanted catch of mammals
- Limit size of trawl net openings
- Limit the length of fishing gear in an area
- Limit the number of fishing vessels or fishing days in an area
- Promote fish and seafood certification (e.g. ecolabels) to reduce consumer demand for fisheries that threaten mammals
- Provide training and tools for safe release of mammals captured by fisheries
- Reduce duration of time fishing gear is in the water
- Retain buoys and lines at the sea floor or river bed when not hauling
- Retain offal on fishing vessels instead of discarding overboard
- Use a smaller mesh size for fishing nets

- Use bindings to keep trawl nets closed until they have sunk below the water surface
- Use different bait species for fishing that are less attractive to mammals
- Use passive listening devices to detect mammals and prompt fishing vessels to move away
- Use sinking lines instead of floating lines
- Use weakened fishing gear.

14.5 Threat: Human intrusions and disturbance

14.5.1 Recreational activities and tourism

Based on the collated evidence, what is the current assessment of the effectiveness of interventions for recreational activities and tourism?	
Likely to be beneficial	• Introduce and enforce regulations for marine and freshwater mammal watching tours • Limit, cease or prohibit feeding of marine and freshwater mammals by tourists • Use volunteers to deter tourists from harassing marine and freshwater mammals at wildlife-viewing sites
Unknown effectiveness	• Inform the public of ways to reduce disturbance to marine and freshwater mammals (e.g. use educational signs) • Train tourist guides to minimize disturbance and promote marine and freshwater mammal conservation
No evidence found (no assessment)	• Create designated areas or access points for recreational activities • Introduce permits or licences for marine and freshwater mammal watching tours • Introduce permits or licences for recreational watersports

Likely to be beneficial

Introduce and enforce regulations for marine and freshwater mammal watching tours

Four studies evaluated the effects of introducing regulations for marine and freshwater mammal watching tours on marine mammals. One study was in each of the North Atlantic Ocean (the Azores), the Cananéia estuary (Brazil), the South Pacific Ocean (Australia) and the Bass Strait (Australia).

COMMUNITY RESPONSE (0 STUDIES)

POPULATION RESPONSE (0 STUDIES)

BEHAVIOUR (4 STUDIES)

Behaviour change (4 studies): Two controlled studies in the North Atlantic Ocean and South Pacific Ocean found that when whale-watching vessels followed approach regulations, fewer sperm whales and humpback whale pods changed their behaviours (e.g. swimming speed, aerial displays) or avoided the vessels compared to when regulations were not followed, but direction of movement and diving patterns or diving behaviours did not differ. One replicated, controlled study in the Cananéia estuary found that when tour boats followed approach regulations, fewer Guiana dolphins displayed negative behaviours (e.g. moving away, diving, groups separating). One study in the Bass Strait found that when boats approached a seal colony to 75 m, more seals remained on shore than when boats approached to 25 m.

Assessment: likely to be beneficial (effectiveness 80%; certainty 60%; harms 0%).

https://www.conservationevidence.com/actions/2838

Limit, cease or prohibit feeding of marine and freshwater mammals by tourists

One study evaluated the effects of setting limits on feeding of marine mammals by tourists. The study was in Shark Bay (Australia).

COMMUNITY RESPONSE (0 STUDIES)

POPULATION RESPONSE (1 STUDY)

Survival (1 study): One controlled, before-and-after study in Shark Bay found that after setting limits on feeding of bottlenose dolphins by tourists, the survival of calves born to females being fed increased and was similar to calves of non-fed mothers.

BEHAVIOUR (0 STUDIES)

Assessment: likely to be beneficial (effectiveness 60%; certainty 40%; harms 0%).

https://www.conservationevidence.com/actions/2845

Use volunteers to deter tourists from harassing marine and freshwater mammals at wildlife-viewing sites

One study evaluated the effects of using volunteers to deter tourists from harassing marine and freshwater mammals at wildlife-viewing sites. The study was at the Ohau Stream waterfall (New Zealand).

COMMUNITY RESPONSE (0 STUDIES)
POPULATION RESPONSE (0 STUDIES)
BEHAVIOUR (0 STUDIES)
OTHER (1 STUDY)
Change in human behaviour (1 study): One randomized, controlled study at the Ohau Stream waterfall found that the presence of an official-looking volunteer resulted in fewer tourists harassing New Zealand fur seals at a waterfall.
Assessment: likely to be beneficial (effectiveness 60%; certainty 40%; harms 0%).

https://www.conservationevidence.com/actions/2844

Unknown effectiveness

Inform the public of ways to reduce disturbance to marine and freshwater mammals (e.g. use educational signs)

One study evaluated the effects of informing the public of ways to reduce disturbance to marine and freshwater mammals. The study was in the South Pacific Ocean (New Zealand).
COMMUNITY RESPONSE (0 STUDIES)
POPULATION RESPONSE (0 STUDIES)
BEHAVIOUR (0 STUDIES)
OTHER (1 STUDY)
Change in human behaviour (1 study): One controlled study in the South Pacific Ocean found that tourist groups that observed information signs approached and disturbed New Zealand fur seals in similar numbers to those that did not.
Assessment: unknown effectiveness (effectiveness 20%; certainty 30%; harms 0%).

https://www.conservationevidence.com/actions/2843

Train tourist guides to minimize disturbance and promote marine and freshwater mammal conservation

One study evaluated the effects of training tourist guides to minimize disturbance and promote marine and freshwater mammal conservation. The study was in the Kenai Fjords (Alaska).
COMMUNITY RESPONSE (0 STUDIES)
POPULATION RESPONSE (0 STUDIES)
BEHAVIOUR (1 STUDY)

Behaviour change (1 study): One before-and-after study in the Kenai Fjords found that fewer harbour seals were disturbed during kayak excursions after training was provided to kayak guides.
Assessment: unknown effectiveness (effectiveness 60%; certainty 27%; harms 0%).

https://www.conservationevidence.com/actions/2840

No evidence found (no assessment)

We have captured no evidence for the following interventions:

- Create designated areas or access points for recreational activities
- Introduce permits or licences for marine and freshwater mammal watching tours
- Introduce permits or licences for recreational watersports.

14.5.2 Work and other activities

Based on the collated evidence, what is the current assessment of the effectiveness of interventions for work and other activities?	
No evidence found (no assessment)	• Introduce regulations for flying drones over marine and freshwater mammals • Introduce regulations for the use of underwater drones in proximity to marine and freshwater mammals

No evidence found (no assessment)

We have captured no evidence for the following interventions:

- Introduce regulations for flying drones over marine and freshwater mammals
- Introduce regulations for the use of underwater drones in proximity to marine and freshwater mammals.

14.6 Threat: Natural system modifications

14.6.1 Dams and water management/use

Based on the collated evidence, what is the current assessment of the effectiveness of interventions for dams and water management/use?	
No evidence found (no assessment)	• Install bypass channels in dams • Maintain water level and flow along regulated rivers • Use automated detection systems to prevent flood gates and locks from closing when mammals are present

No evidence found (no assessment)

We have captured no evidence for the following interventions:

- Install bypass channels in dams
- Maintain water level and flow along regulated rivers
- Use automated detection systems to prevent flood gates and locks from closing when mammals are present.

14.7 Threat: Invasive or problematic species and disease

14.7.1 Invasive or problematic species

Based on the collated evidence, what is the current assessment of the effectiveness of interventions for invasive or problematic species?	
Likely to be beneficial	• Use baited lines instead of nets for shark control
Unknown effectiveness	• Use deterrents to reduce predation on marine and freshwater mammals by native species
No evidence found (no assessment)	• Limit, cease or prohibit ballast water exchange in specific areas • Physically remove invasive or problematic species • Treat ballast water before release • Use biocides or other chemicals to control invasive or problematic species • Use biological control to manage invasive or problematic species

Likely to be beneficial

Use baited lines instead of nets for shark control

Two studies evaluated the effects on marine mammals of using baited lines instead of nets for shark control. One study was in the Indian Ocean (South Africa) and one in the South Pacific Ocean (Australia).

COMMUNITY RESPONSE (0 STUDIES)
POPULATION RESPONSE (1 STUDY)
Survival (1 study): One site comparison study in the South Pacific Ocean found that using baited lines instead of nets increased the survival of entangled common and bottlenose dolphins.
BEHAVIOUR (0 STUDIES)
OTHER (2 STUDIES)
Reduction in entanglements/unwanted catch (2 studies): Two site comparison studies in the Indian Ocean and South Pacific Ocean found that baited lines used for shark control had fewer entanglements of dolphins, whales and dugongs than nets.
Assessment: likely to be beneficial (effectiveness 70%; certainty 50%; harms 10%).

https://www.conservationevidence.com/actions/2856

Unknown effectiveness

Use deterrents to reduce predation on marine and freshwater mammals by native species

One study evaluated the effects of using deterrents to reduce predation by native species on marine mammals. The study was in the North Pacific Ocean (USA).
COMMUNITY RESPONSE (0 STUDIES)
POPULATION RESPONSE (1 STUDY)
Survival (1 study): One controlled study in the North Pacific Ocean found that neither boat motor sounds nor the presence of humans reduced Galapagos shark predation on Hawaiian monk seal pups, although shark presence was low throughout the study.
BEHAVIOUR (0 STUDIES)
Assessment: unknown effectiveness (effectiveness 30%; certainty 27%; harms 0%).

https://www.conservationevidence.com/actions/2855

No evidence found (no assessment)

We have captured no evidence for the following interventions:

- Limit, cease or prohibit ballast water exchange in specific areas
- Physically remove invasive or problematic species
- Treat ballast water before release

- Use biocides or other chemicals to control invasive or problematic species
- Use biological control to manage invasive or problematic species.

14.7.2 Disease

Based on the collated evidence, what is the current assessment of the effectiveness of interventions for disease?	
Likely to be beneficial	• Use drugs to treat parasites
No evidence found (no assessment)	• Carry out surveillance for diseases • Translocate or temporarily bring marine and freshwater mammals into captivity to reduce exposure to disease • Treat disease in wild marine and freshwater mammals • Vaccinate against disease

Likely to be beneficial

Use drugs to treat parasites

Two studies evaluated the effects on marine mammals of using drugs to treat parasites. Both studies were in the North Pacific Ocean (USA).

COMMUNITY RESPONSE (0 STUDIES)

POPULATION RESPONSE (2 STUDIES)

Survival (2 studies): One of two controlled studies (including one before-and-after study) in the North Pacific Ocean found that treating northern fur seal pups with an anti-parasitic drug (ivermectin) reduced mortality rates. The other study found that Hawaiian monk seal pups treated with an anti-parasitic drug (praziquantel) had similar survival rates to untreated pups.

Condition (2 studies): One of two controlled studies (including one before-and-after study) in the North Pacific Ocean found that northern fur seal pups treated with an anti-parasitic drug (ivermectin) had reduced hookworm infections and greater growth rates than untreated pups. The other study

found that Hawaiian monk seal pups treated with an anti-parasitic drug (praziquantel) had similar parasite loads to untreated pups.
BEHAVIOUR (0 STUDIES)

Assessment: likely to be beneficial (effectiveness 50%; certainty 40%; harms 10%).

https://www.conservationevidence.com/actions/2861

No evidence found (no assessment)

We have captured no evidence for the following interventions:

- Carry out surveillance for diseases
- Translocate or temporarily bring marine and freshwater mammals into captivity to reduce exposure to disease
- Treat disease in wild marine and freshwater mammals
- Vaccinate against disease.

14.8 Threat: Pollution

14.8.1 General

Based on the collated evidence, what is the current assessment of the effectiveness of interventions for pollution?	
No evidence found (no assessment)	• Add chemicals or minerals to sediment to remove or neutralize pollutants • Establish pollution emergency plans • Use 'bioremediating' organisms to remove or neutralize pollutants

No evidence found (no assessment)

We have captured no evidence for the following interventions:

- Add chemicals or minerals to sediment to remove or neutralize pollutants
- Establish pollution emergency plans
- Use 'bioremediating' organisms to remove or neutralize pollutants.

14.8.2 Domestic and urban wastewater

Based on the collated evidence, what is the current assessment of the effectiveness of interventions for domestic and urban wastewater?	
No evidence found (no assessment)	• Limit the amount of storm wastewater overflow • Limit, cease or prohibit dumping of sewage sludge • Limit, cease or prohibit dumping of untreated sewage • Set or improve minimum sewage treatment standards

No evidence found (no assessment)

We have captured no evidence for the following interventions:

- Limit the amount of storm wastewater overflow
- Limit, cease or prohibit dumping of sewage sludge
- Limit, cease or prohibit dumping of untreated sewage
- Set or improve minimum sewage treatment standards.

14.8.3 Industrial and military effluents

Based on the collated evidence, what is the current assessment of the effectiveness of interventions for industrial and military effluents?	
No evidence found (no assessment)	• Cease or prohibit the disposal of drill cuttings at sea or in rivers • Cease or prohibit the disposal of mining waste (tailings) at sea or in rivers • Rehabilitate and release marine and freshwater mammals following oil spills • Relocate marine and freshwater mammals following oil spills • Remove or clean-up oil pollution following a spill • Set regulatory ban on marine burial of nuclear waste • Use double hulls to prevent oil spills

No evidence found (no assessment)

We have captured no evidence for the following interventions:

- Cease or prohibit the disposal of drill cuttings at sea or in rivers
- Cease or prohibit the disposal of mining waste (tailings) at sea or in rivers
- Rehabilitate and release marine and freshwater mammals following oil spills
- Relocate marine and freshwater mammals following oil spills
- Remove or clean-up oil pollution following a spill

- Set regulatory ban on marine burial of nuclear waste
- Use double hulls to prevent oil spills.

14.8.4 Aquaculture effluents

Based on the collated evidence, what is the current assessment of the effectiveness of interventions for aquaculture effluents?	
No evidence found (no assessment)	• Introduce and enforce water quality regulations for aquaculture systems • Reduce the amount of antibiotics used in aquaculture systems • Reduce the amount of pesticides used in aquaculture systems

No evidence found (no assessment)

We have captured no evidence for the following interventions:

- Introduce and enforce water quality regulations for aquaculture systems
- Reduce the amount of antibiotics used in aquaculture systems
- Reduce the amount of pesticides used in aquaculture systems.

14.8.5 Agricultural and forestry effluents

Based on the collated evidence, what is the current assessment of the effectiveness of interventions for agricultural and forestry effluents?	
No evidence found (no assessment)	• Create artificial wetlands to reduce the amount of pollutants reaching rivers and the sea • Establish aquaculture to extract the nutrients from run-offs • Establish riparian buffers to reduce the amount of pollutants reaching rivers and the sea • Reduce pesticide, herbicide or fertilizer use • Treat wastewater from intensive livestock holdings

No evidence found (no assessment)

We have captured no evidence for the following interventions:

- Create artificial wetlands to reduce the amount of pollutants reaching rivers and the sea
- Establish aquaculture to extract the nutrients from run-offs
- Establish riparian buffers to reduce the amount of pollutants reaching rivers and the sea
- Reduce pesticide, herbicide or fertilizer use
- Treat wastewater from intensive livestock holdings.

14.8.6 Other pollution

Based on the collated evidence, what is the current assessment of the effectiveness of interventions for other pollution?	
No evidence found (no assessment)	• Limit, cease or prohibit discharge of waste effluents overboard from vessels • Remove and clean-up shoreline waste disposal sites • Use methods to reduce sediment disturbance during dredging (e.g. curtains, screens) • Use non-toxic antifouling coatings on surfaces

No evidence found (no assessment)

We have captured no evidence for the following interventions:

- Limit, cease or prohibit discharge of waste effluents overboard from vessels
- Remove and clean-up shoreline waste disposal sites
- Use methods to reduce sediment disturbance during dredging (e.g. curtains, screens)
- Use non-toxic antifouling coatings on surfaces.

14.8.7 Fishing gear

Based on the collated evidence, what is the current assessment of the effectiveness of interventions for fishing gear?	
Beneficial	• Remove derelict fishing gear from mammals found entangled
No evidence found (no assessment)	• Equip ports with dedicated fishing gear disposal facilities • Establish fishing gear registration programmes • Improve methods for locating fishing gear • Inform fishers of the impacts of derelict fishing gear on mammals to encourage responsible disposal • Offer incentives to fishers for recovering, reusing or recycling fishing gear • Recover lost or discarded fishing gear • Use biodegradable fishing gear

Beneficial

● Remove derelict fishing gear from mammals found entangled

Two studies evaluated the effects of removing derelict fishing gear from mammals found entangled. One study was in the North Pacific Ocean (USA) and one in the North Atlantic Ocean (USA).

COMMUNITY RESPONSE (0 STUDIES)

POPULATION RESPONSE (2 STUDIES)

Reproductive success (1 study): One review in the North Pacific Ocean found that after removing derelict fishing gear from Hawaiian monk seals, along with at least seven other interventions to enhance survival, more than a quarter of the seals reproduced.

Survival (2 studies): One review in the North Pacific Ocean found that removing derelict fishing gear from Hawaiian monk seals, along with at least seven other interventions to enhance survival, resulted in more than a quarter of the seals surviving. One review in the North Atlantic Ocean found that three common bottlenose dolphins survived for at least 1–4 years after they were disentangled from derelict fishing gear and released.

BEHAVIOUR (0 STUDIES)

Assessment: beneficial (effectiveness 70%; certainty 70%; harms 10%).

https://www.conservationevidence.com/actions/2892

No evidence found (no assessment)

We have captured no evidence for the following interventions:

- Equip ports with dedicated fishing gear disposal facilities
- Establish fishing gear registration programmes
- Improve methods for locating fishing gear
- Inform fishers of the impacts of derelict fishing gear on mammals to encourage responsible disposal
- Offer incentives to fishers for recovering, reusing or recycling fishing gear
- Recover lost or discarded fishing gear
- Use biodegradable fishing gear.

14.8.8 Other garbage and solid waste

Based on the collated evidence, what is the current assessment of the effectiveness of interventions for other garbage and solid waste?	
No evidence found (no assessment)	• Install stormwater traps or grids • Limit, cease or prohibit discharge of solid waste overboard from vessels • Remove litter from marine and freshwater environments

No evidence found (no assessment)

We have captured no evidence for the following interventions:

- Install stormwater traps or grids
- Limit, cease or prohibit discharge of solid waste overboard from vessels
- Remove litter from marine and freshwater environments.

14.8.9 Noise pollution

Based on the collated evidence, what is the current assessment of the effectiveness of interventions for noise pollution?	
Trade-off between benefit and harms	• Use acoustic devices to deter marine and freshwater mammals from an area to reduce noise exposure
Unknown effectiveness	• Use methods to dampen underwater noise emissions (e.g. bubble curtains, screens) • Use 'soft start' procedures to deter marine and freshwater mammals to reduce noise exposure
No evidence found (no assessment)	• Delay or cease operations if marine and freshwater mammals are detected within a specified zone • Limit, cease or prohibit the use of sonars • Limit, cease or prohibit the use of underwater explosives • Modify vessels to reduce noise disturbance • Reduce hammer energy during pile driving • Use alternative methods instead of airguns for seismic surveys

Trade-off between benefit and harms

Use acoustic devices to deter marine and freshwater mammals from an area to reduce noise exposure

Four studies evaluated the effects of using acoustic devices to deter marine and freshwater mammals from an area to reduce noise exposure. Two studies were in the North Sea (Germany), one study was in the Great Belt (Denmark) and one was in Faxaflói Bay (Iceland).

COMMUNITY RESPONSE (0 STUDIES)

POPULATION RESPONSE (0 STUDIES)

BEHAVIOUR (4 STUDIES)

Behaviour change (4 studies): Three studies (including two controlled and one before-and-after study) in the North Sea and the Great Belt found that using acoustic devices to deter mammals from an area at a wind farm construction site or pelagic sites reduced the activity and sightings of harbour porpoises at distances of 1–18 km from the devices. One before-and-after study in Faxaflói Bay found that when an acoustic device was deployed from a boat, minke whales swam away from the device, increased their swimming speed, and swam more directly.

Assessment: trade-off between benefit and harms (effectiveness 65%; certainty 60%; harms 20%).

https://www.conservationevidence.com/actions/2896

Unknown effectiveness

Use methods to dampen underwater noise emissions (e.g. bubble curtains, screens)

One study evaluated the effects on marine mammals of using bubble curtains or screens to dampen underwater noise emissions. The study was in the North Sea (Germany).

COMMUNITY RESPONSE (0 STUDIES)

POPULATION RESPONSE (0 STUDIES)

BEHAVIOUR (1 STUDY)

Behaviour change (1 study): One before-and-after, site comparison study in the North Sea found that using bubble curtains or screens during pile driving resulted in harbour porpoise detections within 15 km decreasing less compared to before pile driving than at sites without bubble curtains or screens.

Assessment: unknown effectiveness (effectiveness 50%; certainty 30%; harms 0%).

https://www.conservationevidence.com/actions/2901

Use 'soft start' procedures to deter marine and freshwater mammals to reduce noise exposure

Three studies evaluated the effects of using 'soft start' procedures to deter marine and freshwater mammals to reduce noise exposure. One study was in each of the South Atlantic Ocean (Gabon), the South Pacific Ocean (Australia) and various water bodies (UK).

COMMUNITY RESPONSE (0 STUDIES)

POPULATION RESPONSE (0 STUDIES)

BEHAVIOUR (3 STUDIES)

Behaviour change (3 studies): One study in various water bodies around the UK found that a greater proportion of cetaceans (including whales, dolphins and porpoise) avoided or moved away from vessels during 'soft start' procedures with seismic airguns compared to when airguns were not firing. One study in the South Atlantic Ocean found that during 'soft start' procedures using seismic airguns, a pod of short-finned whales initially moved away but remained within 900 m of the vessel as it passed by. One study in the South Pacific Ocean found that during 'soft-start' procedures with a small experimental airgun array, migrating humpback whales slowed their speed towards the vessel but did not significantly alter their course. *Assessment: unknown effectiveness (effectiveness 55%; certainty 36%; harms 10%).*

https://www.conservationevidence.com/actions/2897

No evidence found (no assessment)

We have captured no evidence for the following interventions:

- Delay or cease operations if marine and freshwater mammals are detected within a specified zone
- Limit, cease or prohibit the use of sonars
- Limit, cease or prohibit the use of underwater explosives
- Modify vessels to reduce noise disturbance
- Reduce hammer energy during pile driving
- Use alternative methods instead of airguns for seismic surveys.

14.8.10 Thermal pollution

Based on the collated evidence, what is the current assessment of the effectiveness of interventions for thermal pollution?	
No evidence found (no assessment)	• Limit, cease or prohibit the discharge of cooling effluents from power stations

No evidence found (no assessment)

We have captured no evidence for the following interventions:

- Limit, cease or prohibit the discharge of cooling effluents from power stations.

14.9 Threat: Climate change and severe weather

Based on the collated evidence, what is the current assessment of the effectiveness of interventions for climate change and severe weather?	
No evidence found (no assessment)	• Establish a network of legally protected areas • Implement rapid response plans for stranded mammals following extreme events • Legally protect areas where climate change impacts are predicted to be less severe • Manage water levels and flow in rivers to maintain deep pools and connectivity

No evidence found (no assessment)

We have captured no evidence for the following interventions:

- Establish a network of legally protected areas
- Implement rapid response plans for stranded mammals following extreme events
- Legally protect areas where climate change impacts are predicted to be less severe
- Manage water levels and flow in rivers to maintain deep pools and connectivity.

14.10 Habitat protection

Based on the collated evidence, what is the current assessment of the effectiveness of interventions for habitat protection?	
Likely to be beneficial	• Cease or prohibit activities that cause disturbance in sensitive areas for marine and freshwater mammals • Legally protect habitat for marine and freshwater mammals
No evidence found (no assessment)	• Cease or prohibit activities that cause disturbance during sensitive periods for marine and freshwater mammals • Enforce existing legislation for habitat protection • Retain or create buffer zones around important habitats

Likely to be beneficial

Cease or prohibit activities that cause disturbance in sensitive areas for marine and freshwater mammals

Two studies evaluated the effects of prohibiting activities that cause disturbance in sensitive areas for marine mammals. One study was in the Kattegat Sea (Denmark) and one in the Indian Ocean (Australia).

COMMUNITY RESPONSE (0 STUDIES)

POPULATION RESPONSE (1 STUDY)

Abundance (1 study): One before-and-after study in the Kattegat sea found that harbour porpoise activity increased at a stony reef after fishing was prohibited and the reef was restored with boulders.

BEHAVIOUR (1 STUDY)

Behaviour change (1 study): One site comparison study in the Indian Ocean found that a beach where human access was fully prohibited had fewer

Australian sea lions showing aggression or retreating compared to a beach where access was partly prohibited.

Assessment: likely to be beneficial (effectiveness 60%; certainty 40%; harms 0%).

https://www.conservationevidence.com/actions/2917

Legally protect habitat for marine and freshwater mammals

Four studies evaluated the effects of legally protecting habitat for marine and freshwater mammals. One study was in each of the North Atlantic Ocean (Portugal), the South Pacific Ocean (New Zealand), the North Sea (UK) and the Port River estuary (Australia).

COMMUNITY RESPONSE (0 STUDIES)

POPULATION RESPONSE (4 STUDIES)

Abundance (2 studies): One before-and-after study in the North Atlantic Ocean found that a population of Mediterranean monk seals increased during eight years after the islands they inhabited were legally protected. One before-and-after study in the North Sea found that a population of bottlenose dolphins was estimated to be a similar size before and after part of its range was protected.

Survival (2 studies): One before-and-after study in the South Pacific Ocean found that the survival rate of Hector's dolphins was higher after a coastal area was legally protected than before. One before-and-after study in the Port River estuary found that after the area became legally protected a similar number of Indo-Pacific bottlenose dolphin strandings were recorded compared to before protection, but the number of strandings caused by humans decreased.

BEHAVIOUR (0 STUDIES)

Assessment: likely to be beneficial (effectiveness 70%; certainty 60%; harms 0%).

https://www.conservationevidence.com/actions/2915

No evidence found (no assessment)

We have captured no evidence for the following interventions:

- Cease or prohibit activities that cause disturbance during sensitive periods for marine and freshwater mammals
- Enforce existing legislation for habitat protection
- Retain or create buffer zones around important habitats.

14.11 Habitat restoration and creation

Based on the collated evidence, what is the current assessment of the effectiveness of interventions for habitat restoration and creation?	
Unknown effectiveness	• Restore habitat for marine and freshwater mammals
No evidence found (no assessment)	• Create artificial habitat for marine and freshwater mammals • Leave anthropogenic structures in place after decommissioning

Unknown effectiveness

Restore habitat for marine and freshwater mammals

One study evaluated the effects of restoring habitat for marine mammals. The study was in the Kattegat sea (Denmark).

COMMUNITY RESPONSE (0 STUDIES)

POPULATION RESPONSE (1 STUDY)

Abundance (1 study): One before-and-after study in the Kattegat sea found that harbour porpoise activity increased at a stony reef after it was restored with boulders and fishing was prohibited.

BEHAVIOUR (0 STUDIES)

Assessment: unknown effectiveness (effectiveness 55%; certainty 30%; harms 0%).

https://www.conservationevidence.com/actions/2920

No evidence found (no assessment)

We have captured no evidence for the following interventions:

- Create artificial habitat for marine and freshwater mammals
- Leave anthropogenic structures in place after decommissioning.

14.12 Species management

14.12.1 Species recovery

Based on the collated evidence, what is the current assessment of the effectiveness of interventions for species recovery?	
Beneficial	• Rescue and release stranded or trapped marine and freshwater mammals
Likely to be beneficial	• Rehabilitate and release injured, sick or weak marine and freshwater mammals
Trade-off between benefit and harms	• Hand-rear orphaned or abandoned marine and freshwater mammal young • Remove individual marine and freshwater mammals exhibiting aggressive behaviours that may limit population recovery
Unknown effectiveness	• Reunite abandoned marine and freshwater mammal young with parents
No evidence found (no assessment)	• Legally protect marine and freshwater mammal species • Place orphaned or abandoned marine and freshwater mammal young with foster parents

Beneficial

Rescue and release stranded or trapped marine and freshwater mammals

Eleven studies evaluated the effects of rescuing and releasing stranded or trapped marine and freshwater mammals. Five studies were in the North Atlantic Ocean (USA), two studies were in the Indian Ocean (Tasmania, South Africa), and one study was in each of the South Atlantic Ocean (Brazil), the

Cachoeira River estuary (Brazil), the North Pacific Ocean (USA) and the Shannon Estuary (Ireland).

COMMUNITY RESPONSE (0 STUDIES)

POPULATION RESPONSE (11 STUDIES)

Reproductive success (2 studies): One review in the North Pacific Ocean found that after rescuing and releasing stranded or trapped Hawaiian monk seals, along with at least seven other interventions to enhance survival, more than a quarter of the seals reproduced. One study in the Shannon Estuary found that a stranded common bottlenose dolphin that was rescued and released was observed with a calf a year later.

Survival (11 studies): Seven studies (including one review) in the North Atlantic Ocean, Indian Ocean and the Shannon Estuary found that 17–100% of rescued and released Atlantic white-sided dolphins, short-beaked common dolphins, common bottlenose dolphins, long-finned pilot whales, short-finned pilot whales, and Cape fur seals survived during post-release monitoring periods, which ranged in length from three weeks to three years. Three studies in the South Atlantic Ocean, the Cachoeira estuary and the Indian Ocean found that a trapped rough-toothed dolphin, two stranded tucuxi dolphins and seven stranded sperm whales were successfully rescued and released, although long-term survival was not reported. One review in the North Pacific Ocean found that rescuing and releasing stranded or trapped Hawaiian monk seals, along with at least seven other interventions to enhance survival, resulted in more than a quarter of the seals surviving.

BEHAVIOUR (0 STUDIES)

Assessment: beneficial (effectiveness 70%; certainty 64%; harms 10%).

https://www.conservationevidence.com/actions/2924

Likely to be beneficial

Rehabilitate and release injured, sick or weak marine and freshwater mammals

Twenty-seven studies evaluated the effects of rehabilitating and releasing injured, sick or weak marine and freshwater mammals. Nine studies were in the North Atlantic Ocean (USA, UK, France), six studies were in the North Pacific Ocean (USA), four studies were in the Gulf of Mexico (USA), two studies were in each of the North Sea (the Netherlands) and the Gulf of Maine (USA), and one study was in each of the Indian River Lagoon (USA), Bohai Bay (China), The Wash estuary (UK), water bodies in Florida (USA), El Dorado Lake (Peru), and the Gulf of California (Mexico).

COMMUNITY RESPONSE (0 STUDIES)
POPULATION RESPONSE (26 STUDIES)
Reproductive success (1 study): One replicated study in the North Pacific Ocean found that more than a quarter of rehabilitated and released Hawaiian monk seals reproduced.
Survival (26 studies): Twenty-one studies (including two controlled studies, four replicated studies and one review) in the North Atlantic Ocean, the Gulf of Maine, the Gulf of Mexico, the North Pacific Ocean, the Indian River Lagoon, The Wash estuary, water bodies in Florida, El Dorado Lake, and the Gulf of California found that 10–100% of dolphins, porpoises, whales, seals, sea lions and manatees released after rehabilitation in captivity survived during post-release monitoring periods, which ranged in length from three days to five years. Five studies (including one replicated study) in the North Sea, the North Atlantic Ocean, Bohai Bay and the North Pacific Ocean found that two of three harbour porpoises, 152 of 188 grey seal pups, a common seal, a west Pacific finless porpoise and 14 of 35 California sea lions were successfully rehabilitated and released but survival after release was not reported. One controlled study in the North Pacific Ocean found that at least a quarter of California sea lions treated for toxic algae poisoning and released back into the wild died or had to be euthanized.
BEHAVIOUR (3 STUDIES)
Behaviour change (3 studies): Two of three controlled studies in the North Atlantic Ocean, the North Pacific Ocean and The Wash estuary found that a harbour porpoise and six harbour seals that were rehabilitated and released had similar movements and/or behaviours to wild mammals. The other study found that California sea lions treated for toxic algae poisoning and released travelled further from the shore, spent less time diving or hauled out and made shorter, shallower dives than wild sea lions without poisoning.
Assessment: likely to be beneficial (effectiveness 60%; certainty 70%; harms 10%).

https://www.conservationevidence.com/actions/2925

Trade-off between benefit and harms

Hand-rear orphaned or abandoned marine and freshwater mammal young

Twelve studies evaluated the effects of hand-rearing orphaned or abandoned marine and freshwater mammal young. Four studies were in the North Pacific Ocean (USA), two studies were in captive facilities (USA), and one study was in each of the North Atlantic Ocean (USA), the Indian River Lagoon (USA), the Salish Sea (USA), the Guerrero Lagoon (USA), the South Atlantic Ocean (Brazil) and water bodies in Florida (USA).

COMMUNITY RESPONSE (0 STUDIES)

POPULATION RESPONSE (11 STUDIES)

Reproductive success (1 study): One replicated study in the South Atlantic Ocean found that most captive-reared Antillean manatees released back into the wild reproduced.

Survival (11 studies): Three studies (including one replicated and controlled study) in the North Pacific Ocean, and the Indian River Lagoon found that a gray whale calf, three Steller sea lion pups, and a common bottlenose dolphin calf that were released after being reared in captivity survived during post-release monitoring periods of between three days to three months. Two replicated studies in the South Atlantic Ocean and water bodies in Florida found that approximately three-quarters of Antillean manatees and two-thirds of Florida manatees that were captive-reared and released were known to survive for at least one year, and some survived for more than seven years. Three studies in the North Atlantic Ocean, the North Pacific Ocean and the Guerrero Lagoon found that three West Indian manatee calves, seven Hawaiian monk seal pups and one Antillean manatee calf that were captive-reared either died before or after release, had to be returned to captivity after release, or survived in the wild only with supplemental feeding. Two studies at captive facilities found that a captive-reared grey whale calf and five pygmy and dwarf sperm whale calves increased in body weight but were either not released or died in captivity. One controlled study in the North Pacific Ocean found that captive-reared, released Pacific harbour seal pups had similar survival estimates to wild pups.

BEHAVIOUR (3 STUDIES)

Behaviour change (3 studies): Two controlled studies (including one replicated study) in the North Pacific Ocean found that captive-reared and released Pacific harbour seal pups and Steller sea lion pups had similar diving behaviour to wild pups. One controlled study in the Salish Sea found that captive-reared and released harbour seal pups travelled greater distances and further from the release site than wild pups born at the same site and in the same season.

Assessment: trade-off between benefit and harms (effectiveness 60%; certainty 50%; harms 20%).

https://www.conservationevidence.com/actions/2926

Remove individual marine and freshwater mammals exhibiting aggressive behaviours that may limit population recovery

One study evaluated the effects of removing individual marine mammals exhibiting aggressive behaviours that may limit population recovery. The study was in the North Pacific Ocean (USA).

COMMUNITY RESPONSE (0 STUDIES)

POPULATION RESPONSE (1 STUDY)

Survival (1 study): One before-and-after study in the North Pacific Ocean found that after removing aggressive male Hawaiian monk seals, the survival of adult female Hawaiian monk seals increased.

Condition (1 study): One before-and-after study in the North Pacific Ocean found that fewer female Hawaiian monk seals were injured after aggressive male Hawaiian monk seals were removed.

BEHAVIOUR (0 STUDIES)

Assessment: trade-off between benefit and harms (effectiveness 60%; certainty 40%; harms 20%).

https://www.conservationevidence.com/actions/2929

Unknown effectiveness

Reunite abandoned marine and freshwater mammal young with parents

One study evaluated the effects of reuniting abandoned marine and freshwater mammal young with parents. The study was in the North Pacific Ocean (USA).

COMMUNITY RESPONSE (0 STUDIES)

POPULATION RESPONSE (1 STUDY)

Reproductive success (1 study): One review in the North Pacific Ocean found that after reuniting Hawaiian monk seal pups with their mothers, along with at least seven other interventions to enhance survival, more than a quarter of the seals reproduced.

Survival (1 study): One review in the North Pacific Ocean found that after reuniting Hawaiian monk seal pups with their mothers, along with at least seven other interventions to enhance survival, more than a quarter of the seals survived.

BEHAVIOUR (0 STUDIES)

Assessment: unknown effectiveness (effectiveness 50%; certainty 25%; harms 0%).

https://www.conservationevidence.com/actions/2927

No evidence found (no assessment)

We have captured no evidence for the following interventions:

- Legally protect marine and freshwater mammal species
- Place orphaned or abandoned marine and freshwater mammal young with foster parents.

14.12.2 Translocation

Based on the collated evidence, what is the current assessment of the effectiveness of interventions for translocation?	
Likely to be beneficial	• Translocate marine and freshwater mammals to re-establish or boost native populations
No evidence found (no assessment)	• Translocate marine and freshwater mammal species before onset of impactful activities

Likely to be beneficial

Translocate marine and freshwater mammals to re-establish or boost native populations

Four studies evaluated the effects of translocating marine mammals to re-establish or boost native populations. The four studies were in the North Pacific Ocean (USA).

COMMUNITY RESPONSE (0 STUDIES)

POPULATION RESPONSE (4 STUDIES)

Reproductive success (2 studies): One replicated study and one review in the North Pacific Ocean found that after translocating Hawaiian monk seals, along with rehabilitation or at least seven other interventions to enhance survival, more than a quarter of the seals reproduced.

Survival (4 studies): Two studies (including one replicated and one controlled study) in the North Pacific Ocean found that 50–83% of translocated, and 52% of rehabilitated and translocated, Hawaiian monk seal pups survived for at least one year. One of the studies and one review in the North Pacific Ocean found that translocated seal pups had similar survival rates to non-translocated pups born at release sites or greater survival rates than non-translocated pups remaining at the original site. One review in the North Pacific Ocean found that translocating Hawaiian monk seals, along with at least seven other interventions to enhance survival, resulted in more than a quarter of the seals surviving.

BEHAVIOUR (1 STUDY)

Behaviour change (1 study): One review in the North Pacific Ocean found that translocated Hawaiian monk seal pups had similar dispersal times to non-translocated seal pups born at release sites.

Assessment: likely to be beneficial (effectiveness 50%; certainty 50%; harms 15%).

https://www.conservationevidence.com/actions/2930

No evidence found (no assessment)

We have captured no evidence for the following interventions:

- Translocate marine and freshwater mammal species before onset of impactful activities.

14.12.3 Captive breeding, rearing and releases (ex-situ conservation)

Based on the collated evidence, what is the current assessment of the effectiveness of interventions for captive breeding, rearing and releases (ex-situ conservation)?	
Trade-off between benefit and harms	• Breed marine and freshwater mammals in captivity
Unknown effectiveness	• Release captive-bred marine and freshwater mammals to re-establish or boost native populations

Trade-off between benefit and harms

Breed marine and freshwater mammals in captivity

Six studies evaluated the effects of breeding marine and freshwater mammals in captivity. Three studies were in the USA, one study was also in China,

Indonesia and Venezuela, and one study was in each of South Africa, Hong Kong and China.

COMMUNITY RESPONSE (0 STUDIES)

POPULATION RESPONSE (6 STUDIES)

Reproductive success (2 studies): One study in Hong Kong found that four of six female Indo-Pacific dolphins successfully conceived during a controlled captive breeding programme and gave birth to a total of nine calves. One study in China found that wild-caught Yangtze finless porpoises successfully reproduced in semi-captive conditions.

Survival (4 studies): Two studies in South Africa and the USA found that a captive-born common bottlenose dolphin, and a captive-born and hand-reared common bottlenose dolphin, survived in captivity for at least two and a half years and four years respectively. One review in the USA found that 80% of common bottlenose dolphins born in captivity over two decades survived, and survival increased with improved husbandry techniques. One review in the USA, China, Indonesia and Venezuela found that most captive-born Amazon river dolphins, narrow-ridged finless porpoises and Irrawaddy dolphins did not survive in captivity.

Condition (1 study): One study in China found that a population of Yangtze finless porpoises breeding in semi-captive conditions had low genetic diversity.

BEHAVIOUR (1 STUDY)

Behaviour change (1 study): One study in the USA found that a captive-born and hand-reared common bottlenose dolphin displayed normal behaviour for the species and joined a dolphin social group in captivity.

Assessment: trade-off between benefit and harms (effectiveness 40%; certainty 40%; harms 20%).

https://www.conservationevidence.com/actions/2932

Unknown effectiveness

Release captive-bred marine and freshwater mammals to re-establish or boost native populations

Two studies evaluated the effects of releasing captive-bred marine and freshwater mammals to re-establish or boost native populations. One study was in the Porto de Pedras estuary (Brazil) and one in water bodies in Florida (USA).

COMMUNITY RESPONSE (0 STUDIES)

POPULATION RESPONSE (2 STUDIES)
Survival (2 studies): Two studies in the Porto de Pedras estuary and water bodies in Florida found that two of three Antillean manatees and two of 14 Florida manatees born in captivity and released into the wild survived for at least one year without further intervention.
BEHAVIOUR (0 STUDIES)
Assessment: unknown effectiveness (effectiveness 30%; certainty 30%; harms 20%).

https://www.conservationevidence.com/actions/2933

14.13 Education and awareness raising

Based on the collated evidence, what is the current assessment of the effectiveness of interventions for education and awareness raising?	
Likely to be beneficial	• Educate the public to improve behaviour towards marine and freshwater mammals
Unknown effectiveness	• Engage policymakers to make policy changes beneficial to marine and freshwater mammals
No evidence found (no assessment)	• Involve local communities in marine and freshwater mammal conservation projects

Likely to be beneficial

Educate the public to improve behaviour towards marine and freshwater mammals

Three studies evaluated the effects of educating the public to improve behaviour towards marine and freshwater mammals. One study was in each of the North Atlantic Ocean (USA), the Sundarbans mangroves (Bangladesh) and the South Pacific Ocean (Peru).

COMMUNITY RESPONSE (0 STUDIES)

POPULATION RESPONSE (0 STUDIES)

BEHAVIOUR (3 STUDIES)

Change in human behaviour (3 studies): Three before-and-after studies in the North Atlantic Ocean, the Sundarbans mangroves and the South Pacific Ocean found that after educational whale-watching tours or an educational exhibition, participants were more willing to change their behaviour to

support marine conservation, to donate money to marine conservation, or to cut their fishing nets to save entangled dolphins.
Assessment: likely to be beneficial (effectiveness 60%; certainty 40%; harms 0%).

https://www.conservationevidence.com/actions/2935

Unknown effectiveness

Engage policymakers to make policy changes beneficial to marine and freshwater mammals

One study evaluated the effects of engaging policymakers to make changes beneficial to marine and freshwater mammals. The study was in the Catazajá wetlands (Mexico).

COMMUNITY RESPONSE (0 STUDIES)

POPULATION RESPONSE (0 STUDIES)

BEHAVIOUR (1 STUDY)

Change in human behaviour (1 study): One study in the Catazajá wetlands reported that engaging policymakers resulted in the designation of a protected area for West Indian manatees.
Assessment: unknown effectiveness (effectiveness 75%; certainty 25%; harms 0%).

https://www.conservationevidence.com/actions/2934

No evidence found (no assessment)

We have captured no evidence for the following interventions:

- Involve local communities in marine and freshwater mammal conservation projects.

15. TERRESTRIAL MAMMAL CONSERVATION

Nick Littlewood, Ricardo Rocha, Rebecca K. Smith, Philip Martin, Sarah Lockhart, Rebecca F. Schoonover, Elspeth Wilman, Andrew J. Bladon, Katie A. Sainsbury, Stuart Pimm & William J. Sutherland

Expert assessors

Jeff Bowman, Ontario Ministry of Natural Resources and Forestry, Canada
Nicolas Caruso, Universidad Nacional del Sur, Argentina
Sebastien Devillard, Claude Bernard University Lyon 1, France
Jeffrey Dunninck, Panthera, USA
Anthony Giordano, SPECIES, USA
Edson Gandiwa, Chinhoyi University of Technology, Zimbabwe
Mohd Azlan Jayasilan bin Abdul Gulam Azad, Universiti Malaysia Sarawak, Malaysia
Claude Fisher, Haute Ecole Specialisee de Suisse Occidentale, Switzerland
Igor Khorozyan, Georg-August University of Göttingen, Germany
Laura Kubasiewicz, Mammal Society, UK
Nick Littlewood, Scotland's Rural College Aberdeen, UK
Lucy Lush, Scottish Wildlife Trust, UK
Silvio Marchini, University of Sao Paolo, Brazil
Catherine McNicol, Gloucestershire Wildlife Trust, UK
William Morgan, University of Cambridge, UK
Hannah Mumby, University of Hong Kong, China
Andres Ordiz, Scandinavian Bear and Wolf Research, Norway
Silviu Petrovan, University of Cambridge, UK Sugoto Roy, IUCN, Switzerland
Viorel Popescu, University of Ohio, USA
Tharmalingam Ramesh, Sálim Ali Centre for Ornithology & Natural History, India
Neil Reid, Queen's University Belfast, UK
Alfredo Romero-Muñoz, Fundacion Cohabitar, Bolivia
Laurentiu Rozylowicz, University of Bucharest, Romania
Euan Ritchie, Deakin University, Australia
Martin Salek, The Czech Academy of Sciences, Czechia
Stephanie Schai-Braun, Universität für Bodenkultur Wien (BOKU), Austria
Henry Schofield, Vincent Wildlife Trust, UK
Ignasi Torre, Museum of Natural Sciencies of Granollers, Spain
Richard Yarnell, Nottingham Trent University, UK
Susanne Vogel, Aarhus University, Denmark
Marco Zaccaroni, University of Florence, Italy
Galo Zapata-Rios, Wildlife Conservation Society, Ecuador
Diana Zlatanova, Sofia University "St. Kliment Ohridski", Bulgaria

Scope of assessment: for native wild terrestrial species across the world.

 https://doi.org/10.11647/OBP.0267.15

Assessed: 2021.

Effectiveness measure is the median % score for effectiveness.

Certainty measure is the median % certainty of evidence for effectiveness, determined by the quantity and quality of the evidence in the synopsis.

Harm measure is the median % score for negative side-effects to the group of species of concern.

This book is meant as a guide to the evidence available for different conservation interventions and as a starting point in assessing their effectiveness. The assessments are based on the available evidence for the target group of species for each intervention. The assessment may therefore refer to different species or habitat to the one(s) you are considering. Before making any decisions about implementing interventions it is vital that you read the more detailed accounts of the evidence in order to assess their relevance for your study species or system.

Full details of the evidence are available at
www.conservationevidence.com

There may also be significant negative side-effects on the target groups or other species or communities that have not been identified in this assessment.

A lack of evidence means that we have been unable to assess whether or not an intervention is effective or has any harmful impacts.

15.1 Threat: Residential and commercial development

Based on the collated evidence, what is the current assessment of the effectiveness of interventions for residential and commercial development?	
Beneficial	• Use collar-mounted devices to reduce predation by domestic animals
Likely to be beneficial	• Keep cats indoors or in outside runs to reduce predation of wild mammals • Prevent mammals accessing potential wildlife food sources or denning sites to reduce nuisance behaviour and human-wildlife conflict • Scare or otherwise deter mammals from human-occupied areas to reduce human-wildlife conflict
Unknown effectiveness	• Install underpasses beneath ski runs • Issue enforcement notices to deter use of non bear-proof garbage dumpsters to reduce human-wildlife conflict • Provide diversionary feeding for mammals to reduce nuisance behaviour and human-wildlife conflict • Provide woody debris in ski run area • Retain wildlife corridors in residential areas
Likely to be ineffective or harmful	• Translocate problem mammals away from residential areas (e.g. habituated bears) to reduce human-wildlife conflict

Based on the collated evidence, what is the current assessment of the effectiveness of interventions for residential and commercial development?	
No evidence found (no assessment)	• Keep dogs indoors or in outside enclosures to reduce threats to wild mammals • Keep domestic cats and dogs well-fed to reduce predation of wild mammals • Protect mammals close to development areas (e.g. by fencing)

Beneficial

Use collar-mounted devices to reduce predation by domestic animals

Five studies evaluated the effects on mammals of using collar-mounted devices to reduce predation by domestic animals. Three studies were in the UK, one was in Australia and one was in the USA.

COMMUNITY RESPONSE (0 STUDIES)

POPULATION RESPONSE (5 STUDIES)

Survival (5 studies): Five replicated studies (including four randomized, controlled studies), in the UK, Australia and the USA, found that bells, a sonic device, and a neoprene flap (which inhibits pouncing) mounted on collars, and a brightly coloured and patterned collar all reduced the rate at which cats predated and returned home with mammals. In one of these studies, an effect was only found in autumn, and not in spring.

BEHAVIOUR (0 STUDIES)

Assessment: beneficial (effectiveness 63%; certainty 70%; harms 0%).

https://www.conservationevidence.com/actions/2332

Likely to be beneficial

Keep cats indoors or in outside runs to reduce predation of wild mammals

One study evaluated the effects on potential prey mammals of keeping cats indoors or in outside runs. This study was in the UK.

COMMUNITY RESPONSE (0 STUDIES)
POPULATION RESPONSE (1 STUDY)
Survival (1 study): One replicated study in the UK found that keeping domestic cats indoors at night reduced the number of dead or injured mammals that were brought home.
BEHAVIOUR (0 STUDIES)
Assessment: likely to be beneficial (effectiveness 60%; certainty 45%; harms 0%).

https://www.conservationevidence.com/actions/2326

Prevent mammals accessing potential wildlife food sources or denning sites to reduce nuisance behaviour and human-wildlife conflict

Two studies evaluated the effects of preventing mammals accessing potential wildlife food sources or denning sites to reduce nuisance behaviour and human-wildlife conflict. One study was in the USA and one was in Switzerland.
COMMUNITY RESPONSE (0 STUDIES)
POPULATION RESPONSE (0 STUDIES)
BEHAVIOUR (0 STUDIES)
OTHER (2 STUDIES)
Human-wildlife conflict (2 studies): A replicated, controlled study in the USA found that electric shock devices prevented American black bears from accessing or damaging bird feeders. A before-and-after study in Switzerland found that electric fencing excluded stone martens from a building.
Assessment: likely to be beneficial (effectiveness 80%; certainty 60%; harms 0%).

https://www.conservationevidence.com/actions/2346

Scare or otherwise deter mammals from human-occupied areas to reduce human-wildlife conflict

Ten studies evaluated the effects of scaring or otherwise deterring mammals from residential areas to reduce human-wildlife conflict. Six studies were in the USA, three were in Canada and one was in Tanzania.
COMMUNITY RESPONSE (0 STUDIES)
POPULATION RESPONSE (0 STUDIES)
BEHAVIOUR (0 STUDIES)
OTHER (10 STUDIES)
Human-wildlife conflict (10 studies): Two of four studies (including one randomized and controlled study) in the USA, found that a range of noise

and pain deterrents did not prevent black bears from returning to urban areas or other human-occupied sites. The other two studies found that such actions did deter them from seeking food at human-occupied sites. Two of three studies, in the USA and Canada, found that chasing nuisance black bears with dogs and chasing elk with people or dogs caused them to stay away longer or remain further from human occupied areas. The other study found that attempts to scare coyotes did not cause them to avoid human occupied areas. A before-and-after study in Canada found that an electric fence prevented polar bear entry to a compound. A study in Canada found that chemical and acoustic repellents did not deter polar bears from baits in most cases. A replicated study in Tanzania found that drones caused African savanna elephants to quickly leave residential areas.
Assessment: likely to be beneficial (effectiveness 50%; certainty 70%; harms 0%).

https://www.conservationevidence.com/actions/2347

Unknown effectiveness

Install underpasses beneath ski runs

One study evaluated the effects on mammals of installing underpasses beneath ski runs. This study was in Australia.
COMMUNITY RESPONSE (0 STUDIES)
POPULATION RESPONSE (0 STUDIES)
BEHAVIOUR (1 STUDY)
Use (1 study): A replicated study in Australia found that boulder-filled crossings beneath ski slopes were used by seven small mammal species.
Assessment: unknown effectiveness (effectiveness 50%; certainty 30%; harms 0%).

https://www.conservationevidence.com/actions/2355

Issue enforcement notices to deter use of non bear-proof garbage dumpsters to reduce human-wildlife conflict

One study evaluated the effects of issuing enforcement notices to deter use of non bear-proof garbage dumpsters to reduce human-wildlife conflict. This study was in the USA.
COMMUNITY RESPONSE (0 STUDIES)
POPULATION RESPONSE (0 STUDIES)
BEHAVIOUR (0 STUDIES)
OTHER (1 STUDY)

Human-wildlife conflict (1 study): A replicated, controlled, before-and-after study in the USA found that issuing enforcement notices requiring appropriate dumpster use did not reduce garbage accessibility to black bears.
Assessment: unknown effectiveness (effectiveness 10%; certainty 30%; harms 0%).

https://www.conservationevidence.com/actions/2345

Provide diversionary feeding for mammals to reduce nuisance behaviour and human-wildlife conflict

Three studies evaluated the effects of providing diversionary feeding for mammals to reduce nuisance behaviour and human-wildlife conflict. Two studies were in the USA and one was in Slovenia.

COMMUNITY RESPONSE (0 STUDIES)

POPULATION RESPONSE (0 STUDIES)

BEHAVIOUR (1 STUDY)

Uptake (1 study): A site comparison study in Slovenia found that 22-63% of the estimated annual energy content of the diet of brown bears comprised provided diversionary food.

OTHER (2 STUDIES)

Human-wildlife conflict (2 studies): Two before-and-after studies (one also a site comparison) in the USA found that diversionary feeding reduced nuisance behaviour by black bears.
Assessment: unknown effectiveness (effectiveness 60%; certainty 20%; harms 0%).

https://www.conservationevidence.com/actions/2323

Provide woody debris in ski run area

One study evaluated the effects on mammals of providing woody debris in ski run areas. This study was in the USA.

COMMUNITY RESPONSE (0 STUDIES)

POPULATION RESPONSE (1 STUDY)

Abundance (1 study): A controlled study in the USA found that placing woody debris on ski slopes did not affect overall small mammal abundance and had mixed effects on individual species abundances.

BEHAVIOUR (0 STUDIES)

Assessment: unknown effectiveness (effectiveness 27%; certainty 27%; harms 0%).

https://www.conservationevidence.com/actions/2356

Retain wildlife corridors in residential areas

One study evaluated the effects on mammals of retaining wildlife corridors in residential areas. This study was in Botswana.

COMMUNITY RESPONSE (0 STUDIES)

POPULATION RESPONSE (0 STUDIES)

BEHAVIOUR (1 STUDY)

Use (1 study): A replicated study in Botswana found that retained wildlife corridors in residential areas were used by 19 mammal species, including African elephants.

Assessment: unknown effectiveness (effectiveness 61%; certainty 30%; harms 0%).

https://www.conservationevidence.com/actions/2354

Likely to be ineffective or harmful

Translocate problem mammals away from residential areas (e.g. habituated bears) to reduce human-wildlife conflict

Eleven studies evaluated the effects of translocating problem mammals (such as bears) away from residential areas to reduce human-wildlife conflict. Six studies were in the USA, two were in Canada, one was Russia, one was in India and one was in Romania.

COMMUNITY RESPONSE (0 STUDIES)

POPULATION RESPONSE (6 STUDIES)

Survival (6 studies): A controlled study in the USA found that grizzly bears translocated away from conflict situations had lower survival rates than did non-translocated bears. A replicated study study in the USA found that fewer than half of black bears translocated from conflict situations survived after one year. Two of three studies (two controlled), in the USA, found that after translocation away from urban sites, white-tailed deer survival was lower than that of non-translocated deer. The third study found that short-term survival was lower but long-term survival was higher than that of non-translocated deer. A study in Russia found that most Amur tigers translocated after attacking dogs or people did not survive for a year after release.

BEHAVIOUR (0 STUDIES)

OTHER (6 STUDIES)

Human-wildlife conflict (6 studies): Five studies (including one controlled and two replicated studies), in the USA and Canada, of brown/grizzly or black

bears translocated away from residential areas or human-related facilities, found that at least some returned to their original capture location and/or continued to cause nuisance. In two of the studies, most returned to their capture area and one black bear returned six times following translocation. A before-and-after study in India found that leopards translocated away from human-dominated areas, attacked more humans and livestock than before-translocation. A controlled study in Romania found that translocated brown bears occurred less frequently inside high potential conflict areas than outside, the opposite to bears that had not been translocated.
Assessment: likely to be ineffective or harmful (effectiveness 31%; certainty 60%; harms 50%).

https://www.conservationevidence.com/actions/2336

No evidence found (no assessment)

We have captured no evidence for the following interventions:

- Keep dogs indoors or in outside enclosures to reduce threats to wild mammals
- Keep domestic cats and dogs well-fed to reduce predation of wild mammals
- Protect mammals close to development areas (e.g. by fencing).

15.2 Threat: Agriculture and aquaculture

15.2.1 All farming systems

Based on the collated evidence, what is the current assessment of the effectiveness of interventions for all farming systems?	
Beneficial	• Create uncultivated margins around intensive arable or pasture fields
Likely to be beneficial	• Establish wild flower areas on farmland • Pay farmers to cover the costs of conservation measures • Plant new or maintain existing hedgerows on farmland • Plant trees on farmland • Provide or retain set-aside areas on farmland
Unknown effectiveness	• Use repellent on slug pellets to reduce non-target poisoning
No evidence found (no assessment)	• Maintain/restore/create habitat connectivity on farmland • Manage hedgerows to benefit wildlife on farmland • Provide refuges during crop harvesting or mowing • Restrict use of rodent poisons on farmland with high secondary poisoning risk

Beneficial

Create uncultivated margins around intensive arable or pasture fields

Nine studies evaluated the effect of creating uncultivated margins around intensive arable, cropped grass or pasture fields on mammals. Six studies were in the UK, two were in Switzerland and one was in the USA.

COMMUNITY RESPONSE (1 STUDY)

Richness/diversity (1 study): One replicated, controlled study in the UK found more small mammal species in uncultivated field margins than in blocks of set-aside.

POPULATION RESPONSE (9 STUDIES)

Abundance (9 studies): One replicated, randomized, controlled study in the USA found more small mammals in uncultivated and unmown field margins than in frequently mown margins. Three of seven replicated, site comparison studies (one randomized), in the UK and Switzerland, found that uncultivated field margins had higher numbers of small mammals, bank voles and brown hares relative to crops (including grassland) and set-aside. The other four studies reported mixed or no effects on bank voles, wood mice and common shrews, small mammals and brown hares. One site comparison study in the UK found that brown hares used grassy field margins more than expected based on their availability.

BEHAVIOUR (0 STUDIES)

Assessment: beneficial (effectiveness 70%; certainty 65%; harms 0%).

https://www.conservationevidence.com/actions/2365

Likely to be beneficial

Establish wild flower areas on farmland

Four studies evaluated the effects of establishing wild flower areas on farmland on small mammals. Two studies were in Switzerland, one in the UK and one in Germany.

COMMUNITY RESPONSE (0 STUDIES)

POPULATION RESPONSE (4 STUDIES)

Abundance (4 studies): Three of four site comparison studies (including three replicated studies), in Switzerland, the UK and Germany, found that sown wildflower areas contained more wood mice, small mammals and common hamsters compared to grass and clover set-aside, grasslands, crop and uncultivated margins, agricultural areas and crop fields.

BEHAVIOUR (0 STUDIES)

Assessment: likely to be beneficial (effectiveness 70%; certainty 60%; harms 0%).

https://www.conservationevidence.com/actions/2359

Pay farmers to cover the costs of conservation measures

Three studies evaluated the effects on mammals of paying farmers to cover the costs of conservation measures. The three studies were in the UK.

COMMUNITY RESPONSE (0 STUDIES)

POPULATION RESPONSE (3 STUDIES)

Abundance (3 studies): A replicated, controlled study, in the UK found that agri-environment scheme enrolment was associated with increased brown hare density in one of two regions studied. A replicated, site comparison study in Northern Ireland, UK found that agri-environment scheme enrolment did not increase numbers of Irish hares. A replicated, controlled study in the UK found that in field margins created through enrolment in an agri-environment scheme, small mammal abundance in spring increased, whereas it remained stable in conventionally managed margins.

BEHAVIOUR (0 STUDIES)

Assessment: likely to be beneficial (effectiveness 57%; certainty 50%; harms 0%).

https://www.conservationevidence.com/actions/2387

Plant new or maintain existing hedgerows on farmland

Three studies evaluated the effects on mammals of planting new or maintaining existing hedgerows on farmland. Two studies were in the UK and one was in Switzerland.

COMMUNITY RESPONSE (0 STUDIES)

POPULATION RESPONSE (3 STUDIES)

Abundance (3 studies): One of two replicated, site comparison studies, in the UK and Switzerland, found that retaining and enhancing hedgerows along with other field boundary features was associated with higher brown hare density in arable sites but not in grassland sites while the other study found that Irish hare numbers did not increase. A replicated, site comparison study in the UK found that establishing hedgerows alongside arable land increased small mammal abundance.

BEHAVIOUR (0 STUDIES)

Assessment: likely to be beneficial (effectiveness 60%; certainty 45%; harms 0%).

https://www.conservationevidence.com/actions/2383

Plant trees on farmland

Four studies evaluated the effects on mammals of planting trees on farmland. Two studies were in the UK, one was in Italy and one was in Australia.

COMMUNITY RESPONSE (0 STUDIES)

POPULATION RESPONSE (2 STUDIES)

Abundance (2 studies): Two replicated studies (including one controlled, and one site comparison study), in the UK, found that farm woodland supported a higher small mammal abundance than on arable land or similar abundance compared to uncultivated field margins and set-aside.

BEHAVIOUR (2 STUDIES)

Use (2 studies): A study in Italy found that tree stands were used more by European hares compared to the wider farmed landscape. A replicated study in Australia found that trees planted on farmland were used by koalas.

Assessment: likely to be beneficial (effectiveness 70%; certainty 60%; harms 0%).

https://www.conservationevidence.com/actions/2386

Provide or retain set-aside areas on farmland

Four studies evaluated the effects on mammals of providing or retaining set-aside areas on farmland. Three studies were in the UK and one was in Switzerland.

COMMUNITY RESPONSE (0 STUDIES)

POPULATION RESPONSE (3 STUDIES)

Abundance (3 studies): Three replicated studies (including two site comparison studies), in the UK and Switzerland, found that set-aside did not enhance small mammal numbers relative to cropland or to uncultivated field margins and farm woodland, or brown hare numbers relative to numbers on farms without set-aside areas.

BEHAVIOUR (1 STUDY)

Use (1 study): A before-and-after study in the UK found that use of uncut set-aside areas by wood mice increased after crop harvesting.

Assessment: likely to be beneficial (effectiveness 40%; certainty 40%; harms 0%).

https://www.conservationevidence.com/actions/2377

Unknown effectiveness

Use repellent on slug pellets to reduce non-target poisoning

One study evaluated the effects on mammals of using repellent on slug pellets

to reduce non-target poisoning. This study was in the UK.
KEY COMMUNITY RESPONSE (0 STUDIES)
POPULATION RESPONSE (0 STUDIES)
BEHAVIOUR (1 STUDY)
Use (1 study): A replicated, controlled study in the UK found that, at some concentrations, food treated with a bitter substance was consumed less by wood mice but not by bank voles or common shrews.
Assessment: unknown effectiveness (effectiveness 31%; certainty 22%; harms 10%).

https://www.conservationevidence.com/actions/2390

No evidence found (no assessment)

We have captured no evidence for the following interventions:

- Maintain/restore/create habitat connectivity on farmland
- Manage hedgerows to benefit wildlife on farmland
- Provide refuges during crop harvesting or mowing
- Restrict use of rodent poisons on farmland with high secondary poisoning risk.

15.2.2 Annual and perennial non-timber crops

Based on the collated evidence, what is the current assessment of the effectiveness of interventions for annual and perennial non-timber crops?	
Beneficial	• Plant crops to provide supplementary food for mammals
Unknown effectiveness	• Create beetle banks on farmland • Establish long-term cover on erodible cropland • Leave cut vegetation in field to provide cover
No evidence found (no assessment)	• Change mowing regime (e.g. timing, frequency, height) • Increase crop diversity for mammals • Leave areas of uncut ryegrass in silage field

Beneficial

Plant crops to provide supplementary food for mammals

Four studies evaluated the effects on mammals of planting crops to provide supplementary food. Two studies were in the USA, one was in the UK and one was in Spain.

COMMUNITY RESPONSE (0 STUDIES)

POPULATION RESPONSE (3 STUDIES)

Abundance (3 studies): Two replicated, controlled studies (including one before-and-after study), in the UK and Spain, found that crops grown to provide food for wildlife resulted in a higher abundance of small mammals in winter, but not in summer and increased European rabbit abundance. A replicated, randomized, controlled study in the USA found that triticale (a cross between wheat and rye) held higher overwintering mule deer abundance relative to barley, annual ryegrass, winter wheat or rye.

BEHAVIOUR (2 STUDIES)

Use (2 studies): A replicated, randomized, controlled study in the USA found that mule deer consumed triticale (a cross between wheat and rye) more than they did barley, annual ryegrass, winter wheat or rye. A replicated, randomized, controlled study in the USA found that supplementary food provided for game species was also consumed by lagomorphs and rodents.

Assessment: beneficial (effectiveness 70%; certainty 62%; harms 0%).

https://www.conservationevidence.com/actions/2394

Unknown effectiveness

Create beetle banks on farmland

One study evaluated the effects on mammals of creating beetle banks on farmland. This study was in the UK.

KEY COMMUNITY RESPONSE (0 STUDIES)

POPULATION RESPONSE (1 STUDY)

Abundance (1 study): One replicated study in the UK found that beetle banks had higher densities of harvest mouse nests than did field margins.

BEHAVIOUR (0 STUDIES)

Assessment: unknown effectiveness (effectiveness 60%; certainty 20%; harms 0%).

https://www.conservationevidence.com/actions/2393

Establish long-term cover on erodible cropland

One study evaluated the effects on mammals of establishing long-term cover on erodible cropland. This study was in the USA.

COMMUNITY RESPONSE (0 STUDIES)

POPULATION RESPONSE (1 STUDY)

Abundance (1 study): A replicated, site comparison study in the USA, found that establishing long-term cover on erodible cropland did not increase the abundance of eastern cottontails.

BEHAVIOUR (0 STUDIES)

Assessment: unknown effectiveness (effectiveness 7%; certainty 27%; harms 0%).

https://www.conservationevidence.com/actions/2402

Leave cut vegetation in field to provide cover

One study evaluated the effects on mammals of leaving cut vegetation in field to provide cover. This study was in the USA

KEY COMMUNITY RESPONSE (0 STUDIES)

POPULATION RESPONSE (1 STUDY)

Abundance (1 study): A controlled, before-and-after study in the USA found that increasing cover, by adding cut vegetation (hay), did not increase rodent abundance.

BEHAVIOUR (0 STUDIES)

Assessment: unknown effectiveness (effectiveness 3%; certainty 30%; harms 0%).

https://www.conservationevidence.com/actions/2401

No evidence found (no assessment)

We have captured no evidence for the following interventions:

- Change mowing regime (e.g. timing, frequency, height)
- Increase crop diversity for mammals
- Leave areas of uncut ryegrass in silage field.

15.2.3 Livestock farming and ranching

Based on the collated evidence, what is the current assessment of the effectiveness of interventions for livestock farming and ranching?	
Beneficial	• Reduce intensity of grazing by domestic livestock
Likely to be beneficial	• Change type of livestock • Exclude livestock from semi-natural habitat (including woodland) • Install mammal crossing points along fences on farmland • Use livestock fences that are permeable to wildlife
Unknown effectiveness	• Use traditional breeds of livestock

Beneficial

Reduce intensity of grazing by domestic livestock

Thirteen studies evaluated the effects on mammals of reducing the intensity of grazing by domestic livestock. Six studies were in the USA, six were in Europe and one was in China.

COMMUNITY RESPONSE (3 STUDIES)

Richness/diversity (3 studies): Two of three site comparison or controlled studies, in the USA and Norway, found that reduced livestock grazing intensity was associated with increased species richness of small mammals whilst one study did not find an increase in species richness.

POPULATION RESPONSE (13 STUDIES)

Abundance (13 studies): Six of nine site comparison or controlled studies (including seven replicated studies), in the USA, Denmark, the UK, China, Netherlands and Norway, found that reductions in livestock grazing intensity were associated with increases in abundances (or proxies of abundances) of small mammals, whilst two studies showed no significant impact of reducing grazing intensity and one study showed mixed results for different species. Two replicated studies (including one controlled and one site comparison study), in the UK and in a range of European countries, found that reducing

grazing intensity did not increase numbers of Irish hares or European hares. A controlled, before-and-after study, in the USA found that exclusion of cattle grazing was associated with higher numbers of elk and mule deer. A replicated, site comparison study in the USA found that an absence of cattle grazing was associated with higher numbers of North American beavers.
BEHAVIOUR (0 STUDIES)
Assessment: beneficial (effectiveness 70%; certainty 65%; harms 0%).

https://www.conservationevidence.com/actions/2408

Likely to be beneficial

Change type of livestock

Two studies evaluated the effect of changing type of livestock on mammals. One study was in the UK and one was in the Netherlands.
COMMUNITY RESPONSE (0 STUDIES)
POPULATION RESPONSE (2 STUDIES)
Abundance (2 studies): One replicated, randomized, paired sites, controlled, before-and-after study in the UK found that sheep and cattle grazing increased field vole abundance relative to sheep-only grazing. One replicated, randomized, paired sites study in the Netherlands found that cattle grazing increased vole abundance relative to horse grazing.
BEHAVIOUR (0 STUDIES)
Assessment: likely to be beneficial (effectiveness 62%; certainty 41%; harms 0%).

https://www.conservationevidence.com/actions/2412

Exclude livestock from semi-natural habitat (including woodland)

Nine studies evaluated the effects of excluding livestock from semi-natural habitat on mammals. Six studies were in the USA, two were in Spain and one was in Australia.
COMMUNITY RESPONSE (2 STUDIES)
Richness/diversity (2 studies): Two replicated, site comparison studies in the USA found more small mammal species on areas from which livestock were excluded.
POPULATION RESPONSE (9 STUDIES)
Abundance (9 studies): Four out of eight studies (including four site comparisons and four controlled studies), in the USA and Spain, found that excluding grazing livestock led to higher abundances of mule deer,

small mammals and, when combined with provision of water, of European rabbits. One study found higher densities of some but not all small mammals species when livestock were excluded and the other three studies found that grazing exclusion did not lead to higher abundances of black-tailed hares, California ground squirrel burrows or of five small mammal species. A site comparison study in Australia found more small mammals where cattle were excluded compared to high intensity cattle-grazing but not compared to medium or low cattle-grazing intensities.

BEHAVIOUR (0 STUDIES)

Assessment: likely to be beneficial (effectiveness 67%; certainty 60%; harms 0%).

https://www.conservationevidence.com/actions/2407

Install mammal crossing points along fences on farmland

Four studies evaluated the effects on mammals of installing mammal crossing points along fences on farmland. Two studies were in Namibia and one each was in the USA and the UK.

KEY COMMUNITY RESPONSE (0 STUDIES)

POPULATION RESPONSE (0 STUDIES)

BEHAVIOUR (4 STUDIES)

Use (4 studies): A study in the USA found that pronghorn antelopes crossed a modified cattle grid which prevented escape of domestic sheep and cows. A controlled, before-and-after study in Namibia found installing swing gates through game fencing reduced the digging of holes by animals under the fence, whilst preventing large predator entry. A study in the UK found that a vertical-sided ditch under an electric fence allowed access by otters. A before-and-after study in Namibia found that tyres installed as crossings through fences were used by wild mammals and reduced fence maintenance requirements.

Assessment: likely to be beneficial (effectiveness 70%; certainty 50%; harms 0%).

https://www.conservationevidence.com/actions/2410

Use livestock fences that are permeable to wildlife

Two studies evaluated the effects on target mammals of using livestock fences that are permeable to wildlife. Both studies were in the USA.

COMMUNITY RESPONSE (0 STUDIES)

POPULATION RESPONSE (0 STUDIES)

BEHAVIOUR (2 STUDIES)

Use (2 studies): A study in the USA found that wild ungulates crossed

a triangular cross-section fence with varying success rates. A replicated, controlled study in the USA found that fences with a lowered top wire were crossed more by elk than were conventional fences.
Assessment: likely to be beneficial (effectiveness 60%; certainty 50%; harms 0%).

https://www.conservationevidence.com/actions/2409

Unknown effectiveness

Use traditional breeds of livestock

One study evaluated the effects of using traditional breeds of livestock on wild mammals. This study was carried out in four European countries.
COMMUNITY RESPONSE (0 STUDIES)
POPULATION RESPONSE (0 STUDIES)
BEHAVIOUR (1 STUDY)
Use (1 study): A replicated, randomized, controlled study in Europe found that European hares did not use areas grazed by traditional livestock breeds more than they used areas grazed by commercial breeds.
Assessment: unknown effectiveness (effectiveness 0%; certainty 10%; harms 0%).

https://www.conservationevidence.com/actions/2411

15.2.4 Reduce human-wildlife conflict

Based on the collated evidence, what is the current assessment of the effectiveness of interventions for reduce human-wildlife conflict?	
Beneficial	• Deter predation of livestock by using shock/ electronic dog-training collars to reduce human-wildlife conflict • Install electric fencing to protect crops from mammals to reduce human-wildlife conflict • Install electric fencing to reduce predation of livestock by mammals to reduce human-wildlife conflict • Install non-electric fencing to exclude predators or herbivores and reduce human-wildlife conflict • Use guardian animals (e.g. dogs, llamas, donkeys) bonded to livestock to deter predators to reduce human-wildlife conflict

Based on the collated evidence, what is the current assessment of the effectiveness of interventions for reduce human-wildlife conflict?	
Likely to be beneficial	• Exclude wild mammals using ditches, moats, walls or other barricades to reduce human-wildlife conflict • Install metal grids at field entrances to prevent mammals entering to reduce human-wildlife conflict • Keep livestock in enclosures to reduce predation by mammals to reduce human-wildlife conflict • Pay farmers to compensate for losses due to predators/wild herbivores to reduce human-wildlife conflict • Provide diversionary feeding to reduce crop damage by mammals to reduce human-wildlife conflict • Use bees to deter crop damage by mammals (e.g. elephants) to reduce human-wildlife conflict • Use chili to deter crop damage by mammals to reduce human-wildlife conflict • Use fire to deter crop damage by mammals to reduce human-wildlife conflict • Use flags to reduce predation of livestock by mammals to reduce human-wildlife conflict • Use lights and sound to deter predation of livestock by mammals to reduce human-wildlife conflict • Use loud noises to deter crop damage (e.g. banger sticks, drums, tins, iron sheets) by mammals to reduce human-wildlife conflict • Use loud noises to deter predation of livestock by mammals to reduce human-wildlife conflict • Use predator scent to deter crop damage by mammals to reduce human-wildlife conflict • Use repellents that taste bad ('contact repellents') to deter crop or property damage by mammals to reduce human-wildlife conflict • Use scent to deter predation of livestock by mammals to reduce human-wildlife conflict

Based on the collated evidence, what is the current assessment of the effectiveness of interventions for reduce human-wildlife conflict?	
	• Use target species distress calls or signals to deter crop damage by mammals to reduce human-wildlife conflict • Use taste-aversion to reduce predation of livestock by mammals to deter human-wildlife conflict
Trade-off between benefit and harms	• Translocate crop raiders away from crops (e.g. elephants) to reduce human-wildlife conflict • Translocate predators away from livestock to reduce human-wildlife conflict
Unknown effectiveness	• Deter predation of livestock by herding livestock using adults instead of children to reduce human-wildlife conflict • Deter predation of livestock by mammals by having people close by to reduce human-wildlife conflict • Dispose of livestock carcasses to deter predation of livestock by mammals to reduce human-wildlife conflict • Drive wild animals away using domestic animals of the same species to reduce human-wildlife conflict • Fit livestock with protective collars to reduce risk of predation by mammals to reduce human-wildlife conflict • Install automatically closing gates at field entrances to prevent mammals entering to reduce human-wildlife conflict • Provide diversionary feeding to reduce predation of livestock by mammals to reduce human-wildlife conflict • Relocate local pastoralist communities to reduce human-wildlife conflict • Use dogs to guard crops to reduce human-wildlife conflict • Use drones to deter crop damage by mammals to reduce human-wildlife conflict • Use light/lasers to deter crop damage by mammals to reduce human-wildlife conflict • Use negative stimuli to deter consumption of livestock feed by mammals to reduce human-wildlife conflict

Based on the collated evidence, what is the current assessment of the effectiveness of interventions for reduce human-wildlife conflict?	
	• Use noise aversive conditioning to deter crop damage by mammals to reduce human-wildlife conflict • Use repellents that smell bad ('area repellents') to deter crop or property damage by mammals to reduce human-wildlife conflict • Use target species scent to deter crop damage by mammals to reduce human-wildlife conflict • Use ultrasonic noises to deter crop damage by mammals to reduce human-wildlife conflict • Use visual deterrents (e.g. scarecrows) to deter predation of livestock by mammals to reduce human-wildlife conflict • Use 'shock collars' to deter crop damage by mammals to reduce human-wildlife conflict
Unlikely to be beneficial	• Use lights and sound to deter crop damage by mammals to reduce human-wildlife conflict
No evidence found (no assessment)	• Establish deviation ponds in fish farms to reduce predation of fish stock by mammals to reduce human-wildlife conflict • Grow unattractive crop in buffer zone around crops (e.g. chili peppers) to reduce human-wildlife conflict • Play predator calls to deter crop damage by mammals to reduce human-wildlife conflict • Use fencing/netting to reduce predation of fish stock by mammals to reduce human-wildlife conflict • Use mobile phone communications to warn farmers of problematic mammals (e.g. elephants) • Use pheromones to deter crop damage by mammals to reduce human-wildlife conflict • Use pheromones to deter predation of livestock by mammals to reduce human-wildlife conflict • Use scarecrows to deter crop damage by mammals to reduce human-wildlife conflict • Use tree nets to deter wild mammals from fruit crops to reduce human-wildlife conflict • Use watchmen to deter crop damage by mammals to reduce human-wildlife conflict

Beneficial

Deter predation of livestock by using shock/electronic dog-training collars to reduce human-wildlife conflict

Five studies evaluated the effects of using shock/electronic dog-training collars to deter predation of livestock to reduce human-wildlife conflict. All five studies were in the USA.

COMMUNITY RESPONSE (0 STUDIES)

POPULATION RESPONSE (0 STUDIES)

BEHAVIOUR (0 STUDIES)

OTHER (5 STUDIES)

Human-wildlife conflict (5 studies): Three of four replicated studies (including two controlled studies), in the USA, found that electric shock collars reduced livestock predation or bait consumption by wolves, whilst one found that they did not reduce wolf bait consumption. One replicated, controlled study in the USA found that electric shock collars reduced the frequency of attacks by captive coyotes on lambs.

Assessment: beneficial (effectiveness 70%; certainty 67%; harms 0%).

https://www.conservationevidence.com/actions/2446

Install electric fencing to protect crops from mammals to reduce human-wildlife conflict

Eleven studies evaluated the effects of installing electric fencing to protect crops from mammals to reduce human-wildlife conflict. Three studies were in Japan, three were in the USA, two were in the UK and one each was in Namibia, India and Guinea-Bissau.

KEY COMMUNITY RESPONSE (0 STUDIES)

POPULATION RESPONSE (0 STUDIES)

BEHAVIOUR (0 STUDIES)

OTHER (11 studies)

Human-wildlife conflict (11 studies): Nine of 11 studies (including three before-and-after studies and three controlled studies), in the USA, the UK, Japan, Namibia, India and Guinea-Bissau, found that electric fences deterred crossings by mammals, ranging in size from European rabbits to elephants. Two studies had mixed results, with some fence designs deterring elephants and black bears.

Assessment: beneficial (effectiveness 80%; certainty 70%; harms 0%).

https://www.conservationevidence.com/actions/2439

Install electric fencing to reduce predation of livestock by mammals to reduce human-wildlife conflict

Eleven studies evaluated the effects of installing electric fencing to reduce predation of livestock by mammals to reduce human-wildlife conflict. Six studies were in the USA (and a further one was presumed to be in the USA) and one each was in Canada, South Africa, Brazil and Spain.

COMMUNITY RESPONSE (0 STUDIES)

POPULATION RESPONSE (0 STUDIES)

BEHAVIOUR (0 STUDIES)

OTHER (11 STUDIES)

Human-wildlife conflict (11 studies): Six out of 10 randomized and/or controlled or before-and-after studies (including eight replicated studies), in the USA (and a further one presumed to be in the USA), Canada, Brazil and Spain, found that electric fences reduced or prevented entry to livestock enclosures or predation of livestock by carnivores. Two studies found that some designs of electric fencing prevented coyotes from entering enclosures and killing or wounding lambs. The other two studies found electric fencing did not reduce livestock predation or prevent fence crossings by carnivores. A before-and-after study in South Africa found that electrifying a fence reduced digging of burrows under the fence that black-backed jackals could pass through.

Assessment: beneficial (effectiveness 70%; certainty 65%; harms 5%).

https://www.conservationevidence.com/actions/2417

Install non-electric fencing to exclude predators or herbivores and reduce human-wildlife conflict

Eight studies evaluated the effects on mammals of installing non-electric fencing to exclude predators or herbivores and reduce human-wildlife conflict. Two studies were in the USA and one each was in Germany, the UK, Spain, China, Tanzania and Kenya.

COMMUNITY RESPONSE (0 STUDIES)

POPULATION RESPONSE (0 STUDIES)

BEHAVIOUR (0 STUDIES)

OTHER (8 STUDIES)

Human-wildlife conflict (8 studies): Four replicated studies (including three before-and-after studies), in USA, China, Tanzania and Kenya, found that non-electric fencing reduced livestock predation by coyotes, Tibetan brown bears, and a range of mammalian predators. A replicated, controlled study in USA found that a high woven wire fence with small mesh, an overhang and an apron (to deter burrowing) was the most effective design at deterring crossings by coyotes. A replicated, controlled study in Germany found that fencing with phosphorescent tape was more effective than fencing with normal yellow tape for deterring red deer and roe deer, but had no effect on crossings by wild boar or brown hare. Two studies (one replicated, before-and-after, site comparison and one controlled study) in the UK and Spain found that fences reduced European rabbit numbers on or damage to crops. *Assessment: beneficial (effectiveness 75%; certainty 70%; harms 2%).*

https://www.conservationevidence.com/actions/2415

Use guardian animals (e.g. dogs, llamas, donkeys) bonded to livestock to deter predators to reduce human-wildlife conflict

Twelve studies evaluated the effects of using guardian animals (e.g. dogs, llamas, donkeys) bonded to livestock to deter mammals from predating these livestock to reduce human-wildlife conflict. Four studies were in the USA, two were in Kenya and one each was in Solvakia, Argentina, Australia, Cameroon, South Africa, and Namibia.

COMMUNITY RESPONSE (0 STUDIES)

POPULATION RESPONSE (0 STUDIES)

BEHAVIOUR (0 STUDIES)

OTHER (12 STUDIES)

Human-wildlife conflict (12 studies): Four of seven studies, (including four site comparison studies), in the USA, Kenya, Solvakia, Australia and Cameroon, found that guardian animals reduced attacks on livestock by predators. The other three studies reported mixed results with reductions in attacks on some but not all age groups or livestock species and reductions for nomadic but not resident pastoralists. Two studies, (including one site comparison study and one before-and-after study), in Argentina and Namibia, found that using dogs to guard livestock reduced the killing of predators by farmers but the number of black-backed jackals killed by farmers and dogs combined increased. A replicated, controlled study in the USA found that fewer sheep guarded by llamas were predated by carnivores in one of two summers whilst a replicated, before-and-after study in South Africa found that using dogs or alpacas to guard livestock reduced attacks by predators. A randomized, replicated, controlled study in USA found that dogs bonded with livestock reduced contact between white-tailed deer and domestic cattle. *Assessment: beneficial (effectiveness 70%; certainty 67%; harms 10%).*

https://www.conservationevidence.com/actions/2433

Likely to be beneficial

Exclude wild mammals using ditches, moats, walls or other barricades to reduce human-wildlife conflict

Two studies evaluated the effects of excluding wild mammals using ditches, moats, walls or other barricades to reduce human-wildlife conflict. One study was in Cameroon and Benin and one was in Cameroon.

COMMUNITY RESPONSE (0 STUDIES)

POPULATION RESPONSE (0 STUDIES)

BEHAVIOUR (0 STUDIES)

OTHER (2 STUDIES)

Human-wildlife conflict (2 studies): Two studies (including one before-and-after study and one site comparison), in Cameroon and Benin and in Cameroon, found that fewer livestock were predated when they were kept in enclosures, especially when these were reinforced.

Assessment: likely to be beneficial (effectiveness 70%; certainty 50%; harms 0%).

https://www.conservationevidence.com/actions/2420

Install metal grids at field entrances to prevent mammals entering to reduce human-wildlife conflict

Two studies evaluated the effects on mammal incursions of installing metal grids at field entrances to reduce human-wildlife conflict. Both of these studies were in the USA.

COMMUNITY RESPONSE (0 STUDIES)

POPULATION RESPONSE (0 STUDIES)

BEHAVIOUR (0 STUDIES)

OTHER (2 STUDIES)

Human-wildlife conflict (2 studies): One of two replicated studies (including one controlled study), in the USA, found that deer guards (horizontal, ground-level metal grids) reduced entry into enclosures by white-tailed deer whilst the other found that they did not prevent crossings by mule deer or elk.

Assessment: likely to be beneficial (effectiveness 50%; certainty 45%; harms 0%).

https://www.conservationevidence.com/actions/2440

Keep livestock in enclosures to reduce predation by mammals to reduce human-wildlife conflict

One study evaluated the effects of keeping livestock in enclosures to reduce predation by mammals to reduce human-wildlife conflict. This study was in Portugal.

COMMUNITY RESPONSE (0 STUDIES)

POPULATION RESPONSE (0 STUDIES)

BEHAVIOUR (0 STUDIES)

OTHER (1 STUDY)

Human-wildlife conflict (1 study): A replicated study in Portugal found fewer wolf attacks on cattle on farms where cattle were confined for at least some of the time compared to those with free-ranging cattle.

Assessment: likely to be beneficial (effectiveness 70%; certainty 50%; harms 0%).

https://www.conservationevidence.com/actions/2438

Pay farmers to compensate for losses due to predators/wild herbivores to reduce human-wildlife conflict

Five studies evaluated the effects on mammals of paying farmers compensation for losses due to predators or wild herbivores to reduce human-wildlife conflict. Three studies were in Kenya and one each was in Italy and Sweden.

COMMUNITY RESPONSE (0 STUDIES)

POPULATION RESPONSE (5 STUDIES)

Abundance (2 studies): Two studies, in Italy and Sweden, found that compensating livestock owners for losses to predators led to increasing populations of wolves and wolverines.

Survival (3 studies): Three before-and-after studies (including two replicated studies), in Kenya, found that when pastoralists were compensated for livestock killings by predators, fewer lions were killed.

BEHAVIOUR (0 STUDIES)

Assessment: likely to be beneficial (effectiveness 80%; certainty 60%; harms 0%).

https://www.conservationevidence.com/actions/2414

Provide diversionary feeding to reduce crop damage by mammals to reduce human-wildlife conflict

Six studies evaluated the effects of providing diversionary feeding to reduce crop damage by mammals to reduce human-wildlife conflict. Three studies were in Canada and one was in each of France, Spain and Austria.

COMMUNITY RESPONSE (0 STUDIES)

POPULATION RESPONSE (0 STUDIES)
BEHAVIOUR (0 STUDIES)
OTHER (6 STUDIES)
Human-wildlife conflict (6 studies): Three of six studies (including four controlled and one before-and-after study) in Canada, France, Spain and Austria found that diversionary feeding reduced damage by red squirrels to pine trees and European rabbits to grape vines, and resulted in fewer red deer using vulnerable forest stands. Two studies found that diversionary feeding did not reduce damage by voles to apple trees or wild boar to grape vines. One study found mixed results on damage by voles to crabapple trees depending on the food provided.
Assessment: likely to be beneficial (effectiveness 50%; certainty 50%; harms 0%).

https://www.conservationevidence.com/actions/2457

Use bees to deter crop damage by mammals (e.g. elephants) to reduce human-wildlife conflict

Three studies evaluated the effects on elephants of using bees to deter crop damage to reduce human-wildlife conflict. All three studies were in Kenya.
COMMUNITY RESPONSE (0 STUDIES)
POPULATION RESPONSE (0 STUDIES)
BEHAVIOUR (0 STUDIES)
OTHER (3 STUDIES)
Human-wildlife conflict (3 studies): Three replicated studies (including one controlled study), in Kenya, found that beehive fences reduced crop raiding by African elephants.
Assessment: likely to be beneficial (effectiveness 70%; certainty 60%; harms 0%).

https://www.conservationevidence.com/actions/2489

Use chili to deter crop damage by mammals to reduce human-wildlife conflict

Seven studies evaluated the effects on elephants of using chili to deter crop damage to reduce human-wildlife conflict. Four studies were in Zimbabwe, two were in Kenya and one was in India.
KEY COMMUNITY RESPONSE (0 STUDIES)
POPULATION RESPONSE (0 STUDIES)
BEHAVIOUR (0 STUDIES)
OTHER (7 STUDIES)

Human-wildlife conflict (7 studies): Five of seven studies (including four replicated and two before-and-after studies), in Zimbabwe, Kenya and India, found that chill-based deterrents (chili-spray, chili smoke, chili fences and chili extract in a projectile, in some cases along with other deterrents) repelled elephants at least initially, whist two studies found that chili smoke (and in one case chili fences) did not reduce crop raiding.
Assessment: likely to be beneficial (effectiveness 50%; certainty 55%; harms 0%).

https://www.conservationevidence.com/actions/2492

Use fire to deter crop damage by mammals to reduce human-wildlife conflict

Two studies evaluated the effects on mammals of using fire to deter crop damage by mammals to reduce human-wildlife conflict. One study was in Zimbabwe and one was in India.
COMMUNITY RESPONSE (0 STUDIES)
POPULATION RESPONSE (0 STUDIES)
BEHAVIOUR (0 STUDIES)
OTHER (2 STUDIES)
Human-wildlife conflict (2 studies): A replicated study in Zimbabwe found that a combination of large fires and people with drums and dogs repelled African elephants from crops faster than did a combination of people with dogs and slingshots, drums and burning sticks. A study in India found that fire reduced the chance of Asian elephants damaging crops.
Assessment: likely to be beneficial (effectiveness 55%; certainty 40%; harms 4%).

https://www.conservationevidence.com/actions/2499

Use flags to reduce predation of livestock by mammals to reduce human-wildlife conflict

Five studies evaluated the effects on mammals of using flags to reduce predation of livestock by mammals to reduce human-wildlife conflict. Three studies were in the USA, one was in Italy and one was in Canada.
COMMUNITY RESPONSE (0 STUDIES)
POPULATION RESPONSE (0 STUDIES)
BEHAVIOUR (0 STUDIES)
OTHER (5 STUDIES)
Human-wildlife conflict (5 studies): Three studies (including two before-and-after studies and a controlled study), in Italy, Canada and the USA, found that flags hanging from fence lines (fladry) deterred crossings by wolves

but not by coyotes. A further replicated, controlled study in the USA found that electric fences with fladry were not crossed by wolves. A replicated, controlled, before-and-after study in the USA found that fladry did not reduce total deer carcass consumption by a range of carnivores.

Assessment: likely to be beneficial (effectiveness 50%; certainty 50%; harms 0%).

https://www.conservationevidence.com/actions/2421

Use lights and sound to deter predation of livestock by mammals to reduce human-wildlife conflict

Three studies evaluated the effects of using lights and sound to deter predation of livestock by mammals to reduce human-wildlife conflict. All three studies were in the USA.

COMMUNITY RESPONSE (0 STUDIES)

POPULATION RESPONSE (0 STUDIES)

BEHAVIOUR (0 STUDIES)

OTHER (3 STUDIES)

Human-wildlife conflict (3 studies): Three replicated studies (including one controlled study), in the USA, found that devices emitting sounds and lights deterred predators from predating sheep or consuming bait.

Assessment: likely to be beneficial (effectiveness 70%; certainty 46%; harms 0%).

https://www.conservationevidence.com/actions/2449

Use loud noises to deter crop damage (e.g. banger sticks, drums, tins, iron sheets) by mammals to reduce human-wildlife conflict

Ten studies evaluated the effects of using loud noises to deter crop damage by mammals to reduce human-wildlife conflict. Three studies were in the USA, two were in Zimbabwe and Kenya and one each was in the UK, Namibia, and India.

COMMUNITY RESPONSE (0 STUDIES)

POPULATION RESPONSE (0 STUDIES)

BEHAVIOUR (0 STUDIES)

OTHER (10 STUDIES)

Human-wildlife conflict (10 studies): Five of six studies (including two controlled, one replicated and two before-and-after studies), in the USA, Namibia, Kenya and India, found that loud noises activated when an animal was in the vicinity reduced or partially reduced crop damage or crop visits by white-tailed deer, black-tailed deer (when combined with using electric shock collars) and elephants. The other study found that using loud noises (along with chili fences and chili smoke) did not reduce crop-raiding by African elephants. Three studies (including two controlled studies), in the UK and the USA, found that regularly sounding loud noises did not repel European rabbits or white-tailed deer. Two replicated studies, in Zimbabwe, found that, from among a range of deterrents, African elephants were repelled faster from crop fields when scared by firecrackers or by a combination of deterrents that included drums.

Assessment: likely to be beneficial (effectiveness 50%; certainty 55%; harms 0%).

https://www.conservationevidence.com/actions/2460

Use loud noises to deter predation of livestock by mammals to reduce human-wildlife conflict

Three studies evaluated the effects of using loud noises to deter predation of livestock by mammals to reduce human-wildlife conflict. Two studies were in the USA and one was in Mexico.

COMMUNITY RESPONSE (0 STUDIES)

POPULATION RESPONSE (0 STUDIES)

BEHAVIOUR (0 STUDIES)

OTHER (3 STUDIES)

Human-wildlife conflict (3 studies): Three replicated studies (including two controlled studies), in the USA and Mexico, found that loud noises at least temporarily deterred sheep predation or food consumption by coyotes and (combined with visual deterrents) deterred livestock predation by large predators.

Assessment: likely to be beneficial (effectiveness 60%; certainty 50%; harms 0%).

https://www.conservationevidence.com/actions/2435

Use predator scent to deter crop damage by mammals to reduce human-wildlife conflict

Three studies evaluated the effects of using predator scent to deter crop damage by mammals to reduce human-wildlife conflict. All three studies were in the USA.

COMMUNITY RESPONSE (0 STUDIES)

POPULATION RESPONSE (0 STUDIES)

BEHAVIOUR (0 STUDIES)

OTHER (3 STUDIES)

Human-wildlife conflict (3 studies): Two of three replicated, randomized, controlled studies (including two before-and-after studies), in the USA, found that coyote scent reduced food consumption by mountain beavers and white-tailed deer. The third study found that it did not reduce trail use by white-tailed deer.

Assessment: likely to be beneficial (effectiveness 50%; certainty 47%; harms 0%).

https://www.conservationevidence.com/actions/2505

Use repellents that taste bad ('contact repellents') to deter crop or property damage by mammals to reduce human-wildlife conflict

Twelve studies evaluated the effects of using repellents that taste bad ('contact repellents') to deter crop or property damage by mammals to reduce human-wildlife conflict. Nine studies were in the USA, two were in the UK and one was in Italy.

COMMUNITY RESPONSE (0 STUDIES)

POPULATION RESPONSE (0 STUDIES)

BEHAVIOUR (0 STUDIES)

OTHER (12 STUDIES)

Human-wildlife conflict (12 studies): Five of 11 controlled studies (including 10 replicated studies), in the USA, Italy and the UK, of a range of contact repellents, found that they reduced herbivory or consumption of baits. The other six studies reported mixed results with at least some repellents at some concentrations deterring herbivory, sometimes for limited periods. A replicated, controlled study in the USA found that a repellent did not prevent chewing damage by coyotes.

Assessment: likely to be beneficial (effectiveness 50%; certainty 55%; harms 0%).

https://www.conservationevidence.com/actions/2509

Use scent to deter predation of livestock by mammals to reduce human-wildlife conflict

Three studies evaluated the effects of using scent to deter predation of livestock by mammals to reduce human-wildlife conflict. Two studies were in the USA and one was in Botswana.

COMMUNITY RESPONSE (0 STUDIES)

POPULATION RESPONSE (0 STUDIES)

BEHAVIOUR (0 STUDIES)

OTHER (3 STUDIES)

Human-wildlife conflict (3 studies): Two of three studies (including one

replicated, before-and-after study), in the USA and Botswana, found that applying scent marks from unfamiliar African wild dogs and grey wolves restricted movements of these species. The other study found that applying scent marks from coyotes did not restrict their movements.
Assessment: likely to be beneficial (effectiveness 50%; certainty 40%; harms 5%).

https://www.conservationevidence.com/actions/2450

Use target species distress calls or signals to deter crop damage by mammals to reduce human-wildlife conflict

Five studies evaluated the effects of using target species distress calls or signals to deter crop damage by these species to reduce human-wildlife conflict. Two studies were in the USA and one each was in Namibia, Australia and Sri Lanka.

COMMUNITY RESPONSE (0 STUDIES)
POPULATION RESPONSE (0 STUDIES)
BEHAVIOUR (0 STUDIES)
OTHER (5 STUDIES)
Human-wildlife conflict (5 studies): Two of five replicated studies (including four controlled studies), in the USA, Namibia, Australia and Sri Lanka, found that white-tailed deer and Asian elephants were deterred or repelled from areas by playing their respective distress calls. Two studies found that, in most cases, elephants and white-tailed deer were not deterred from entering or remaining at sites when distress calls were played. The fifth study found mixed results but, overall, eastern grey kangaroo foot-thumping noises did not increase numbers leaving a site.
Assessment: likely to be beneficial (effectiveness 40%; certainty 50%; harms 0%).

https://www.conservationevidence.com/actions/2488

Use taste-aversion to reduce predation of livestock by mammals to deter human-wildlife conflict

Nine studies evaluated the effects of using taste-aversion to reduce predation of livestock by mammals to deter human-wildlife conflict. Six studies were in the USA, two were in Canada and one was at an unnamed location.

COMMUNITY RESPONSE (0 STUDIES)
POPULATION RESPONSE (0 STUDIES)
BEHAVIOUR (0 STUDIES)
OTHER (9 STUDIES)
Human-wildlife conflict (9 studies): Three of seven replicated studies (including

three controlled studies), in the USA, Canada and at an unnamed location, found that coyotes killed fewer sheep, rabbits or turkeys after taste-aversion treatment. The other four studies found that taste-aversion treatment did not reduce killing by coyotes of chickens, sheep or rabbits. A replicated, before-and-after study in the USA found that taste-aversion treatment reduced egg predation by mammalian predators whilst a replicated, controlled, paired sites study in the USA found no such effect.

Assessment: likely to be beneficial (effectiveness 40%; certainty 50%; harms 0%).

https://www.conservationevidence.com/actions/2429

Trade-off between benefit and harms

Translocate crop raiders away from crops (e.g. elephants) to reduce human-wildlife conflict

Two studies evaluated the effects on mammals of translocating crop-raiding animals away from crops to reduce human-wildlife conflict. One study was in Kenya and one was in the USA.

COMMUNITY RESPONSE (0 STUDIES)

POPULATION RESPONSE (1 STUDY)

Survival (1 study): A controlled study in Kenya found that translocated crop-raiding African elephants had a lower survival rate after release than did non-translocated elephants at the same site.

BEHAVIOUR (0 STUDIES)

OTHER (1 STUDY)

Human-wildlife conflict (1 study): A study in the USA found that most American black bears translocated from sites of crop damage were not subsequently recaptured at sites of crop damage.

Assessment: trade-off between benefit and harms (effectiveness 40%; certainty 40%; harms 37%).

https://www.conservationevidence.com/actions/2485

Translocate predators away from livestock to reduce human-wildlife conflict

Eleven studies evaluated the effects on mammals of translocating predators away from livestock to reduce human-wildlife conflict. Four studies were in the USA two were in Botswana, one each was in Canada, Zimbabwe and Namibia, one was in Venezuela and Brazil and one covered multiple locations in North and Central America and Africa.

COMMUNITY RESPONSE (0 STUDIES)
POPULATION RESPONSE (8 STUDIES)
Reproductive success (2 studies): Two studies, in Zimbabwe and Namibia, found that predators translocated away from livestock bred in the wild after release.
Survival (8 studies): Four of eight studies (including three replicated studies and a systematic review), in the USA, Canada, Zimbabwe, South America, Botswana and Namibia, found that translocating predators reduced their survival or that most did not survive more than 6–12 months after release. Three studies found that translocated predators had similar survival to that of established animals or persisted in the wild and one study could not determine the effect of translocation on survival.
BEHAVIOUR (0 STUDIES)
OTHER (6 STUDIES)
Human-wildlife conflict (6 studies): Four of six studies (including a review and a systematic review), in the USA, South America and in North and Central America and Africa, found that some translocated predators continued to predate livestock or returned to their capture sites. One study found that translocated predators were not subsequently involved in livestock predation and one study could not determine the effect of translocation on livestock predation.
Assessment: trade-off between benefit and harms (effectiveness 40%; certainty 56%; harms 30%).

https://www.conservationevidence.com/actions/2436

Unknown effectiveness

Deter predation of livestock by herding livestock using adults instead of children to reduce human-wildlife conflict

One study evaluated the effects on predatory mammal activities of herding livestock using adults instead of children to reduce human-wildlife conflict. This study was in Cameroon.
COMMUNITY RESPONSE (0 STUDIES)
POPULATION RESPONSE (0 STUDIES)
BEHAVIOUR (0 STUDIES)
OTHER (1 STUDY)
Human-wildlife conflict (1 study): A site comparison study in Cameroon found that using adults to herd livestock reduced losses through predation

relative to that of livestock herded solely by children.
Assessment: unknown effectiveness (effectiveness 70%; certainty 30%; harms 0%).

https://www.conservationevidence.com/actions/2445

Deter predation of livestock by mammals by having people close by to reduce human-wildlife conflict

One study evaluated the effects of deterring predation of livestock by mammals by having people close by to reduce human-wildlife conflict. This study was in Kenya.

COMMUNITY RESPONSE (0 STUDIES)

POPULATION RESPONSE (0 STUDIES)

BEHAVIOUR (0 STUDIES)

OTHER (1 STUDY)

Human-wildlife conflict (1 study): One study in Kenya recorded fewer attacks by predators on livestock in bomas when people were also present but the presence of people did not reduce predator attacks on grazing herds.
Assessment: unknown effectiveness (effectiveness 50%; certainty 30%; harms 0%).

https://www.conservationevidence.com/actions/2444

Dispose of livestock carcasses to deter predation of livestock by mammals to reduce human-wildlife conflict

One study evaluated the effects of disposing of livestock carcasses to deter predation of livestock by mammals to reduce human-wildlife conflict. This study was in the USA.

COMMUNITY RESPONSE (0 STUDIES)

POPULATION RESPONSE (0 STUDIES)

BEHAVIOUR (0 STUDIES)

OTHER (1 STUDY)

Human-wildlife conflict (1 study): One site comparison study in the USA found that burying or removing sheep carcasses reduced predation on livestock by coyotes, but burning carcasses did not alter livestock predation rates.
Assessment: unknown effectiveness (effectiveness 60%; certainty 35%; harms 0%).

https://www.conservationevidence.com/actions/2432

Drive wild animals away using domestic animals of the same species to reduce human-wildlife conflict

One study evaluated the effects of using domestic animals to drive away wild mammals to reduce human-wildlife conflict. This study was in India.

COMMUNITY RESPONSE (0 STUDIES)
POPULATION RESPONSE (0 STUDIES)
BEHAVIOUR (0 STUDIES)
OTHER (1 STUDY)
Human-wildlife conflict (1 study): One study in India found that using domestic elephants to drive wild Asian elephants away from villages did not reduce the probability of elephants damaging crops.
Assessment: unknown effectiveness (effectiveness 0%; certainty 20%; harms 0%).

https://www.conservationevidence.com/actions/2513

Fit livestock with protective collars to reduce risk of predation by mammals to reduce human-wildlife conflict

One study evaluated the effects of fitting livestock with protective collars to reduce human-wildlife conflict on rates of livestock killings by predators. This study was in South Africa.
COMMUNITY RESPONSE (0 STUDIES)
POPULATION RESPONSE (0 STUDIES)
BEHAVIOUR (0 STUDIES)
OTHER (1 STUDY)
Human-wildlife conflict (1 study): A replicated, before-and-after study in South Africa found that livestock protection collars reduced predation on livestock by carnivores.
Assessment: unknown effectiveness (effectiveness 60%; certainty 30%; harms 0%).

https://www.conservationevidence.com/actions/2448

Install automatically closing gates at field entrances to prevent mammals entering to reduce human-wildlife conflict

One study evaluated the effects on mammal movements of installing automatically closing gates at field entrances to reduce human-wildlife conflict. This study was in USA.
KEY COMMUNITY RESPONSE (0 STUDIES)
POPULATION RESPONSE (0 STUDIES)
BEHAVIOUR (0 STUDIES)
OTHER (1 STUDY)
Human-wildlife conflict (1 study): A replicated, controlled study, in the USA

found that vehicle-activated bump gates prevented white-tailed deer from entering enclosures.

Assessment: unknown effectiveness (effectiveness 70%; certainty 38%; harms 0%).

https://www.conservationevidence.com/actions/2441

Provide diversionary feeding to reduce predation of livestock by mammals to reduce human-wildlife conflict

Two studies evaluated the effects of providing diversionary feeding to reduce predation of livestock by mammals to reduce human-wildlife conflict. One study was in the USA and one was in Canada.

COMMUNITY RESPONSE (0 STUDIES)

POPULATION RESPONSE (1 STUDY)

Reproductive success (1 study): A controlled study in the USA found that diversionary feeding of predators did not increase overall nest success rates for ducks.

BEHAVIOUR (0 STUDIES)

OTHER (2 STUDIES)

Human-wildlife conflict (2 studies): One of two studies (one controlled, one before-and-after study) in the USA and Canada found that diversionary feeding reduced striped skunk predation on duck nests. The other study found that diversionary feeding of grizzly bears did not reduce predation on livestock.

Assessment: unknown effectiveness (effectiveness 21%; certainty 20%; harms 0%).

https://www.conservationevidence.com/actions/2437

Relocate local pastoralist communities to reduce human-wildlife conflict

One study evaluated the effects on mammals of relocating local pastoralists to reduce human-wildlife conflict. This study was in India.

COMMUNITY RESPONSE (0 STUDIES)

POPULATION RESPONSE (1 STUDY)

Abundance (1 study): A study in India found that after most pastoralists were relocated outside of an area, Asiatic lion numbers increased.

BEHAVIOUR (0 STUDIES)

Assessment: unknown effectiveness (effectiveness 60%; certainty 10%; harms 10%).

https://www.conservationevidence.com/actions/2413

Use dogs to guard crops to reduce human-wildlife conflict

One study evaluated the effects on mammals of using dogs to guard crops to reduce human-wildlife conflict. This study was in Zimbabwe.

COMMUNITY RESPONSE (0 STUDIES)

POPULATION RESPONSE (0 STUDIES)

BEHAVIOUR (0 STUDIES)

OTHER (1 STUDY)

Human-wildlife conflict (1 study): A replicated study in Zimbabwe found that people with dogs took longer to repel African elephants from crops compared to scaring them by using combinations of people, dogs, slingshots, drums, burning sticks, large fires and spraying with capsicum.

Assessment: unknown effectiveness (effectiveness 27%; certainty 20%; harms 0%).

https://www.conservationevidence.com/actions/2512

Use drones to deter crop damage by mammals to reduce human-wildlife conflict

One study evaluated the effects on mammals of using drones to deter crop damage by mammals to reduce human-wildlife conflict. This study was in Tanzania.

KEY COMMUNITY RESPONSE (0 STUDIES)

POPULATION RESPONSE (0 STUDIES)

BEHAVIOUR (0 STUDIES)

OTHER (1 STUDY)

Human-wildlife conflict (1 study): A replicated study in Tanzania found that drones repelled African savanna elephants from crops within one minute.

Assessment: unknown effectiveness (effectiveness 72%; certainty 35%; harms 0%).

https://www.conservationevidence.com/actions/2481

Use light/lasers to deter crop damage by mammals to reduce human-wildlife conflict

Two studies evaluated the effects of using light or lasers to deter crop damage by mammals to reduce human-wildlife conflict. Both studies were in the USA.

COMMUNITY RESPONSE (0 STUDIES)

POPULATION RESPONSE (0 STUDIES)

BEHAVIOUR (0 STUDIES)

OTHER (2 STUDIES)

Human-wildlife conflict (2 studies): A replicated, randomized, controlled study in the USA found that red lasers did not disperse white-tailed deer from fields at night whilst a study in India found that spotlights directed at the eyes of Asian elephants did reduce the probability of crop damage.
Assessment: unknown effectiveness (effectiveness 40%; certainty 23%; harms 0%).

https://www.conservationevidence.com/actions/2496

Use negative stimuli to deter consumption of livestock feed by mammals to reduce human-wildlife conflict

One study evaluated the effects of using negative stimuli to deter consumption of livestock feed by mammals to reduce human-wildlife conflict. This study was in the USA.

KEY COMMUNITY RESPONSE (0 STUDIES)

POPULATION RESPONSE (0 STUDIES)

BEHAVIOUR (0 STUDIES)

OTHER (1 STUDY)

Human-wildlife conflict (1 study): A replicated, controlled study in the USA found that white-tailed deer presence at cattle feeders was usually reduced by a device that produced a negative stimulus.
Assessment: unknown effectiveness (effectiveness 50%; certainty 20%; harms 0%).

https://www.conservationevidence.com/actions/2486

Use noise aversive conditioning to deter crop damage by mammals to reduce human-wildlife conflict

One study evaluated the effects of using noise aversive conditioning to deter crop damage by mammals to reduce human-wildlife conflict. This study was in the USA.

COMMUNITY RESPONSE (0 STUDIES)

POPULATION RESPONSE (0 STUDIES)

BEHAVIOUR (0 STUDIES)

OTHER (1 STUDY)

Human-wildlife conflict (1 study): A replicated, controlled study in USA found that noise aversive conditioning reduced bait consumption by white-tailed deer.
Assessment: unknown effectiveness (effectiveness 55%; certainty 30%; harms 0%).

https://www.conservationevidence.com/actions/2461

Use repellents that smell bad ('area repellents') to deter crop or property damage by mammals to reduce human-wildlife conflict

One study evaluated the effects of using repellents that smell bad ('area repellents') to deter crop or property damage by mammals to reduce human-wildlife conflict. This study was in the UK.

KEY COMMUNITY RESPONSE (0 STUDIES)

POPULATION RESPONSE (0 STUDIES)

BEHAVIOUR (0 STUDIES)

OTHER (1 STUDY)

Human-wildlife conflict (1 study): A randomized, replicated, controlled study in the UK found that a repellent reduced use of treated areas by moles.

Assessment: unknown effectiveness (effectiveness 60%; certainty 21%; harms 0%).

https://www.conservationevidence.com/actions/2511

Use target species scent to deter crop damage by mammals to reduce human-wildlife conflict

One study evaluated the effects on mammals of using target species scent to deter crop damage to reduce human-wildlife conflict. This study was in South Africa.

COMMUNITY RESPONSE (0 STUDIES)

POPULATION RESPONSE (0 STUDIES)

BEHAVIOUR (0 STUDIES)

OTHER (1 STUDY)

Human-wildlife conflict (1 study): A replicated, controlled study in South Africa found that African elephants were not deterred from feeding by the presence of secretions from elephant temporal glands.

Assessment: unknown effectiveness (effectiveness 0%; certainty 20%; harms 0%).

https://www.conservationevidence.com/actions/2506

Use ultrasonic noises to deter crop damage by mammals to reduce human-wildlife conflict

One study evaluated the effects of using ultrasonic noises to deter crop damage by mammals to reduce human-wildlife conflict. This study was in Australia.

COMMUNITY RESPONSE (0 STUDIES)

POPULATION RESPONSE (0 STUDIES)

BEHAVIOUR (0 STUDIES)

OTHER (1 STUDY)

Human-wildlife conflict (1 study): A replicated, controlled, paired sites study in Australia found that ultrasonic devices did not repel eastern gray kangaroos.

Assessment: unknown effectiveness (effectiveness 0%; certainty 20%; harms 0%).

https://www.conservationevidence.com/actions/2479

Use visual deterrents (e.g. scarecrows) to deter predation of livestock by mammals to reduce human-wildlife conflict

Two studies evaluated the effects of using visual deterrents, such as scarecrows, to deter predation of livestock by mammals to reduce human-wildlife conflict. One study was in Kenya and one was in Mexico.

COMMUNITY RESPONSE (0 STUDIES)

POPULATION RESPONSE (0 STUDIES)

BEHAVIOUR (0 STUDIES)

OTHER (2 STUDIES)

Human-wildlife conflict (2 studies): A study in Kenya recorded more livestock predation at bomas with scarecrows than those without scarecrows whereas a replicated, controlled study in Mexico found that a combination of visual and sound deterrents reduced livestock predation.

Assessment: unknown effectiveness (effectiveness 30%; certainty 30%; harms 10%).

https://www.conservationevidence.com/actions/2427

Use 'shock collars' to deter crop damage by mammals to reduce human-wildlife conflict

One study evaluated the effects on mammals of using 'shock collars' to deter crop damage to reduce human-wildlife conflict. This study was in the USA.

COMMUNITY RESPONSE (0 STUDIES)

POPULATION RESPONSE (0 STUDIES)

BEHAVIOUR (0 STUDIES)

OTHER (1 STUDY)

Human-wildlife conflict (1 study): A replicated, controlled study in the USA found that electric shock collars (combined with loud noise) reduced damage caused by black-tailed deer to tree seedlings.

Assessment: unknown effectiveness (effectiveness 60%; certainty 20%; harms 0%).

https://www.conservationevidence.com/actions/2508

Unlikely to be beneficial

Use lights and sound to deter crop damage by mammals to reduce human-wildlife conflict

Two studies evaluated the effects of using both lights and sound to deter crop damage by mammals to reduce human-wildlife conflict. Both studies were in the USA.

COMMUNITY RESPONSE (0 STUDIES)

POPULATION RESPONSE (0 STUDIES)

BEHAVIOUR (0 STUDIES)

OTHER (2 STUDIES)

Human-wildlife conflict (2 studies): Two replicated paired sites, controlled studies (one also randomized), in the USA, found that frightening devices, emitting lights and sound, did not reduce crop intrusions by white-tailed deer or food consumption by elk and mule deer.

Assessment: unlikely to be beneficial (effectiveness 20%; certainty 40%; harms 0%).

https://www.conservationevidence.com/actions/2456

No evidence found (no assessment)

We have captured no evidence for the following interventions:

- Establish deviation ponds in fish farms to reduce predation of fish stock by mammals to reduce human-wildlife conflict
- Grow unattractive crop in buffer zone around crops (e.g. chili peppers) to reduce human-wildlife conflict
- Play predator calls to deter crop damage by mammals to reduce human-wildlife conflict
- Use fencing/netting to reduce predation of fish stock by mammals to reduce human-wildlife conflict
- Use mobile phone communications to warn farmers of problematic mammals (e.g. elephants)
- Use pheromones to deter crop damage by mammals to reduce human-wildlife conflict
- Use pheromones to deter predation of livestock by mammals to reduce human-wildlife conflict
- Use scarecrows to deter crop damage by mammals to reduce human-wildlife conflict

- Use tree nets to deter wild mammals from fruit crops to reduce human-wildlife conflict
- Use watchmen to deter crop damage by mammals to reduce human-wildlife conflict.

15.3 Threat: Energy production and mining

Based on the collated evidence, what is the current assessment of the effectiveness of interventions for energy production and mining?	
Likely to be beneficial	• Restore former mining sites
Unknown effectiveness	• Translocate mammals away from sites of proposed energy developments • Use repellents to reduce cable gnawing
No evidence found (no assessment)	• Use electric fencing to deter mammals from energy installations or mines

Likely to be beneficial

Restore former mining sites

Twelve studies evaluated the effects of restoring former mining sites on mammals. Eleven studies were in Australia and one was in the USA.

COMMUNITY RESPONSE (8 STUDIES)

Species richness (8 studies): A review in Australia found that seven of 11 studies indicated that rehabilitated areas had lower mammal species richness compared to unmined areas. Four of five replicated, site comparison studies, in Australia, found that mammal species richness was similar in restored mine areas compared to unmined areas or higher in restored areas (but similar when considering only native species). One study found that species richness was lower in restored compared to in unmined areas. A replicated, controlled study in Australia found that thinning trees and burning vegetation as part of mine restoration did not increase small mammal species richness. A replicated, site comparison study in Australia found that restored mine areas were recolonized by a range of mammal species within 10 years.

POPULATION RESPONSE (5 STUDIES)

Abundance (5 studies): A review of rehabilitated mine sites in Australia found that only two of eight studies indicated that rehabilitated areas had equal or higher mammal densities compared to those in unmined areas. One of three replicated, site comparison studies, in the USA and Australia, found that small mammal density was similar on restored mines compared to on unmined land. One study found that for three of four species (including all three native species studied) abundance was lower in restored compared to unmined sites and one study found mixed results, including that abundances of two out of three focal native species were lower in restored compared to unmined sites. A replicated, controlled study in Australia found that thinning trees and burning vegetation as part of mine restoration did not increase small mammal abundance.

BEHAVIOUR (2 STUDIES)

Use (2 studies): A replicated, site comparison study in Australia found that most restored former mine areas were not used by koalas while another replicated site comparison study in Australia found quokka activity to be similar in revegetated mined sites compared to in unmined forest.

OTHER (1 STUDY)

Genetic diversity (1 study): A site comparison study in Australia found that in forest on restored mine areas, genetic diversity of yellow-footed antechinus was similar to that in unmined forest.

Assessment: likely to be beneficial (effectiveness 40%; certainty 50%; harms 0%).

https://www.conservationevidence.com/actions/2490

Unknown effectiveness

Translocate mammals away from sites of proposed energy developments

Two studies evaluated the effects of translocating mammals away from sites of proposed energy developments. One study was in Brazil and one was in Australia.

COMMUNITY RESPONSE (0 STUDIES)

POPULATION RESPONSE (0 STUDIES)

BEHAVIOUR (2 STUDIES)

Behaviour change (2 studies): A study in Brazil found that lesser anteaters translocated away from a hydroelectric development site remained close to release sites while a study in Australia found that at least one out of eight chuditchs translocated from a site to be mined returned to its site of capture. *Assessment: unknown effectiveness (effectiveness 50%; certainty 23%; harms 5%).*

https://www.conservationevidence.com/actions/2517

Use repellents to reduce cable gnawing

One study evaluated the effects of using repellents to reduce cable gnawing. This study was in the USA.

COMMUNITY RESPONSE (0 STUDIES)

POPULATION RESPONSE (0 STUDIES)

BEHAVIOUR (0 STUDIES)

OTHER (1 STUDY)

Human-wildlife conflict (1 study): A randomized, replicated, controlled study in the USA found that repellents only deterred cable gnawing by northern pocket gophers when encased in shrink-tubing. *Assessment: unknown effectiveness (effectiveness 40%; certainty 30%; harms 0%).*

https://www.conservationevidence.com/actions/2502

No evidence found (no assessment)

We have captured no evidence for the following interventions:

- Use electric fencing to deter mammals from energy installations or mines.

15.4 Threat: Transportation and service corridors

15.4.1 Roads and railroads

Based on the collated evidence, what is the current assessment of the effectiveness of interventions for roads and railroads?	
Beneficial	• Install barrier fencing along roads • Install barrier fencing and underpasses along roads • Install overpasses over roads/railways • Install rope bridges between canopies
Likely to be beneficial	• Install fences around existing culverts or underpasses under roads/railways • Install ledges in culverts under roads/railways • Install one-way gates or other structures to allow wildlife to leave roadways • Install pole crossings for gliders/flying squirrels • Install signage to warn motorists about wildlife presence • Install tunnels/culverts/underpass under railways • Install tunnels/culverts/underpass under roads • Install wildlife exclusion grates/cattle grids • Install wildlife warning reflectors along roads • Modify vegetation along railways to reduce collisions by reducing attractiveness to mammals • Provide food/salt lick to divert mammals from roads or railways

Based on the collated evidence, what is the current assessment of the effectiveness of interventions for roads and railroads?	
Unknown effectiveness	• Close roads in defined seasons • Dig trenches around culverts under roads/ railways • Install acoustic wildlife warnings along roads • Install barrier fencing along railways • Install traffic calming structures to reduce speeds • Install wildlife crosswalks • Modify culverts to make them more accessible to mammals • Modify the roadside environment to reduce collisions by reducing attractiveness of road verges to mammals • Reduce legal speed limit • Use road lighting to reduce vehicle collisions with mammals
Unlikely to be beneficial	• Fit vehicles with ultrasonic warning devices • Use chemical repellents along roads or railways
No evidence found (no assessment)	• Modify vegetation along roads to reduce collisions with mammals by enhancing visibility for drivers • Remove roadkill regularly to reduce kill rate of predators/scavengers • Retain/maintain road verges as small mammal habitat • Use alternative de-icers on roads • Use reflective collars or paint on mammals to reduce collisions with road vehicles • Use wildlife decoy to reduce vehicle collisions with mammals

Beneficial

● Install barrier fencing along roads

Twelve studies evaluated the effects on mammals of installing barrier fencing along roads. Eight studies were in the USA, one each was in Canada, Germany and Brazil and one spanned the USA, Canada and Sweden.

COMMUNITY RESPONSE (0 STUDIES)

POPULATION RESPONSE (9 STUDIES)

Survival (9 studies): Three controlled studies, in the USA, Germany and Brazil, found that roadside fencing or equivalent barrier systems reduced the numbers of mammals, including wildcats and coypu, killed by vehicles on roads. Two before-and-after studies, in the USA, found that roadside fencing with one-way gates to allow escape from the road, reduced the number of collisions between vehicles and deer. A study in the USA found that a 2.7-m-high fence did not reduce road-kills of white-tailed deer compared to a 2.2-m-high fence. A controlled, before-and-after study in the USA found that barrier fencing with designated crossing points did not significantly reduce road deaths of mule deer. A replicated, controlled, before-and-after study in Canada found that electric fences, (along with an underpass beneath one highway), reduced moose-vehicle collisions. A review of fencing studies from USA, Canada and Sweden, found that longer fencing along roadsides led to a greater reduction of collisions between large mammals and cars than did shorter fence sections.

BEHAVIOUR (5 STUDIES)

Behaviour change (5 studies): A controlled, before-and-after study in the USA found that 2.3-m-high fencing in good condition prevented most white-tailed deer accessing a highway. A replicated, controlled, before-and-after study in Canada found that electric fences reduced moose access to highways. Three studies (two replicated), in the USA, found that higher fences (2.4–2.7 m) prevented more white-tailed deer from entering highways than did fences that were 2.2 m high, 1.2 m high with outriggers or 1.2–1.8 m high.

Assessment: beneficial (effectiveness 70%; certainty 70%; harms 0%).

https://www.conservationevidence.com/actions/2567

Install barrier fencing and underpasses along roads

Fifty-five studies evaluated the effects on mammals of installing barrier fencing and underpasses along roads. Twenty-seven were in the USA, nine were in Canada, seven were in Australia, two each were in Spain, Portugal, the UK and Sweden, one each was in Denmark, Germany and Croatia and one was a review covering Australia, Europe and North America.

COMMUNITY RESPONSE (0 STUDIES)

POPULATION RESPONSE (15 STUDIES)

Survival (15 studies): Eleven of 15 studies (including 12 before-and-after studies and two site comparisons), in the USA, Australia, Sweden and Canada, found that installing underpasses and associated roadside barrier fencing reduced collisions between vehicles and mammals. Three studies found that the roadkill rate was not reduced and one study found that vehicle-mammal collisions continued to occur after installation.

BEHAVIOUR (52 STUDIES)

Use (52 studies): Seventeen of 18 studies (including 10 before-and-after studies) in the USA, Canada and Sweden, which reported exclusively on ungulates, found that underpasses installed along with roadside barrier fencing were used by a range of ungulate species. These were mule deer, mountain goat, pronghorn, white-tailed deer, elk, moose and Florida Key deer. The other study found that underpasses were not used by moose whilst one of the studies that did report use by ungulates further reported that they were not used by white-tailed deer. Further observations from these studies included that elk preferred more open, shorter underpasses to those that were enclosed or longer, underpass use was not affected by traffic levels and that mule deer used underpasses less than they used overpasses. Thirty-four studies (including four before-and-after studies, seven replicated studies, three site comparisons and two reviews), in the USA, Canada, Australia, Spain, Portugal, the UK, Denmark, Germany, Croatia and across multiple continents, that either studied mammals other than ungulates or multiple species including ungulates, found that underpasses in areas with roadside fencing were used by mammals. Among these studies, one found that small culverts were used by mice and voles more than were larger underpasses, one found that bandicoots used underpasses less after they were lengthened and one found that culverts were used by grizzly bears less often than were overpasses.

Assessment: beneficial (effectiveness 72%; certainty 70%; harms 0%).

https://www.conservationevidence.com/actions/2571

● Install overpasses over roads/railways

Twenty-two studies evaluated the effects on mammals of installing overpasses over roads or railways. Seven studies were in Canada, three were in Spain, three were in Australia, two were in Sweden, one each was in the Netherlands, Germany, Croatia and the USA, and three (including two reviews) were conducted across multiple countries.

COMMUNITY RESPONSE (0 STUDIES)

POPULATION RESPONSE (4 STUDIES)

Survival (4 studies): Four studies (including three before-and-after studies), in Canada, Sweden and Australia, found that overpasses (in combination with roadside fencing) reduced collisions between vehicles and mammals. In two of these studies, data from overpasses and underpasses were combined for analysis.

BEHAVIOUR (21 STUDIES)

Use (21 studies): Nineteen studies, in North America, Europe and Australia, found that overpasses were used by mammals. A wide range of mammals was reported using overpasses, including rodents and shrews, rabbits and hares, carnivores, ungulates, bears, marsupials and short-beaked echidna. A review of crossing structures in Australia, Europe and North America found that overpasses were used by a range of mammals, particularly larger mammal species. A global review of crossing structures (including overpasses) found that all studies reported that the majority of crossings were used by wildlife.

Assessment: beneficial (effectiveness 90%; certainty 80%; harms 0%).

https://www.conservationevidence.com/actions/2526

Install rope bridges between canopies

Ten studies evaluated the effects on mammals of install rope bridges between canopies. Eight studies were in Australia, one was in Brazil and one in Peru.

COMMUNITY RESPONSE (0 STUDIES)

POPULATION RESPONSE (1 STUDY)

Survival (1 study): A study in Australia found that arboreal marsupials using rope bridges did not suffer high predation rates when doing so.

BEHAVIOUR (9 STUDIES)

Use (9 studies): Nine studies (including three replicated studies and a site comparison), in Australia, Brazil and Peru found that rope bridges were used by a range of mammals. Seven of these studies found between three and 25 species using rope bridges, one found that that they were used by squirrel gliders and one that they were used by mountain brushtail possums and common ringtail possums but not by koalas and squirrel gliders. One of the studies found that crossing rates were higher over the canopy bridges than at ground level.

Assessment: beneficial (effectiveness 80%; certainty 70%; harms 0%).

https://www.conservationevidence.com/actions/2556

Likely to be beneficial

Install fences around existing culverts or underpasses under roads/railways

Four studies evaluated the effects on mammals of installing fences around existing culverts under roads/railways. Two studies were in the USA one was in Portugal and one was in South Africa.

COMMUNITY RESPONSE (0 STUDIES)

POPULATION RESPONSE (3 STUDIES)

Survival (3 studies): Two out of three before-and-after studies (including a controlled and a site comparison study), in the USA, Portugal and South Africa, found that installing or enhancing roadside fencing alongside existing culverts reduced mammal road mortality whilst one study found that such fences did not alter mammal road mortality.

BEHAVIOUR (1 STUDY)

Use (1 study): A replicated, randomized, controlled, before-and-after study in the USA found that fences installed to funnel animals to existing culverts did not increase culvert use by bobcats.

Assessment: likely to be beneficial (effectiveness 50%; certainty 46%; harms 0%).

https://www.conservationevidence.com/actions/2525

Install ledges in culverts under roads/railways

Three studies evaluated the effects on mammals of installing ledges in culverts under roads or railways. Two studies were in the USA and one was in Portugal.

COMMUNITY RESPONSE (0 STUDIES)

POPULATION RESPONSE (0 STUDIES)

BEHAVIOUR (3 STUDIES)

Use (3 studies): A replicated, controlled study in Portugal found that under-road culverts with ledges were used more than culverts without ledges by two of five mammal species. A before-and-after study in the USA found that installing ledges within under-road culverts did not increase the number or diversity of small mammal species crossing through them, and only one of six species used ledges. A study in the USA found that ledges in under-road culverts were used by nine of 12 small mammal species and ledges with access ramps were used more often than those without.

Assessment: likely to be beneficial (effectiveness 60%; certainty 45%; harms 0%).

https://www.conservationevidence.com/actions/2523

Install one-way gates or other structures to allow wildlife to leave roadways

Seven studies evaluated the effects on mammals of installing one-way gates or other structures to allow wildlife to leave roadways. All seven studies were in the USA.

COMMUNITY RESPONSE (5 STUDIES)

Survival (5 studies): Two before-and-after studies (one replicated), in the USA, found that barrier fencing with one-way gates reduced deer-vehicle collisions. One of two studies (one before-and-after and one replicated, controlled), in the USA, found that barrier fencing with escape gates along roads with one or more underpasses reduced moose-vehicle collisions, whilst the other found no reduction in total mammal road casualty rates. A replicated, controlled, before-and-after study in USA found that earth escape ramps reduced mammal road mortalities.

POPULATION RESPONSE (0 STUDIES)

BEHAVIOUR (4 STUDIES)

Use (4 studies): One of two studies (one replicated) in the USA, found that one-way gates allowed mule deer to escape when trapped along highways with barrier fencing, whilst the other found that a small proportion used one-way gates. A replicated, controlled, before-and-after study in the USA found that earth escape ramps were used more often than were one-way escape gates to enable deer to escape highways with barrier fencing. A replicated, controlled study in the USA found that barrier fencing with escape gates and underpasses facilitated road crossings by a range of mammals.

Assessment: likely to be beneficial (effectiveness 60%; certainty 57%; harms 0%).

https://www.conservationevidence.com/actions/2558

Install pole crossings for gliders/flying squirrels

Seven studies evaluated the effects on gliders/flying squirrels of installing pole crossings. Six studies were in Australia and one was in the USA.

COMMUNITY RESPONSE (0 STUDIES)

POPULATION RESPONSE (1 STUDY)

Survival (1 study): A study in Australia found that arboreal marsupials using artificial road crossing structures did not suffer high predation rates when doing so.

BEHAVIOUR (6 STUDIES)

Use (6 studies): Six studies (five replicated), in Australia and the USA, found that poles were used for crossing roads by squirrel gliders, sugar gliders and Carolina northern flying squirrels.

Assessment: likely to be beneficial (effectiveness 80%; certainty 60%; harms 0%).

https://www.conservationevidence.com/actions/2546

Install signage to warn motorists about wildlife presence

Six studies evaluated the effects on mammals of installing signage to warn motorists about wildlife presence. Four studies were in the USA one was in Australia and one was in Canada.

COMMUNITY RESPONSE (0 STUDIES)

POPULATION RESPONSE (6 STUDIES)

Abundance (1 study): A before-and-after study in Australia found that when wildlife signs were installed along with speed restrictions, rumble strips, reflective wildlife deterrents, wildlife escape ramps and an educational pamphlet, a small population of eastern quoll re-established in the area.

Survival (6 studies): Three of five studies (including four controlled and three before-and-after studies), in the USA and Canada, found that warning signs did not reduce collisions between vehicles and deer. The other two studies found that warning signs did reduce collisions between vehicles and deer. A before-and-after study in Australia found that wildlife signs along with speed restrictions, rumble strips, reflective wildlife deterrents, wildlife escape ramps and an educational pamphlet, reduced collisions between vehicles and Tasmanian devils but not eastern quolls.

BEHAVIOUR (0 STUDIES)

OTHER (2 STUDIES)

Human behaviour change (2 studies): Two controlled studies (one also replicated, before-and-after), in the USA, found that signs warning of animals on the road reduced vehicles speeds.

Assessment: likely to be beneficial (effectiveness 50%; certainty 50%; harms 0%).

https://www.conservationevidence.com/actions/2608

Install tunnels/culverts/underpass under railways

Six studies evaluated the effects on mammals of installing tunnels, culverts or underpass under railways. Two studies were in Spain, one was in each of Australia, Canada and the Netherlands and one reviewed literature from a range of countries.

COMMUNITY RESPONSE (0 STUDIES)

POPULATION RESPONSE (1 STUDY)

Survival (1 study): A review found that most studies recorded no evidence of predation in or around passages under railways or roads of mammals using those passages.

BEHAVIOUR (5 STUDIES)

Use (5 studies): Five studies, in Spain, Australia, Canada and the Netherlands, found that tunnels, culverts and underpasses beneath railways were used by a range of mammals including rodents, rabbits and hares, carnivores, marsupials, deer and bears. One of these studies found that existing culverts were used more than were specifically designed wildlife tunnels.

Assessment: likely to be beneficial (effectiveness 80%; certainty 60%; harms 0%).

https://www.conservationevidence.com/actions/2519

Install tunnels/culverts/underpass under roads

Twenty-five studies evaluated the effects on mammals of installing tunnels, culverts or underpass under roads. Eight studies were in the USA, four were in Australia, four were in Canada, two were in Spain, one each was in Germany, the Netherlands and South Korea and three were reviews with wide geographic coverage.

COMMUNITY RESPONSE (0 STUDIES)

POPULATION RESPONSE (3 STUDIES)

Survival (3 studies): A study in South Korea found that road sections with higher underpass density did not have fewer wildlife-vehicle collisions. A review found that most studies recorded no evidence of predation of mammals using crossings under roads. A controlled, before-and-after, site comparison study in Australia found that overwinter survival of mountain pygmy-possums increased after an artificial rocky corridor, which included two underpasses, was installed.

BEHAVIOUR (23 STUDIES)

Use (23 studies): Seventeen of 20 studies (including seven replicated studies and two reviews), in the USA, Canada, Australia, Spain, the Netherlands, and across multiple continents, found that crossing structures beneath roads were used by mammals whilst two studies found mixed results depending on species and one study found that culverts were rarely used as crossings by mammals. One of the studies found that crossing structures were used by two of four species more than expected compared to their movements through adjacent habitats. A controlled, before-and-after, site comparison study in Australia found that an artificial rocky corridor, which included two underpasses, was used by mountain pygmy-possums. A replicated study in Germany found that use of tunnels by fallow deer was affected by tunnel colour and design. A study in the USA found that a range of mammals used culverts, including those with shelves fastened to the sides.

Behaviour change (1 study): A controlled, before-and-after, site comparison study in Australia found that after an artificial rocky corridor, which included two underpasses, was installed, dispersal of mountain pygmy-possums increased.

Assessment: likely to be beneficial (effectiveness 62%; certainty 60%; harms 0%).

https://www.conservationevidence.com/actions/2514

Install wildlife exclusion grates/cattle grids

Three studies evaluated the effects on mammals of installing wildlife exclusion grates or cattle grids. All three studies were in the USA.

COMMUNITY RESPONSE (0 STUDIES)

POPULATION RESPONSE (0 STUDIES)

BEHAVIOUR (3 STUDIES)

Behaviour change (3 studies): Two of three studies (including two replicated, before-and-after studies), in the USA, found that steel grates largely prevented crossings by deer whilst two found that they did not prevent crossings by deer and elk or black bears. In one of the studies, only one of three designs prevented crossings.

Assessment: likely to be beneficial (effectiveness 47%; certainty 41%; harms 0%).

https://www.conservationevidence.com/actions/2594

Install wildlife warning reflectors along roads

Fifteen studies evaluated the effects on mammals of installing wildlife warning reflectors along roads. Nine studies were in the USA, three were in Austalia, two were in Germany and one was in Denmark.

COMMUNITY RESPONSE (0 STUDIES)

POPULATION RESPONSE (10 STUDIES)

Abundance (1 study): A before-and-after study in Australia found that when warning reflectors were installed (along with speed restrictions, reflective wildlife signs, rumble strips, wildlife escape ramps and an educational pamphlet) a small population of eastern quoll re-established in the area.

Survival (10 studies): Five of eight controlled or before-and-after studies in the USA and Germany found that wildlife warning reflectors did not reduce collisions between vehicles and deer. Two studies found that vehicle-deer collisions were reduced by reflectors and one found that collisions were reduced in rural areas but increased in suburban areas. A before-and-after study in Australia found that when warning reflectors were installed (along with speed restrictions, reflective wildlife signs, rumble strips, wildlife escape ramps and an educational pamphlet) vehicle collisions with Tasmanian devils, but not eastern quolls, decreased. A review of two studies in Australia found mixed responses of mammal road deaths to wildlife warning reflectors.

BEHAVIOUR (5 STUDIES)

Behaviour change (5 studies): Three of four studies (including three controlled studies), in the USA, Denmark and Germany, found that wildlife warning reflectors did not cause deer to behave in ways that made collisions with vehicles less likely (such as by avoiding crossing roads). The other study found that deer initially responded to wildlife reflectors with alarm and flight but then became habituated. A replicated, controlled study in Australia found that one of four reflector model/colour combinations increased fleeing behaviour of bush wallabies when lights approached. The other combinations had no effect and none of the combinations affected red kangaroos.

Assessment: likely to be beneficial (effectiveness 40%; certainty 60%; harms 5%).

https://www.conservationevidence.com/actions/2591

Modify vegetation along railways to reduce collisions by reducing attractiveness to mammals

Two studies evaluated the effects of modifying vegetation along railways to reduce collisions by reducing attractiveness to wildlife. Both studies were in Norway.

COMMUNITY RESPONSE (0 STUDIES)

POPULATION RESPONSE (2 STUDIES)

Survival (2 studies): Two site comparison studies in Norway found that clearing vegetation from alongside railways reduced moose-train collisions.

BEHAVIOUR (0 STUDIES)

Assessment: likely to be beneficial (effectiveness 70%; certainty 40%; harms 0%).

https://www.conservationevidence.com/actions/2603

Provide food/salt lick to divert mammals from roads or railways

Three studies evaluated the effects of providing food or salt licks to divert mammals from roads. One study was in the USA, one was in Norway and one was a review of studies from across North America and Europe.

COMMUNITY RESPONSE (0 STUDIES)

POPULATION RESPONSE (2 STUDIES)

Survival (2 studies): A replicated, controlled study in the USA found that intercept feeding reduced mule deer road deaths along two of three highways in one of two years. A replicated, site comparison study in Norway found that intercept feeding reduced moose collisions with trains.

BEHAVIOUR (1 STUDY)

Behaviour change (1 study): A review of feeding wild ungulates in North

America, and Europe found that feeding diverted ungulates away from roads in one of three studies.

Assessment: likely to be beneficial (effectiveness 60%; certainty 52%; harms 0%).

https://www.conservationevidence.com/actions/2617

Unknown effectiveness

Close roads in defined seasons

One study evaluated the effects on mammals of closing roads in defined seasons. This study was in the USA.

COMMUNITY RESPONSE (0 STUDIES)

POPULATION RESPONSE (0 STUDIES)

BEHAVIOUR (1 STUDY)

Use (1 study): A site comparison study in the USA found that closing roads to traffic during the hunting season increased use of those areas by mule deer.

Assessment: unknown effectiveness (effectiveness 60%; certainty 15%; harms 0%).

https://www.conservationevidence.com/actions/2626

Dig trenches around culverts under roads/railways

One study evaluated the effects on mammals of digging trenches around culverts under roads and/or railways. This study was in South Africa.

COMMUNITY RESPONSE (0 STUDIES)

POPULATION RESPONSE (1 STUDY)

Survival (1 study): A replicated, randomized, controlled, before-and-after study in South Africa found that digging trenches alongside culverts did not reduce mammal mortality on roads.

BEHAVIOUR (0 STUDIES)

Assessment: unknown effectiveness (effectiveness 10%; certainty 20%; harms 0%).

https://www.conservationevidence.com/actions/2524

Install acoustic wildlife warnings along roads

Two studies evaluated the effects on mammals of installing acoustic wildlife warnings along roads. One study was in Demark and one was in Australia.

COMMUNITY RESPONSE (0 STUDIES)

POPULATION RESPONSE (0 STUDIES)

BEHAVIOUR (2 STUDIES)

Behaviour change (2 studies): A before-and-after study in Denmark found that sound from acoustic road markings did not alter fallow deer behaviour.

A controlled study in Australia found that Roo-Guard® sound emitters did not deter tammar wallabies from food and so were not considered suitable for keeping them off roads.

Assessment: unknown effectiveness (effectiveness 5%; certainty 37%; harms 0%).

https://www.conservationevidence.com/actions/2592

Install barrier fencing along railways

One study evaluated the effects on mammals of installing barrier fencing along railways. This study was in Norway.

COMMUNITY RESPONSE (0 STUDIES)

POPULATION RESPONSE (1 STUDY)

Survival (1 study): A before-and-after study in Norway found that fencing eliminated moose collisions with trains, except at the fence end.

BEHAVIOUR (0 STUDIES)

Assessment: unknown effectiveness (effectiveness 80%; certainty 30%; harms 0%).

https://www.conservationevidence.com/actions/2590

Install traffic calming structures to reduce speeds

One study evaluated the effects on mammals of installing traffic calming structures to reduce speeds. This study was in Australia.

COMMUNITY RESPONSE (0 STUDIES)

POPULATION RESPONSE (1 STUDY)

Abundance (1 study): A before-and-after study in Australia found that following installation of barriers to create a single lane, rumble strips, reflective wildlife signs, reflective wildlife deterrents, wildlife escape ramps and production of an educational pamphlet, a small population of eastern quoll population re-established in the area.

Survival (1 study): A before-and-after study in Australia found that following installation of barriers to create a single lane, rumble strips, reflective wildlife signs, reflective wildlife deterrents, wildlife escape ramps and production of an educational pamphlet, vehicle collisions with Tasmanian devils, but not eastern quolls decreased.

BEHAVIOUR (0 STUDIES)

Assessment: unknown effectiveness (effectiveness 60%; certainty 30%; harms 0%).

https://www.conservationevidence.com/actions/2598

Install wildlife crosswalks

One study evaluated the effects on mammals of installing wildlife crosswalks.

This study was in the USA.
COMMUNITY RESPONSE (0 STUDIES)
POPULATION RESPONSE (1 STUDY)
Survival (1 study): A replicated, before-and-after, site comparison study in the USA found that designated crossing points with barrier fencing did not significantly reduce road deaths of mule deer.
BEHAVIOUR (0 STUDIES)
Assessment: unknown effectiveness (effectiveness 0%; certainty 20%; harms 0%).

https://www.conservationevidence.com/actions/2593

Modify culverts to make them more accessible to mammals

One study evaluated the effects of modifying culverts to make them more accessible to mammals. This study was in the USA.
COMMUNITY RESPONSE (0 STUDIES)
POPULATION RESPONSE (0 STUDIES)
BEHAVIOUR (1 STUDY)
Use (1 study): A replicated, site comparison study in the USA found that modified culverts (with a dry walkway, open-air central section and enlarged entrances) were used more by bobcats to make crossings than were unmodified culverts.
Assessment: unknown effectiveness (effectiveness 70%; certainty 30%; harms 0%).

https://www.conservationevidence.com/actions/2522

Modify the roadside environment to reduce collisions by reducing attractiveness of road verges to mammals

One study evaluated the effects of modifying the roadside environment to reduce collisions by reducing attractiveness of road verges to mammals. This study was in Canada.
COMMUNITY RESPONSE (0 STUDIES)
POPULATION RESPONSE (0 STUDIES)
BEHAVIOUR (1 STUDY)
(1 study): A replicated, before-and-after, site comparison study in Canada found that draining roadside salt pools and filling them with rocks reduced the number and duration of moose visits.
Assessment: unknown effectiveness (effectiveness 60%; certainty 25%; harms 0%).

https://www.conservationevidence.com/actions/2600

Reduce legal speed limit

One study evaluated the effects on mammals of reducing the legal speed limit. This study was in Canada.

COMMUNITY RESPONSE (0 STUDIES)

POPULATION RESPONSE (1 STUDY)

Survival (1 study): A controlled, before-and-after study in Canada found that speed limit reductions and enforcement did not reduce vehicle collisions with bighorn sheep or elk.

BEHAVIOUR (0 STUDIES)

Assessment: unknown effectiveness (effectiveness 5%; certainty 27%; harms 0%).

https://www.conservationevidence.com/actions/2596

Use road lighting to reduce vehicle collisions with mammals

Two studies evaluated the effects on mammals of using road lighting to reduce vehicle collisions with mammals. Both studies were in the USA.

COMMUNITY RESPONSE (0 STUDIES)

POPULATION RESPONSE (2 STUDIES)

Survival (2 studies): One of two studies (one controlled and one before-and-after), in the USA, found that road lighting reduced vehicle collisions with moose. The other study found that road lighting did not reduce vehicle collisions with mule deer.

BEHAVIOUR (0 STUDIES)

Assessment: unknown effectiveness (effectiveness 50%; certainty 30%; harms 2%).

https://www.conservationevidence.com/actions/2614

Unlikely to be beneficial

Fit vehicles with ultrasonic warning devices

Three studies evaluated the effects on mammals of fitting vehicles with ultrasonic warning devices. Two studies were in the USA and one was in Australia.

COMMUNITY RESPONSE (0 STUDIES)

POPULATION RESPONSE (1 STUDY)

Survival (1 study): A replicated, controlled study in Australia found that Shu Roo warning whistles did not reduce animal-vehicle collisions for eastern grey kangaroos or red kangaroos

BEHAVIOUR (3 STUDIES)

Behaviour change (3 studies): Three controlled studies (two replicated), in the USA and Australia, found that ultrasonic warning devices did not deter mule deer, eastern grey kangaroos, red kangaroos or white-tailed deer from roads. *Assessment: unlikely to be beneficial (effectiveness 0%; certainty 55%; harms 0%).*

https://www.conservationevidence.com/actions/2606

Use chemical repellents along roads or railways

Five studies evaluated the effects on mammals of using chemical repellents along roads or railways. Two studies were in Canada and one each was in Germany, Norway and Denmark.

COMMUNITY RESPONSE (0 STUDIES)

POPULATION RESPONSE (2 STUDIES)

Survival (2 studies): Two studies (one before-and-after, one site comparison), in Germany and Norway, found that chemical-based repellents did not reduce collisions between ungulates and road vehicles or trains.

BEHAVIOUR (4 STUDIES)

Behaviour change (4 studies): Two of four studies (including three replicated, controlled studies), in Germany, Canada, and Denmark, found that chemical repellents, trialled for potential to deter animals from roads, did not deter ungulates. The other two studies found mixed results with repellents temporarily deterring some ungulate species in one study and one of three deterrents deterring caribou in the other.

Assessment: unlikely to be beneficial (effectiveness 20%; certainty 50%; harms 0%).

https://www.conservationevidence.com/actions/2615

No evidence found (no assessment)

We have captured no evidence for the following interventions:

- Modify vegetation along roads to reduce collisions with mammals by enhancing visibility for drivers
- Remove roadkill regularly to reduce kill rate of predators/scavengers
- Retain/maintain road verges as small mammal habitat
- Use alternative de-icers on roads
- Use reflective collars or paint on mammals to reduce collisions with road vehicles
- Use wildlife decoy to reduce vehicle collisions with mammals.

15.4.2 Utility and service lines

Based on the collated evidence, what is the current assessment of the effectiveness of interventions for utility and service lines?	
Likely to be beneficial	• Install crossings over/under pipelines

Likely to be beneficial

Install crossings over/under pipelines

Three studies evaluated the effects on mammals of installing crossings over/under pipelines. Two studies were in the USA and one was in Canada.

COMMUNITY RESPONSE (0 STUDIES)

POPULATION RESPONSE (0 STUDIES)

BEHAVIOUR (3 STUDIES)

Use (3 studies): A study in USA found that buried pipeline sections were used more frequently than their availability as crossing points by caribou. A study in USA found that pipeline sections elevated specifically to permit mammal crossings underneath were not used by moose or caribou more than were other elevated sections. A controlled study in Canada found that a range of large mammal species used wildlife crossings over pipelines.

Assessment: likely to be beneficial (effectiveness 60%; certainty 40%; harms 0%).

https://www.conservationevidence.com/actions/2627

15.4.3 Shipping lanes

Based on the collated evidence, what is the current assessment of the effectiveness of interventions for shipping lanes?	
Likely to be beneficial	• Install overpasses over waterways • Provide mammals with escape routes from canals
No evidence found (no assessment)	• Install barrier fencing along waterways

Likely to be beneficial

Install overpasses over waterways

Two studies evaluated the effects on mammals of installing overpasses over waterways. One study was in the USA and one was in Spain.

COMMUNITY RESPONSE (0 STUDIES)

POPULATION RESPONSE (0 STUDIES)

BEHAVIOUR (2 STUDIES)

Use (2 studies): Two studies (one replicated, one a site comparison) in the USA and Spain, found that bridges and overpasses over waterways were used by desert mule deer, collared peccaries and coyotes and by a range of large and medium-sized mammals.

Assessment: likely to be beneficial (effectiveness 70%; certainty 40%; harms 0%).

https://www.conservationevidence.com/actions/2628

Provide mammals with escape routes from canals

Five studies evaluated the effects on mammals of providing mammals with escape routes from canals. Two studies were in Germany and one each was in the USA, the Netherlands and Argentina.

COMMUNITY RESPONSE (0 STUDIES)

POPULATION RESPONSE (2 STUDIES)

Survival (2 studies): One of two studies (one before-and-after), in Germany and the USA, found that ramps and ladders reduced mule deer drownings whilst the other study found that ramps and shallow-water inlets did not reduce mammal drownings.

BEHAVIOUR (3 STUDIES)

Use (3 studies): Three studies (one replicated) in Germany, the Netherlands and Argentina, found that ramps and other access or escape routes out of water were used by a range of medium-sized and large mammals species.

Assessment: likely to be beneficial (effectiveness 60%; certainty 45%; harms 0%).

https://www.conservationevidence.com/actions/2638

No evidence found (no assessment)

We have captured no evidence for the following interventions:

- Install barrier fencing along waterways.

15.5 Threat: Biological resource use

15.5.1 Hunting and collecting terrestrial animals

Based on the collated evidence, what is the current assessment of the effectiveness of interventions for hunting and collecting terrestrial animals?	
Likely to be beneficial	• Prohibit or restrict hunting of a species • Provide/increase anti-poaching patrols • Set hunting quotas based on target species population trends
Unknown effectiveness	• Ban exports of hunting trophies • Ban private ownership of hunted mammals • Incentivise species protection through licensed trophy hunting • Prohibit or restrict hunting of particular sex/ breeding status/age animals • Site management for target mammal species carried out by field sport practitioners • Use wildlife refuges to reduce hunting impacts
No evidence found (no assessment)	• Commercially breed for the mammal production trade • Make introduction of non-native mammals for sporting purposes illegal • Promote mammal-related ecotourism • Promote sustainable alternative livelihoods • Use selective trapping methods in hunting activities

Likely to be beneficial

Prohibit or restrict hunting of a species

Five studies evaluated the effects of prohibiting or restricting hunting of a mammal species. One study each was in Norway, the USA, South Africa, Poland and Zimbabwe.

COMMUNITY RESPONSE (0 STUDIES)

POPULATION RESPONSE (5 STUDIES)

Abundance (2 studies): Two studies (including one before-and-after study), in the USA and Poland, found that prohibiting hunting led to population increases of tule elk and wolves.

Survival (3 studies): A before-and-after study in Norway found that restricting or prohibiting hunting did not alter the number of brown bears killed. A study in Zimbabwe reported that banning the hunting, possession and trade of Temminck's ground pangolins did not eliminate hunting of the species. A before-and-after study in South Africa found that increasing legal protection of leopards, along with reducing human-leopard conflict by promoting improved animal husbandry, was associated with increased survival.

BEHAVIOUR (0 STUDIES)

Assessment: likely to be beneficial (effectiveness 50%; certainty 45%; harms 0%).

https://www.conservationevidence.com/actions/2597

Provide/increase anti-poaching patrols

Seven studies evaluated the effects of providing or increasing anti-poaching patrols on mammals. Two studies were in Thailand and one each was in Brazil, Iran, Lao People's Democratic Republic, South Africa and Tajikistan.

COMMUNITY RESPONSE (0 STUDIES)

POPULATION RESPONSE (7 STUDIES)

Abundance (6 studies): Two studies, in Thailand and Iran, found more deer and small mammals and more urial sheep and Persian leopards close to ranger stations (from which anti-poaching patrols were carried out) than further from them. One of three before-and-after studies, in Brazil, Thailand and Lao People's Democratic Republic, found that ranger patrols increased mammal abundance. The other two studies found that patrols did not increase tiger abundance. A site comparison study in Tajikistan found more snow leopard, argali, and ibex where anti-poaching patrols were conducted.

Survival (1 study): A study in South Africa found that anti-poaching patrols did not deter African rhinoceros poaching.

BEHAVIOUR (0 STUDIES)

Assessment: likely to be beneficial (effectiveness 65%; certainty 60%; harms 0%).

https://www.conservationevidence.com/actions/2618

Set hunting quotas based on target species population trends

Three studies evaluated the effects of setting hunting quotas for mammals based on target species population trends. One study each was in Canada, Spain and Norway.

COMMUNITY RESPONSE (0 STUDIES)

POPULATION RESPONSE (3 STUDIES)

Abundance (2 studies): Two studies, in Spain and Norway, found that restricting hunting and basing quotas on population targets enabled population increases for Pyrenean chamois and Eurasian lynx.

Survival (1 study): A before-and-after study in Canada found that setting harvest quotas based on population trends, and lengthening the hunting season, did not decrease the number of cougars killed by hunters.

BEHAVIOUR (0 STUDIES)

Assessment: likely to be beneficial (effectiveness 64%; certainty 42%; harms 0%).

https://www.conservationevidence.com/actions/2607

Unknown effectiveness

Ban exports of hunting trophies

One study evaluated the effects of banning exports of hunting trophies on wild mammals. This study was in Cameroon.

COMMUNITY RESPONSE (0 STUDIES)

POPULATION RESPONSE (1 STUDY)

Abundance (1 study): A before-and-after study in Cameroon found similar hippopotamus abundances before and after a ban on exporting hippopotamus hunting trophies.

BEHAVIOUR (0 STUDIES)

Assessment: unknown effectiveness (effectiveness 10%; certainty 27%; harms 0%).

https://www.conservationevidence.com/actions/2625

Ban private ownership of hunted mammals

One study evaluated the effects of banning private ownership of hunted mammals. This study was in Sweden.

COMMUNITY RESPONSE (0 STUDIES)
POPULATION RESPONSE (1 STUDY)
Survival (1 study): A before-and-after study in Sweden found that fewer brown bears were reported killed after the banning of private ownership of hunted bears.
BEHAVIOUR (0 STUDIES)
Assessment: unknown effectiveness (effectiveness 70%; certainty 30%; harms 0%).

https://www.conservationevidence.com/actions/2602

Incentivise species protection through licensed trophy hunting

One study evaluated the effects on mammals of incentivising species protection through licensed trophy hunting. This study was in Nepal.
COMMUNITY RESPONSE (0 STUDIES)
POPULATION RESPONSE (1 STUDY)
Abundance (1 study): A study in Nepal found that after trophy hunting started, bharal abundance increased, though the sex ratio of this species, and of Himalayan tahr, became skewed towards females.
BEHAVIOUR (0 STUDIES)
Assessment: unknown effectiveness (effectiveness 60%; certainty 20%; harms 20%).

https://www.conservationevidence.com/actions/2610

Prohibit or restrict hunting of particular sex/ breeding status/age animals

Two studies evaluated the effects of prohibiting or restricting hunting of particular sex, breeding status or age animals. Both studies were in the USA.
COMMUNITY RESPONSE (0 STUDIES)
POPULATION RESPONSE (2 STUDIES)
Reproduction (2 studies): Two replicated, before-and-after studies, in the USA, found that limiting hunting of male deer did not increase the numbers of young deer/adult female.
Population structure (1 study): A replicated, before-and-after study in the USA found that limiting hunting of older male elk resulted in an increased ratio of male:female elk.
BEHAVIOUR (0 STUDIES)
Assessment: unknown effectiveness (effectiveness 50%; certainty 35%; harms 0%).

https://www.conservationevidence.com/actions/2609

Site management for target mammal species carried out by field sport practitioners

One study evaluated the effects of site management for a target mammal species being carried out by field sport practitioners. This study was in Ireland.

COMMUNITY RESPONSE (0 STUDIES)

POPULATION RESPONSE (1 STUDY)

Abundance (1 study): A replicated, site comparison study in the Republic of Ireland found that sites managed for the sport of coursing Irish hares held more of this species than did the wider countryside.

BEHAVIOUR (0 STUDIES)

Assessment: unknown effectiveness (effectiveness 65%; certainty 20%; harms 0%).

https://www.conservationevidence.com/actions/2605

Use wildlife refuges to reduce hunting impacts

Two studies evaluated the effects on mammal species of using wildlife refuges to reduce hunting impacts. One study was in Canada and one was in Mexico.

COMMUNITY RESPONSE (0 STUDIES)

POPULATION RESPONSE (2 STUDIES)

Abundance (2 studies): One of two replicated site comparison studies in Canada and Mexico found more moose in areas with limited hunting than in more heavily hunted areas. The other study found mixed results with only one of five species being more numerous in a non-hunted refuge.

BEHAVIOUR (0 STUDIES)

Assessment: unknown effectiveness (effectiveness 50%; certainty 35%; harms 0%).

https://www.conservationevidence.com/actions/2612

No evidence found (no assessment)

We have captured no evidence for the following interventions:

- Commercially breed for the mammal production trade
- Make introduction of non-native mammals for sporting purposes illegal
- Promote mammal-related ecotourism
- Promote sustainable alternative livelihoods
- Use selective trapping methods in hunting activities.

15.5.2 Logging and wood harvesting

Based on the collated evidence, what is the current assessment of the effectiveness of interventions for logging and wood harvesting?	
Likely to be beneficial	• Thin trees within forest • Use patch retention harvesting instead of clearcutting • Use selective harvesting instead of clearcutting
Unknown effectiveness	• Allow forest to regenerate naturally following logging • Apply fertilizer to trees • Clear or open patches in forests • Fell trees in groups, leaving surrounding forest unharvested • Gather coarse woody debris into piles after felling • Leave coarse woody debris in forests • Leave standing deadwood/snags in forests • Plant trees following clearfelling • Provide supplementary feed to reduce tree damage • Remove competing vegetation to allow tree establishment in clearcut areas • Retain dead trees after uprooting • Retain understorey vegetation within plantations • Retain undisturbed patches during thinning operations • Retain wildlife corridors in logged areas • Use thinning of forest instead of clearcutting
No evidence found (no assessment)	• Control firewood collection in remnant native forest and woodland • Coppice trees • Harvest timber outside mammal reproduction period • Retain riparian buffer strips during timber harvest • Use tree tubes/small fences/cages to protect individual trees

Likely to be beneficial

Thin trees within forest

Twelve studies evaluated the effects on mammals of thinning trees within forests. Six studies were in Canada and six were in the USA.

COMMUNITY RESPONSE (2 STUDIES)

Species richness (2 studies): A replicated, site comparison study the USA found that in thinned tree forest stands, there was similar mammal species richness compared to in unthinned stands. A replicated, controlled study in Canada found that thinning of regenerating lodgepole pine did not increase small mammal species richness 12–14 years later.

POPULATION RESPONSE (8 STUDIES)

Abundance (8 studies): Three of eight replicated, controlled and replicated, site comparison studies, in the USA and Canada, found that thinning trees within forests lead to higher numbers of small mammals. Two studies showed increases for some, but not all, small mammal species with a further study showing an increase for one of two squirrel species in response to at least some forest thinning treatments. The other two studies showed no increases in abundances of small mammals or northern flying squirrels between 12 and 14 years after thinning.

BEHAVIOUR (4 STUDIES)

Use (4 studies): Three of four controlled and comparison studies (three also replicated, one randomized) in Canada found that thinning trees within forests did not lead to greater use of areas by mule deer, moose or snowshoe hares. The other study found that a thinned area was used more by white-tailed deer than was unthinned forest.

Assessment: likely to be beneficial (effectiveness 40%; certainty 42%; harms 0%).

https://www.conservationevidence.com/actions/2650

Use patch retention harvesting instead of clearcutting

Three studies evaluated the effects on mammals of using patch retention harvesting instead of clearcutting. Two studies were in Canada and one was in Australia.

COMMUNITY RESPONSE (0 STUDIES)

POPULATION RESPONSE (3 STUDIES)

Abundance (3 studies): Two replicated, controlled, before-and-after studies and a replicated, site comparison study in Canada and Australia found that retaining patches of unharvested trees instead of clearcutting whole forest stands increased or maintained numbers of some but not all small mammals.

Higher abundances where tree patches were retained were found for southern red-backed voles, bush rat and for female agile antechinus. No benefit of retaining forest patches was found on abundances of deer mouse, meadow vole and male agile antechinus.
BEHAVIOUR (0 STUDIES)
Assessment: likely to be beneficial (effectiveness 50%; certainty 40%; harms 0%).

https://www.conservationevidence.com/actions/2639

Use selective harvesting instead of clearcutting

Eight studies evaluated the effects on mammals of using selective harvesting instead of clearcutting. Four studies were in Canada, three were in the USA and one was a review of studies in North America.
COMMUNITY RESPONSE (1 STUDY)
Richness/diversity (1 study): A replicated, site comparison study in Canada found that harvesting trees selectively did not result in higher small mammal species richness compared to clearcutting.
POPULATION RESPONSE (7 STUDIES)
Abundance (7 studies): One of six replicated, controlled or replicated, site comparison studies in the USA and Canada found more small mammals in selectively harvested forest stands than in fully harvested, regenerating stands. Three studies found that selective harvesting did not increase small mammal abundance relative to clearcutting. The other two studies found mixed results with one of four small mammal species being more numerous in selectively harvested stands or in selectively harvested stands only in some years. A systematic review in North American forests found that partially harvested forests had more red-backed voles but not deer mice than did clearcut forests.
BEHAVIOUR (1 STUDY)
Use (1 study): A site comparison study in the USA found that partially harvested forest was not used by snowshoe hares more than was largely clearcut forest.
Assessment: likely to be beneficial (effectiveness 40%; certainty 40%; harms 0%).

https://www.conservationevidence.com/actions/2637

Unknown effectiveness

Allow forest to regenerate naturally following logging

One study evaluated the effects on mammals of allowing forest to regenerate naturally following logging. This study was in Canada.

COMMUNITY RESPONSE (0 STUDIES)
POPULATION RESPONSE (1 STUDY)
Abundance (1 study): A replicated, site comparison study in Canada found that, natural forest regeneration increased moose numbers relative to more intensive management in the short- to medium-term but not in the longer term.
BEHAVIOUR (0 STUDIES)
Assessment: unknown effectiveness (effectiveness 45%; certainty 20%; harms 0%).

https://www.conservationevidence.com/actions/2634

Apply fertilizer to trees

Three studies evaluated the effects on mammals of applying fertilizer to trees. All three studies were in Canada.
COMMUNITY RESPONSE (0 STUDIES)
POPULATION RESPONSE (0 STUDIES)
BEHAVIOUR (3 STUDIES)
Use (3 studies): One of three replicated studies (including one controlled study and two site comparison studies), in Canada, found that thinned forest stands to which fertilizer was applied were used more by snowshoe hares in winter but not in summer over the short-term. The other studies found that forest stands to which fertilizer was applied were not more used by snowshoe hares in the longer term or by mule deer or moose.
Assessment: unknown effectiveness (effectiveness 30%; certainty 30%; harms 0%).

https://www.conservationevidence.com/actions/2649

Clear or open patches in forests

Four studies evaluated the effects on mammals of clearing or opening patches in forests. Two studies were in the USA, one was in Bolivia and one was in Canada.
COMMUNITY RESPONSE (0 STUDIES)
POPULATION RESPONSE (4 STUDIES)
Abundance (4 studies): Two of four replicated studies (including three controlled studies and a site comparison study), in Bolivia, the USA and Canada, found that creating gaps or open patches within forests did not increase small mammal abundance relative to uncut forest. One study found that it did increase small mammal abundance and one found increased abundance for one of four small mammal species.
BEHAVIOUR (0 STUDIES)
Assessment: unknown effectiveness (effectiveness 32%; certainty 30%; harms 0%).

https://www.conservationevidence.com/actions/2641

Fell trees in groups, leaving surrounding forest unharvested

Three studies evaluated the effects on mammals of felling trees in groups, leaving surrounding forest unharvested. Two studies were in Canada and one was in the UK.

COMMUNITY RESPONSE (0 STUDIES)

POPULATION RESPONSE (3 STUDIES)

Abundance (2 studies): One of two replicated studies (including one controlled study and one site comparison study), in Canada, found that felling groups of trees within otherwise undisturbed stands increased the abundance of one of four small mammal species relative to clearcutting. The other study found that none of four small mammal species monitored showed abundance increases.

Survival (1 study): A study in the UK found that when trees were felled in large groups with surrounding forest unaffected, there was less damage to artificial hazel dormouse nests than when trees were felled in small groups or thinned throughout.

BEHAVIOUR (0 STUDIES)

Assessment: unknown effectiveness (effectiveness 40%; certainty 30%; harms 0%).

https://www.conservationevidence.com/actions/2648

Gather coarse woody debris into piles after felling

Two studies evaluated the effects on mammals of gathering coarse woody debris into piles after felling. Both studies were in Canada.

COMMUNITY RESPONSE (1 STUDY)

Richness/diversity (1 study): A randomized, replicated, controlled study in Canada found higher mammal species richness where coarse woody debris was gathered into piles.

POPULATION RESPONSE (2 STUDIES)

Abundance (2 studies): One of two randomized, replicated, controlled studies in Canada found higher counts of San Bernardino long-tailed voles where coarse woody debris was gathered into piles. The other study found higher small mammal abundance at one of three plots where debris was gathered into piles.

BEHAVIOUR (0 STUDIES)

Assessment: unknown effectiveness (effectiveness 70%; certainty 31%; harms 0%).

https://www.conservationevidence.com/actions/2653

Leave coarse woody debris in forests

Three studies evaluated the effects on mammals of leaving coarse woody debris in forests. One study was in Canada, one was in the USA and one was in Malaysia.

COMMUNITY RESPONSE (1 STUDY)

Richness/diversity (1 study): A replicated, site comparison study, in Malaysia found more small mammal species groups in felled forest areas with woody debris than without.

POPULATION RESPONSE (3 STUDIES)

Abundance (3 studies): One out of three replicated studies (two controlled, one site comparison, one before-and-after) in Canada, the USA and Malaysia found that retaining or adding coarse woody debris did not increase numbers or frequency of records of small mammals. The other study found that two of three shrew species were more numerous in areas with increased volumes of coarse woody debris than areas without coarse woody debris.

BEHAVIOUR (0 STUDIES)

Assessment: unknown effectiveness (effectiveness 50%; certainty 30%; harms 0%).

https://www.conservationevidence.com/actions/2647

Leave standing deadwood/snags in forests

One study evaluated the effects on mammals of leaving standing deadwood or snags in forests. This study was in the USA.

COMMUNITY RESPONSE (0 STUDIES)

POPULATION RESPONSE (1 STUDY)

Abundance (1 study): A replicated, controlled study in the USA found that increasing the quantity of standing deadwood in forests increased the abundance of one of three shrew species, compared to removing deadwood.

BEHAVIOUR (0 STUDIES)

Assessment: unknown effectiveness (effectiveness 55%; certainty 25%; harms 0%).

https://www.conservationevidence.com/actions/2646

Plant trees following clearfelling

One study evaluated the effects on mammals of planting trees following clearfelling. This study was in Canada.

COMMUNITY RESPONSE (0 STUDIES)

POPULATION RESPONSE (0 STUDIES)

BEHAVIOUR (1 STUDY)

Use (1 study): A replicated, site comparison study in Canada found that forest

stands subject to tree planting and herbicide treatment after logging were used more by American martens compared to naturally regenerating stands.
Assessment: unknown effectiveness (effectiveness 50%; certainty 20%; harms 0%).

https://www.conservationevidence.com/actions/2631

Provide supplementary feed to reduce tree damage

One study evaluated the effects of providing supplementary feed on the magnitude of tree damage caused by mammals. This study was in USA.

COMMUNITY RESPONSE (0 STUDIES)

POPULATION RESPONSE (0 STUDIES)

BEHAVIOUR (0 STUDIES)

OTHER (1 STUDY)

Human-wildlife conflict (1 study): A replicated, randomized, paired sites, controlled, before-and-after study in USA found that supplementary feeding reduced tree damage by black bears.
Assessment: unknown effectiveness (effectiveness 70%; certainty 30%; harms 0%).

https://www.conservationevidence.com/actions/2629

Remove competing vegetation to allow tree establishment in clearcut areas

Three studies evaluated the effects on mammals of removing competing vegetation to allow tree establishment in clearcut areas. Two studies were in Canada and one was in the USA.

COMMUNITY RESPONSE (0 STUDIES)

POPULATION RESPONSE (0 STUDIES)

BEHAVIOUR (3 STUDIES)

Use (3 studies): One of three studies (including two controlled studies and one site comparison study), in the USA and Canada, found that where competing vegetation was removed to allow tree establishment in clearcut areas, American martens used the areas more. One study found mixed results for moose and one found no increase in site use by snowshoe hares.
Assessment: unknown effectiveness (effectiveness 40%; certainty 30%; harms 0%).

https://www.conservationevidence.com/actions/2644

Retain dead trees after uprooting

One study evaluated the effects on mammals of retaining dead trees after uprooting. This study was in the USA.

COMMUNITY RESPONSE (0 STUDIES)
POPULATION RESPONSE (0 STUDIES)
BEHAVIOUR (1 STUDY)
Use (1 study): A replicated, controlled study in the USA found that areas where trees were uprooted but left on site were used more by desert cottontails than were cleared areas.
Assessment: unknown effectiveness (effectiveness 60%; certainty 20%; harms 0%).

https://www.conservationevidence.com/actions/2642

Retain understorey vegetation within plantations

One study evaluated the effects on mammals of retaining understorey vegetation within plantations. This study was in Chile.
COMMUNITY RESPONSE (1 STUDY)
Richness/diversity (1 study): A replicated, controlled, before-and-after study in Chile found that areas with retained understorey vegetation had more species of medium-sized mammal, compared to areas cleared of understorey vegetation.
POPULATION RESPONSE (0 STUDIES)
BEHAVIOUR (1 STUDY)
Use (1 study): A replicated, controlled, before-and-after study in Chile found that areas with retained understorey vegetation had more visits from medium-sized mammals, compared to areas cleared of understorey vegetation.
Assessment: unknown effectiveness (effectiveness 62%; certainty 30%; harms 0%).

https://www.conservationevidence.com/actions/2645

Retain undisturbed patches during thinning operations

Two studies evaluated the effects on mammals of retaining undisturbed patches during thinning operations. Both studies were in the USA.
COMMUNITY RESPONSE (0 STUDIES)
POPULATION RESPONSE (0 STUDIES)
BEHAVIOUR (2 STUDIES)
Use (2 studies): Two randomized, replicated, controlled studies (one also before-and-after) in the USA found that snowshoe hares and tassel-eared squirrels used retained undisturbed forest patches more than thinned areas.
Assessment: unknown effectiveness (effectiveness 70%; certainty 30%; harms 0%).

https://www.conservationevidence.com/actions/2640

Retain wildlife corridors in logged areas

Two studies evaluated the effects on mammals of retaining wildlife corridors in logged areas. One study was in Australia and one was in Canada.

COMMUNITY RESPONSE (0 STUDIES)

POPULATION RESPONSE (0 STUDIES)

BEHAVIOUR (2 STUDIES)

Use (2 studies): A replicated study in Australia found that corridors of trees, retained after harvesting, supported seven species of arboreal marsupial. A replicated, controlled study in Canada found that lines of woody debris through clearcut areas that were connected to adjacent forest were not used more by red-backed voles than were isolated lines of woody debris.

Assessment: unknown effectiveness (effectiveness 50%; certainty 30%; harms 0%).

https://www.conservationevidence.com/actions/2651

Use thinning of forest instead of clearcutting

One study evaluated the effects on mammals of using thinning of forest instead of clearcutting. This study was in the USA.

COMMUNITY RESPONSE (0 STUDIES)

POPULATION RESPONSE (0 STUDIES)

BEHAVIOUR (1 STUDY)

Use (1 study): A replicated, controlled study in the USA found that thinned forest areas were used more by desert cottontails than were fully cleared or uncleared areas.

Assessment: unknown effectiveness (effectiveness 60%; certainty 20%; harms 0%).

https://www.conservationevidence.com/actions/2643

No evidence found (no assessment)

We have captured no evidence for the following interventions:

- Control firewood collection in remnant native forest and woodland
- Coppice trees
- Harvest timber outside mammal reproduction period
- Retain riparian buffer strips during timber harvest
- Use tree tubes/small fences/cages to protect individual trees.

15.6 Threat: Human intrusions and disturbance

Based on the collated evidence, what is the current assessment of the effectiveness of interventions for human intrusions and disturbance?	
Beneficial	• Exclude or limit number of visitors to reserves or protected areas
Likely to be beneficial	• Use conditioned taste aversion to reduce human-wildlife conflict in non-residential sites • Use non-lethal methods to deter carnivores from attacking humans • Use prescribed burning • Use signs or access restrictions to reduce disturbance to mammals
Trade-off between benefit and harms	• Translocate mammals that have habituated to humans (e.g. bears)
Unknown effectiveness	• Habituate mammals to visitors • Treat mammals to reduce conflict caused by disease transmission to humans
No evidence found (no assessment)	• Provide paths to limit extent of disturbance to mammals • Set maximum number of people/vehicles approaching mammals • Set minimum distances for approaching mammals • Use voluntary agreements with locals to reduce disturbance

Beneficial

Exclude or limit number of visitors to reserves or protected areas

Five studies evaluated the effects on mammals of excluding or limiting the number of visitors to reserves or protected areas. Three studies were in the USA, one was in Ecuador and one was in Thailand.

COMMUNITY RESPONSE (0 STUDIES)

POPULATION RESPONSE (2 STUDIES)

Abundance (1 study): A site comparison study in Ecuador found that a road with restricted access had a higher population of medium-sized and large mammals compared to a road with unrestricted access.

Survival (1 study): A before-and-after study in the USA found that temporarily restricting visitor access resulted in fewer bears being killed to protect humans.

BEHAVIOUR (3 STUDIES)

Use (3 studies): Three studies (one a before-and-after study), in the USA and Thailand, found that restricting human access to protected areas resulted in increased use of these areas by grizzly bears and leopards.

Assessment: beneficial (effectiveness 80%; certainty 70%; harms 0%).

https://www.conservationevidence.com/actions/2330

Likely to be beneficial

Use conditioned taste aversion to reduce human-wildlife conflict in non-residential sites

Two studies evaluated the effects on mammals of using conditioned taste aversion to reduce human-wildlife conflict in non-residential sites. Both studies were in the USA.

COMMUNITY RESPONSE (0 STUDIES)

POPULATION RESPONSE (0 STUDIES)

BEHAVIOUR (0 STUDIES)

OTHER (2 STUDIES)

Human-wildlife conflict (2 studies): Two studies, in the USA, found that lacing foodstuffs with substances that induce illness led to these foods being avoided by coyotes and black bears.

Assessment: likely to be beneficial (effectiveness 75%; certainty 50%; harms 0%).

https://www.conservationevidence.com/actions/2384

Use non-lethal methods to deter carnivores from attacking humans

Eight studies evaluated the effects of using non-lethal methods to deter carnivores from attacking humans. Three studies were in the USA, two were in Australia, one was in the USA and Canada, one was in Austria and one was in Bangladesh.

COMMUNITY RESPONSE (0 STUDIES)

POPULATION RESPONSE (1 STUDY)

Survival (1 study): A study in Bangladesh found that when domestic dogs accompanied people to give advance warning of tiger presence, fewer tigers were killed by people.

BEHAVIOUR (0 STUDIES)

OTHER (8 STUDIES)

Human-wildlife conflict (8 studies): Two studies, in the USA and Canada, found that pepper spray caused all or most American black bears and grizzly bears to flee or cease aggressive behaviour. One of these studies also showed that tear gas repelled half of American black bears. Two studies in the USA and Austria found that grizzly/brown bears were repelled by rubber bullets or by a range of deterrents including rubber bullets, chasing, shouting and throwing items. A study in the USA found that hikers wearing bear bells were less likely to be approached or charged by grizzly bears than were hikers without bells. A replicated, controlled study in Australia found that ultrasonic sound deterrent units did not affect feeding location choices of dingoes. A study in Bangladesh found that domestic dogs accompanying people gave advance warning of tiger presence, enabling people to take precautionary actions. A study in Australia found that a motorised water pistol caused most dingoes to change direction or speed or move ≥5 m away, but sounding a horn did not.

Assessment: likely to be beneficial (effectiveness 62%; certainty 60%; harms 0%).

https://www.conservationevidence.com/actions/2385

Use prescribed burning

Thirty-seven studies evaluated the effects on mammals of using prescribed burning. Twenty-five studies were in the USA, three each were in Canada and South Africa, two each were in Spain and Tanzania and one each was in France and Auatralia.

COMMUNITY RESPONSE (2 STUDIES)

Richness/diversity (2 studies): A replicated, randomized, controlled study in the USA found similar small mammal species richness after prescribed burning compared to in unburned forest. A replicated, site comparison study in Australia found that prescribed burns early in the dry season resulted in higher small mammal species richness relative to wildfires later in the season.

POPULATION RESPONSE (16 STUDIES)

Abundance (11 studies): Five of 10 replicated studies (of which eight were controlled and two were site comparisons), in the USA, Spain and Australia, found that prescribed burning did not increase abundances of small mammals. Three studies found mixed effects, on cottontail rabbits and small mammals and two found that burning increased numbers of European rabbits and small mammals. A systematic review in the USA found that two mammal species showed positive responses (abundance or reproduction) to prescribed burning while three showed no response.

Reproductive success (1 study): A before-and-after, site comparison study in South Africa found that 92% of Cape mountain zebra foals were produced in the three years post-fire compared to 8% in the three years pre-fire.

Condition (1 study): A replicated, controlled study, in the USA, found that prescribed burning did not reduce bot fly infestation rates among rodents and cottontail rabbits.

Occupancy/range (3 studies): Two of three studies (including two site comparisons and one controlled study), in the USA and Canada, found that prescribed burning resulted in larger areas being occupied by black-tailed prairie dog colonies and smaller individual home ranges of Mexican fox squirrels. The third study found that prescribed burning did not increase occupancy rates of beaver lodges.

BEHAVIOUR (22 STUDIES)

Use (21 studies): Ten of 21 studies (including eight controlled studies and eight site comparisons with a further four being before-and-after studies), in the USA, Canada, South Africa, Tanzania and France, found that prescribed burning increased use of areas (measured either as time spent in areas or consumption of food resources) by bighorn sheep, mule deer, pronghorn antelope, elk, plains bison, Cape mountain zebrasand mouflon. Six studies found mixed effects, with responses differing among different ages or sexes of white-tailed deer, bison and elk, differing among different large herbivore species or varying over time for elk, while swift foxes denned more but did not hunt more in burned areas. The other five studies showed that prescribed burning did not increase use or herbivory by elk, black-tailed deer, white-tailed deer or mixed species groups of mammalian herbivores.

Behaviour change (1 study): A site comparison study in Tanzania found that vigilance of Thomson's gazelles did not differ between those on burned and unburned areas.

Assessment: likely to be beneficial (effectiveness 49%; certainty 50%; harms 5%).

https://www.conservationevidence.com/actions/2388

Use signs or access restrictions to reduce disturbance to mammals

One study evaluated the effects of using signs or access restrictions to reduce disturbance to mammals. This study was in the USA.

COMMUNITY RESPONSE (0 STUDIES)

POPULATION RESPONSE (0 STUDIES)

BEHAVIOUR (1 STUDY)

Use (1 study): A replicated, paired sites, site comparison study in the USA found that removing or closing roads increased use of those areas by black bears.

Assessment: likely to be beneficial (effectiveness 80%; certainty 40%; harms 0%).

https://www.conservationevidence.com/actions/2325

Trade-off between benefit and harms

Translocate mammals that have habituated to humans (e.g. bears)

Two studies evaluated the effects of translocating mammals that have habituated to humans. One study was in the USA and one was in the USA and Canada.

COMMUNITY RESPONSE (0 STUDIES)

POPULATION RESPONSE (0 STUDIES)

BEHAVIOUR (0 STUDIES)

OTHER (2 STUDIES)

Human-wildlife conflict (2 studies): A study in the USA found that almost half of the translocated 'nuisance' black bears returned to their capture locations. A review of studies in the USA and Canada found that black bears translocated away from sites of conflict with humans were less likely to return to their capture site if translocated as younger bears, over greater distances, or across geographic barriers.

Assessment: trade-off between benefit and harms (effectiveness 50%; certainty 49%; harms 20%).

https://www.conservationevidence.com/actions/2341

Unknown effectiveness

Habituate mammals to visitors

One study evaluated the effects of habituating mammals to visitors. This study was in the USA.

COMMUNITY RESPONSE (0 STUDIES)

POPULATION RESPONSE (0 STUDIES)

BEHAVIOUR (0 STUDIES)

OTHER (1 STUDY)

Human-wildlife conflict (1 study): A study in the USA found that brown bears that were highly habituated to humans showed less aggression towards human visitors than did non-habituated bears.

Assessment: unknown effectiveness (effectiveness 50%; certainty 20%; harms 10%).

https://www.conservationevidence.com/actions/2340

Treat mammals to reduce conflict caused by disease transmission to humans

One study evaluated the effects of treating mammals to reduce conflict caused by disease transmission to humans. This study was in Germany.

COMMUNITY RESPONSE (0 STUDIES)

POPULATION RESPONSE (0 STUDIES)

BEHAVIOUR (0 STUDIES)

OTHER (1 STUDY)

Human-wildlife conflict (1 study): A controlled, before-and-after study in Germany found that following a worming programme, proportions of red foxes infested with small fox tapeworm fell.

Assessment: unknown effectiveness (effectiveness 50%; certainty 30%; harms 0%).

https://www.conservationevidence.com/actions/2342

No evidence found (no assessment)

We have captured no evidence for the following interventions:

- Provide paths to limit extent of disturbance to mammals
- Set maximum number of people/vehicles approaching mammals
- Set minimum distances for approaching mammals
- Use voluntary agreements with locals to reduce disturbance.

15.7 Threat: Natural system modifications

Based on the collated evidence, what is the current assessment of the effectiveness of interventions for natural system modifications?	
Likely to be beneficial	• Burn at specific time of year • Provide artificial waterholes in dry season • Remove mid-storey vegetation in forest • Remove trees and shrubs to recreate open areas of land • Thin trees to reduce wildfire risk
Unknown effectiveness	• Provide supplementary food after fire
Unlikely to be beneficial	• Remove burnt trees and branches after wildfire • Remove understorey vegetation in forest
No evidence found (no assessment)	• Provide shelter structures after fire • Use fencing to protect water sources for use by wild mammals

Likely to be beneficial

Burn at specific time of year

Two studies evaluated the effects on mammals of burning at a specific time of year. One study was in Australia, and one was in the USA.

COMMUNITY RESPONSE (0 STUDIES)

POPULATION RESPONSE (1 STUDY)

Abundance (1 study): A replicated, randomized, controlled, before-and-after study in the USA found that carrying out prescribed burns in autumn did not increase small mammal abundances or biomass relative to burning in summer.

Survival (1 study): A randomized, replicated, controlled study in Australia found that in forest burned early in the dry season, northern brown bandicoot survival rate declined less than in forests burned late in the dry season.

BEHAVIOUR (0 STUDIES)

Assessment: likely to be beneficial (effectiveness 50%; certainty 40%; harms 0%).

https://www.conservationevidence.com/actions/2416

Provide artificial waterholes in dry season

Three studies evaluated the effects on mammals of providing artificial waterholes in the dry season. One study was in South Africa, one was in Tanzania and one was in Jordan.

COMMUNITY RESPONSE (1 STUDY)

Richness/diversity (1 study): A site comparison study in Tanzania found that artificial waterholes were used by a similar number of large mammal species as was a natural waterhole.

POPULATION RESPONSE (0 STUDIES)

BEHAVIOUR (2 STUDIES)

Use (2 studies): A study in South Africa found that areas around artificial waterholes were used more by eight out of 13 mammalian herbivore species than was the wider landscape. A study in Jordan found that artificial waterholes were used by striped hyenas.

Assessment: likely to be beneficial (effectiveness 60%; certainty 40%; harms 10%).

https://www.conservationevidence.com/actions/2484

Remove mid-storey vegetation in forest

One study evaluated the effects on mammals of removing mid-storey vegetation in forest. This study was in the USA.

COMMUNITY RESPONSE (1 STUDY)

Richness/diversity (1 study): A randomized, replicated, controlled study in the USA found that after removing mid-storey vegetation, mammal species richness increased.

POPULATION RESPONSE (1 STUDY)

Abundance (1 study): A randomized, replicated, controlled study in the

USA found that after removing mid-storey vegetation, mammal abundance increased.

BEHAVIOUR (0 STUDIES)

Assessment: likely to be beneficial (effectiveness 65%; certainty 40%; harms 0%).

https://www.conservationevidence.com/actions/2480

Remove trees and shrubs to recreate open areas of land

Two studies evaluated the effects on mammals of removing trees and shrubs to recreate open areas of land. Both studies were in the USA.

COMMUNITY RESPONSE (0 STUDIES)

POPULATION RESPONSE (1 STUDY)

Abundance (1 study): A controlled study in the USA found that where Ashe juniper trees were removed, there were higher abundances of three rodent species.

BEHAVIOUR (1 STUDY)

Use (1 study): A before-and-after, site comparison study in the USA found that removing trees increased use of areas by Rocky Mountain bighorn sheep.

Assessment: likely to be beneficial (effectiveness 70%; certainty 45%; harms 0%).

https://www.conservationevidence.com/actions/2483

Thin trees to reduce wildfire risk

Three studies evaluated the effects on mammals of thinning trees to reduce wildfire risk. All three studies were in the USA.

COMMUNITY RESPONSE (0 STUDIES)

POPULATION RESPONSE (2 STUDIES)

Abundance (2 studies): A replicated, controlled, before-and-after study in the USA found that reducing tree density increased abundances of two of four small mammal species. A systematic review in the USA found that, in thinned forests, two mammal species were recorded in higher densities compared to in unmanaged forests, while three species showed no effect.

BEHAVIOUR (1 STUDY)

(1 study): A replicated, controlled study in the USA found that thinning followed by prescribed burning did not increase use of forest areas by North American elk in most season, stand age and sex comparisons.

Assessment: likely to be beneficial (effectiveness 50%; certainty 50%; harms 0%).

https://www.conservationevidence.com/actions/2477

Unknown effectiveness

Provide supplementary food after fire

One study evaluated the effects on mammals of providing supplementary food after fire. This study was in the USA.

COMMUNITY RESPONSE (0 STUDIES)

POPULATION RESPONSE (1 STUDY)

Survival (1 study): A replicated, randomized, controlled study in the USA found that supplementary feeding did not increase survival of hispid cotton rats following prescribed fire.

BEHAVIOUR (0 STUDIES)

Assessment: unknown effectiveness (effectiveness 0%; certainty 21%; harms 0%).

https://www.conservationevidence.com/actions/2494

Unlikely to be beneficial

Remove burnt trees and branches after wildfire

One study evaluated the effects on mammals of removing burnt trees and branches after wildfire. This study was in Spain.

COMMUNITY RESPONSE (0 STUDIES)

POPULATION RESPONSE (1 STUDY)

Abundance (1 study): A replicated, randomized, controlled study in Spain found that removing burned trees and branches after wildfire did not increase European wild rabbit numbers compared to removing burned trees but leaving branches in place.

BEHAVIOUR (0 STUDIES)

Assessment: unlikely to be beneficial (effectiveness 0%; certainty 40%; harms 0%).

https://www.conservationevidence.com/actions/2478

Remove understorey vegetation in forest

Three studies evaluated the effects on mammals of removing understorey vegetation in forest. All three studies were in the USA.

COMMUNITY RESPONSE (0 STUDIES)

POPULATION RESPONSE (3 STUDIES)

Abundance (3 studies): Three replicated, randomized, controlled studies (two also before-and-after), in the USA, found that compared to prescribed burning, mechanically removing understorey vegetation growth in forests did

not increase abundances of white-footed mice, shrews or four rodent species.
BEHAVIOUR (0 STUDIES)

Assessment: unlikely to be beneficial (effectiveness 10%; certainty 50%; harms 0%).

https://www.conservationevidence.com/actions/2482

No evidence found (no assessment)

We have captured no evidence for the following interventions:

- Provide shelter structures after fire
- Use fencing to protect water sources for use by wild mammals.

15.8 Threat: Invasive alien and other problematic species

15.8.1 Invasive non-native/alien species/diseases

Based on the collated evidence, what is the current assessment of the effectiveness of interventions for invasive non-native/alien species/ diseases?	
Beneficial	• Remove/control non-native mammals
Likely to be beneficial	• Remove/control non-native mammals within a fenced area • Use conditioned taste aversion to prevent non-target species from entering traps • Use drugs to treat parasites • Use reward removal to prevent non-target species from entering traps
Unknown effectiveness	• Modify traps used in the control/eradication of non-native species to avoid injury of non-target mammal • Remove/control non-native invertebrates • Remove/control non-native plants
No evidence found (no assessment)	• Control non-native prey species to reduce populations and impacts of non-native predators • Control non-native/problematic plants to restore habitat • Provide artificial refuges for prey to evade/escape non-native predators • Reintroduce top predators to suppress and reduce the impacts of smaller non-native predator and prey species • Remove/control non-native amphibians (e.g. cane toads) • Remove/control non-native species that could interbreed with native species

Beneficial

Remove/control non-native mammals

Twenty-five studies evaluated the effects on non-controlled mammals of removing or controlling non-native mammals. Twenty-one studies were in Australia, and one was in each of France, the UK, Equador and the USA.

COMMUNITY RESPONSE (0 STUDIES)

POPULATION RESPONSE (24 STUDIES)

Abundance (21 studies): Ten of 18 controlled, before-and-after or site comparison studies, in Australia, found that after controlling red foxes, abundances, densities or trapping frequencies increased for rock-wallaby spp., eastern grey kangaroo, woylie,, brush-tail possum, tammar wallaby, chuditch and quenda. Seven studies found mixed results with increases in some species but not others, increases followed by declines or increases only where cats as well as foxes were controlled. The other study found no increase in bush rat numbers with fox control. One of three replicated, before-and-after studies (including two controlled studies), in Australia, France and Ecuador, found that control of invasive rodents increased numbers of lesser white-toothed shrews and greater white-toothed shrews. One study found that Santiago rice rat abundance declined less with rodent control and one found mixed results, with increased numbers of short-tailed mice at one out of four study sites.

Survival (1 study): A replicated, controlled study in Australia found that controlling red foxes increased survival of juvenile eastern grey kangaroos.

Occupancy/range (3 studies): Three studies (two before-and-after, one controlled), in the UK and Australia, found that after controlling non-native American mink, red foxes and European rabbits, there were increases in ranges or proportions of sites occupied by water vole, common brushtail possum, long-nosed potoroo and southern brown bandicoot and four native small mammal species.

BEHAVIOUR (1 STUDY)

Behaviour change (1 study): A before-and-after study in the USA found that following removal of feral cats, vertebrate prey increased as a proportion of the diet of island foxes.

Assessment: beneficial (effectiveness 67%; certainty 70%; harms 10%).

https://www.conservationevidence.com/actions/2504

Likely to be beneficial

Remove/control non-native mammals within a fenced area

One study evaluated the effects on native mammals of removing or controlling non-native mammals within a fenced area. This study was in Australia.

COMMUNITY RESPONSE (1 STUDY)

Richness/diversity (1 study): A site comparison study in Australia found that in a fenced area where invasive cats, red foxes and European rabbits were removed, native mammal species richness was higher than outside the fenced area.

POPULATION RESPONSE (1 STUDY)

Abundance (1 study): A site comparison study in Australia found that in a fenced area where invasive cats, red foxes and European rabbits were removed, native mammals overall and two out of four small mammal species were more abundant than outside the fenced area.

BEHAVIOUR (0 STUDIES)

Assessment: likely to be beneficial (effectiveness 65%; certainty 40%; harms 0%).

https://www.conservationevidence.com/actions/2528

Use conditioned taste aversion to prevent non-target species from entering traps

One study evaluated the effects on mammals of using conditioned taste aversion to prevent non-target species from entering traps. This study was in the USA.

COMMUNITY RESPONSE (0 STUDIES)

POPULATION RESPONSE (0 STUDIES)

BEHAVIOUR (1 STUDY)

Behaviour change (1 study): A replicated, controlled study in the USA found that using bait laced with lithium chloride reduced the rate of entry of San Clemente Island foxes into traps set for feral cats.

Assessment: likely to be beneficial (effectiveness 60%; certainty 40%; harms 0%).

https://www.conservationevidence.com/actions/2536

Use drugs to treat parasites

Seven studies evaluated the effects on mammals of using drugs to treat parasites. Three studies were in the USA, two were in Spain, one was in Germany and one was in Croatia.

COMMUNITY RESPONSE (0 STUDIES)

POPULATION RESPONSE (7 STUDIES)

Survival (1 study): A randomized, replicated, controlled study the USA found that medical treatment of Rocky Mountain bighorn sheep against lungworm did not increase lamb survival.

Condition (6 studies): Three of four before-and-after studies (one controlled), in Germany, the USA and Croatia, found that after administering drugs to mammals, parasite burdens were reduced in roe deer and in wild boar piglets and numbers of white-tailed deer infected were reduced. A third study found that levels of lungworm larvae in bighorn sheep faeces were reduced one month after drug treatment but not after three to seven months. One of these studies also found that the drug treatment resulted in increased body weight in roe deer fawns. A replicated, controlled, before-and-after study in Spain found that higher doses of ivermectin treated sarcoptic mange in Spanish ibex faster than lower doses, and treatment was more effective in animals with less severe infections. A replicated, before-and-after study in Spain found that after injecting Spanish ibex with ivermectin to treat sarcoptic mange a mange-free herd was established.

BEHAVIOUR (0 STUDIES)

Assessment: likely to be beneficial (effectiveness 55%; certainty 60%; harms 0%).

https://www.conservationevidence.com/actions/2587

Use reward removal to prevent non-target species from entering traps

One study evaluated the effects on mammals of using reward removal to prevent non-target species from entering traps. This study was in the USA.

COMMUNITY RESPONSE (0 STUDIES)

POPULATION RESPONSE (0 STUDIES)

BEHAVIOUR (1 STUDY)

Behaviour change (1 study): A replicated, controlled study in the USA found that when reward removal was practiced, the rate of San Clemente Island fox entry into traps set for feral cats was reduced.

Assessment: likely to be beneficial (effectiveness 60%; certainty 40%; harms 0%).

https://www.conservationevidence.com/actions/2537

Unknown effectiveness

Modify traps used in the control/eradication of non-native species to avoid injury of non-target mammal

One study evaluated the effects of modifying traps used in the control or eradication of non-native species to avoid injury of non-target mammals. This study was in the USA.

COMMUNITY RESPONSE (0 STUDIES)

POPULATION RESPONSE (1 STUDY)

Condition (1 study): A before-and-after study in the USA found that modifying traps used for catching non-native mammals reduced moderate but not severe injuries among incidentally captured San Nicolas Island foxes.

BEHAVIOUR (0 STUDIES)

Assessment: unknown effectiveness (effectiveness 40%; certainty 22%; harms 0%).

https://www.conservationevidence.com/actions/2535

Remove/control non-native invertebrates

One study evaluated the effects on mammals of removing or controlling non-native invertebrates. This study was in the USA.

COMMUNITY RESPONSE (0 STUDIES)

POPULATION RESPONSE (1 STUDY)

Abundance (1 study): A replicated, controlled, before-and-after study the USA found that after the control of red imported fire ants, capture rates of northern pygmy mice increased.

BEHAVIOUR (0 STUDIES)

Assessment: unknown effectiveness (effectiveness 67%; certainty 25%; harms 0%).

https://www.conservationevidence.com/actions/2501

Remove/control non-native plants

Two studies evaluated the effects on mammals of removing or controlling non-native invasive plants. Both studies were in the USA.

COMMUNITY RESPONSE (1 STUDY)

Richness/diversity (1 study): A replicated study in the USA found that control of introduced saltcedar did not change small mammal species richness.

POPULATION RESPONSE (1 STUDY)

Abundance (1 study): A site comparison study in the USA found that partial removal of velvet mesquite did not increase abundances of six mammal species.

BEHAVIOUR (0 STUDIES)

Assessment: unknown effectiveness (effectiveness 8%; certainty 25%; harms 0%).

https://www.conservationevidence.com/actions/2529

No evidence found (no assessment)

We have captured no evidence for the following interventions:

- Control non-native prey species to reduce populations and impacts of non-native predators
- Control non-native/problematic plants to restore habitat
- Provide artificial refuges for prey to evade/escape non-native predators
- Reintroduce top predators to suppress and reduce the impacts of smaller non-native predator and prey species
- Remove/control non-native amphibians (e.g. cane toads)
- Remove/control non-native species that could interbreed with native species.

15.8.2 Problematic native species/diseases

Based on the collated evidence, what is the current assessment of the effectiveness of interventions for problematic native species/diseases?	
Beneficial	• Use vaccination programme
Likely to be beneficial	• Cull disease-infected animals
Trade-off between benefit and harms	• Remove or control predators
Unknown effectiveness	• Control ticks/fleas/lice in wild mammal populations • Establish populations isolated from disease • Provide diversionary feeding for predators • Remove or control competitors • Sterilize predators • Train mammals to avoid problematic species • Treat disease in wild mammals

Based on the collated evidence, what is the current assessment of the effectiveness of interventions for problematic native species/diseases?	
No evidence found (no assessment)	• Eliminate highly virulent diseases early in an epidemic by culling all individuals (healthy and infected) in a defined area • Sterilize non-native domestic or feral species (e.g. cats and dogs)

Beneficial

Use vaccination programme

Seven studies evaluated the effects on mammals of using vaccination programmes. Three studies were in the UK and one study was in each of Belgium, Spain, Poland and Ethiopia.

COMMUNITY RESPONSE (0 STUDIES)

POPULATION RESPONSE (7 STUDIES)

Abundance (1 study): A before-and-after study in Poland found that following an anti-rabies vaccination programme, red fox numbers increased.

Condition (6 studies): Five studies (including three replicated, three controlled and two before-and-after studies) in Belgium, Spain and the UK found that following vaccination, rabies was less frequent in red foxes, numbers of Eurasian badgers infected with tuberculosis was reduced and European rabbits developed immunity to myxomatosis and rabbit haemorrhagic disease. One of the studies also found that vaccination reduced the speed and extent of infection in infected Eurasian badgers. A study in Ethiopia found that following vaccination of Ethiopian wolves, a rabies outbreak halted.

BEHAVIOUR (0 STUDIES)

Assessment: beneficial (effectiveness 80%; certainty 70%; harms 0%).

https://www.conservationevidence.com/actions/2582

Likely to be beneficial

Cull disease-infected animals

One study evaluated the effects on mammals of culling disease-infected animals. This study was in Tasmania.

COMMUNITY RESPONSE (0 STUDIES)

POPULATION RESPONSE (1 STUDY)

Condition (1 study): A before-and-after, site comparison study in Tasmania found that culling disease-infected Tasmanian devils resulted in fewer animals with large tumours associated with late stages of the disease.

BEHAVIOUR (0 STUDIES)

Assessment: likely to be beneficial (effectiveness 60%; certainty 40%; harms 15%).

https://www.conservationevidence.com/actions/2586

Trade-off between benefit and harms

Remove or control predators

Ten studies evaluated the effects on non-controlled mammals of removing or controlling predators. Seven studies were in North America, one was in Finland, one in Portugal and one in Mexico.

COMMUNITY RESPONSE (0 STUDIES)

POPULATION RESPONSE (10 STUDIES)

Abundance (6 studies): Three of six studies (including three controlled, one before-and-after and one replicated, paired sites study), in Finland Portugal, Mexico and the USA, found that removing predators increased abundances of pronghorns, moose and European rabbits and Iberian hares. One of these studies also found that mule deer abundance did not increase. The other three studies found that removing predators did not increase mountain hare, caribou or desert bighorn sheep abundance.

Reproductive success (2 studies): Two replicated, before-and-after studies (one also controlled), in the USA, found that predator removal was associated with increased breeding productivity of white-tailed deer and less of a productivity decline in pronghorns. However, one of these studies also found that there was no change in breeding productivity of mule deer.

Survival (5 studies): Two of five before-and-after studies (including two controlled studies and one replicated study), in the USA, Canada and the USA and Canada combined, found that controlling predators did not increase survival of caribou calves, or of calf or adult female caribou. Two studies found that moose calf survival and woodland caribou calf survival increased with predator control. The other study found mixed results with increases in white-tailed deer calf survival in some but not all years with predator control.

BEHAVIOUR (0 STUDIES)

Assessment: trade-off between benefit and harms (effectiveness 50%; certainty 50%; harms 46%).

https://www.conservationevidence.com/actions/2613

Unknown effectiveness

Control ticks/fleas/lice in wild mammal populations

Two studies evaluated the effects of controlling ticks, fleas or lice in wild mammal populations. Both studies were in the USA.

COMMUNITY RESPONSE (0 STUDIES)

POPULATION RESPONSE (2 STUDIES)

Condition (2 studies): A replicated, paired sites, controlled study in the USA found that a grain-bait insecticide product did not consistently reduce flea burdens on Utah prairie dogs. A controlled study the USA found that treating wolves with ivermectin cleared them of infestations of biting dog lice.

BEHAVIOUR (0 STUDIES)

Assessment: unknown effectiveness (effectiveness 50%; certainty 30%; harms 0%).

https://www.conservationevidence.com/actions/2589

Establish populations isolated from disease

One study evaluated the effects on mammals of establishing populations isolated from disease. The study was in sub-Saharan Africa.

COMMUNITY RESPONSE (0 STUDIES)

POPULATION RESPONSE (1 STUDY)

Condition (1 study): A site comparison study throughout sub-Saharan Africa found that fencing reduced prevalence of canine distemper but not of rabies, coronavirus or canine parvovirus in African wild dogs.

BEHAVIOUR (0 STUDIES)

Assessment: unknown effectiveness (effectiveness 25%; certainty 20%; harms 0%).

https://www.conservationevidence.com/actions/2588

Provide diversionary feeding for predators

One study evaluated the effects on potential prey mammals of providing diversionary feeding for predators. This study was in Canada.

COMMUNITY RESPONSE (0 STUDIES)

POPULATION RESPONSE (1 STUDY)

Survival (1 study): A controlled, before-and-after study in Canada found that diversionary feeding of predators appeared to increase woodland caribou calf survival.

BEHAVIOUR (0 STUDIES)

Assessment: unknown effectiveness (effectiveness 67%; certainty 20%; harms 0%).

https://www.conservationevidence.com/actions/2578

Remove or control competitors

Two studies evaluated the effects on non-controlled mammals of removing or controlling competitors. One study was across Norway and Sweden and one was in Norway.

COMMUNITY RESPONSE (0 STUDIES)

POPULATION RESPONSE (1 STUDY)

Reproductive success (1 study): A replicated, controlled study in Norway and Sweden found that red fox control, along with supplementary feeding, was associated with an increase in arctic fox litters.

BEHAVIOUR (1 STUDY)

Use (1 study): A controlled study in Norway found that where red foxes had been controlled arctic foxes were more likely to colonize.

Assessment: unknown effectiveness (effectiveness 70%; certainty 33%; harms 12%).

https://www.conservationevidence.com/actions/2575

Sterilize predators

One study evaluated the effects on potential prey mammals of sterilizing predators. This study was in the USA and Canada.

COMMUNITY RESPONSE (0 STUDIES)

POPULATION RESPONSE (1 STUDY)

Survival (1 study): A before-and-after study in the USA and Canada found that sterilising some wolves (combined with trapping and removing others) did not increase caribou survival.

BEHAVIOUR (0 STUDIES)

Assessment: unknown effectiveness (effectiveness 0%; certainty 15%; harms 0%).

https://www.conservationevidence.com/actions/2573

Train mammals to avoid problematic species

Two studies evaluated the effects of training mammals to avoid problematic species. Both studies were in Australia.

COMMUNITY RESPONSE (0 STUDIES)

POPULATION RESPONSE (1 STUDY)

Survival (1 study): A controlled study in Australia found that training greater bilbies to avoid introduced predators did not increase their post-release survival.

BEHAVIOUR (2 STUDIES)

Behaviour change (2 studies): One of two controlled studies in Australia found that greater bilbies trained to avoid introduced predators showed more predator avoidance behaviour, the second study found no difference in behaviour between trained and untrained bilbies.

Assessment: unknown effectiveness (effectiveness 40%; certainty 30%; harms 0%).

https://www.conservationevidence.com/actions/2580

Treat disease in wild mammals

Three studies evaluated the effects on mammals of treating disease in the wild. Two studies were in the USA and one was in Germany.

COMMUNITY RESPONSE (0 STUDIES)

POPULATION RESPONSE (2 STUDIES)

Condition (2 studies): A replicated study in Germany found that medical treatment of mouflons against foot rot disease healed most infected animals. A before-and-after study in the USA found that management which included vaccination of Yellowstone bison did not reduce prevalence of brucellosis.

BEHAVIOUR (1 STUDY)

Uptake (1 study): A study in the USA found that a molasses-based bait was readily consumed by white-tailed deer, including when it contained a dose of a disease vaccination.

Assessment: unknown effectiveness (effectiveness 50%; certainty 32%; harms 0%).

https://www.conservationevidence.com/actions/2581

No evidence found (no assessment)

We have captured no evidence for the following interventions:

- Eliminate highly virulent diseases early in an epidemic by culling all individuals (healthy and infected) in a defined area
- Sterilize non-native domestic or feral species (e.g. cats and dogs).

15.8.3 Other

Based on the collated evidence, what is the current assessment of the effectiveness of interventions for other sources of non-native, invasive or other problematic species?	
Likely to be beneficial	• Use fencing to exclude grazers or other problematic species • Use fencing to exclude predators or other problematic species

Likely to be beneficial

Use fencing to exclude grazers or other problematic species

Three studies evaluated the effects on mammals of using fencing to exclude grazers or other problematic species. One study was in each of the USA, Australia and Spain.

COMMUNITY RESPONSE (1 STUDY)

Richness/diversity (1 study): A controlled, before-and-after study in Australia found that after fencing to exclude introduced herbivores, native mammal species richness increased.

POPULATION RESPONSE (3 STUDIES)

Abundance (3 studies): Two controlled studies (including one replicated, paired sites study) in Spain and Australia found that using fences to exclude large or introduced herbivores increased the abundance of Algerian mice and native mammals. A replicated, paired sites study in the USA found that in areas fenced to exclude livestock grazing and off-road vehicles, abundance of black-tailed hares was lower compared to in unfenced areas.

BEHAVIOUR (0 STUDIES)

Assessment: likely to be beneficial (effectiveness 50%; certainty 46%; harms 10%).

https://www.conservationevidence.com/actions/2495

Use fencing to exclude predators or other problematic species

Ten studies evaluated the effects on mammals of using fencing to exclude predators or other problematic species. Four studies were in Australia, four were in the USA and two were in Spain.

COMMUNITY RESPONSE (1 STUDY)

Richness/diversity (1 study): A site comparison study in Australia found that fencing which excluded feral cats, foxes and rabbits increased small mammal species richness.

POPULATION RESPONSE (10 STUDIES)

Abundance (4 studies): Two of three studies (including two replicated, controlled studies), in Spain, Australia and the USA, found that abundances of European rabbits and small mammals were higher within areas fenced to exclude predators or other problematic species, compared to in unfenced areas. The third study found that hispid cotton rat abundance was not higher with predator fencing. A replicated, controlled study in Spain found that translocated European rabbit abundance was higher in fenced areas that excluded both terrestrial carnivores and raptors than in areas only accessible to raptors.

Reproductive success (1 study): A replicated, controlled study in USA found that predator exclosures increased the number of white-tailed deer fawns relative to the number of adult females.

Survival (7 studies): Four of six studies (including four replicated, controlled studies) in Spain, Australia and the USA, found that fencing to exclude predators did not increase survival of translocated European rabbits, hispid cotton rats, southern flying squirrels or western barred bandicoots. The other two studies found that persistence of populations of eastern barred bandicoots and long-haired rats was greater inside than outside fences. A controlled, before-and-after study in the USA found that electric fencing reduced coyote incursions into sites frequented by black-footed ferrets.

BEHAVIOUR (0 STUDIES)

Assessment: likely to be beneficial (effectiveness 50%; certainty 50%; harms 0%).

https://www.conservationevidence.com/actions/2497

15.9 Threat: Pollution

Based on the collated evidence, what is the current assessment of the effectiveness of interventions for pollution?	
Likely to be beneficial	• Leave headlands in fields unsprayed
Unknown effectiveness	• Reduce pesticide or fertilizer use • Translocate mammals away from site contaminated by oil spill
No evidence found (no assessment)	• Establish riparian buffers

Likely to be beneficial

Leave headlands in fields unsprayed

Two studies evaluated the effects on mammals of leaving headlands in fields unsprayed. One study was in the UK and one was in the Netherlands.

COMMUNITY RESPONSE (0 STUDIES)

POPULATION RESPONSE (0 STUDIES)

BEHAVIOUR (2 STUDIES)

Use (2 studies): Two replicated studies (one also controlled) in the UK and the Netherlands, found that crop edge headlands that were not sprayed with pesticides were used more by mice than were sprayed crop edges.

Assessment: likely to be beneficial (effectiveness 60%; certainty 50%; harms 0%).

https://www.conservationevidence.com/actions/2540

Unknown effectiveness

Reduce pesticide or fertilizer use

Three studies evaluated the effects on mammals of reducing pesticide, herbicide or fertilizer use. Two studies were in the UK, one was in Italy and one was in Argentina.

COMMUNITY RESPONSE (1 STUDY)

Richness/diversity (1 study): A replicated, site comparison study in Argentina found that farming without pesticides or fertilizers did not increase small mammal species richness in field margins.

POPULATION RESPONSE (2 STUDIES)

Abundance (2 studies): One of two site comparison studies, in the UK and Italy, found that reducing pesticide or fertilizer use, by farming organically, increased wood mouse abundance. The other study found that it did not increase European hare abundance.

BEHAVIOUR (1 STUDY)

Use (1 study): A replicated, site comparison study in Argentina found that farming without pesticides or fertilizers did not increase small mammal use of field margins.

Assessment: unknown effectiveness (effectiveness 38%; certainty 23%; harms 0%).

https://www.conservationevidence.com/actions/2539

Translocate mammals away from site contaminated by oil spill

One study evaluated the effects of translocating mammals away from a site contaminated by oil spill. This study was in the USA.

COMMUNITY RESPONSE (0 STUDIES)

POPULATION RESPONSE (1 STUDY)

Survival (1 study): A study in the USA found that after being translocated in a trial of responses to a hypothetical pollution incident, most sea-otters survived for the duration of monitoring.

BEHAVIOUR (1 STUDY)

Behaviour change (1 study): A study in the USA found that after being translocated in a trial of responses to a hypothetical pollution incident, most sea-otters did not return to their capture location.

Assessment: unknown effectiveness (effectiveness 65%; certainty 20%; harms 10%).

https://www.conservationevidence.com/actions/2542

No evidence found (no assessment)

We have captured no evidence for the following interventions:

- Establish riparian buffers.

15.10 Threat: Climate change and severe weather

Based on the collated evidence, what is the current assessment of the effectiveness of interventions for climate change and severe weather?	
Unknown effectiveness	• Apply water to vegetation to increase food availability during drought • Translocate animals from source populations subject to similar climatic conditions
No evidence found (no assessment)	• Protect habitat along elevational gradients • Provide dams/water holes during drought • Remove flood water • Retain/provide migration corridors

Unknown effectiveness

Apply water to vegetation to increase food availability during drought

One study evaluated the effects on mammals of applying water to vegetation to increase food availability during drought. This study was in the USA.

COMMUNITY RESPONSE (0 STUDIES)

POPULATION RESPONSE (0 STUDIES)

BEHAVIOUR (1 STUDY)

Use (1 study): A controlled, before-and-after study in the USA found that watering scrub during drought increased its use by adult Sonoran pronghorns for feeding.

Assessment: unknown effectiveness (effectiveness 70%; certainty 34%; harms 0%).

https://www.conservationevidence.com/actions/2555

Translocate animals from source populations subject to similar climatic conditions

One study evaluated the effects of translocating mammals from source populations subject to similar climatic conditions. This study was in the USA.

COMMUNITY RESPONSE (0 STUDIES)

POPULATION RESPONSE (1 STUDY)

Reproductive success (1 study): A study in the USA found that bighorn sheep translocated from populations subject to a similar climate to the recipient site reared more offspring than did those translocated from milder climatic areas.

BEHAVIOUR (0 STUDIES)

Assessment: unknown effectiveness (effectiveness 55%; certainty 25%; harms 0%).

https://www.conservationevidence.com/actions/2553

No evidence found (no assessment)

We have captured no evidence for the following interventions:

- Protect habitat along elevational gradients
- Provide dams/water holes during drought
- Remove flood water
- Retain/provide migration corridors.

15.11 Habitat protection

Based on the collated evidence, what is the current assessment of the effectiveness of interventions for habitat protection?	
Beneficial	• Legally protect habitat for mammals
Unknown effectiveness	• Build fences around protected areas • Increase resources for managing protected areas • Increase size of protected area
No evidence found (no assessment)	• Encourage habitat protection of privately-owned land • Retain buffer zones around core habitat

Beneficial

● Legally protect habitat for mammals

Seven studies evaluated the effects of legally protecting habitat for mammals. One study each was in Zambia, the USA, Tanzania, Brazil, Nepal and India and one was a systematic review of sites with a wide geographic spread.

COMMUNITY RESPONSE (0 STUDIES)

POPULATION RESPONSE (7 STUDIES)

Abundance (7 studies): A systematic review of protected areas across the globe found that 24 of 31 studies reported an increase in mammal populations in protected areas relative to unprotected areas. Three studies (including two site comparison studies), in Zambia, the USA and Nepal, found that populations of red lechwe, black bears and one-horned rhinoceros grew following site protection or were higher than in adjacent non-protected sites. One of three site comparison studies, in Tanzania, Brazil and India, found

that populations of more mammal species increased inside protected areas than in adjacent unprotected areas. One study found that populations of only three of 11 species were higher on protected than on unprotected land whilst the third study found that 13 of 16 species were less abundant in a protected area than in a nearby unprotected area.
BEHAVIOUR (0 STUDIES)
Assessment: beneficial (effectiveness 70%; certainty 65%; harms 0%).

https://www.conservationevidence.com/actions/2559

Unknown effectiveness

Build fences around protected areas

Two studies evaluated the effects on mammals of building fences around protected areas. One study was in Kenya and one was in Mozambique.
COMMUNITY RESPONSE (1 STUDY)
Richness/diversity (1 study): A before-and-after study in Kenya found that after a fence was built around a protected area, mammal species richness initially increased in both study sites, but subsequently declined at one of the sites.
POPULATION RESPONSE (2 STUDIES)
Abundance (2 studies): A paired sites study in Mozambique found that inside a fenced sanctuary there were more mammal scats than outside the sanctuary. A before-and-after study in Kenya found that after a fence was built around a protected area, mammal abundance initially increased in both study sites, but it subsequently declined at one of the sites.
BEHAVIOUR (0 STUDIES)
Assessment: unknown effectiveness (effectiveness 50%; certainty 23%; harms 20%).

https://www.conservationevidence.com/actions/2561

Increase resources for managing protected areas

One study evaluated the effects on mammals of increasing resources for managing protected areas. This study was in Tanzania.
COMMUNITY RESPONSE (1 STUDY)
Species richness (1 study): A site comparison study in Tanzania found that mammal species richness was higher in a well-resourced national park, than in a less well-resourced forest reserve.
POPULATION RESPONSE (1 STUDY)

Abundance (1 study): A site comparison study Tanzania found that there were greater occupancy rates or relative abundances of most mammal species in a well-resourced national park than in a less well-resourced forest reserve.
BEHAVIOUR (0 STUDIES)
Assessment: unknown effectiveness (effectiveness 80%; certainty 30%; harms 0%).

https://www.conservationevidence.com/actions/2564

Increase size of protected area

One study evaluated the effects on mammals of increasing the size of a protected area. This study was in South Africa.
COMMUNITY RESPONSE (0 STUDIES)
POPULATION RESPONSE (0 STUDIES)
BEHAVIOUR (1 STUDY)
Behaviour change (1 study): A before-and-after study in South Africa found that expanding a fenced reserve resulted in the home range of a reintroduced group of lions becoming larger but the core range becoming smaller.
Assessment: unknown effectiveness (effectiveness 55%; certainty 30%; harms 0%).

https://www.conservationevidence.com/actions/2563

No evidence found (no assessment)

We have captured no evidence for the following interventions:
- Encourage habitat protection of privately-owned land
- Retain buffer zones around core habitat.

15.12 Habitat restoration and creation

Based on the collated evidence, what is the current assessment of the effectiveness of interventions for habitat restoration and creation?	
Beneficial	• Provide artificial dens or nest boxes on trees
Likely to be beneficial	• Create or maintain corridors between habitat patches • Manage vegetation using livestock grazing • Provide artificial refuges/breeding sites • Remove vegetation by hand/machine • Restore or create forest • Restore or create grassland • Restore or create savannas • Restore or create shrubland
Unknown effectiveness	• Apply fertilizer to vegetation to increase food availability • Manage vegetation using grazing by wild herbivores • Manage wetland water levels for mammal species • Provide more small artificial breeding sites rather than fewer large sites • Restore or create wetlands
Unlikely to be beneficial	• Remove vegetation using herbicides
No evidence found (no assessment)	• Remove topsoil that has had fertilizer added to mimic low nutrient soil • Replant vegetation

Beneficial

Provide artificial dens or nest boxes on trees

Thirty studies evaluated the effects on mammals of providing artificial dens or nest boxes on trees. Fourteen studies were in Australia, nine were in the USA, three were in the UK, one was in each of Canada, Lithuania, South Africa and Japan.

COMMUNITY RESPONSE (0 STUDIES)

POPULATION RESPONSE (6 STUDIES)

Abundance (5 studies): Three of five controlled studies (three also replicated) in the USA, the UK, Canada and Lithuania, found that provision of artificial dens or nest boxes increased abundances of gray squirrels and common dormice. The other two studies found that northern flying squirrel and Douglas squirrel abundances did not increase.

Condition (1 study): A replicated, randomized, paired sites, controlled, before-and-after study in Canada found that nest boxes provision did not increase body masses of northern flying squirrel or Douglas squirrel.

BEHAVIOUR (27 STUDIES)

Use (27 studies): Twenty-seven studies, in Australia, the USA, the UK, Canada, South Africa and Japan found that artificial dens or nest boxes were used by a range of mammal species for roosting and breeding.

Assessment: beneficial (effectiveness 65%; certainty 70%; harms 0%).

https://www.conservationevidence.com/actions/2584

Likely to be beneficial

Create or maintain corridors between habitat patches

Four studies evaluated the effects on mammals of creating or maintaining corridors between habitat patches. One study was in each of Canada, the USA, Norway and the Czech Republic.

COMMUNITY RESPONSE (0 STUDIES)

POPULATION RESPONSE (0 STUDIES)

BEHAVIOUR (4 STUDIES)

Use (4 studies): Four studies (three replicated) in Canada, the USA, Norway and the Czech Republic found that corridors between habitat patches were used by small mammals. Additionally, North American deermice moved further through corridors with increased corridor width and connectivity and root voles moved further in corridors of intermediate width.

Assessment: likely to be beneficial (effectiveness 70%; certainty 50%; harms 0%).

https://www.conservationevidence.com/actions/2576

Manage vegetation using livestock grazing

Six studies evaluated the effects on mammals of managing vegetation using livestock grazing. Four studies were in the USA, one was in Norway and one was in Mexico.

COMMUNITY RESPONSE (0 STUDIES)

POPULATION RESPONSE (1 STUDY)

Abundance (1 study): A replicated, controlled, before-and-after study in the USA found that introduction of livestock grazing increased the abundance of Stephens' kangaroo rat after two years.

BEHAVIOUR (5 STUDIES)

Use (4 studies): One of four studies (three replicated controlled studies and a before-and-after study), in the USA and Norway, found that sheep-grazed pasture was used by feeding reindeer more than was ungrazed pasture. One found mixed effects on Rocky Mountain elk use of grazed plots and another found no response of Rocky Mountain elk to spring cattle grazing. The forth study found cattle grazing to increase the proportion of rough fescue biomass utilized by elk in the first, but not second winter after grazing.

Behaviour change (1 study): A replicated, paired sites study in Mexico found that in pastures grazed by cattle, Tehuantepec jackrabbits spent more time feeding than they did in pastures not grazed by cattle.

Assessment: likely to be beneficial (effectiveness 50%; certainty 50%; harms 0%).

https://www.conservationevidence.com/actions/2545

Provide artificial refuges/breeding sites

Eight studies evaluated the effects on mammals of providing artificial refuges/ breeding sites. Two studies were in each of the USA, Spain and Portugal and one was in each of Argentina and Australia.

COMMUNITY RESPONSE (0 STUDIES)

POPULATION RESPONSE (4 STUDIES)

Abundance (3 studies): Two studies (one controlled), in Spain and Portugal, found that artificial warrens increased European rabbit abundance. A replicated, randomized, controlled, before-and-after study in Argentina found that artificial refuges did not increase abundances of small vesper mice or Azara's grass mice.

Survival (1 study): A study in USA found that artificial escape dens increased swift fox survival rates.

BEHAVIOUR (4 STUDIES)

Use (4 studies): Four studies (two replicated), in Australia, Spain, Portugal and the USA, found that artificial refuges, warrens or nest structures were used by fat-tailed dunnarts, European rabbits, and Key Largo woodrats and Key Largo cotton mice.

Assessment: likely to be beneficial (effectiveness 65%; certainty 55%; harms 0%).

https://www.conservationevidence.com/actions/2583

Remove vegetation by hand/machine

Twenty studies evaluated the effects on mammals of removing vegetation by hand or machine. Eleven studies were in the USA, and one each was in Canada, South Africa, Israel, Norway, Portugal, France, Spain, the Netherlands and Thailand.

COMMUNITY RESPONSE (1 STUDY)

Richness/diversity (1 study): A site comparison study in the USA found that mechanically clearing trees within woodland reduced small mammal diversity.

POPULATION RESPONSE (12 STUDIES)

Abundance (11 studies): Eight of 11 site comparison or controlled studies (nine of which were replicated), in the USA, Israel, Portugal, Spain and the Netherlands, found that clearing woody vegetation or herbaceous and grassland vegetation benefitted target mammals. Population or density increases were recorded for small mammals, European rabbits and Stephens' kangaroo rat while black-tailed prairie dog and California ground squirrel colonies were larger or denser and Utah prairie dog colonies established better than in uncleared areas. Two studies found mixed results of clearing woody vegetation, with hazel dormouse abundance declining, then increasing and small mammal abundance increasing, then declining in both cleared and uncleared plots alike. One study found no effect of scrub clearance from sand dunes on habitat specialist small mammals.

Survival (1 study): A replicated, site comparison study in the USA found that mechanical disturbance of woody vegetation within forest (combined with reseeding, follow-up herbicide application and further seeding) increased overwinter survival of mule deer fawns.

BEHAVIOUR (8 STUDIES)

Use (8 studies): Four of seven studies (of which six were site comparisons or controlled), in the USA, Canada, Norway, France and Thailand, found that areas cleared of woody vegetation or herbaceous and grassland vegetation were utilized more by mule deer, reindeer, mouflon and gaur. One study found that clearing woody vegetation promoted increased use by white-tailed deer in some but not all plots, one found that it did not increase use by mule deer and one found that carrying out a second clearance on previously cleared plots did not increase use by white-tailed deer. A before-and-after study in South Africa found that clearing woody vegetation from shrubland increased wildebeest and zebra abundance following subsequent burning but not when carried out without burning whilst other mammals did not show consistent responses.

Assessment: likely to be beneficial (effectiveness 65%; certainty 60%; harms 0%).

https://www.conservationevidence.com/actions/2550

Restore or create forest

Five studies evaluated the effects on mammals of restoring or creating forest. Two studies were in the USA and one each were in Colombia, Italy and Australia.

COMMUNITY RESPONSE (2 STUDIES)

Richness/diversity (2 studies): Two site comparison studies (one replicated) in the USA and Colombia found that mammal species richness in restored forest was similar to that in established forest.

POPULATION RESPONSE (2 STUDIES)

Abundance (2 studies): One of two replicated studies (one a site comparison) in Australia and Italy found that replanted or regrowing forest supported a higher abundance of hazel dormice than did coppiced forest. The other study found only low numbers of common brushtail possums or common ringtail possums by 7–30 years after planting.

BEHAVIOUR (1 STUDY)

Usage (1 study): A replicated, site comparison study in the USA found that restored riparian forest areas were visited more by carnivores than were remnant forests when restored areas were newly established, but not subsequently, whilst restored areas were not visited more frequently by black-tailed deer.

Assessment: likely to be beneficial (effectiveness 55%; certainty 42%; harms 0%).

https://www.conservationevidence.com/actions/2570

Restore or create grassland

Three studies evaluated the effects on mammals of restoring or creating grassland. One study each was in Portugal, the USA and Hungary.

COMMUNITY RESPONSE (1 STUDY)

Richness/diversity (1 study): A replicated, site comparison study in Hungary found that grassland restored on former cropland hosted a similar small mammal species richness compared to native grassland.

POPULATION RESPONSE (3 STUDIES)

Abundance (2 studies): A controlled, before-and-after study in Portugal found that sowing pasture grasses into areas cleared of scrub did not increase European rabbit densities. A replicated, site comparison study in Hungary found that grassland restored on former cropland hosted a similar abundance of small mammals compared to native grassland.

Survival (1 study): A replicated, site comparison study in the USA found that seeding with grassland species as part of a suite of actions including mechanical disturbance and herbicide application increased overwinter survival of mule deer fawns.

BEHAVIOUR (0 STUDIES)

Assessment: likely to be beneficial (effectiveness 40%; certainty 40%; harms 0%).

https://www.conservationevidence.com/actions/2566

Restore or create savannas

Two studies evaluated the effects on mammals of restoring or creating savannas. One study was in Senegal and one was in the USA.

COMMUNITY RESPONSE (1 STUDY)

Richness/diversity (1 study): A replicated, randomized, paired sites, controlled study in the USA found that restoring savannas by removing trees increased small mammal diversity.

POPULATION RESPONSE (2 STUDIES)

Abundance (2 studies): A study in Senegal found that in a population of dorcas gazelle translocated into a fenced enclosure where vegetation had been restored, births outnumbered deaths. A replicated, randomized, paired sites, controlled study in the USA found that restoring savannas by removing trees did not, in most cases, change small mammal abundance.

BEHAVIOUR (0 STUDIES)

Assessment: likely to be beneficial (effectiveness 59%; certainty 50%; harms 0%).

https://www.conservationevidence.com/actions/2568

Restore or create shrubland

Three studies evaluated the effects on mammals of restoring or creating shrubland. Two studies were in the USA and one was in Mexico.

COMMUNITY RESPONSE (2 STUDIES)

Richness/diversity (2 studies): Two site comparison studies, in the USA and Mexico, found that following desert scrub or shrubland restoration, mammal species richness was similar to that in undisturbed areas.

POPULATION RESPONSE (1 STUDY)

Abundance (1 study): A site comparison study in the USA found that restored desert scrub hosted similar small mammal abundance compared to undisturbed desert scrub.

BEHAVIOUR (1 STUDY)

Use (1 study): A replicated, site comparison study in the USA found that restoring shrubland following tree clearance did not increase usage of areas by mule deer compared to tree clearance alone.

Assessment: likely to be beneficial (effectiveness 45%; certainty 40%; harms 0%).

https://www.conservationevidence.com/actions/2569

Unknown effectiveness

Apply fertilizer to vegetation to increase food availability

Two studies evaluated the effects on mammals of applying fertilizer to vegetation to increase food availability. One study was in Canada and one was in the USA.

COMMUNITY RESPONSE (0 STUDIES)

POPULATION RESPONSE (0 STUDIES)

BEHAVIOUR (2 STUDIES)

Use (2 studies): Two replicated, controlled studies, in Canada and the USA, found that applying fertilizer increased the use of vegetation by pronghorns and Rocky Mountain elk.

Assessment: unknown effectiveness (effectiveness 60%; certainty 30%; harms 0%).

https://www.conservationevidence.com/actions/2577

Manage vegetation using grazing by wild herbivores

Two studies evaluated the effects on mammals of managing vegetation using grazing by wild herbivores. One study was in the USA and one was in South Africa.

COMMUNITY RESPONSE (0 STUDIES)

POPULATION RESPONSE (2 STUDIES)

Abundance (2 studies): A site comparison study in the USA found that areas with higher numbers of wild herbivore grazers hosted more small mammals than did areas grazed by fewer wild herbivores. A study in South Africa found that grazing by Cape mountain zebras did not lead to a higher population of bontebok.

BEHAVIOUR (0 STUDIES)

Assessment: unknown effectiveness (effectiveness 43%; certainty 30%; harms 0%).

https://www.conservationevidence.com/actions/2548

Manage wetland water levels for mammal species

One study evaluated the effects of managing wetland water levels for mammal species. This study was in the USA.

COMMUNITY RESPONSE (0 STUDIES)

POPULATION RESPONSE (1 STUDY)

Abundance (1 study): A replicated, site comparison study in the USA found that managing wetland water levels to be higher in winter increased the abundance of muskrat houses.

BEHAVIOUR (0 STUDIES)

Assessment: unknown effectiveness (effectiveness 60%; certainty 21%; harms 0%).

https://www.conservationevidence.com/actions/2574

Provide more small artificial breeding sites rather than fewer large sites

One study evaluated the effects on mammals of providing more small artificial breeding sites rather than fewer larger sites. This study was in Spain.

COMMUNITY RESPONSE (0 STUDIES)

POPULATION RESPONSE (1 STUDY)

Abundance (1 study): A replicated, controlled study in Spain found that smaller artificial warrens supported higher rabbit densities than did larger artificial warrens.

BEHAVIOUR (0 STUDIES)

Assessment: unknown effectiveness (effectiveness 60%; certainty 30%; harms 0%).

https://www.conservationevidence.com/actions/2595

Restore or create wetlands

Four studies evaluated the effects on mammals of restoring or creating wetlands. Three studies were in the USA and one was in the UK.

COMMUNITY RESPONSE (2 STUDIES)

Community composition (1 study): A site comparison study in the USA found that the composition of mammal species present differed between a created and a natural wetland.

Richness/diversity (2 studies): Two site comparison studies (one replicated) in the USA, found that mammal species richness did not differ between created and natural wetlands.

POPULATION RESPONSE (2 STUDIES)

Abundance (1 study): A before-and-after study in the USA found that following marshland restoration, muskrat abundance increased.

Survival (1 study): A replicated, controlled, before-and-after study in the UK, found that water voles persisted better in wetlands that were partially restored using mechanical or manual methods than they did in wetlands undergoing complete mechanical restoration.

BEHAVIOUR (0 STUDIES)

Assessment: unknown effectiveness (effectiveness 59%; certainty 30%; harms 0%).

https://www.conservationevidence.com/actions/2572

Unlikely to be beneficial

Remove vegetation using herbicides

Six studies evaluated the effects on mammals of removing vegetation using herbicides. All six studies were in the USA.

COMMUNITY RESPONSE (0 STUDIES)

POPULATION RESPONSE (4 STUDIES)

Abundance (2 studies): Two controlled studies (one replicated) in the USA found that applying herbicide did not increase numbers of translocated Utah prairie dogs or alter mule deer densities in areas of tree clearance.

Survival (1 study): A replicated, site comparison study in the USA found that applying herbicide, along with mechanical disturbance and seeding, increased overwinter survival of mule deer fawns.

Condition (1 study): A replicated, controlled study in the USA found that applying herbicide did not reduce bot fly infestation rates of rodents and cottontail rabbits.

BEHAVIOUR (2 STUDIES)

Use (2 studies): Two replicated, controlled studies in the USA found that applying herbicide increased forest use by female, but not male, white-tailed deer and increased pasture use by cottontail rabbits in some, but not all, sampling seasons.

Assessment: unlikely to be beneficial (effectiveness 30%; certainty 42%; harms 19%).

https://www.conservationevidence.com/actions/2565

No evidence found (no assessment)

We have captured no evidence for the following interventions:

- Remove topsoil that has had fertilizer added to mimic low nutrient soil
- Replant vegetation.

15.13 Species management

15.13.1 Translocate mammals

Based on the collated evidence, what is the current assessment of the effectiveness of interventions for translocate mammals?	
Beneficial	• Release translocated mammals into fenced areas • Translocate to re-establish or boost populations in native range
Likely to be beneficial	• Hold translocated mammals in captivity before release • Provide supplementary food during/after release of translocated mammals • Translocate predators for ecosystem restoration • Use holding pens at release site prior to release of translocated mammals
Trade-off between benefit and harms	• Translocate mammals to reduce overpopulation
Unknown effectiveness	• Airborne translocation of mammals using parachutes • Use tranquilizers to reduce stress during translocation

Beneficial

● Release translocated mammals into fenced areas

Twenty-four studies evaluated the effects of releasing translocated mammals into fenced areas. Nine studies were in Australia, six studies were in South Africa, two studies were in the USA and one study was in each of India,

China, Spain, Hungary, Namibia and South Africa and France.

COMMUNITY RESPONSE (0 STUDIES)

POPULATION RESPONSE (22 STUDIES)

Abundance (5 studies): Five studies (one replicated) in the USA, Australia and South Africa found that following translocation into fenced areas, 18 African elephant populations, tule elk, brushtail possum and elk and bison increased in number and following eradiation of invasive species a population of translocated and released captive-bred burrowing bettongs increased. A replicated, controlled study in Spain found that the abundance of translocated European rabbits was higher in areas fenced to exclude predators than unfenced areas.

Reproductive success (7 studies): Two replicated, controlled studies in France and Spain found that after translocation, reproductive success of common hamsters and European rabbits was higher inside than outside fenced areas or warrens. Four studies (one replicated, controlled) in China and South Africa found that following translocation into a fenced area, Père David's deer, lions, translocated and captive-bred African wild dogs and one of two groups of Cape buffalo reproduced. A study in Australia found that four of five mammal populations released into a predator-free enclosure and one population released into a predator-reduced enclosure reproduced, whereas two populations released into an unfenced area with ongoing predator management did not survive to breed.

Survival (13 studies): Two replicated, controlled studies in Spain and France found that after translocation, survival rates of common hamsters and European rabbits were higher inside than outside fenced areas or warrens. A study in Australia found that four of five mammal populations released into a predator-free enclosure and one population released into a predator-reduced enclosure survived, whereas two populations released into an unfenced area with ongoing predator management did not persist. Five studies in India, China, South Africa, Namibia and South Africa and Australia found that following translocation into fenced areas, most black rhinoceroses and greater Indian rhinoceroses, Père David's deer, most oribi and offspring of translocated golden bandicoots survived for between one and 10 years. Two studies in Australia found that only two of five translocated numbats survived over seven months and western barred bandicoots did not persist. A study in South Africa found that translocated and captive-bred African wild dogs released into fenced reserves in family groups had high survival rates. A study in Australia found that following release into fenced areas, a translocated population of red-tailed phascogales survived longer than a released captive-bred population. A replicated, controlled study in South Africa found that after translocation to a fenced reserve with holding pens, survival of released lions was higher than that of resident lions.

Condition (3 studies): A replicated, before-and-after study in Australia found that eastern bettongs translocated into fenced predator proof enclosures increased in body weight post-release, with and without supplementary food. A replicated study in South Africa found that following translocation into fenced reserves, stress hormone levels of African elephants declined over time. A study in Australia found that golden bandicoots descended from a population translocated into a fenced area free from non-native predators, maintained genetic diversity relative to the founder and source populations.

BEHAVIOUR (2 STUDIES)

Use (2 studies): A site comparison study in Australia found that following translocation into a predator-free fenced area, woylies developed home ranges similar in size to those of an established population outside the enclosure. A study in Hungary found that one fifth of translocated European ground squirrels released into a fenced area with artificial burrows remained in the area after release.

Assessment: beneficial (effectiveness 80%; certainty 70%; harms 0%).

https://www.conservationevidence.com/actions/2467

Translocate to re-establish or boost populations in native range

Sixty-four studies evaluated the effects of translocating mammals to re-establish or boost populations in their native range. Twenty studies were in the USA, eight in Italy, four in Canada and South Africa, three in the Netherlands and Spain, two in each of the USA and Canada, Zimbabwe, Sweden, Australia and the USA and Mexico and one in each of Uganda, the UK, Brazil, France, Portugal, Africa, Europe, North America, Botswana, Nepal, Chile, Slovakia, Ukraine, Slovakia and Poland and one global study.

COMMUNITY RESPONSE (0 STUDIES)

POPULATION RESPONSE (62 STUDIES)

Abundance (22 studies): Two studies (incuding one controlled and one before-and-after, site comparison study) in Spain and Canada found that translocating animals increased European rabbit abundance or American badger population growth rate at release sites. Fourteen studies (one replicated) in South Africa, the USA, the Netherlands, Italy, France and Spain found that following translocation, populations of warthogs, Eurasian beavers, red squirrels, roe deer, Alpine ibex, Iberian ibex, Cape mountain zebra, 22 species of grazing mammals, black bears, brown bear, bobcats and most populations of river otters increased. Two reviews in South Africa and Australia found that reintroductions (mainly through translocations) led to increasing populations for four of six species of large carnivores and that over half of translocations were classified as successful. One replicated study in the USA and Mexico found that translocating desert bighorn sheep did not increase the population size. Two studies (one replicated) and a review in USA and Canada, the USA and Australia found that translocated American martens, and sea otters at four of seven sites, established populations and that translocated and released captive-bred macropod species established populations in 44 of 72 cases. A study in Italy found that following the translocation of red deer, the density of Apennine chamois in the area almost halved. A worldwide review found that translocating ungulates was more successful when larger numbers were released, and small populations grew faster if they contained more mature individuals and had an equal ratio of males and females.

Reproductive success (16 studies): A controlled study in Italy found that wild-caught translocated Apennine chamois reproduced in similar numbers to released captive-bred chamois. Fourteen studies (four replicated) in Canada, the USA, Zimbabwe, South Africa, the UK, Italy, the Netherlands and Slovakia found that translocated black and white rhinoceroses, warthogs, common dormice, European ground squirrels, cougars, bobcats, brown bears, sea otters, river otters and some Eurasian otters reproduced. A study in the Netherlands found that translocated beavers were slow to breed.

Survival (39 studies): Four of five studies (including three controlled, two replicated and one before-and-after, site comparison study) in the USA, Canada and Chile found that wild-born translocated long-haired field mice, female elk, cougars and American badgers had lower survival rates than non-translocated resident animals. One found that translocated Lower Keys marsh rabbits had similar survival rates to non-translocated resident animals. Five of four studies (two replicated, four controlled) and two reviews in Canada, Canada and the USA, the USA, Italy, Sweden and Africa, Europe, and North America found that wild-born translocated swift foxes, European otters, black-footed ferret kits and a mix of carnivores had higher survival rates than released captive-bred animals. One study found that wild-born translocated Apennine chamois had a similar survival rate to released captive-bred animals. Twenty of twenty-one studies (including two replicated and one before-and-after study) and a review in Nepal, France, Italy, Portugal, Ukraine, Slovakia and Poland, Canada, USA, Brazil, Uganda, South Africa, Zimbabwe and Botswana found that following translocation, populations of or individual mammals survived between two months and at least 25 years. The other two studies found that two of 10 translocated white rhinoceroses died within three days of release and an American marten population did not persist. A review in Australia found that over half of translocations, for which the outcome could be determined, were classified as successful. Two of three studies (one replicated) and one review in Sweden, the UK, the Netherlands and the USA and Mexico found that following release of wild-caught translocated and captive-bred animals, European otters and common dormice survived three months to seven years. The review found that most black-footed ferret releases were unsuccessful at maintaining a population. A replicated study in the USA found that following translocation of bighorn sheep, 48–98% of their offspring survived into their first winter.

Condition (3 studies): Three studies (including one replicated, controlled study) in the USA and Italy found that following translocation, populations of elk had similar levels of genetic diversity to non-translocated populations, descendants of translocated swift fox had genetic diversity at least as high as that of the translocated animals and brown bear genetic diversity declined over time.

BEHAVIOUR (9 STUDIES)

Use (7 studies): A study in Italy found that following translocation, Alpine ibex used similar habitats to resident animals. Two of four studies (including one randomized, controlled study) in the USA, Netherlands and Botswana found that following translocation (and in one case release of some captive-bred animals), most Eurasian otters settled and all three female grizzly bears established ranges at their release site. The other two studies found that most nine-banded armadillos and some white rhinoceroses (when released into areas already occupied by released animals) dispersed from their release site. Two studies (one replicated) in Spain found that following translocation, Iberian ibex expanded their range and roe deer increased their distribution six-fold.

Behaviour change (2 studies): A replicated controlled study in Chile found that following translocation, long-haired field mice travelled two- to four-times further than non-translocated mice. A controlled study in Italy found that wild-caught translocated Apennine chamois moved further from the release site than released captive-bred animals.

Assessment: beneficial (effectiveness 75%; certainty 70%; harms 15%).

https://www.conservationevidence.com/actions/2397

Likely to be beneficial

Hold translocated mammals in captivity before release

Fifteen studies evaluated the effects of holding translocated mammals in captivity before release. Four studies were in the USA, two were in Australia and one was in each of India, Canada, Switzerland, Croatia and Slovenia, the USA and Canada, the UK, France, Spain and South Africa.

COMMUNITY RESPONSE (0 STUDIES)

POPULATION RESPONSE (13 STUDIES)

Abundance (2 studies): Two studies (one replicated, before-and-after study) in Croatia and Slovenia and the USA found that following translocation, with time in captivity prior to release, Eurasian lynx established an increasing population and Allegheny woodrat numbers in four of six sites increased over the first two years.

Reproductive success (4 studies): Four studies in Croatia and Slovenia, Spain, the USA and Canada and Australia found that following translocation, with time in captivity prior to release, Eurasian lynx established a breeding population, and swift foxes, European otters and red-tailed phascogales reproduced.

Survival (10 studies): Two studies (one controlled) in the UK and USA found that being held for longer in captivity before release increased survival rates of translocated European hedgehogs and, along with release in spring increased the survival rate of translocated Canada lynx in the first year. Four of six studies in India, the USA and Canada, the USA, France, South Africa and Australia found that following translocation, with time in captivity prior to release, most swift foxes and greater Indian rhinoceroses survived for at least 12-20 months, 48% of Eurasian lynx survived for 2–11 years and red-tailed phascogales survived for at least six years. The other two studies found that most kangaroo rats and all rock hyraxes died within 5-87 days. A replicated, controlled study in Canada found that translocated swift foxes that had been held in captivity prior to release had higher post-release survival rates than did released captive-bred animals.

Condition (3 studies): A randomised, controlled study in Australia found that holding translocated eastern bettongs in captivity before release did not increase their body mass after release compared to animals released directly into the wild. A controlled study the UK found that being held for longer in captivity before release, reduced weight loss after release in translocated European hedgehogs. A study in Spain found that offspring of translocated

European otters that were held in captivity before release, had similar genetic diversity to donor populations.

Occupancy/range (2 studies): A study in the USA found that most translocated and captive-bred mountain lions that had been held in captivity prior to release established home ranges in the release area. A study in Croatia and Slovenia and review in Switzerland found that following translocation, with time in captivity prior to release, the range of Eurasian lynx increased over time.

BEHAVIOUR (0 STUDIES)

Assessment: likely to be beneficial (effectiveness 60%; certainty 60%; harms 0%).

https://www.conservationevidence.com/actions/2458

Provide supplementary food during/after release of translocated mammals

Sixteen studies evaluated the effects of providing supplementary food during/after release of translocated mammals. Four studies were in the UK, two were in each of the USA, France, Australia and Argentina, and one was in each of Italy, Spain, Ireland and South Africa.

COMMUNITY RESPONSE (0 STUDIES)

POPULATION RESPONSE (15 STUDIES)

Abundance (2 studies): A controlled study in Spain found that providing supplementary food during translocation did not increase European rabbit abundance. A study in France found that following supplementary feeding in a holding pen prior to release, a translocated deer population increased over six years.

Reproductive success (4 studies): Three studies (one replicated) in the USA, Italy and Ireland found that having been provided with supplementary food in holding pens prior to release, translocated black-tailed prairie dogs, a pair of Eurasian badgers and most female red squirrels reproduced in the wild. A study in the UK found that some translocated pine martens released from holding pens and then provided with supplementary food and nest boxes bred in the first year after release.

Survival (10 studies): Six of 10 studies (including one replicated and one controlled study) in the UK, France, Italy, Ireland, South Africa, USA, Argentina and Australia found that at sites with supplementary food in holding pens before (and in two cases after) release, translocated populations of black-tailed prairie dogs, approximately half of female roe deer and over half of red squirrels, Eurasian badgers, pine martens and released rehabilitated or captive reared giant anteaters survived for between one month and at least two years. Four studies found that at translocation release sites with provision of supplementary food, in most cases artificial refuges and in one case water, no red squirrels, rock hyraxes or burrowing bettongs survived over 2-5 months and most translocated Tipton and Heermann's kangaroo rat spp. died within five days. A controlled study in France found that translocated European rabbits provided with supplementary food in holding pens for three days prior to release had higher female (but not male) survival rates immediately following release compared to those released directly. A controlled study in the UK found that survival of translocated and rehabilitated European hedgehogs that were provided with supplementary food after release varied with release method.

Condition (2 studies): One of three studies (including one replicated, one controlled and two before-and-after studies) in the UK and Australia found that translocated common dormice gained weight after being provided with supplementary food. One found that translocated eastern bettongs did not have increased body weights after provision of supplementary food in fenced enclosures prior to release. The other found that translocated and rehabilitated European hedgehogs provided with food after release all lost body mass, with effects varying with release method.

BEHAVIOUR (2 STUDIES)

Use (1 study): A controlled study in Australia found that supplementary feeding stations were visited by translocated burrowing bettongs.

Behaviour change (1 study): A controlled study in Argentina found that after being provided with supplementary food and kept in holding pens, released captive-bred giant anteaters were less nocturnal than wild-born rehabilitated and released individuals.

Assessment: likely to be beneficial (effectiveness 67%; certainty 60%; harms 0%).

https://www.conservationevidence.com/actions/2470

Translocate predators for ecosystem restoration

Two studies evaluated the effects of translocating predators for ecosystem restoration. These studies were in the USA.

COMMUNITY RESPONSE (0 STUDIES)

POPULATION RESPONSE (2 STUDIES)

Abundance (2 studies): A before-and-after study in the USA found that following reintroduction of wolves, populations of beavers and bison increased. A before-and-after study in the USA found that after the translocation of wolves to the reserve, adult elk numbers approximately halved.

Reproductive success (1 study): A before-and-after study in the USA found that after the translocation of wolves to the reserve, elk calf:cow ratios approximately halved.

BEHAVIOUR (0 STUDIES)

Assessment: likely to be beneficial (effectiveness 70%; certainty 55%; harms 10%).

https://www.conservationevidence.com/actions/2431

Use holding pens at release site prior to release of translocated mammals

Thirty-five studies evaluated the effects of using holding pens at the release site prior to release of translocated mammals. Ten studies were in the USA, seven were in South Africa, four were in the UK, three studies were in France, two studies were in each of Canada, Australia and Spain and one was in each of Kenya, Zimbabwe, Italy, Ireland and India.

COMMUNITY RESPONSE (0 STUDIES)

POPULATION RESPONSE (31 STUDIES)

Abundance (4 studies): Three of four studies (two replicated, one before-and-after study) in South Africa, Canada, France and Spain found that following release from holding pens at release sites (in some cases with other associated actions), populations of roe deer, European rabbits and lions increased in size. The other study found that elk numbers increased at two of four sites.

Reproductive success (10 studies): A replicated study in the USA found that translocated gray wolves had similar breeding success in the first two years after release when adult family groups were released together from holding pens or when young adults were released directly into the wild. Seven of nine studies (including two replicated and one controlled study) in Kenya, South Africa, the USA, Italy, Ireland, Australia and the UK found that following release from holding pens at release sites (in some cases with other associated actions), translocated populations of roan, California ground squirrels, black-tailed prairie dogs, lions, four of four mammal populations, most female red squirrels and some pine martens reproduced successfully. Two studies found that one of two groups of Cape buffalo and one pair out of 18 Eurasian badgers reproduced.

Survival (26 studies): Two of seven studies (five controlled, three replicated studies) in Canada, the USA, France, the UK found that releasing animals from holding pens at release sites (in some cases with associated actions) resulted in higher survival for water voles and female European rabbits compared to those released directly into the wild. Four studies found that translocated swift foxes, gray wolves, Eurasian lynx and Gunnison's prairie dogs released from holding pens had similar survival rates to those released directly into the wild. One study found that translocated American martens released from holding pens had lower survival than those released directly into the wild. Two of four studies (three controlled) in South Africa, Spain, and the USA found that translocated African wild dogs and European rabbits that spent longer in holding pens at release sites had a higher survival rate after release. One study found mixed effects for swift foxes and one found no effect of time in holding pens for San Joaquin kit foxes. Eleven studies (one replicated) in Kenya, South Africa, the USA, France, Italy, Ireland, India, Australia and the UK found that after release from holding pens at release sites (in some cases with other associated actions), translocated populations or individuals survived between one month and six years, and four of four mammal populations survived. Two studies in the UK and South Africa found that no released red squirrels or rock hyraxes survived over five months or 18 days respectively. One of two controlled studies (one replicated, one before-and-after) in South Africa and the USA found that following release from holding pens, survival of translocated lions was higher than that of resident animals, whilst that of translocated San Joaquin kit foxes was lower than that of resident animals. A study in Australia found that translocated bridled nailtail wallabies kept in holding pens prior to release into areas where predators had been controlled had similar annual survival to that of captive-bred animals.

Condition (1 study): A controlled study in the UK found that translocated common dormice held in pens before release gained weight after release whereas those released directly lost weight.

BEHAVIOUR (5 STUDIES)

Behaviour change (5 studies): Three studies (one replicated) in the USA and Canada found that following release from holding pens, fewer translocated sea otters and gray wolves returned to the capture site compared to those released immediately after translocation, and elk remained at all release sites. Two studies in Zimbabwe and South Africa found that following release from holding pens, translocated lions formed new prides.

Assessment: likely to be beneficial (effectiveness 72%; certainty 60%; harms 10%).

https://www.conservationevidence.com/actions/2434

Trade-off between benefit and harms

Translocate mammals to reduce overpopulation

Three studies evaluated the effects of translocating mammals to reduce overpopulation. Two studies were in the USA and one was in Australia.

COMMUNITY RESPONSE (0 STUDIES)

POPULATION RESPONSE (3 STUDIES)

Abundance (1 study): A before-and-after study in the USA found that adult elk numbers approximately halved after the translocation of wolves to the reserve.

Reproductive success (1 study): A before-and-after study in the USA found that elk calf:cow ratios approximately halved after the translocation of wolves to the reserve.

Survival (2 studies): A study in Australia found that koalas translocated to reduce overpopulation had lower survival than individuals in the source population. A study in the USA found that following translocation to reduce over-abundance, white-tailed deer had lower survival rates compared to non-translocated deer at the recipient site.

Occupancy/range (1 study): A study in the USA found that following translocation to reduce over-abundance at the source site, white-tailed deer had similar home range sizes compared to non-translocated deer at the recipient site.

BEHAVIOUR (0 STUDIES)

Assessment: trade-off between benefit and harms (effectiveness 66%; certainty 47%; harms 20%).

https://www.conservationevidence.com/actions/2430

Unknown effectiveness

Airborne translocation of mammals using parachutes

One study evaluated the effects of airborne translocation of mammals using parachutes. This study was in the USA.

COMMUNITY RESPONSE (0 STUDIES)

POPULATION RESPONSE (1 STUDY)

Survival (1 study): A study in the USA found that at least some North American beavers translocated using parachutes established territories and survived over one year after release.

BEHAVIOUR (0 STUDIES)

Assessment: unknown effectiveness (effectiveness 50%; certainty 24%; harms 15%).

https://www.conservationevidence.com/actions/2466

Use tranquilizers to reduce stress during translocation

One study evaluated the effects on mammals of using tranquilizers to reduce stress during translocation. This study was in France.

COMMUNITY RESPONSE (0 STUDIES)

POPULATION RESPONSE (1 STUDY)

Survival (1 study): A controlled study in France found that using tranquilizers to reduce stress during translocation did not increase post-release survival of European rabbits.

BEHAVIOUR (0 STUDIES)

Assessment: unknown effectiveness (effectiveness 10%; certainty 15%; harms 10%).

https://www.conservationevidence.com/actions/2465

15.13.2 Captive-breeding

Based on the collated evidence, what is the current assessment of the effectiveness of interventions for captive-breeding?	
Beneficial	• Breed mammals in captivity
Likely to be beneficial	• Use artificial insemination
Unknown effectiveness	• Clone rare species • Place captive young with captive foster parents • Preserve genetic material for use in future captive breeding programs

Beneficial

Breed mammals in captivity

Three studies evaluated the effects of breeding mammals in captivity. One study was across Europe, one was in the USA and one was global.

COMMUNITY RESPONSE (0 STUDIES)
POPULATION RESPONSE (3 STUDIES)
Abundance (1 study): A review of captive-breeding programmes across the world found that the majority of 118 captive-bred mammal populations increased.
Reproductive success (2 studies): A review of a captive breeding programme across Europe found that the number of European otters born in captivity tended to increase over 15 years. A study in the USA found that wild-caught Allegheny woodrats bred in captivity.
Survival (1 study): A review of a captive breeding programme across Europe found that the number of European otters born in captivity that survived tended to increase over 15 years.
BEHAVIOUR (0 STUDIES)
Assessment: beneficial (effectiveness 80%; certainty 70%; harms 0%).

https://www.conservationevidence.com/actions/2471

Likely to be beneficial

Use artificial insemination

Three studies evaluated the effects on mammals of using artificial insemination. One study was in the USA, one was in Brazil and one was in China.
COMMUNITY RESPONSE (0 STUDIES)
POPULATION RESPONSE (3 STUDIES)
Reproductive success (3 studies): A study in the USA found that following artificial insemination, fewer than half of female black-footed ferrets gave birth. A study in Brazil found that following artificial insemination, a captive female Amazonian brown brocket deer gave birth. A replicated study in China found that following artificial insemination, a lower proportion of captive female giant pandas became pregnant than after natural mating.
BEHAVIOUR (0 STUDIES)
Assessment: likely to be beneficial (effectiveness 40%; certainty 40%; harms 9%).

https://www.conservationevidence.com/actions/2473

Unknown effectiveness

Clone rare species

One study evaluated the effects of cloning rare species. This study was in Iran.
COMMUNITY RESPONSE (0 STUDIES)

POPULATION RESPONSE (1 STUDY)

Reproductive success (1 study): A controlled study in Iran found that immature eggs of domestic sheep have potential to be used for cloning of Esfahan mouflon.

BEHAVIOUR (0 STUDIES)

Assessment: unknown effectiveness (effectiveness 40%; certainty 20%; harms 10%).

https://www.conservationevidence.com/actions/2474

Place captive young with captive foster parents

Two studies evaluated the effects of placing captive young mammals with captive foster parents. One study was in the USA and one was in Sweden and Norway.

COMMUNITY RESPONSE (0 STUDIES)

POPULATION RESPONSE (2 STUDIES)

Survival (2 studies): A replicated, controlled study in the USA found that most captive coyote pups placed with foster parents were successfully reared. A replicated study in Sweden and Norway found that captive grey wolf pups placed with foster parents had higher survival rates than pups that stayed with their biological mother.

Condition (1 study): A replicated study in Sweden and Norway found that captive grey wolf pups placed with foster parents weighed less than pups that stayed with their biological mother.

BEHAVIOUR (0 STUDIES)

Assessment: unknown effectiveness (effectiveness 70%; certainty 32%; harms 10%).

https://www.conservationevidence.com/actions/2472

Preserve genetic material for use in future captive breeding programs

Two studies evaluated the effects of preserving genetic material for use in future captive breeding programs. One study was in Mexico and one was in the USA.

COMMUNITY RESPONSE (0 STUDIES)

POPULATION RESPONSE (2 STUDIES)

Survival (2 studies): A study in Mexico found that a series of non-traditional techniques, combined with natural mating, produced five aoudad embryos that could be cryogenically preserved. A study in USA, found that artificial insemination using preserved genetic material increased genetic diversity and lowered inbreeding in a captive black-footed ferret population.

BEHAVIOUR (0 STUDIES)

Assessment: unknown effectiveness (effectiveness 60%; certainty 35%; harms 0%).

https://www.conservationevidence.com/actions/2475

15.13.3 Release captive-bred mammals

Based on the collated evidence, what is the current assessment of the effectiveness of interventions for release captive-bred mammals?	
Beneficial	• Provide supplementary food during/after release of captive-bred mammals • Release captive-bred individuals to re-establish or boost populations in native range • Release captive-bred mammals into fenced areas • Use holding pens at release site prior to release of captive-bred mammals
Likely to be beneficial	• Captive rear in large enclosures prior to release • Provide live natural prey to captive mammals to foster hunting behaviour before release
Unknown effectiveness	• Train captive-bred mammals to avoid predators

Beneficial

Provide supplementary food during/after release of captive-bred mammals

Fifteen studies evaluated the effects of providing supplementary food during/after release of captive-bred mammals. Four studies were in Australia, two were in each of the USA, China and Argentina, and one was in each of Poland, the UK, Oman and Saudi Arabia.

COMMUNITY RESPONSE (0 STUDIES)

POPULATION RESPONSE (14 STUDIES)

Abundance (5 studies): Four studies (one replicated, one before-and-after study) and one review in Poland, Oman, China and Australia found that following provision of supplementary food (and in one case water) to released captive-bred animals, populations of European bison increased more than six-fold over 20 years, Arabian oryx increased over 14 years, eastern-barred bandicoots increased for the first five years before declining, Père David's deer increased more than six-fold over 12 years and Przewalski's horses (enclosed in winter) increased over 11 years.

Reproductive success (9 studies): Eight studies (including two replicated and one before-and-after study) and one review in Poland, the UK, China, the USA, Australia and Saudi Arabia found that following the provision of supplementary food (and in one case water or artificial nests) after release of captive-bred animals, some from holding pens, European bison, European otters, Père David's deer, eastern-barred bandicoots, Przewalski's horses and some captive-bred red wolves successfully reproduced, Arabian gazelles started breeding in the year following releases and sugar gliders established a breeding population.

Survival (6 studies): Four of six studies (one controlled, before-and-after study) in the UK, USA, Argentina and Australia found that following the provision of supplementary food (and in one case water or nest boxes) after release of captive-bred animals, many from holding pens, 19% of red wolves survived for at least seven years, Eurasian otters survived for at least two years, over half the giant anteaters (some rehabilitated) survived for at least six months and hare-wallabies survived at least two months. Two of the studies found that red-tailed phascogales survived for less than a year and most Mexican wolves survived less than eight months.

BEHAVIOUR (1 STUDY)

Behaviour change (1 study): A controlled study in Argentina found that after being provided with supplementary food and kept in holding pens, released captive-bred giant anteaters were less nocturnal in their activity patterns than released wild-born rehabilitated individuals.

Assessment: beneficial (effectiveness 72%; certainty 70%; harms 10%).

https://www.conservationevidence.com/actions/2527

Release captive-bred individuals to re-establish or boost populations in native range

Thirty-one studies evaluated the effects of releasing captive-bred mammals to establish or boost populations in their native range. Seven studies were in the USA, three were in Australia and Italy, two studies were in each of Canada, Sweden, Saudi Arabia, the UK, the Netherlands and South Africa and one study was in each of France, Africa, Europe, and North America, Estonia, the USA and Mexico, Poland and China.

COMMUNITY RESPONSE (0 STUDIES)

POPULATION RESPONSE (30 STUDIES)

Abundance (7 studies): Five of five studies (one replicated) and two reviews in Saudi Arabia, Australia, the USA, South Africa, France, the Netherlands and China found that following release of captive-bred (or in one case captive-reared, or including translocated) animals, populations of mountain gazelles, Corsican red deer, Père David's deer, Eurasian otters and swift foxes increased. The two reviews found that following release of mainly translocated but some captive-bred large carnivores, populations of four of six species increased, and over half of mammal release programmes were considered successful.

Reproductive success (5 studies): Four studies (one replicated) in Saudi Arabia, the UK and the Netherlands found that released captive-bred (and in some cases some wild-born translocated) mountain gazelles, dormice and some Eurasian otters reproduced successfully and female Arabian oryx reproduced successfully regardless of prior breeding experience. A controlled study in Italy found that released captive-born Apennine chamois reproduced in similar numbers to wild-caught translocated chamois.

Survival (24 studies): Four of three controlled studies (two replicated) and two reviews in Canada, Canada and the USA, Sweden, Italy and across the world found that released captive-bred swift foxes, European otters and mammals from a review of 49 studies had lower post-release survival rates than did wild-born translocated animals. The other study found that released captive-born Apennine chamois survived in similar numbers to wild-caught translocated chamois. Three studies (one replicated) in the USA and Canada found that released captive-born Key Largo woodrats, Vancouver Island marmots and swift fox pups had lower survival rates than wild-born, wild-living animals. One of the studies also found that Vancouver Island marmots released at two years old were more likely to survive than those released as yearlings. Eleven studies (three replicated) in Italy, Sweden, the UK, Estonia, Poland, Saudi Arabia, Australia and the USA found that following the release of captive-bred (and in some cases some wild-born translocated) animals, Arabian oryx, populations of European otters, European mink and mountain gazelle survived for 2-11 years, roe deer and over a third of brush-tailed rock-wallabies, black-footed ferrets and brown hares survived for 0.5-24 months and dormice populations survived three months to over seven years. A review in Australia found that release programmes for macropod species resulted in successful establishment of populations in 61% of cases and that

40% survived over five years, and another review in Australia found that over half of programmes were considered successful. Two studies and a review in the USA, USA and Mexico and South Africa found that over 40% of released captive-bred American black bears were killed or had to be removed, only one of 10 oribi survived over two years and that most black-footed ferret releases were unsuccessful at maintaining a population.
BEHAVIOUR (3 STUDIES)
Use (3 studies): Two studies in the USA and Australia found that following release, most captive-bred and translocated mountain lions that had been held in captivity prior to release and most released captive-bred brush-tailed rock-wallabies established stable home ranges. A controlled study in Italy found that released captive-born Apennine chamois remained closer to the release site than released wild-caught translocated chamois.
Assessment: beneficial (effectiveness 70%; certainty 67%; harms 10%).

https://www.conservationevidence.com/actions/2476

Release captive-bred mammals into fenced areas

Fourteen studies evaluated the effects of releasing captive-bred mammals into fenced areas. Nine studies were in Australia and one each was in Jordan, South Africa, the USA, Saudi Arabia and Senegal.
COMMUNITY RESPONSE (0 STUDIES)
POPULATION RESPONSE (14 STUDIES)
Abundance (5 studies): Four studies (one replicated) and a review in Australia, Jordan and Senegal found that after releasing captive-bred animals into fenced areas, a population of burrowing bettongs increased, a population of Arabian oryx increased six-fold in 12 years, a population of dorcas gazelle almost doubled over four years, three populations of eastern barred bandicoot initially increased and abundance of eastern barred bandicoots increased.
Reproductive success (6 studies): Four studies and a review in South Africa, Australia, Saudi Arabia and Senegal found that following release of captive-bred animals into fenced areas (in some cases with other associated management), African wild dogs, three populations of eastern barred bandicoot, dorcas gazelle and most female black-footed rock-wallabies reproduced, and Arabian gazelles started breeding in the year following the first releases. A study in Australia found that four of five mammal populations released into a predator-free enclosure and one released into a predator-reduced enclosure reproduced, whereas two populations released into an unfenced area with ongoing predator management did not survive to reproduce.

Survival (10 studies): A study in Australia found that four of five mammal populations released into a predator-free enclosure and one population released into a predator-reduced enclosure survived, whereas two populations released into an unfenced area with ongoing predator management did not. Six studies (one controlled before-and-after study and two replicated studies) in Australia and the USA found that following release of captive-bred animals into fenced areas (in some cases with other associated management), a burrowing bettong population, three eastern barred bandicoot populations and over half of black-footed rock-wallabies survived between one and eight years, most captive-bred hare-wallabies survived at least two months, at least half of black-footed ferrets survived more than two weeks, and bandicoots survived at five of seven sites up to three years after the last release. One study in Australia found that following release into fenced areas, a captive-bred population of red-tailed phascogales survived for less than a year. A study in South Africa found that captive-bred African wild dogs released into fenced reserves in family groups had high survival rates. A randomized, controlled study in Australia found that captive-bred eastern barred bandicoots released into a fenced reserve after time in holding pens had similar post-release survival compared to bandicoots released directly from captivity.
Condition (1 study): A randomized, controlled study in Australia found that captive-bred eastern barred bandicoots released into a fenced reserve after time in holding pens had similar post-release body weight compared to those released directly from captivity.
BEHAVIOUR (0 STUDIES)
Assessment: beneficial (effectiveness 77%; certainty 70%; harms 0%).

https://www.conservationevidence.com/actions/2521

Use holding pens at release site prior to release of captive-bred mammals

Thirty-one studies evaluated the effects of using holding pens at the release site prior to release of captive-bred mammals. Seven studies were in Australia, and in the USA, four were in the UK, three in Argentina, two in each of Israel, Saudi Arabia and China and one in each of Canada, Namibia, South Africa and Germany.
COMMUNITY RESPONSE (0 STUDIES)
POPULATION RESPONSE (30 STUDIES)
Abundance (2 studies): A study in Saudi Arabia found that a population of captive-bred Arabian sand gazelles kept in holding pens prior to release

nearly doubled in size over four years. A before-and-after study in China found that following release of captive-bred animals from a pre-release enclosure into the semi-wild (free-roaming in summer, enclosed in winter and provided with food), Przewalski's horses increased in number.

Reproductive success (10 studies): Eight studies (one replicated) and one review in the UK, Saudi Arabia, the USA, Israel and Australia found that following the use of holding pens prior to release (and in some cases provision of supplementary food), captive-bred Eurasian otters, Arabian sand gazelles, eastern-barred bandicoots, some swift foxes, some red wolves and over 33% of Persian fallow deer reproduced, Arabian gazelles started breeding in the first year and the reproductive success of female Asiatic wild ass increased over 10 years. A study in Australia found that after being kept in a holding pen, all four mammal populations released into an invasive-species-free fenced enclosure reproduced.

Survival (23 studies): One of three studies (two controlled, one replicated) in the UK, Canada and Australia found that using holding pens prior to release of captive-bred (and some translocated) animals resulted in greater post-release survival for water voles compared to animals released directly into the wild. The other two studies found similar survival rates for eastern barred bandicoots and swift foxes compared to animals released directly into the wild. A replicated study in the USA found that captive-bred Allegheny woodrats kept in holding pens prior to release, had higher early survival rates than those not kept in holding pens, but overall survival rates tended to be lower than wild resident woodrats. Three studies in South Africa, USA and Argentina found that released captive-bred (and some translocated) African wild dogs, riparian brush rabbits and guanacos that spent longer in, and in one case in larger, holding pens had a higher survival rate. Three studies (one controlled) in Australia and the USA found that captive-bred animals kept in holding pens prior to release had similar (bridled nailtail wallabies) or lower (black-footed ferret kits) annual survival rate after release to that of wild-born translocated animals and lower (black-footed ferrets) survival rates than resident animals. Ten studies (including one controlled, before-and-after study) and one review in Saudi Arabia, the USA, Argentina, China, Israel, Australia and Germany found that following the use of holding pens prior to release of captive-bred animals (or in some cases captive-reared/rehabilitated, or with provision of supplementary food), four of four mammal populations, 19% of red wolves, Asiatic wild ass, Persian fallow deer, most Arabian sand gazelles, most swift foxes, eastern-barred bandicoots and European mink survived at least 1-10 years, over half of giant anteaters, hare-wallabies and Père David's deer survived for at least 1.5-6 months. Three studies in Namibia, the USA and Australia found that that following the use of holding pens prior to release of captive-bred or reared animals (some provided with nest boxes and/or supplementary food), red-tailed phascogales, most Mexican wolves and African wild dogs survived less than 6-12 months.

Condition (4 studies): A randomized, controlled study in Australia found that eastern barred bandicoots released after time in holding pens lost a similar proportion of body weight and recovered to a similar weight compared to bandicoots released directly. A controlled study in the UK found that common dormice lost weight after being put into holding pens whereas wild translocated dormice gained weight. A controlled, before-and-after study in Australia found that captive-bred rufous hare-wallabies placed in holding pens prior to release lost body condition in holding pens. A before-and-after study in Australia found that captive-bred brush-tailed rock-wallabies placed in a holding pen prior to release maintained good health.

BEHAVIOUR (1 STUDY)

Behaviour change (1 study): A controlled study in Argentina found that after being kept in holding pens and provided with supplementary food, released captive-bred giant anteaters were less nocturnal in their activity patterns than released wild-born rehabilitated individuals.

Assessment: beneficial (effectiveness 70%; certainty 65%; harms 10%).

https://www.conservationevidence.com/actions/2510

Likely to be beneficial

Captive rear in large enclosures prior to release

Four studies evaluated the effects of captive rearing mammals in large enclosures prior to release. Two studies were in the USA, one was in Mexico and one was in Australia.

COMMUNITY RESPONSE (0 STUDIES)

POPULATION RESPONSE (3 STUDIES)

Reproductive success (1 study): A study in Mexico found that peninsular pronghorn taken from the wild and kept in a large enclosure bred successfully and the population increased, providing stock suitable for reintroductions.

Survival (2 studies): A replicated, controlled study in USA found that black-footed ferrets reared in outdoor pens had higher post-release survival rates than did ferrets raised indoors. A controlled study in Australia found that

Tasmanian devils reared free-range in large enclosures did not have greater post-release survival rates than animals from intensively managed captive-rearing facilities.
Condition (1 study): A controlled study in Australia found that Tasmanian devils reared free-range in large enclosures did not gain more body weight post-release compared to animals from intensively managed captive-rearing facilities.
BEHAVIOUR (1 STUDY)
Behaviour change (1 study): A controlled study in USA found that captive-bred black-footed ferrets raised in large enclosures dispersed shorter distances post-release than did ferrets raised in small enclosures.
Assessment: likely to be beneficial (effectiveness 50%; certainty 46%; harms 0%).

https://www.conservationevidence.com/actions/2507

Provide live natural prey to captive mammals to foster hunting behaviour before release

Three studies evaluated the effects of providing live natural prey to captive mammals to foster hunting behaviour before release. One study was in Spain, one was in the USA and one was in Botswana.
COMMUNITY RESPONSE (0 STUDIES)
POPULATION RESPONSE (2 STUDIES)
Survival (2 studies): Two studies in Spain and Botswana found that a rehabilitated Iberian lynx and wild-born but captive-reared orphaned cheetahs and leopards that were provided with live natural prey in captivity survived for between at least three months and 19 months after release.
BEHAVIOUR (1 STUDY)
Behaviour change (1 study): A controlled study in the USA found that captive-bred black-footed ferrets fed on live prairie dogs took longer to disperse after release but showed greater subsequent movements than did ferrets not fed with live prairie dogs.
Assessment: likely to be beneficial (effectiveness 65%; certainty 40%; harms 0%).

https://www.conservationevidence.com/actions/2518

Unknown effectiveness

Train captive-bred mammals to avoid predators

Two studies evaluated the effects of training captive-bred mammals to avoid predators. One study was in Australia and one was in the USA.

COMMUNITY RESPONSE (0 STUDIES)

POPULATION RESPONSE (1 STUDY)

Survival (1 study): A randomized, controlled study in the USA found that training captive-born juvenile black-tailed prairie dogs, by exposing them to predators, increased post-release survival.

BEHAVIOUR (1 STUDY)

Behaviour change (1 study): A before-and-after study in Australia found that rufous hare-wallabies could be conditioned to become wary of potential predators.

Assessment: unknown effectiveness (effectiveness 70%; certainty 37%; harms 0%).

https://www.conservationevidence.com/actions/2520

15.13.4 Release captive-bred/translocated mammals

Based on the collated evidence, what is the current assessment of the effectiveness of interventions for release captive-bred/translocated mammals?	
Beneficial	• Release translocated/captive-bred mammals in areas with invasive/problematic species eradication/control • Release translocated/captive-bred mammals in family/social groups • Release translocated/captive-bred mammals into area with artificial refuges/breeding sites
Likely to be beneficial	• Release translocated/captive-bred mammals at a specific time (e.g. season, day/night) • Release translocated/captive-bred mammals in larger unrelated groups • Release translocated/captive-bred mammals to areas outside historical range • Release translocated/captive-bred mammals to islands without invasive predators

Beneficial

● Release translocated/captive-bred mammals in areas with invasive/problematic species eradication/control

Twenty-two studies evaluated the effects of releasing translocated or captive-bred mammals in areas with eradication or control of invasive or problematic species. Sixteen studies were in Australia, four were in the USA, and one in the UK.

COMMUNITY RESPONSE (0 STUDIES)

POPULATION RESPONSE (21 STUDIES)

Abundance (4 studies): A replicated study in Australia found that increasing amounts of regular predator control increased population numbers of released captive-bred eastern barred bandicoots. Two studies in Australia found that following eradication or control of invasive species, a population of translocated and released captive-bred burrowing bettongs increased and a population of translocated western barred bandicoots increased over four years. A study in Australia found that following the release of captive-bred bridled nailtail wallabies and subsequent predator controls, numbers increased over a three years, but remained low compared to the total number released.

Reproductive success (2 studies): A study in Australia found that four of five captive-bred mammal populations released into a predator-free enclosure and one population released into a predator-reduced enclosure produced a second generation, whereas two populations released into an unfenced area with ongoing predator management did not survive to reproduce. A study in Australia found that most female captive-reared black-footed rock-wallabies released into a large predator-free fenced area reproduced.

Survival (18 studies): Ten studies (one controlled, three replicated, two before-and-after studies) in Australia, and the UK found that following the eradication/control of invasive species (and in some cases release into a fenced area), a translocated population of woylies, western barred bandicoots and red-tailed phascogales survived over four years, released captive-bred eastern barred bandicoots survived up to three years at five of seven sites, offspring of translocated golden bandicoots survived three years, over half of released captive-reared black-footed rock-wallabies survived over two years, captive-bred water voles survived for at least 20 months or over 11 months at over half of release sites, most released captive-bred hare-wallabies survived at least two months, most captive-bred eastern barred bandicoots survived for over three weeks. A replicated study in Australia found that after the control of invasive species, four translocated populations of burrowing bettongs died out within four months. A review of studies in Australia found that in seven studies where red fox control was carried out before or after the release of captive-bred eastern-barred bandicoots, survival varied. A study in Australia found that four of five captive-bred mammal populations released into a predator-free enclosure and one population released into a predator-reduced enclosure survived, whereas two populations released into an unfenced area with ongoing predator management did not. A study in Australia found that captive-bred bridled nailtail wallabies released from holding pens in areas where predators had been controlled had similar annual survival rates to that of wild-born translocated animals. Two studies (one replicated) in the USA found that where predators were managed, at least half of released captive-bred black-footed ferrets survived more than two weeks, but that post-release mortality was higher than resident wild ferrets. A before-and-after study in the USA found following the onset of translocations of black bears away from an elk calving site, survival of the offspring of translocated elk increased.

Condition (2 studies): A study Australia found that wild-born golden bandicoots, descended from a translocated population released into a predator-free enclosure, maintained genetic diversity relative to the founder and source populations. A replicated, before-and-after study in Australia found that one to two years after release into predator-free fenced reserves, translocated eastern bettongs weighed more and had improved nutritional status compared to before release.

BEHAVIOUR (1 STUDY)

Behaviour change (1 study): A replicated, before-and-after study in the USA found that translocated Utah prairie dogs released after the control of native predators into an area with artificial burrows showed low site fidelity and different pre- and post-release behaviour.

Assessment: beneficial (effectiveness 69%; certainty 62%; harms 10%).

https://www.conservationevidence.com/actions/2469

Release translocated/captive-bred mammals in family/ social groups

Twenty-six studies evaluated the effects of releasing translocated or captive-bred mammals in family or social groups. Eleven were in the USA, seven were in South Africa and one was in each of Poland, Zimbabwe, along the USA–Canada border, Russia, Italy, Canada, China and India.

COMMUNITY RESPONSE (0 STUDIES)

POPULATION RESPONSE (22 STUDIES)

Abundance (4 studies): A study in the USA found that a translocated population of Rocky Mountain bighorn sheep released in groups increased at a similar rate to that of a population newly established through natural recolonization. A replicated, controlled study in the USA found that after translocating black-tailed prairie dogs in social groups to areas with artificial burrows, colonies increased in size over four years. A replicated study in Canada found that following translocation of elk, most of which had been kept in holding pens in groups, numbers increased at two of four sites. A study in the USA found that following the release of captive-reared bighorn sheep in groups, the overall population declined over 14 years.

Reproductive success (11 studies): A study in the USA found that captive-reared bighorn sheep released in groups had similar population recruitment rates compared to wild-reared sheep. A replicated, paired study in the USA found that black-tailed prairie dogs translocated as family groups had higher reproductive success than those translocated in non-family groups. A replicated study in the USA found that translocated gray wolves had similar breeding success when adult family groups were released together from holding pens or when young adults were released directly into the wild. Six of eight studies (one replicated) in Poland, Russia, South Africa, the USA and the USA–Canada border found that when translocated and/or captive-bred animals were released in social or family groups, cheetahs, European bison, lions, African wild dogs, most European beavers and some swift foxes reproduced successfully. One study found that one of two translocated Cape buffalo groups released after being held in a holding pen formed a single herd and reproduced, while the other scattered and escaped the reserve. One study found that no Gunnison's prairie dogs reproduced during the first year.

Survival (19 studies): One of three studies (one controlled, before-and-after) in the USA found that when translocated or captive-bred animals were released in family or social groups, captive-reared bighorn sheep had similar survival compared to wild-reared sheep, whereas two found lower survival compared to wild white-tailed deer and San Joaquin kit foxes. Three replicated studies (one controlled, one paired) in the USA found that when translocated as a social or family group, black-tailed prairie dogs had higher and white-tailed deer and gray wolves had similar survival rates to those translocated as unrelated groups or individuals. Ten studies (one replicated) in Poland, Russia, Italy, South Africa, the USA, USA–Canada border, China and India found that when translocated and/or captive-bred animals were released in social or family groups, a population of Przewalski's horses and European bison persisted 5-11 years, lions, most swift foxes and European beavers and half or more cheetahs survived for at least one year, and one-horned rhinoceroses and over half of Gunnison's prairie dogs and Eurasian badgers survived at least 1-6 months. Three studies in the USA and South Africa found that when translocated or captive-bred animals were released in family or social groups (some provided with artificial refuges and/or supplementary food), most Mexican wolves did not survive over eight months and all rock hyraxes died within 90 days. A study in South Africa found that translocated and captive-bred African wild dogs released in family groups into fenced reserves had high survival rates.

Condition (1 study): A study in China found that following the release of captive-bred Przewalski's horses in groups, the population had a lower genetic diversity than two captive populations.

BEHAVIOUR (4 STUDIES)

Behaviour change (4 studies): Two replicated, controlled (one before-and-after) studies in the USA found that when translocated as a social or family group, white-tailed deer had similar average dispersal distances and Utah prairie dogs had similar release site fidelity and post-release behaviour compared to those translocated as unrelated groups. One found that deer translocated together did not stay together, whether they had previously been part of the same social group or not. A study in Zimbabwe found that a translocated lion family joined with immigrant lions and formed a new pride. A study in South Africa found that translocated lions that were released in groups that had already been socialised and formed into prides, established stable home ranges.

Assessment: beneficial (effectiveness 70%; certainty 65%; harms 5%).

https://www.conservationevidence.com/actions/2463

Release translocated/captive-bred mammals into area with artificial refuges/breeding sites

Seventeen studies evaluated the effects of releasing translocated or captive-bred mammals into areas with artificial refuges or breeding sites. Five studies were in the USA, three were in Australia, three were in Spain, two were in the UK and one was in each of Ireland, South Africa, Hungary and Slovakia, the Czech Republic and Poland.

COMMUNITY RESPONSE (0 STUDIES)

POPULATION RESPONSE (15 STUDIES)

Abundance (5 studies): Two of three studies (two replicated, two controlled) in Spain and the USA found that translocation release sites with artificial burrows provided had higher abundances of European rabbits and densities of California ground squirrels compared to those without. The other study found that abundance of European rabbits following translocation was similar with and without artificial burrows provided. A replicated, controlled study in the USA found that after translocating black-tailed prairie dogs to areas with artificial burrows, colonies increased in size. A before-and-after study in Spain found that translocating European rabbits into areas with artificial refuges to supplement existing populations did not alter rabbit abundance, although two of three populations persisted for at least three years.

Reproductive success (4 studies): Three studies in Australia, Ireland and the UK found that released captive-bred sugar gliders, most translocated female red squirrels and some translocated pine martens provided with nest boxes and supplementary food reproduced. A study of 12 translocation projects in Slovakia, the Czech Republic and Poland found that translocated European ground squirrels released initially into enclosures or burrows with retention caps reproduced after release, whereas those without enclosures or burrows dispersed from release sites.

Survival (9 studies): Five of eight studies in Australia, the USA, UK, Ireland and South Africa found that at release sites with artificial refuges, and in some cases food provided, a population of captive-bred sugar gliders survived at least three years, two of three populations of red-tailed phascogales survived for more than four years, most translocated black bears survived at least one year and over half translocated red squirrels and pine martens survived 8-12 months. Three studies found that at release sites with artificial refuges, food and in one case water provided, no translocated red squirrels survived more than five months, all translocated rock hyraxes died within three months and most translocated Tipton and Heermann's kangaroo rat spp. died within five days. A randomised, replicated, controlled study in Hungary found that translocated European ground squirrels released into plugged artificial burrows had higher recapture rates than those released into unplugged artificial burrows.

BEHAVIOUR (3 STUDIES)

Use (2 studies): Two studies in Australia found that released captive-bred sugar gliders used artificial nest boxes provided.

Behaviour change (1 study): A replicated, before-and-after study in the USA found that translocated Utah prairie dogs released into an area with artificial burrows, after the control of native predators, tended to leave the release site and spent more time being vigilant than before.
Assessment: beneficial (effectiveness 62%; certainty 62%; harms 0%).

https://www.conservationevidence.com/actions/2453

Likely to be beneficial

Release translocated/captive-bred mammals at a specific time (e.g. season, day/night)

Seven studies evaluated the effects of releasing translocated or captive-bred mammals at a specific time (season or day/night). Three studies were in the USA and one each was in the UK, Canada, Ireland and Hungary.
COMMUNITY RESPONSE (0 STUDIES)
POPULATION RESPONSE (7 STUDIES)
Survival (7 studies): Four of five studies in the UK, Canada and the USA found that translocated common dormice, black bears and Canadian lynx and captive-bred swift foxes released in a specific season had higher survival rates than those released during another season. The other study found that red squirrels translocated in autumn and winter had similar survival rates. A randomised, replicated, controlled study in Hungary found that translocated European ground squirrels released during the morning had higher recapture rates than those released during the afternoon. A study in the USA found that most translocated kangaroo rats released at dusk in artificial burrows supplied with food died within five days of release.
Condition (1 study): A study in the UK found that common dormice translocated during summer lost less weight than those translocated during spring.
BEHAVIOUR (2 STUDIES)
Behaviour change (2 studies): Two studies in the UK and USA found that common dormice translocated during spring and black bears translocated during winter travelled shorter distances or settled closer to the release site than those translocated during summer.
Assessment: likely to be beneficial (effectiveness 70%; certainty 60%; harms 10%).

https://www.conservationevidence.com/actions/2447

Release translocated/captive-bred mammals in larger unrelated groups

Five studies evaluated the effects of releasing translocated or captive-bred mammals in larger unrelated groups. Two studies were in South Africa, one was in Namibia and South Africa, one was in the USA and one was in Australia.

COMMUNITY RESPONSE (0 STUDIES)

POPULATION RESPONSE (5 STUDIES)

Reproductive success (3 studies): A replicated, paired sites study in the USA found that black-tailed prairie dogs translocated in larger groups had higher reproductive success than smaller groups. A study in South Africa found that Cape buffalo translocated to a fenced reserve as a larger group formed a single herd and reproduced, whilst a smaller group separated. A study in South Africa found that rehabilitated and captive-bred cheetahs released in groups (unrelated and family) and as individuals reproduced.

Survival (4 studies): A replicated, paired sites study in the USA found that black-tailed prairie dogs translocated in larger groups had higher initial daily survival rate than smaller groups. Two studies (one controlled) in Namibia and South Africa and Australia found that releasing translocated black rhinoceroses and burrowing bettongs in larger groups did not increase survival. A study in South Africa found that most adult rehabilitated and captive-bred cheetahs released in groups (unrelated and family) and as individuals survived at least one year.

BEHAVIOUR (2 STUDIES)

Behaviour change (2 studies): A replicated, paired sites study in the USA found that black-tailed prairie dogs translocated in larger groups attracted more immigrants than smaller groups. A study in South Africa found that Cape buffalo translocated as a larger group formed a single herd and stayed in the fenced reserve, whilst a smaller group scattered and escaped the reserve.

Assessment: likely to be beneficial (effectiveness 67%; certainty 54%; harms 0%).

https://www.conservationevidence.com/actions/2462

Release translocated/captive-bred mammals to areas outside historical range

Seven studies evaluated the effects of releasing translocated or captive-bred mammals to areas outside their historical range. Three studies were in Australia, one study was in each of Kenya, France and South Africa, and one was a review of studies in Andorra, Spain and France.

COMMUNITY RESPONSE (0 STUDIES)
POPULATION RESPONSE (7 STUDIES)
Abundance (5 studies): Three of four studies in Kenya, Australia, France, and South Africa found that after translocating mammals to areas outside their historical range, populations increased for Alpine marmots, most of 22 herbivorous species and bridled nailtail wallabies (including captive and enclosure bred animals). A study in Kenya found that a population of translocated roan persisted for more than six years but did not increase. A review of studies in Andorra, Spain and France found that following translocation to areas outside their native range, alpine marmots had similar densities and family group sizes to those of populations in their native range.
Reproductive success (1 study): A study in Kenya found that a population of roan translocated into an area outside their native range persisted and bred for more than six years.
Survival (3 studies): A study in Australia found that captive-bred, translocated and enclosure born bridled nailtail wallabies released into areas outside their historical range had annual survival rates of 40–88% over four years. A study in Australia found that most captive-bred Tasmanian devils released into an area outside their native range survived over four months. A study in Australia found that half the captive-bred and wild-caught translocated eastern barred bandicoots released to a red fox-free island outside their historical range survived for at least two months.
BEHAVIOUR (0 STUDIES)
Assessment: likely to be beneficial (effectiveness 70%; certainty 52%; harms 15%).

https://www.conservationevidence.com/actions/2443

Release translocated/captive-bred mammals to islands without invasive predators

Six studies evaluated the effects of releasing translocated or captive-bred mammals to islands without invasive predators. The six studies were in Australia.
COMMUNITY RESPONSE (0 STUDIES)
POPULATION RESPONSE (7 STUDIES)
Abundance (2 studies): A study in Australia found that following release of captive-bred dibblers on to an island free of introduced predators, numbers increased. A replicated study in Australia found that following release of captive-bred and wild-born brush-tailed bettong onto islands free of foxes or cats, numbers increased on two of four islands.

Reproductive success (3 studies): A study in Australia found that captive-bred proserpine rock-wallabies released on an island without introduced predators established a breeding population. Two studies in Australia found that following release on to islands without invasive predators, captive-bred rufous hare-wallabies and captive-bred dibblers.
Survival (3 studies): A review of 28 translocation studies in Australia found that 67% of marsupial populations translocated to islands without predators survived more than five years, compared to 0% translocated to islands with predators and 20% translocated to the mainland. A study in Australia found that most captive-bred rufous hare-wallabies released on an island without non-native predators survived more than a year. A replicated study in Australia found that wild-born golden bandicoots descended from translocated populations released onto two predator-free islands persisted for 2–3 years.
Condition (1 study): A replicated study in Australia found that wild-born golden bandicoots descended from translocated populations that had been released onto two predator-free islands, maintained genetic diversity relative to founder and source populations.
BEHAVIOUR (0 STUDIES)
Assessment: likely to be beneficial (effectiveness 80%; certainty 60%; harms 0%).

https://www.conservationevidence.com/actions/2464

15.13.5 Other

Based on the collated evidence, what is the current assessment of the effectiveness of interventions for other sources of non-native, invasive or other problematic species?	
Likely to be beneficial	• Place orphaned or abandoned wild young with captive foster parents • Place orphaned or abandoned wild young with wild foster parents • Provide supplementary water to increase reproduction/survival • Rehabilitate injured, sick or weak mammals
Trade-off between benefit and harms	• Hand-rear orphaned or abandoned young in captivity • Provide supplementary food to increase reproduction/survival

Based on the collated evidence, what is the current assessment of the effectiveness of interventions for other sources of non-native, invasive or other problematic species?	
Unknown effectiveness	• Cease/reduce payments to cull mammals • Graze herbivores on pasture, instead of sustaining with artificial foods
No evidence found (no assessment)	• Temporarily hold females and offspring in fenced area to increase survival of young

Likely to be beneficial

Place orphaned or abandoned wild young with captive foster parents

Two studies evaluated the effects of placing orphaned or abandoned wild young with captive foster parents. One study was in Canada and one was in the USA.

COMMUNITY RESPONSE (0 STUDIES)

POPULATION RESPONSE (1 STUDY)

Survival (1 study): A controlled study in the USA found that stranded sea otter pups reared in captivity by foster mothers had higher post-release survival than did unfostered pups reared mostly alone, and similar survival to wild pups.

BEHAVIOUR (2 STUDIES)

Behaviour change (2 studies): A study in Canada found that a captive white-tailed deer adopted a wild orphaned fawn. A controlled study in the USA found that stranded sea otter pups reared in captivity by foster mothers began foraging earlier than did unfostered pups reared mostly alone.

Assessment: likely to be beneficial (effectiveness 70%; certainty 40%; harms 7%).

https://www.conservationevidence.com/actions/2364

Place orphaned or abandoned wild young with wild foster parents

Three studies evaluated the effects of placing orphaned or abandoned wild young with wild foster parents. One study was in the USA, one was in South Africa and one was in Botswana.

COMMUNITY RESPONSE (0 STUDIES)

POPULATION RESPONSE (3 STUDIES)

Survival (3 studies): Two studies (one controlled) in the USA and Botswana, found that orphaned young black bears and African wild dogs had greater or equal survival compared to animals released alone or young of wild mammals with their biological parents. A study in South Africa found that an orphaned cheetah cub was not accepted by a family of cheetahs.
BEHAVIOUR (0 STUDIES)
Assessment: likely to be beneficial (effectiveness 52%; certainty 42%; harms 10%).

https://www.conservationevidence.com/actions/2343

Provide supplementary water to increase reproduction/survival

Six studies evaluated the effects on mammals of providing supplementary water to increase reproduction/survival. Two studies were in Australia and one each was in Oman, Portugal, Saudi Arabia and the USA and Mexico.
COMMUNITY RESPONSE (0 STUDIES)
POPULATION RESPONSE (5 STUDIES)
Abundance (2 studies): A replicated study in the USA and Mexico found that providing supplementary water was associated with increases in desert bighorn sheep population size. A study in Oman found that a released captive-bred Arabian oryx population initially provided with supplementary water and food increased over 14 years.
Reproduction (2 studies): A study in Saudi Arabia found that released captive-bred Arabian gazelles initially provided with supplementary water and food after release into a fenced area started breeding in the first year. A study in Australia found that most female released captive-reared black-footed rock-wallabies provided with supplementary water after release into a large predator-free fenced area reproduced in the first two years.
Survival (2 studies): A controlled, before-and-after study in Australia found that most released captive-bred hare-wallabies provided with supplementary water, along with supplementary food and predator control, survived at least two months after release into a fenced peninsula. A study in Australia found that over half of released captive-reared black-footed rock-wallabies provided with supplementary water after release into a large predator-free fenced area survived for at least two years.
BEHAVIOUR (1 STUDY)
Use (1 study): A replicated study in Portugal found that artificial waterholes were used by European rabbits and stone martens.
Assessment: likely to be beneficial (effectiveness 80%; certainty 60%; harms 0%).

https://www.conservationevidence.com/actions/2396

Rehabilitate injured, sick or weak mammals

Thirteen studies evaluated the effects of rehabilitating injured, sick or weak mammals. Four studies were in the UK, three were in Spain, two were in Argentina and one each was in Uganda, Australia, the USA and Brazil.

COMMUNITY RESPONSE (0 STUDIES)

POPULATION RESPONSE (12 STUDIES)

Survival (11 studies): Five studies, in the UK and Spain, found that varying proportions of European hedgehogs released after being rehabilitated in captivity survived during post-release monitoring periods, which ranged from two weeks to 136 days. Five studies, in Australia, Spain, the USA and Brazil, found that four koalas, an Iberian lynx, a gray wolf, a puma and two brown bears released following rehabilitation in captivity survived for varying durations during monitoring periods, which ranged in length from three months to up to seven years. A study in Argentina found that over half of released rehabilitated and captive-reared giant anteaters survived for at least six months.

Condition (2 studies): A study in Uganda found that a snare wound in a white rhinoceros healed after treatment and rehabilitation. A study in the UK found that two of three rehabilitated European hedgehogs lost 12-36% of their body weight after release into the wild.

BEHAVIOUR (1 STUDY)

Behaviour change (1 study): A controlled study in Argentina found that released wild-born rehabilitated giant anteaters were more nocturnal in their activity patterns than captive-bred individuals.

Assessment: likely to be beneficial (effectiveness 70%; certainty 55%; harms 10%).

https://www.conservationevidence.com/actions/2352

Trade-off between benefit and harms

Hand-rear orphaned or abandoned young in captivity

Six studies evaluated the effects of hand-rearing orphaned mammals. Two were in the USA, one each was in Australia, South Africa and India and one was in six countries across North America, Europe and Asia.

COMMUNITY RESPONSE (0 STUDIES)

POPULATION RESPONSE (5 STUDIES)

Reproductive success (1 study): One study in India found that three hand-reared orphaned or abandoned greater one-horned rhinoceroses gave birth in the wild.

Survival (5 studies): Five studies (including one controlled and one replicated) in Australia, the USA, India and in six countries across North America, Europe and Asia, found that some hand-reared orphaned or abandoned ringtail possums, white-tailed deer, sea otters, bears and greater one-horned rhinoceroses survived for periods of time after release.

BEHAVIOUR (1 STUDY)

Behaviour change (1 study): A study in South Africa found that a hand-reared, orphaned serval established a home range upon release.

Assessment: trade-off between benefit and harms (effectiveness 50%; certainty 40%; harms 20%).

https://www.conservationevidence.com/actions/2358

Provide supplementary food to increase reproduction/ survival

Twenty-four studies evaluated the effects on mammals of providing supplementary food to increase reproduction/survival. Nine studies were in the USA, two were in Canada, two were in South Africa, two were in Poland, and one each was in Sweden, the Netherlands, eSwatini, Spain, Portugal, Slovenia, Austria, Norway and Sweden and one was across North America and Europe.

COMMUNITY RESPONSE (0 STUDIES)

POPULATION RESPONSE (18 STUDIES)

Abundance (8 studies): Four of eight studies (incuding four controlled, two site comparisons and five before-and-after studies) in the USA, Canada, South Africa, Poland and Austria found that supplementary feeding increased the abundance or density of bank voles, red squirrels, striped mice, brown hyena and black-backed jackals. One study found a temporary increased in prairie vole abundance. The other three studies found supplementary feeding not to increase abundance or density of white-footed mice, northern flying squirrels, Douglas squirrels or Eurasian otters.

Reproduction (8 studies): Four of five controlled studies (three also replicated) in the USA, South Africa, Norway and Sweden, Sweden and Spain, found that supplementary food increased the proportion of striped mice that were breeding, the number of arctic fox litters and the size of prairie vole litters. However, there was no increase in the number of arctic fox cubs in each litter or the proportion of female Iberian lynx breeding. One of two replicated studies (one site comparison and one controlled), in the Netherlands and the USA, found that supplementary feeding increased the number of young wild boar produced and recruited in to the population. The other study found that the number of mule deer produced/adult female did not increase. A review of studies across North America and Europe found that supplementary feeding increased ungulate reproductive rates in five of eight relevant studies.

Survival (9 studies): Four of eight studies (including seven controlled studies and two before-and-after studies) in the USA, Canada, Poland and Spain, found that supplementary feeding increased survival of mule deer, bank voles, northern flying squirrels and eastern cottontail rabbits. Five studies found no increase in survival for white-tailed deer, Douglas squirrels, mule deer, Rocky Mountain bighorn sheep lambs or Iberian lynx. A review of studies across North America and Europe found that supplementary feeding increased ungulate survival in four out of seven relevant studies.

Condition (4 studies): One of three studies (including two controlled and two before-and-after studies) in Poland, the USA, and Canada, found that supplementary food lead to weight gain or weight recovery in bank voles. One study found no body mass increase with supplementary feeding in northern flying squirrels and Douglas squirrels. The third study found mixed results, with supplementary feeding increasing weight gains in some cotton rats, depending on their sex, weight and the time of year. A review of studies from across North America and Europe found that different proportions of studies found supplementary feeding to improve a range of measures of ungulate condition.

BEHAVIOUR (6 STUDIES)

Use (2 studies): A replicated, controlled study in Sweden found that supplementary food increased occupancy of Arctic fox dens. A replicated study in Portugal found that artificial feeding stations were used by European rabbits.

Behaviour (4 studies): Two of three replicated studies (two also controlled), in eSwatini, Slovenia and the USA, found that supplementary feeding led to reduced home range sizes or shorter movements of red deer and elk. The third study found home ranges and movement distances to be similar between fed and unfed multimammate mice. One replicated study in Poland found that supplementary feeding of ungulates altered brown bear behaviour.

Assessment: trade-off between benefit and harms (effectiveness 70%; certainty 70%; harms 20%).

https://www.conservationevidence.com/actions/2367

Unknown effectiveness

Cease/reduce payments to cull mammals

One study evaluated the effects of ceasing or reducing payments to cull mammals. This study was in Sweden and Norway.

COMMUNITY RESPONSE (0 STUDIES)

POPULATION RESPONSE (1 STUDY)

Survival (1 study): A before-and-after study in Sweden and Norway found that fewer brown bears were reported killed after the removal of financial hunting incentives.

BEHAVIOUR (0 STUDIES)

Assessment: unknown effectiveness (effectiveness 61%; certainty 39%; harms 0%).

https://www.conservationevidence.com/actions/2349

Graze herbivores on pasture, instead of sustaining with artificial foods

One study evaluated the effects of grazing mammalian herbivores on pasture, instead of sustaining with artificial foods. This study was in South Africa.

COMMUNITY RESPONSE (0 STUDIES)

POPULATION RESPONSE (1 STUDY)

Reproductive success (1 study): A site comparison study in South Africa found that a population of roan antelope grazed on pasture had a higher population growth rate than populations provided solely with imported feed.

BEHAVIOUR (0 STUDIES)

Assessment: unknown effectiveness (effectiveness 65%; certainty 35%; harms 0%).

https://www.conservationevidence.com/actions/2398

No evidence found (no assessment)

We have captured no evidence for the following interventions:

- Temporarily hold females and offspring in fenced area to increase survival of young.

15.14 Education and awareness raising

Based on the collated evidence, what is the current assessment of the effectiveness of interventions for education and awareness raising?	
Likely to be beneficial	• Encourage community-based participation in land management • Provide education programmes to improve behaviour towards mammals and reduce threats • Publish data on ranger performance to motivate increased anti-poacher efforts • Train and support local staff to help reduce persecution of mammals
Unknown effectiveness	• Use campaigns and public information to improve behaviour towards mammals and reduce threats
No evidence found (no assessment)	• Provide science-based films, radio programmes, or books about mammals to improve behaviour towards mammals and reduce threats

Likely to be beneficial

Encourage community-based participation in land management

Two studies evaluated the effects of encouraging community-based participation in management of mammals to reduce mammal persecution. One study was in Pakistan and one was in India.

COMMUNITY RESPONSE (0 STUDIES)

POPULATION RESPONSE (1 STUDY)
Abundance (1 study): A study in Pakistan found that involving local communities with park management was associated with an increasing population of Himalayan brown bears.
BEHAVIOUR (0 STUDIES)
OTHER (1 STUDY)
Human behaviour change (1 study): A study in Namibia found that fewer farmers who engaged in community-based management of land, through membership of a conservancy, removed large carnivores from their land than did non-conservancy members.
Assessment: likely to be beneficial (effectiveness 70%; certainty 40%; harms 0%).

https://www.conservationevidence.com/actions/2395

Provide education programmes to improve behaviour towards mammals and reduce threats

Two studies evaluated the effects of providing education programmes to improve behaviour towards mammals and reduce threats. One study was in South Africa and one was in the USA.
COMMUNITY RESPONSE (0 STUDIES)
POPULATION RESPONSE (1 STUDY)
Abundance (1 study): A before-and-after study in South Africa found that educating ranchers on ways of reducing livestock losses, along with stricter hunting policies, increased leopard density.
Survival (1 study): A before-and-after study in South Africa found that educating ranchers on ways of reducing livestock losses, along with stricter hunting policies, reduced leopard mortalities.
BEHAVIOUR (0 STUDIES)
OTHER (1 STUDY)
Human behaviour change (1 study): A replicated, controlled, before-and-after study in the USA found that visiting households to educate about the danger of garbage to black bears did not increase use of wildlife-resistant dumpsters.
Assessment: likely to be beneficial (effectiveness 55%; certainty 40%; harms 0%).

https://www.conservationevidence.com/actions/2423

Publish data on ranger performance to motivate increased anti-poacher efforts

One study evaluated the effects on poaching incidents of publishing data on ranger performance to motivate increased anti-poacher efforts. This study was in Ghana.

COMMUNITY RESPONSE (0 STUDIES)

POPULATION RESPONSE (1 STUDY)

Survival (1 study): A replicated, before-and-after, site comparison study in Ghana found that when data were publishing on staff performance, poaching incidents decreased on these sites and on sites from which performance data were not published.

BEHAVIOUR (0 STUDIES)

OTHER (1 STUDY)

Human behaviour change (1 study): A replicated, before-and-after, site comparison study in Ghana found that publishing data on staff performance lead to an increase in anti-poaching patrols.

Assessment: likely to be beneficial (effectiveness 60%; certainty 40%; harms 0%).

https://www.conservationevidence.com/actions/2426

Train and support local staff to help reduce persecution of mammals

One study evaluated the effects of training and supporting local staff to help reduce persecution of mammals. This study was in Kenya.

COMMUNITY RESPONSE (0 STUDIES)

POPULATION RESPONSE (1 STUDY)

Survival (1 study): A replicated, before-and-after study in Kenya found that employing local tribesmen to dissuade pastoralists from killing lions and to assist with livestock protection measures, alongside compensating for livestock killed by lions, reduced lion killings by pastoralists.

BEHAVIOUR (0 STUDIES)

Assessment: likely to be beneficial (effectiveness 75%; certainty 40%; harms 0%).

https://www.conservationevidence.com/actions/2425

Unknown effectiveness

Use campaigns and public information to improve behaviour towards mammals and reduce threats

Two studies evaluated the effects of using campaigns and public information to improve behaviour towards mammals and reduce threats. One study was in the USA and one was in Lao People's Democratic Republic.

COMMUNITY RESPONSE (0 STUDIES)

POPULATION RESPONSE (0 STUDIES)

BEHAVIOUR (0 STUDIES)

OTHER (2 STUDIES)

Human behaviour change (2 studies): A randomized, replicated, controlled, before-and-after study in the USA found that displaying education signs did not reduce the percentage of garbage containers that were accessible to black bears. A controlled, before-and-after study in Lao People's Democratic Republic found that a social marketing campaign promoting a telephone hotline increased reporting of illegal hunting.

Assessment: unknown effectiveness (effectiveness 40%; certainty 30%; harms 0%).

https://www.conservationevidence.com/actions/2422

No evidence found (no assessment)

We have captured no evidence for the following interventions:

- Provide science-based films, radio programmes, or books about mammals to improve behaviour towards mammals and reduce threats.

This book need not end here...

At Open Book Publishers, we are changing the nature of the traditional academic book. The title you have just read will not be left on a library shelf, but will be accessed online by hundreds of readers each month across the globe. OBP publishes only the best academic work: each title passes through a rigorous peer-review process. We make all our books free to read online so that students, researchers and members of the public who can't afford a printed edition will have access to the same ideas.
This book and additional content is available at:https://www.openbookpublishers.com/product/1031

Customise

Personalise your copy of this book or design new books using OBP and third-party material. Take chapters or whole books from our published list and make a special edition, a new anthology or an illuminating coursepack. Each customised edition will be produced as a paperback and a downloadable PDF. Find out more at:
https://www.openbookpublishers.com/section/59/1

Donate

If you enjoyed this book, and feel that research like this should be available to all readers, regardless of their income, please think about donating to us. We do not operate for profit and all donations, as with all other revenue we generate, will be used to finance new Open Access publications:
https://www.openbookpublishers.com/section/13/1/support-us

You may also be interested in:

Global Warming in Local Discourses
How Communities around the World Make Sense of Climate Change

Edited by Michael Brüggemann and Simone Rödder

https://doi.org/10.11647/OBP.0212

Terrestrial Mammal Conservation
Global Evidence for the Effects of Interventions for Terrestrial Mammals Excluding Bats and Primates

N.A. Littlewood, R. Rocha, R.K. Smith, W.J. Sutherland et al.

https://doi.org/10.11647/OBP.0234

Right Research
Modelling Sustainable Research Practices in the Anthropocene

Edited by Chelsea Miya, Oliver Rossier and Geoffrey Rockwell

https://doi.org/10.11647/OBP.0213

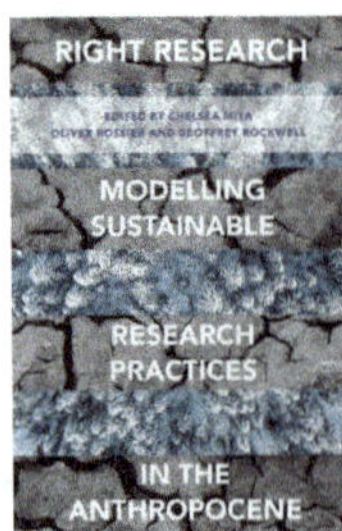

www.ingramcontent.com/pod-product-compliance
Lightning Source LLC
LaVergne TN
LVHW052011200926
848481LV00005B/1460